Speaking of Apes

A Critical Anthology of
Two-Way Communication with Man

TOPICS IN CONTEMPORARY SEMIOTICS

Series Editors: Thomas A. Sebeok and Jean Umiker-Sebeok
                    Indiana University

SPEAKING OF APES
Edited by Thomas A. Sebeok and Jean Umiker-Sebeok

Speaking of Apes

A Critical Anthology of
Two-Way Communication with Man

Edited by
Thomas A. Sebeok
and
Jean Umiker-Sebeok

Indiana University
Bloomington, Indiana

Plenum Press · New York and London

Library of Congress Cataloging in Publication Data

Main entry under title:

Speaking of apes.

(Topics in contemporary semiotics) Bibliography: p.
Includes index.
1. Chimpanzees–Psychology. 2. Human-animal communication. 3. Language and languages. 4. Animal communication. I. Sebeok, Thomas Albert, 1920- II. Umiker-Sebeok, Jean.
QL737. P96S63   599'.884   79-17714
ISBN 0-306-40279-3

© 1980 Plenum Press, New York
A Division of Plenum Publishing Corporation
227 West 17th Street, New York, N.Y. 10011

Printed in the United States of America

Oskar Pfungst
(1874-1932)

CONTRIBUTORS

URSULA BELLUGI, The Salk Institute, San Diego, California

T. G. BEVER, Department of Psychology, Columbia University, New York, New York

SALLY BOYSEN, Yerkes Regional Primate Research Center, Emory University, Atlanta, Georgia

JACOB BRONOWSKI, Late of The Salk Institute, San Diego, California

ROGER BROWN, Department of Psychology, Harvard University, Cambridge, Massachusetts

NOAM CHOMSKY, Department of Linguistics and Philosophy, Massachusetts Institute of Technology, Cambridge, Massachusetts

ROGER S. FOUTS, Department of Psychology, University of Oklahoma, Norman, Oklahoma

BEATRICE T. GARDNER, Department of Psychology, University of Nevada, Reno, Nevada

R. ALLEN GARDNER, Department of Psychology, University of Nevada, Reno, Nevada

ALICE F. HEALY, Department of Psychology, Yale University, New Haven, Connecticut

H. HEDIGER, Emeritus, University of Zurich, Switzerland

JANE H. HILL, Department of Anthropology, Wayne State University, Detroit, Michigan

WINTHROP N. KELLOGG, Late of the Department of Psychology, Florida State University, Tallahassee, Florida

ERIC H. LENNEBERG, Late of the Department of Psychology, Cornell University, Ithaca, New York

JOHN LIMBER, Department of Psychology, University of New Hampshire, Durham, New Hampshire

WILLIAM A. MALMI, Department of Anthropology, University of California, Concord, California

PETER MARLER, The Rockefeller University, New York, New York

DAVID MCNEILL, Department of Behavioral Sciences, University of Chicago, Chicago, Illinois

GEORGES MOUNIN, Université de Provence, Aix-en-Provence, France

RANDALL L. RIGBY, HHB 2/17 FA, APO SF 96251

DUANE M. RUMBAUGH, Department of Psychology, Georgia State University, Atlanta, Georgia

E. SUE SAVAGE-RUMBAUGH, Yerkes Regional Primate Research Center, Emory University, Atlanta, Georgia

THOMAS A. SEBEOK, Research Center for Language and Semiotic Studies, Indiana University, Bloomington, Indiana

H. S. TERRACE, Department of Psychology, Columbia University, New York, New York

JEAN UMIKER-SEBEOK, Research Center for Language and Semiotic Studies, Indiana University, Bloomington, Indiana

ACKNOWLEDGMENTS

The editors wish to thank the following for permission to reprint copyrighted materials in this book:

"Communication and Language in the Home-Raised Chimpanzee," by Winthrop N. Kellogg. *Science 162*, 423–427. Copyright 1968 by the American Association for the Advancement of Science. Reprinted by permission of the American Association for the Advancement of Science.

"A Word Between Us," by Eric H. Lenneberg. In *Communication*, edited by John D. Roslansky, p. 111–131. Amsterdam: North-Holland, 1969. Reprinted with permission of the publisher and Mrs. Lenneberg.

"The First Sentences of Child and Chimpanzee," by Roger Brown. *Psycholinguistics. Selected Papers by Roger Brown*, pp. 208–231. New York: Free Press. Copyright 1970 by The Free Press, a Division of Macmillan Publishing Co., Inc. Reprinted with permission of Macmillan Publishing Co., Inc. and the author.

"Language, Name and Concept," by Jacob Bronowski and Ursula Bellugi. *Science 168*, 669–673. Copyright 1970 by the American Association for the Advancement of Science. Reprinted with permission of the American Association for the Advancement of Science, Dr. Bellugi, and Mrs. Bronowski.

"Of Language Knowledge, Apes, and Brains," by Eric H. Lenneberg. *Journal of Psycholinguistic Research 1*, 1–29. Copyright 1971 by Plenum Publishing Corp. Reprinted with permission of the editor, publisher, and Mrs. Lenneberg.

"Can Chimpanzees Learn a Phonemic Language?" by Alice F. Healy. *Journal of Psycholinguistic Research 2*, 167–170. Copyright 1973 by Plenum Publishing Corp. Reprinted with permission of the editor, publisher, and author.

"Sentence Structure in Chimpanzee Communication," by David McNeill. In *The Growth of Competence*, edited by Kevin Connolly and Jerome Bruner, pp. 75–94. Copyright 1974 by Academic Press, Ltd. Reprinted with permission of the publisher and author.

"Language, Communication, Chimpanzees," by Georges Mounin, *Current Anthropology 17*, 1–7. Copyright 1976 by the University of Chicago Press. Reprinted with permission of the publisher and author.

"What Might Be Learned from Studying Language in the Chimpanzee?" by H. S. Terrace and T. G. Bever. In *Origins and Evolution of Language and Speech*, edited by S. R. Harnad, H. D. Steklis, and J. Lancaster, pp. 579–588 (*Annals of the New York Academy of Sciences 280*), 1976. Reprinted with permission of the New York Academy of Sciences and the authors.

"Chimpanzees and Language Evolution," by William A. Malmi. In *Origins and Evolution of Language and Speech*, edited by S. R. Harnad, H. D. Steklis, and J. Lancaster, pp. 598–603. (*Annals of the New York Academy of Sciences 280*), 1976. Reprinted with permission of the New York Academy of Sciences.

"Language in Child and Chimp?" by John Limber. *American Psychologist 32*, 280–295. Copyright 1977 by the American Psychological Association. Reprinted with permission of the American Psychological Association and the author.

"Primate Vocalization: Affective or Symbolic?" by Peter Marler. In *Progress in Ape Research*, edited by Geoffrey H. Bourne, pp. 85–96. New York: Academic Press, Inc. Copyright 1977 by Academic Press, Inc. Reprinted with permission of the publisher and author.

"Language Behavior of Apes," by Duane M. Rumbaugh. In *Behavioral Primatology*, edited by A. M. Schrier, pp. 105–138. Hillsdale, N. J.: Lawrence Erlbaum. Copyright 1977 by Lawrence Erlbaum Associates, Inc. Reprinted with permission of the publisher and author.

"Man–Chimpanzee Communication," by Roger S. Fouts and Randall L. Rigby. In *How Animals Communicate*, edited by Thomas A. Sebeok, pp. 1034–1054. Bloomington: Indiana University Press. Reprinted with permission of the publisher and authors.

"Comparative Psychology and Language Acquisition," by R. Allen and Beatrice T. Gardner. In *Psychology: The State of the Art*, edited by Kurt Salzinger and Florence L. Denmark, pp. 37–76. (*Annals of the New York Academy of Sciences 309*), 1978. Reprinted with permission of the New York Academy of Sciences and the authors.

"Apes and Language," by Jane H. Hill. *Annual Review of Anthropology 7*, 89–112. Copyright 1978 by Annual Reviews, Inc. Reprinted with permission of Annual Reviews, Inc. and the author.

"Linguistically-Mediated Tool Use and Exchange by Chimpanzees (*Pan Troglodytes*)," by E. Sue Savage-Rumbaugh, Duane M. Rumbaugh and Sally Boysen. *The Behavioral and Brain Science 4*, 539–554, 1978. Copyright 1979 by Cambridge University Press. Reprinted with permission of the publisher and authors.

"Is Problem-Solving Language?" by H. S. Terrace. *Journal of the Experimental Analysis of Behavior 31*, 161–175, 1979. Copyright 1979 by the Society for the Experimental Analysis of Behavior. Reprinted with permission of the Society for the Experimental Analysis of Behavior and author.

"Looking in the Destination for What Should Have Been Sought in the Source," by Thomas A. Sebeok. In *The Sign & Its Masters*. Austin: University of Texas Press, 1979. Reprinted with permission of the publisher and author.

"Human Language and Other Semiotic Systems," by Noam Chomsky, *Semiotica 25*, 31–44, 1979. Reprinted with permission of the editor and author.

"Do You Speak Yerkish? The Newest Colloquial Language with Chimpanzees," by H. Hediger, translation of "Sprechen Sie Yerkisch?, Die Neueste Umgangssprache mit Schimpansen," *Das Tier 81498*, 18–19 (January, 1979). Copyright 1979 by the author. Reprinted with permission of the publisher and author.

CONTENTS

MAN–CHIMPANZEE COMMUNICATION 261

Roger S. Fouts and Randall L. Rigby

INTRODUCTION: QUESTIONING APES

JEAN UMIKER-SEBEOK AND THOMAS A. SEBEOK

The peculiarity of the case is just that there are so many sources of possible deception in most of the observations that the whole lot of them may *be worthless. . . . I am also constantly baffled as to what to think of this or that particular story, for the sources of error in any one observation are seldom fully knowable.*

William James (in Murphy & Ballou 1960:310, 320).

VARIETIES OF INTERSPECIFIC COMMUNICATION

A common definition of communication involves nothing more than the moving of information in time from one place to another. From this point of view, it is of little consequence whether the trafficked ware consists of random noise, the 10^{10} bits of information inherent in a strand of DNA programmed to rebuild itself, or some configuration of signs, as a sonata by Mozart, charged with high aesthetic value. If information is available, it can be communicated.

Semiosis, or a triadic cooperative production involving a sign, its object, and its interpretant (Peirce 1935–1966:5.484), is as much a criterial attribute of all life as is the ability to metabolize. Whatever form life may assume, two basic activities that sustain interaction with its environment are the packing and discharging of nutrients and gases in order to grow or derive energy, and the processing of signs in order to maneuver and operate information so as to maximize the chances of the survival of its kind.

Just as there are different sorts of strategies for metabolic activity, there are also various kinds of communication devices. Plants, for example, are characterized by an autotrophic reaction: the organism derives energy from sunlight. Animals employ heterotrophic reaction: they usually educe nourishment from organic compounds outside their own body. Phytosemiosic techniques have hardly begun to be scientifically explored, so our knowledge of them is still very much at a rudimentary level. The investigation of the enormous variety of zoosemiosic systems, on the

---

JEAN UMIKER-SEBEOK AND THOMAS A. SEBEOK ● Research Center for Language and Semiotic Studies, Indiana University, Bloomington, Indiana 47405.

other hand, has long interested humanity, and, during the last quarter of the century, has emerged as the cardinal area of concentration in the biological study of animal behavior (Sebeok 1978b).

Broadly speaking, two principal modes of communicative behavior can be demarcated in animals for purposes of research: intraspecific and interspecific. Each species of animal possesses a unique code. By adverting to this code, an amalgam of partially inherited and partially learned components, the conspecific membership can exchange messages that tend biologically and/or socially to promote the welfare of that population. This vast, heterogeneous category of substratal communicative events (Sebeok 1977) will concern us hereafter only incidentally.

Interspecific communication, by contrast, always entails more than one specific code, the number depending on the complexity of the ecosystem occupied by animals belonging to different species that congregate there to their mutual advantage (Sebeok 1978b:21). In mixed populations, where members of every participant species necessarily coexist in some sort of affiliation with individuals of another—in relationships technically known by terms such as parasitism, commensalism, mutualism, or, more generally, symbiosis—some messages flowing to and fro are registered against a higher order, as it were, xenogeneic code, a quantum of knowledge of which is necessarily shared by concerned communicants belonging to separate but allied species, say, the oxpecker (*Buphagus africanus*) calling out to the rhinoceros prior to alighting upon its skin, or to the zebra signifying by its posture (legs spread, tail raised, ears drooping) its own readiness to be cleansed. The insects, ticks, or other parasites infesting the quadruped also constitute an essential part of this particular communicative network. The permutations in mutually modifying response patterns of this second kind can become very elaborate indeed when the members of one species are barraged by the displays of one or more others that, in turn, react to displays of *theirs*.

The intricacy of the animal's environment may be heightened further still in the interesting special case when one of the participating species in the system chances to be man. It is upon some varieties of this type of communicative dyad that we shall focus in what follows.

The circumstances under which man may encounter animals, necessitating that each party learn—even if never entirely master—the essential elements of the reciprocal's own code, are many and varied. A few of the most common situations in which man/animal interactions take place (cf. Hediger 1969, Chap. 4) can be identified and briefly exemplified:

1. Man preys upon, even annihilates, an animal species for different reasons—some (such as antelopes) may be hunted down as game, certain carnivores (such as the East African crocodile) condemned as "vermin," primates may be overused in medical research, marsupials may be killed for their hide, cetaceans may be

exploited for their oil (cf. Ehrenfeld 1972). In effect, every time a population of animals is exterminated, a unique communicative code is concurrently and irreversibly eliminated.

2. Man becomes the casualty of an animal's depredations, as of the Southeast Asian *Anopheles balabacensis*, one of perhaps 50 species of this genus, members of which may come into noisome taction with man (more people die each year, even now, of mosquito-borne disease than from any other single cause; see Gillett 1973).

3. Man coexists with an animal in some sort of partnership, for example, in a purely guest/host relationship (as aquarium fishes with their master); or in a nexus of mutual dependency (such as a Seeing Eye dog working in the service of a blind man, or a cormorant catching fish for a Japanese fisherman and being rewarded with food according to the size of the catch).

4. Parasitism may work in either direction:

a. The activities of man in relation to the reindeer (Zeuner 1963:46–48) can, for instance, be described as those of a social parasite. We are the exploiting species, taking advantage of the reindeer's greed, sated only when it licks up salty matter (as human urine) that attracts and binds the animal to human camps. The Masai custom of drinking the blood of their cattle without killing them is another good example of parasitism where the human guest lives, especially in times of famine, on body fluids of the host.

b. It has justly been observed that each of us has about as many organisms on our skin surface as there are people on earth. The mite *Demodex,* crab lice, fleas and bedbugs are but a few samples of the teeming miniature parasitic population sharing our ecology (e.g., Rosebury 1969).

5. An animal may accept a human as a conspecific. As early as 1910, Heinroth described the attachment of incubator-hatched greylag goslings to human beings. These goslings reject any goose and gander as parent objects, opting instead to look upon humans as their exclusive parents. Many other hand-reared birds were later found to have transferred their adult sexual behavior toward their human caretaker. Morris and Morris (1966a:182) have recounted attempts by a female panda, Chi-Chi, to mate with her keepers, and the ardent sexual advances of a male dolphin, Peter, toward his female trainer, Margaret Howe, were recorded in her published protocol (Lilly 1967:274f.).

6. An animal may define a human as a part of its inanimate environment, as when men are driving through a wildlife park in a closed vehicle. In certain circumstances, small mammals (such as bats, house-mice), and even larger ones (e.g., koala bears) grasp a human limb and, mistaking it for an insensate substrate, will climb it.

7. Taming, defined as the reduction, or possibly total elimination, of an ani-

mal's flight reaction from man, may be deliberately induced. This is an indispensable precondition both for training and domestication. The latter is a state wherein not only the care and feeding but most particularly the breeding of the animal—or communication of genetic information from one generation to the next—have all come (at least to a degree) under human control. (When the biologically altered domesticated animal breaks out of control, it is referred to as "feral," as opposed to "wild.")

8. Man trains an animal. Such management may take one of two counterpolar forms: a. A rat forced to swim under water to escape drowning is taught to take the alley in a submerged Y-maze when the correct decision is indicated by the brighter of the two alleys; a porpoise is brought under behavioral control to locate and retrieve underwater objects. Such efforts are called *apprentissage,* loosely rendered as "scientific training."   b. A horse is taught to perform a comedy act for purposes of exhibition (Bouissac 1976, Chap. 8); a porpoise is taught to play basketball. Such efforts are called *dressage,* or circus (viz., oceanarium) training.

Apprentissage and *dressage* are fundamentally distinct ways of shaping behavior, although from a semiotic point of view they constitute complementary measures, in particular as regards their pragmatic import. This variance was intuitively appreciated by Hediger as early as in his 1935 dissertation and later advanced substantially in several of his published writings (e.g., 1968:120f., 1974). For instance, Hediger insightfully emphasized that *apprentissage* entails a reduction of the animal–man nexus to as close to zero as may be feasible. *Dressage,* on the other hand, requires a maximum intensification of the ligature, with the richest possible emotional involvement. The signs employed fluctuate along this dimension. Apropos *dressage,* the Brelands (1966:108) related an interesting informal observation concerning the emotional component of a parrot's vocalization: in the exhibition in question, the bird picks up a toy telephone, holds it up to his ear, and says "Hello!"—after which he receives a peanut. It was noted that, every time the bird said "Hello," "the pupils of his eyes contracted and dilated remarkably." This sign is emitted solely in an emotionally charged situation, for the pupil-size cue may not occur if the bird is talking merely for peanuts (for kindred observations in domestic cats, see Hess 1975:116f).

TWO MODES OF ANIMAL TRAINING: APPRENTISSAGE VERSUS DRESSAGE

Some of the searches for primate language capacity, represented in this book and elsewhere, clearly constitute a striving toward pure *apprentissage;* others can more readily be characterized as leaning toward *dressage;* but most of these projects fall somewhere between these two ideal types of semiosic construction. This is due to both the practical and theoretical constraints upon such investigations. On the practical side, it was found that experimental ape subjects performed poorly if deprived of a congenial social environment. Gill and Rumbaugh report, for example, that

during early training we maintained that Lana should have little interaction with the experimenter so that her strict training schedule would not be disputed. We soon found, however, that the social aspect of language training was of great importance. Accordingly we modified our initial decision and began allowing increased social contact between Lana and the experimenters. Lana did much better in an informal social setting with the training procedures modified as needed. (Rumbaugh 1977:161)

Premack, too, describes the "most efficacious training phase" as the one where Sarah's trainer worked inside her cage, sitting on one side of the table with Sarah on an adjacent side (1976:24). "The whole arrangement was intimate," he writes:

Sarah's hand could be guided when needed and her head turned gently in a desired direction to assure attention. Astute cajoling often served to stretch out the lesson. Trainers held her hand, patting and encouraging her with affectionate tones. Sometimes she reciprocated, gently chucking the trainer under the chin. (*ibid*.:25)

Sarah was rewarded with tidbits. Although Premack, so far as we know, did not test to see whether Sarah would have performed as well without such material reinforcements, he emphasizes the importance of the rapport between trainer and chimpanzee by supposing that "social approval . . . would have proved to be a more potent reward" than the food (*ibid*.:27).

Trainers, even those educated in the tradition of the experimental technique of *apprentissage,* also appear to derive a great deal of satisfaction from a close, emotional involvement with their ape charges. It is not uncommon for young apes to elicit parental feelings of attachment in their caretakers, as evidenced in the following passage by Ann Premack, who played an active role in training Sarah, as well as in planning the chimpanzee's overall teaching program:

People who raise chimps have high expectations for them as they have for their own children, and when the chimps don't perform at these levels, the "parents" are often bitter. . . . Aside from a human baby, I can think of no creature which can arouse stronger feelings of tenderness than an infant chimpanzee. It has huge round eyes and a delicate head and is far more alert than a human infant of the same age. When you pick up a young chimp, it encircles your body with its long, trembling arms and legs, and the effect is devastating—you want to take it home! (1976:16–17)[1]

In contrast to experiments where a chimpanzee being given linguistic training is raised as much as possible as a regular member of a human family (e.g., Hayes 1951; Kellogg & Kellogg 1933; Temerlin 1975), reports of a trainer's subjective reactions to his or her ape subject in more recent ape language projects are rare,

[1]The appealing features of infant chimpanzees described by Mrs. Premack are comparable with the releasers of caretaking responses displayed by human babies. See, e.g., Guthrie (1976) and Morris (1977).

probably because they are thought to be irrelevant to the scientific aims of training and testing. Since every trainer, even one brought in from the outside for purposes of blind testing, must, if he or she is to work directly with the animal subject, go through usually several "get acquainted" sessions, it would appear that the avoidance to allude to, let alone analyze, such encounters, which seem to be in some ways as fundamental a part of *apprentissage* as of the more explicitly emotive technique of *dressage,* is due more to received notions concerning scientific method and *apprentissage* in general than to the actuality of each case. A more candid expression of these human affective responses, such as that provided by Mrs. Premack, where she describes her earliest associations between herself and Sarah with memories of her childhood experiences with a retarded sister, and the slow process of accommodation between herself and the chimpanzee, would no doubt furnish a richer picture of the actual circumstances in which humans are pursuing their overt stated goal of *apprentissage.*

Tradition dictates that the ape's "learning" begins only during the next stage—during the first actual training sessions—and that this type of encounter warrants a fuller description than the preceding one, in which trainer and subject first strive to establish a social rapport, i.e., they learn to interpret and respond to one another's signs—come to share, in sum, a common semiotic code—in such a way that mutual understanding will thereafter become possible. As alluded to above, even scholars such as Premack and Rumbaugh, who are careful to achieve as emotion-free a context as possible for their work, admit that "interpersonal" communication of affect is a *sine qua non* of a successful training or testing session, and yet, strangely, they are quick to deny that the exchange of such information interferes with the attainment of more or less pure *apprentissage.* The implication is, of course, that communication may be neatly divided into two types—affective and symbolic—and that the two are independent and may be separated one from the other at will. However, speaking of ape communication in general (this would presumably apply as well to intraspecific communication among humans), Marler notes that "[i]t is important not to underestimate the potential richness of the information content of affective signaling" and, further, citing Premack himself, "as long as there is some concordance between the preferences and aversions of communicants then a remarkable amount of information can be transmitted by an affective system" (this volume:226).

It is ironic that Premack should defend the affective communication system of apes (and presumably humans) and then proceed to more or less ignore the importance of the affective communication taking place between his experimental chimpanzees and their trainers. This is especially puzzling because, in his descriptions of interactions between Sarah and her trainers, there is frequent allusion to the human's emission of nonverbal signs. Describing Sarah's extralinguistic antics, for example, he writes that "[W]hen she miscalculated in her gymnastics and fell 6–8 feet to a concrete floor, she rose without a whimper, often to come over to investigate *the grimacing trainer whose face was wreathed in empathetic pain"* (1976:27; emphasis added). After 16 months of training in the intimate conditions described above, Sarah became sexually mature and sometimes dangerous, so that training had to be carried on through a small door in her cage, human and chimpanzee on opposite sides. This arrangement proved less successful than the preceding one. "It was now difficult to pat Sarah and hold her hand, and she rejected far more lessons than she had in the first phase" (*ibid.*:28).

The emotional commitment of Sarah's trainers to her success is even palpable in the following passage, where Premack is denying that the Clever Hans phenomenon played a part in his project: "There were numerous informal contraindications of social clues, the most vital of which were the clever innovations made by the subjects themselves. *These innovations delighted and surprised the trainers, for they were in no way traceable to the training"* (*ibid.*:29; emphasis added). It is reasonable to assume that the emotions of delight and surprise were expressed at least nonverbally in the presence of the chimpanzees, and could in fact have served as added reinforcing cues. Whether in the Premack project or other similar ones, this sort of personal involvement in the outcome of training and testing could be due to the trainer's relationship to the project itself, as well as to the appealing chimpanzee or gorilla subjects which figure in it. A graduate student, for example, may strive for successful performances as a way of pleasing his employer, the principal investigator, or both himself and his professor, should the former be developing a doctoral thesis from his work with apes. The principal investigators themselves, of course, require success in order to obtain continued financial support for their project, as well as personal recognition and career advancement.[2] Whatever the reason, it is immediately clear to an outside visitor to any of the ape "language" projects, or to anyone who talks at length with those involved in them, that the animals are surrounded by a dedicated group of enthusiastic workers, one that constitutes a tightly knit social community with a solid core of shared beliefs and goals in opposition to outside visitors, as well as against groups elsewhere which are competing for scarce research resources. In fact, it is difficult to imagine a skeptic being taken on as a member of such a "team." The fervor of the team members increases with the amount of time which individuals have spent on the project, which, in some cases, now exceeds ten years. The total dedication required for projects of this sort may partially explain why it seems to be the rule rather than the exception that the work becomes a family affair, the principal investigators—

---

[2]Barber (1976) reviews studies of cases of intentional fudging of data both by investigators, often in intense competition among each other for funds and prestige (pp. 36–44), and by experimenters (pp. 60–63), who, if they do not have a stake in the research, may fail to follow required procedures or fudge data in order to carry out assigned tasks and hand in materials which will be acceptable to the investigator.

husbands and wives (e.g., the Gardners, Premacks, and Rumbaughs)—both becoming deeply involved, with graduate students and younger colleagues serving as the subjects' uncles and aunts, or members of an extended family.

Whatever the sociological and psychological context of the anthropoid ape "language" projects, and the degree to which they approach *apprentissage* as opposed to *dressage,* those involved are for the most part unaware of the literature on *dressage.* How many psychologists, we wonder, have ever even glanced at Hachet-Souplet's eye-opening chapters (1897:79–91) on "le dressage du singe," reporting training procedures traceable back to at least the 13th century and reaching a stage of modern sophistication by the 18th?[3] This knowledgeable circus buff and onlooker upon its world of wonders puts in doubt even educated monkeys' supposed ability to imitate human movements, and tells us how that illusion is built up by dint of their quadrumanous configuration. Part of the difficulty in properly assessing attempts at *apprentissage,* such as in the ape "language" projects, or apparently miraculous performances such as talking animals (Sebeok 1979a, this volume), is that those who are called in to serve as expert judges of such events are usually those trained only in *apprentissage,* whereas it is precisely the person who is thoroughly familiar with *dressage* (e.g., Hediger, who is the world's leading expert on the psychology of animals in captivity, especially man–animal communication in zoos and circuses—see 1968, 1969, 1974), or other performance-oriented activities, such as conjury (e.g., Christopher 1970; Randi 1975), or a combination of scientific methodology and conjury (e.g., see Diaconis 1978; M. Gardner 1957, 1966) who are best able to detect the operation of subtle affective cues—whether unwitting and therefore self-deceptive, or aimed at misleading others—in a supposedly "controlled," emotion-free laboratory experiment.

GREAT EXPECTATIONS

All the studies assembled in this book fall into the eighth category of man–animal interactions, or man training animal, in which, however, further far-reaching distinctions must now be introduced. Two variables of direct concern in the present context are: the performance intended to be induced, and the organism selected for the training. As for the former, the class of behaviors most particularly to the point are those involving attempts to *humanize* an animal, that is, to attribute to it hominine traits, indeed, "to endow the animal with psychical capabilities like those of men, and to say that it acts from similar motives" (Katz 1937:19). This trend of thought, associated, among others, with the romantic indulgences of the school of

---

[3]Cf. *Haney's Art of Training Animals* (Burroughs, 1869), and William F. Pinchbeck's *The Expositor, Or Many Mysteries Unravelled* (privately published, 1805).

Romanes, and known, of course, as anthropomorphism, was recently adverted to by Griffin (1976:68–71), who, particularly in reference to nonverbal communication, suggests "that it is more likely than not that thoughts or mental experiences in people and animals share important properties without being completely identical" (p. 70). Griffin cautions "that enthusiastic observers of animals are constantly in danger of interpreting their behavior in more complex terms than is necessary or correct" (p. 72), and proffers the Clever Hans effect as "an outstanding example" (*ibid.*). More recently, the special issue of *The Behavioral and Brain Sciences* devoted to "Cognition and Consciousness in Nonhuman Species" and featuring articles by Griffin (1978b), Premack and Woodruff (1978), and Savage-Rumbaugh, Rumbaugh, and Boysen (reprinted in this volume), plus a lengthy section of commentaries by a wide range of discussants, brings this problem sharply into focus, especially from the perspective of input from projects designed to teach apes languagelike skills to the establishment of a cognitive ethology.

Several separate but intertwined issues appear to be at stake here—namely, the risk that the beliefs, aspirations, and expectations of investigators and experimenters will lead to (1) inaccurate observations and/or recordings of ape behaviors; (2) the overinterpretation of ape performances; or (3) the unintended modification of an animal's behavior in the direction of the desired results.

OBSERVATIONAL AND RECORDING ERRORS

The first of these issues is especially important for the analysis of projects attempting to teach apes the American sign language of the deaf (ASL), since their methods of observation and informal testing are such that it is unclear exactly what should be counted and recorded, in which case, as several studies have shown (see Barber 1976:74–75), experimenters' expectancies can affect how they interpret or score responses of their subjects. But this issue is certainly relevant as well to researches using artificial symbol systems. Barber (*ibid.*:57–59), in fact, reviewing studies which show that experimenters frequently make errors in recording events, and that these mistakes are generally in the direction of the experimenters' expectations or desires, illustrates the relevance of this particular type of pitfall to any experimental procedure.

Beginning with the seminal work of the Gardners with the chimpanzee Washoe, all subsequent ASL research with either chimpanzees—i.e., that led by Fouts in Oklahoma, and by Terrace and Bever in New York—or with gorillas— i.e., the work by Patterson at Stanford—has been open to question as to the accuracy of observations made of spontaneous signing by ape subjects in more or less informal, conversational settings. The Gardners report (this volume:300) that both in their original work with Washoe and with later subjects (Kojo, Pili, Tatu, and

Dar), the criterion of reliability of reporting sign acquisition was that a sign be reported on three independent occasions by three different observers, at which point it is listed as an acquired sign. Such a sign was counted as a "reliable item of vocabulary" only when, in addition, it had been noted to occur "spontaneously" and "appropriately" at least once on each of 15 consecutive days.

The Gardners claim that these are much more stringent criteria than those used by students of language acquisition by children, e.g., by asking the mother to list new words—a technique, by the way, no longer favored in modern child language studies—or by taking random samples of a child's speech on different occasions. However, such comparisons are not altogether appropriate. In contrast with the situation of a mother with her language-learning child, for example, the assistants teaching the chimpanzees are brought into contact with the animals for the sole purpose of inculcating in them certain carefully programmed communication skills. They are thus more narrowly and intensely focused on a restricted range of the learner's behavior than is a mother in a home setting. In addition, they have frequent discussions with one another and with project leaders about signs they believe they have seen, doubtless even reading one another's notes and research diaries.[4] In this way, one observer may prime another to inadvertently create situations in which a newly reported sign might be likely to recur by, for example, using a certain tool when the sign for that object has been reported by another observer. This outcome is especially likely owing to the fact that trainers were instructed by the Gardners to record the context in which a new sign was observed as well as the sign itself. Should the ape actually produce a sign in this sort of context, it would not be considered the result of outright prompting, molding, or the Clever Hans effect, all of which the Gardners deny played a role in the reporting of new signs, and yet such innocent provision of opportunities for corroboration of other observers' records would certainly influence the overall course of the study.

Other methods employed in the ape ASL projects make such distortions even more likely. To take the Gardners' project as an example, "old" signs were, first of all, continually reassessed to see if they had become a permanent part of the chimpanzee's repertory. Reassessment consisted of deliberately introducing the appropriate context for signs in the chimpanzee's vocabulary which had not yet appeared spontaneously during the day. New signs which had not yet met the criterion were elicited, for example, by presenting a picture of a cow, since it would be unlikely that the chimpanzees would see a real one 15 days in a row, and, finally, new signs were taught through prompting and molding.

A second weakness in the reporting procedures in this type of enterprise is that a trainer may, after conferring with his colleagues about new signs, simply be more prone than he would have been to interpret a certain gesture on the part of the ape

---

[4]This procedure is also followed by Fouts, Patterson, and Terrace.

as the new sign reported yesterday by someone else. If a sign has come up in conversation, it is only to be expected that this knowledge will predispose observers to interpret any gesture which remotely resembles that sign configuration as the sign itself, in the manner of a *pars pro toto*. Thus, Clifford and Bull (1978:161) have shown that when a group of witnesses gets together to reconstruct an event it may produce a more complete account of details, but "at the cost of increased inaccuracy with groups obtaining 40 per cent more errors of commission than individuals. . . . Group interactions may pressure witnesses into offering inferences rather than perceptions, or even into prefabrications." It is thus misleading to emphasize either the independence of observations or the spontaneity of the apes' performances, since the actual working conditions of such projects considerably reduce both such elements.[5]

That reports of ape signs may be incorrect is even more likely when one considers that trainers were commonly asked to keep daily diaries, for which they had to rely on their memory of sign exchanges occurring while many other events competed for their attention. Anyone who has ever attempted to reconstruct a dialogue that took place even a few minutes before knows how difficult a task this is, even if one only tries to capture the basic form of each utterance, ignoring all the paralinguistic and nonverbal elements that infused and surrounded it.[6]

There are several other methodological considerations which cast further doubt upon the objectivity of the recordkeeping procedures involving signing apes. The Gardners report that, as their subjects matured and the frequency of signs increased, maintaining a complete inventory of signs and contextual descriptions became impossible. "Often," they write, "the brisk pace of an extended conversational interchange precludes verbatim recording." It thus became necessary to employ a different method of recording, whereby, during a specified period, "everything signed by the chimpanzee," as well as contextual descriptions, were recorded. These sign protocols required a team of two persons, one whispering an immediate spoken transcription of signing, together with notes about context, into a tape recorder, the other "performing the usual role of teacher, caretaker, playmate, and interlocutor" (this volume:311). The Gardners patterned this method after the work of Roger Brown and his colleagues with language-learning children, but in so doing overlooked the critical difference between the medium of commu-

---

[5]On this point, cf. Ristau and Robbins (1979:291).

[6]See Clifford and Bull (1978) for a full review of the many linguistic and cognitive factors which influence the perceptions and memory of eyewitnesses, often leading to inaccurate accounts of events. The material they present fully supports C. S. Peirce's theory of the logical nature of perception and hypothesis (see Sebeok & Umiker-Sebeok 1980). In light of the immense amount of evidence brought forward by Clifford and Bull, among others, persons attempting to teach languagelike skills to apes would do well to heed the warning, given those involved in juridical proceedings, about the danger of accepting the fictions of the law court—namely, that eyewitnesses see and hear correctly and so testify, and that, if they do not testify with accuracy, cross-examination will straighten them out.

nication being studied in each case. The majority of the assistants working with signing apes are native speakers of English, having learned ASL only for limited use during their participation in the project. These observers are thus placed in a situation where they must transform one sign system, ASL, which is for them a secondary form of communication, into another, speech, which requires not only a simple translation within a single channel, but a transmutation from the visual to the acoustic mode. It seems reasonable to conclude that there would be a far greater rate of misinterpretations in this type of situation than even in one where a speaker of, say, English, were called upon to make a simultaneous translation of the utterances made by a language-learning child in a spoken language—say, Hungarian— in which that observer is not fluent!

The simultaneous recording of children's speech, even when in one's own native linguistic code, is itself difficult, especially when undertaken within the rapid, noise-filled context of a young child's living quarters. Errors are easily introduced at every stage of the process, from the original observations to the final transcription of the tape itself. It is puzzling that critics of the ASL ape projects have overlooked this fundamental shortcoming of this method of data collection. Despite the not inconsiderable number of sources of misjudgment, they seem to take for granted the reliability of observers, based largely, one would assume, on their trust in the spot checks made by so-called expert witnesses who are proficient in ASL and who testify that they have in fact observed the apes making what they took to be genuine ASL signs. We shall return to the question of the reliability of such expert testimony below.

Further possible sources of error can be uncovered if one imagines what kind of communication could be taking place between the pair of humans involved in such team recording sessions as described above. Reports indicate that trainers and observers enter these situations with the tacit understanding that it is the chimpanzee's, not one another's, behavior that is of interest. But surely at least some of the comments of the recorder will be heard by the ape's interlocutor, thus inevitably altering the latter's own interpretations of the ape's signs and in this way influencing the direction the human–ape conversation may take. If the session is to yield usable data, the ape's interlocutor must make sure that signs are visible to the recorder, and this, coupled with the feedback produced by overhearing a slightly delayed interpretation of the ongoing interaction, and the knowledge that one is performing for an audience, cannot help but make sign exchanges less than completely spontaneous. The construction of reality in such situations would involve communication between all three participants, in complex two-directional forms between both humans and between each human and the ape. By observing the human interlocutor's reactions to the ape's behavior, the reporter is as much cued as to the meaning of the exchange as by watching the chimpanzee's performance, just as the human interlocutor is influenced by the spoken interpretation of the interaction by the reporter.[7]

This is not to say that some researchers have not been aware of the springes mentioned above. Terrace and his colleagues, for example, have attempted to check the reliability of their trainers' observations of Nim's signs by having outside observers, some of them, unlike the students, fluent in ASL, witness sessions between the chimpanzee and one of his trainers, and then compare the two resulting accounts of each human involved. They report (Terrace, Pettito, Sanders, & Bever, n.d.:13) that agreement between the teachers' reports and the transcripts of independent observers was between 77 and 94%. This sounds impressive until one reads (n.d.:fn. 33) that these levels of agreement were attained only following a discussion between the teacher and the independent observer, during which they attempted to reconcile whatever differences had been found in their respective transcripts. The two human participants, in other words, renegotiated the reality of the events in question, leaving us to wonder how much the figures quoted represent what communication took place between the chimpanzee and the teacher, as opposed to between the two humans.[8] The honesty of Terrace and his colleagues in reporting this and other procedural steps which others, we suspect, omit from their accounts, is to be applauded. It makes it possible for the reader to assess more accurately the reliability and import of the data presented.

To continue with the example of the use of independent observers of Nim's sign behavior, other questions come to mind concerning the collection of data itself, prior to their evaluation. Nim's classroom consisted of a bare, 8'-square room, one wall of which contained a large one-way mirror, beneath which there was a portal used for photographing and videotaping the chimpanzee's behavior. The independent observer sat in an adjacent room, viewing the chimpanzee and the teacher through the one-way mirror. This situation did not preclude the possibility that the observer, assuming that he was not deaf (of which we are not informed), could hear the teacher's spoken account of the interaction and thus receive acoustic as well as visual cues from the latter concerning the meaning of Nim's gestures. We are not told that sound could *not* transude from the classroom to the observation room,

---

[7]This type of situation presents an interesting combination of eristic and heuristic dialogue (Perelman & Olbrechts-Tyteca 1969:39): (1) two of the participants, the trainer and the ape, communicate directly with one another, as specific individuals; (2) two of the participants communicate with the third person, the observer, only indirectly, through one another; and (3) all three actors address a "universal" audience—that is, the reader/viewer of a written or filmed record of their interaction.

[8]This sort of communication might be compared with discussions that take place between a mother and a stranger concerning the meaning of the utterance of the mother's child. The influence of linguistic interpretations over perceptions works in both directions, especially if the stranger is an "authority" on language acquisition, child psychology, or the like. The extent to which students of language acquisition bias a mother regarding her own child's linguistic performances is still unknown.

although this seems all the more probable in view of the existence of the portal and the fact that, in order to assure proper recordings of the teacher's and observer's spoken commentaries, all possible sources of noise in the acoustic channel had probably been eliminated. In any case, this one example exposes a few among the many conceivable flaws in the laudable plan to introduce "independent" observers in order to check the reliability of project personnel.

OCKHAM'S RAZOR

The second issue at stake here is the methodological principle of economy of explanation, commonly known as "Ockham's razor," which is invoked in several alternative formulations of the influential 14th century philosopher, as: "What can be done with fewer assumptions is done in vain with more." The principal justification for parsimony is the elimination of pseudo-explanatory entities. Criteria for judgment in accounting for any fact, according to Ockham, must derive either from manifest experience (the empiricist criterion of evidence) or from compelling reasoning.

Probably the most common form of criticism leveled by one ape "language" investigator against another is that simpler, more parsimonious explanations are available to account for a rival's data, whether or not they were gathered through participant observation method or forced-choice testing.[9] Commenting on the Gardners' use of spontaneous productions to judge knowledge of language, Lenneberg (this volume:81) warns that all Washoe's spontaneous signs should be disregarded for purposes of assessing her linguistic competence, not only because of the peril of the Clever Hans effect, but also because, in our attempts to determine the appropriateness of such messages, "we are simply testing our own ingenuity to assign interpretations to productions that might, for all we know, have been emitted randomly. It would always be possible to claim that the ape had 'intended' to tell us something. . . ." Overly rich interpretations of chimpanzee and gorilla ASL productions abound in the published reports. Savage-Rumbaugh *et al.* (this volume:375) have repeatedly argued that, given the weak acquisition criteria employed by those working with signing apes—namely, the ability to name pictures or to produce an iconic gesture repeatedly, under the same or similar circumstances—"it is impossible to tell whether the chimpanzee is simply imitating or echoing, in a performative sense, the action or object, or whether the animal is

---

[9]As Griffin (1978b:555) comments, "The leading groups of experimenters training apes to use symbolic communication share a general faith in the significance of the endeavor, but they are often critical of each other's specific experimental procedures. If one accepts the sharpest of these mutual criticisms, it is tempting to dismiss all of the results, as inconclusive."

indeed attempting to relay a message." Since, as they further allege, ASL is a highly iconic system of communication, they conclude that there is nothing to prevent us saying parsimoniously that the apes are merely producing "short-circuited iconic sequences."[10] Unless it can be shown, they go on to say (p. 378), that the ape could competently use a sign apart from the original context in which it was acquired (i.e., in a situation where the actions and signs of the trainer cannot remind the animal of the hand movements used in the sign), the mere announcement that signs occurred in "new contexts" is not sufficient to rule out the possibility that the animal's behavior can be accounted for by the simpler explanations of deferred imitation or error interpreted in a novel manner by the experimenters. The records kept on Washoe, according to Premack, were meant to preserve sense or meaning rather than word order, and, as such, "reflect dispositions of the recording organisms no less than of the organism recorded" (1976:331).

Much has been made of various experimental apes' so-called "innovations," several examples of which—including generalizations of signs and construction of new lexical items—are provided by Hill in this volume (p. 339–340). Unfortunately, these instances are generally reported in such a way that not enough is learned of the context of occurrence to enable us to rule out the possibility of either trainer suggestion or overinterpretation. Terrace *et al.* (n.d.:6), to take one example, note that there is available a more parsimonious explanation for Fouts's report that Washoe was creating a new compound lexical item when she produced "water" plus "bird" in the presence of a swan and was asked, "what that?" Since Fouts provides no evidence that Washoe characterized the swan as a *bird that inhabits water,* we may just as well assume that Washoe, who was very familiar with the question form *what that?,* was merely responding by first identifying a body of water, then a bird (cf. Martin Gardner's comment on this example, cited in Sebeok 1979b:57–58).

If a sign or other response produced by an ape appears to be inappropriate, to take a second type of example, human trainers appear all too willing to stretch their imaginations in order to make the animal's performance "fit" conversationally, reminding us of the clever interpretations made of the so-called telepathically-communicated images drawn by such allegedly psychic persons as Uri Geller, where gullible observers, wishing to believe in his extraordinary powers eagerly project their own concept onto the sometimes amazingly crude but suggestive scribbles made by the performer (Randi 1975).[11] Thus, anomalous chimpanzee or gorilla signs may be read as jokes, insults, metaphors, or the like, much as the "not infrequent offenses against the very elements of counting and the fundamental arithmetical

---

[10]On the Rumbaughs' own use of richly interpreted anecdotes in their work with Lana and other chimpanzees, see Mellgren and Fouts (1978) and Schubert (1978).
[11]Cf. Randi (1978).

processes" made by Clever Hans "were regarded in part as intentional jokes and by an authority in pedagogy as a 'sign of independence and stubbornness which might also be called humor'" (Pfungst 1965:145). Patterson (1978a:456) reports that Koko

seems to relish the effects of her practical jokes, often responding exactly opposite to what I ask her to do. One day, during a videotaping session, I asked Koko to place a toy animal under a bag, and she responded by taking the toy and stretching to hold it up to the ceiling.

Or, to take another example from the same source, it was interpreted as a deliberate joke (the gorilla was "grinning") when Koko, in response to persistent attempts to get her to sign "drink," made "a perfect drink sign—in her ear." If Koko was not used to making this sign, one wonders how the observer knew that this variation was in fact a play on an established signifier–signified relationship (see Sebeok, this volume)? While accepting that the sign play of apes is genuine, Hill nevertheless comments (this volume:341) that

there remain serious questions about whether their [the apes'] metaphorical abilities are truly linguistic. . . . No matter how suggestive is the circumstantial evidence, the semantics of the chimpanzees' symbol systems cannot be said to be the same as the semantics of the signs into which they "translate," since the system is so truncated.[12]

Premack (1976:29) claims that Sarah's symbolic innovations were the "most vital" type of informal evidence that the animal's performances in general were not the result of social cuing. The examples which he provides (*ibid.*:30), however, are all ambiguous, based on a trainer's hunch about what the chimpanzee meant by a deviation from the regular trained responses.

The determination of trainers to show their animal in the best light has even led to the denial that errors are mistakes, again bringing to mind one of the assumptions which sets research on psychic phenomena apart from normal scientific procedures. A so-called psychic's mistakes on tests of his powers are frequently used to prove that those powers are real, based on the assumption that if the performer were using mere tricks he would be correct every time, while his occasional mistakes prove that his correct responses are based not on conjury but a special yet unpredictable "force." "Thus," as Randi (1975:11) points out, "the performer who can consistently turn out effects that defy explanation by ordinary means is considered a fraud, and the one who 'hits and misses' or who has periods of impotency is judged to be the real goods." Turning to the chimpanzee experiments, we find Premack using Sarah's mistakes on matching sets with more than five items as proof that Sarah was not responding to social cues (1976:30–31). He argues that since the

---

[12]Cf. Lenneberg (this volume) regarding the difficulty in telling whether Washoe is making metaphors or mistakes.

chimpanzee had done well on matching sets that did not involve numerosity, the trainer would expect her to do as well on those that did, and thus would be likely to cue her toward success. This, however, is a mighty assumption. It is at least as likely that the trainer would reasonably doubt that a chimpanzee could handle so many items at once, despite her earlier performances. As far as we can tell, Premack never tried to determine what the trainer actually expected or did in the training situation.

Ristau and Robbins have noted that examples of gratuitous interpretations of conversational exchanges with Lana are also common:

[A]t times the interpretations of Lana's productions (Chaps. 9 and 12 of Rumbaugh 1977) involve assumptions about reasons for her errors that either the authors cannot possibly know or that rely on information the authors have not offered the reader. For example, they cite an error as the "result of Lana not attending to the question posed," giving no indication of what clues led them to this conclusion (p. 243)." (1979:270)

In another example (Rumbaugh 1977:243–244), of two seemingly irrelevant contributions by Lana to a conversation, one is simply ignored, the other excused, without evidence, as a "typing error" (Ristau & Robbins 1979:271).

Another example of a chimpanzee being in a "no-forfeit" situation is again provided by Premack (1976:31–32). He notes that Sarah had difficulty learning the concepts "is," "if–then," and the quantifiers "all," "none," "one," etc. He assumed that, "if social cues were the basis of Sarah's learning, she should have had no more difficulty learning complex predicates than simple ones" (p. 31). The logic of this argument is difficult to follow. According to his training procedures, complex predicates were broken down into simpler parts, so that Sarah had to go through more preliminary training sets to reach the full form. Yet, he says, because the difficulty of each step is about the same (his reason for breaking difficult patterns down), the animal should make no greater number of errors learning a complex pattern than a simple one! But it would seem that the very fact that there were more steps involved would mean a greater number of errors, even if the chimpanzee performed about as reliably on each step as when learning a simple predicate. Premack further confuses the issue by saying that Sarah's difficulty on some of the complex forms was *his* fault, not hers, since he did not break the form down properly. Is he saying, then, that Sarah did *not* make more mistakes on complex forms?

The danger of overinterpretation of an ape's signing or use of artificial symbols is thus present even when the method of testing controls the input to the animal. On the level of single lexical items, Premack (1976:36) claims that all that the Gardners' double-blind vocabulary tests have shown is that Washoe "learned to associate different responses with different objects," and that, "in view of the chimpanzee's capacity . . . is there any serious reason to doubt that it can learn to associate different responses with different objects?" Likewise, Rumbaugh (this volume:242) criticizes the Gardners' testing procedure, saying that it

did not provide for a way to determine the degree to which Washoe's signs were essentially generalized responses to stimuli similar to those used in prior training. Her competence in the use of her signs syntactically in order to generate and transmit truly novel messages, information, or requests was not tapped by this or any other test. Instances of her using signs in these ways were reported separately; however, they may have been fortuitous productions.

The tests of Washoe's comprehension of multi-sign sequences have also been criticized from the point of view that in most cases non-syntactic or nonlinguistic explanations are readily available to account for the chimpanzee's responses. Terrace *et al*. (n.d.:62), for example, note that the Gardners accepted as correct any response to a wh-question which they thought was lexically appropriate, so that, for example, if Washoe signed "blue" in response to the question "what color?" when the object being used was a red ball, her answer was counted as correct because it was a color. Washoe clearly learned to respond to such category questions by producing a response which fell within the correct category, but many of her specific responses were inappropriate. Certainly, as Terrace and his collaborators maintain, such responses cannot be used to show, as the Gardners claim, that Washoe's performance at the time was comparable to that of the human child at Stage III in Brown's scheme of language development (see Brown 1968).

Even when Washoe's responses to wh-questions were appropriate, Rumbaugh (this volume:244) wonders

whether this level of mastery constitutes "grammatical" control, for grammar subsumes word-order effects. The constituents of Washoe's responses to the wh-questions (e.g., Q. "Who you?" A. "Linn"; Q. "Whose that?" A. "Shoes you"; Q. "Where we go?" A. "You me out") were by and large appropriate, but the report (Gardner & Gardner 1975) and its analyses did not address questions regarding the logic inherent in the organization of responses that entailed *more* than one word—a relevant issue in the evaluation of "grammatical control."

For one reason or another, investigators appear to find it difficult to avoid the use of tests in which the animal subject can rely upon nonlinguistic strategies to accomplish a task in a way which is considered correct. Take, for example, Fouts and Rigby's description (this volume:277) of their testing of comprehension of sign sequences by a chimpanzee named Ally. Ally was first taught to pick out one of five objects from a box and put it in one of three places, e.g., "Put baby in purse." Next, new objects were put in the box and a new place to put them added. "With five items to choose from and three places to put them, chance would produce one correct response in fifteen trials," they claim. But Ally's performance "was far above the chance level." At closer look, however, one wonders what such results actually mean. Since the sentence frame, "Put x in y," did not vary, only the object (slot x filler) and the location (slot y filler) elements were not redundant. All Ally had to do was to recall his prior associations between a restricted number of signs and their objects and perform the general task, learned in the training phase of the experiment, of placing the object in the designated place. The order, *object* plus *place,* was maintained throughout the test, so that Ally would know from this non-linguistic cue that (1) the first nonredundant element was something in the box, and (2) it was to be moved toward the second nonredundant element in the string of signs. At most, then, this test was one of Ally's ability to name, not of his knowledge of sentence structure. Compare the similar analysis, made by Terrace and Bever (this volume:183–184), of Sarah's alleged sensitivity to the sentence structure of apparently more complex sequences such as "Sarah insert banana pail apple dish."

A number of kindred attacks have been made on Premack's interpretations of Sarah's use of language tokens in forced-choice tests. After analyzing his treatment of Sarah's training with "name of" concept, for example, the Gardners (this volume:316) conclude that "since Sarah's program of training and testing concentrated on each 'linguistic concept' for days and weeks at a time, she could have solved Premack's entire battery of problems by rote memory alone." Savage-Rumbaugh *et al.* (this volume:379) also affirm that some of Sarah's performances—notably the match-to-sample paradigm tests designed to demonstrate Sarah's symbolic representational capacity—could be accounted for on the basis of simple recall, their own experimental chimpanzee, Lana, having performed much more difficult recall problems. "Questions as to whether or not such a [symbolic] capacity is necessary to solve the various problems could and should be raised for the entire gamut of tests presented to Sarah," they write (*ibid.*:380). They agree with Mounin (this volume:174) that "the suspicion lingers" that only an illusion of language is created by the translation of Sarah's token sequences into English, while, stripped of this linguistic surface, the animal's behavior could reasonably be explained in terms of simple response–reward associations which have little to do with linguistic competence.[13]

Terrace has, in fact, applied the law of parsimony to a large number of Premack's testing procedures, and, in each case, "the homogeneous nature of the questions posed during any one session, along with the restricted range of possible answers, increases the likelihood that nonlinguistic contextual cues contributed to the performance of Premack's subjects" (this volume:392). Sometimes, as he shows, it was even possible for the experimental animals to solve a problem without knowing the critical word involved. Simple discrimination learning and memory could account for many of Sarah's responses which hitherto have been cited by

---

[13]One of the reasons Pfungst (1965:227) concluded that Mr. von Osten, the owner of Clever Hans, was not deceiving the public was because von Osten's training technique, which the horse's master thought of as "instruction," took much longer to accomplish than would have a straightforward stimulus–response training, without the illusion of intervening mathematical or linguistic concepts.

Premack as examples of knowledge of sentence structure, communicative intent, pronouns, prepositions, and predicates. For example, Premack (1976:32) argues that Sarah was responding to sentential complexity rather than nonlinguistic factors of a certain test because, for one thing, she took longer to handle the complex sentences than simple ones presented (although length of time was not recorded—this was just Premack's impression, it seems), and she would do a smaller number of the former in any given training session. This might be said to prove, rather, that Sarah was solving problems on the basis of memory and that it took her longer to remember more training associations for tests involving complex sentences, since she had to go through a greater number of training steps for these.

Terrace and Bever (this volume:185), noting that the Lana project's use of a computer provided Lana with access to a much larger set of alternative symbols than did Premack's, nevertheless point out that Lana was also trained in long sessions devoted to a single type of problem, requiring constant primary reinforcement in such a way that a sequence, glossed by Rumbaugh and his associates as "Please machine give Lana M & M" or "Please machine give Lana piece of raisin," "can be analyzed as a complex X–R chain." Such two-step, chained associations would be the equivalent, they say, of the learning of a rat in a double-T maze. The Gardners (this volume:317) have also criticized the interpretations made of Lana's performances, claiming that all Lana had to do to get a reward was to produce one of a very small number of correct color sequences on her computer console. "There is no reason to suppose," therefore, "that Lana's productions have any semantic content." The large number of drill sessions with the same small set of keys and stimulus arrays, they add, would not "have placed any great strain on the rote memory of a chimpanzee" (cf. Fouts and Rigby this volume). Several commentators have, moreover, suggested simpler explanations of the tool use and exchange by the chimpanzees Austin and Sherman, described by Savage-Rumbaugh *et al.* (this volume). Jolly, for example, remarks that "the cleverness . . . is the experimenters' not the chimps'" (1978:580). "At the circus we applaud with awe as the human acrobat brachiates above ground," she continues (*ibid.*), "[t]hen we chortle when bediapered chimps waddle in on their hind legs and sit down to tea. Could the authors now devise an intellectual equivalent of letting the chimpanzees loose on the trapeze?" (cf. Lockard 1978; Mellgren & Fouts 1978).

Harlow (cited by Hill, this volume:337) concluded his review of the Premack project by describing it as "a series of learned tasks ranging in difficulty from simple discrimination to relatively simple matching-from-sample problems." While Harlow compares Sarah's performances with "more complex learning accomplished by macaques," Terrace *et al.* (n.d.:59–60) compare the multi-sign utterances of both Sarah and Lana with the problem-solving of pigeons trained by Straub and others to respond to nonsense symbols in a fixed order (see also Chomsky, this volume; and Ristau & Robbins 1979:272). Apparently, pigeons can learn to peck arrays of four colors in a particular sequence, irrespective of the physical position of the colors. If, as seems likely, they say, pigeons can also be trained to respond to ABC*X* problems, where *X* could refer to a variety of objects or actions, such performances could be compared with the chimpanzees' "Mary give Sarah apple," or "Please machine give apple," but why refer to the sequence, "green, white, red, blue" as "Trainer give R-42 grain"? Schubert (1978:597) also questions the assignment of English labels by Rumbaugh and his collaborators to the computer lexigrams manipulated by their subjects, Lana, Austin, Sherman, and Ally, arguing that, in the case of some of these translations, such as "straw" in the tool exchange experiment described in this volume, the experimenters "cash in on the overtones of the word" (*ibid.*:598) in an unintentionally misleading way.

Comparison with pigeons[14] points up the degree to which the very fact of choosing apes for this type of research introduces substantial distortion. We are much more likely to anthropomorphize an ape's performance, accepting English glosses as appropriate, than we are that of a pigeon, rat, or woodpecker (see below:27–28, 425). It is up to those doing this sort of work to prove that the sequences of symbols produced by an ape are evidence of linguistic relationships. Simply assigning English glosses is appealing, but insufficient as well as misleading.[15] The same may be said of our tendency to assume that, because a chimpanzee or gorilla produces hand configurations which resemble those used by deaf users of ASL, the animal is employing signs as do his human counterparts. In what is certainly the most skeptical approach yet made to this type of data, Terrace and his associates (n.d.), analyzing more than 20,000 combinations of two or more signs produced by Nim, reached the conclusion that each instance of what appeared to be grammatical competence on the chimpanzee's part could be adequately explained by simpler, nonlinguistic processes, including especially the imitation of a trainer's preceding utterances.

CLEVER HANS EFFECT

The third issue at stake here is the possibility of the human observer unintentionally modifying the animal's behavior to produce the desired results. It may never be feasible to exclude the fallible human half of the dyad under conditions of *apprentissage,* let alone the looser protocols of *dressage.* Hediger deems the task

---

[14]Cf. Brown's discussion (this volume) of Premack's work regarding the "pigeon ping-pong" problem— i.e., you can teach two pigeons to bat a ball, but is it ping-pong?

[15]Lenneberg (1975), using normal high school students, replicated Premack's study as closely as possible and found that, although the human subjects quickly outperformed Premack's chimpanzees, they could not correctly translate into English any of the sentences which they themselves had completed. Thinking that they were merely solving puzzles, the subjects failed to connect the plastic tokens with language.

of elimination of the Clever Hans effect analogous to squaring the circle, "if only for the reason that every experimental method is necessarily a human method and must thus *per se* constitute a human influence on the animal. . . . The concept of an experiment with animals—be it psychological, physiological or pharmacological—without some direct or indirect contact between human being and animals is basically untenable" (1974:29). When this statement is considered in the light of Niels Bohr's complementarity principle, the unavoidable interaction between the observer and the system observed is seen to impose the ultimate limitation on the knowability of the "real" state of affairs. The quantum of action couples the observer with the system observed such that an observation necessarily changes the observed system.

If Hediger is right, as we believe he is, then the door always remains wide open, as well, to the subtle intrusion of a host of both investigator and experimenter effects (Barber 1976) in the course of participant observation as a method to decipher animal behavior, for the technique necessarily hinges on a particular hermeneutic: the interpretation of signs by and of actors in the game. This holds *a fortiori* for animals, such as all apes, which can scarcely survive in the absence of an intense level of social intercourse with their handler before they are capable of undergoing the rigors of training for any task. We have already noted (p. 5) the often profound emotional, even familial, ties that exist between the apes learning ASL or an artificial language and their trainers, to the extent that the apes' performances uniformly declined—or the animals refused to cooperate altogether—when faced with a novel interlocutor in a double-blind testing situation. The temperamental response of a subject has been an all too convenient excuse used by investigators for not relying more heavily on outside experimenters in direct tests of ape language capacities. As in some parapsychological research, the basic rule that subjects do not select their own experimenter is broken when chimpanzees or gorillas are allowed to exclude certain individuals or classes of persons from testing situations.[16] In any case, it would be interesting to know precisely *why* the animals have such difficulty performing for strangers.

A related issue concerns the personal preferences exhibited by the apes for

---

[16]Randi (1975) has shown how important it is to Uri Geller and other so-called psychics that skeptical observers be excluded from tests of their special powers. He also demonstrated how it is the extraordinary personal control Geller has over the perceptions and expressions of sympathetic observers that makes it possible for him to appear to them as a true psychic. Kreskin (1973:26–27), in his discussion of ESP as entertainment, explains how he goes about capturing an audience:

> Rapport with the audience is built up through verbal contact, and to a lesser extent, body movement. The latter is not studied but does coordinate with patter to command attention. I attempt to keep all eyes on me. I then go about creating a climate for suggestible responses, literally playing it by ear and "feel" until I can sense that the audience is ready for communication and response. The main task is to instill faith, establish a "faith–prestige" relationship early. It may take fifteen or twenty minutes but the audience is seldom aware that the program is rapidly changing from the establishment of rapport and the conditioning of conjuring to an area far removed from magic. When volunteers come up onstage, they are unknowingly ready for response.

certain project members. All the animals used so far in the ape "language" projects have shown some preference for certain trainers, performing better for these "favorites" than for the rest, but there has been no systematic attempt to account for these differential responses on the part of the animals. As Hill (this volume:348) has pointed out, project reports have not "attended to the possibility that the chimpanzees have input into the communication dynamics, and hence into the society being constructed. The chimpanzees have often been seen as exclusively receptive; the interest in their output has been in its structural complexity and lexical variety." But, as Hill has noted, it is clear from many anecdotes, as well as from the very design features of some of the testing and training protocols, that there is a great deal of input from the animals in these experiments, and this input should be considered seriously as a component of every procedure followed.

The ape subjects are not the only ones to show a preference for certain project personnel, of course. Investigators appear to favor certain subjects—notably such "star" performers as Washoe, Koko, and Sarah—and their reports tend to emphasize the results obtained with these animals over those stemming from the behavior of less gifted individuals. Thus, what Barber calls the "Investigator Data Analysis Effect" (1976:19ff.) comes into play when, for example, an investigator obtains negative results with some animals and fails to report them, excluding from his analysis subjects which do not conform to his predictions and thereby distorting the reliability of his conclusions as to the significance of his data. A particularly disturbing example of this type of pitfall is found in Premack's work. In his book-length report (1976), for example, an early chapter, devoted to a discussion of the experimental subjects used and containing a series of *ad hoc* justifications for the poor performances of several chimpanzee subjects, is followed by others in which not only are the performances of Sarah, his prize pupil, highlighted, but it is often overpainted exactly which subjects are being discussed.[17]

Turning to the flow of messages from experimenter to ape, it should be noted that some trainers were more successful at teaching the apes than were others. Terrace and his colleagues (n.d.:14–15), for example, report that some of Nim's teachers were much better at their job than were others, so much so that they conclude that the rate of learning by the animal could be said to tell as much about the human involved as the chimpanzee. It is regrettable that, once again, individual differences in success in teaching subjects have received so little attention from investigators in this field, especially since numerous studies have shown that "experimenters differing in such characteristics as sex, age, ethnic identity, prestige, anxiety, friendliness, dominance, and warmth at times obtained divergent

[17]Ristau and Robbins (1979:271–272) discuss another type of Investigator Data Analysis Effect, this one in relation to the Lana project. Lana was obliged to end all her sentences by pushing a "period" key on her computer console. Investigators assumed that when the chimpanzee depressed this key if a grammatically incorrect sentence was in progress, she meant "erase," whereas she might just as well merely have intended to end that sentence, knowing that a reward would be forthcoming only if she did so.

results when testing similar [human] subjects," and that such an "Experimenter Personal Attributes Effect has been found on a wide variety of tasks including intelligence tests, projective tests, verbal conditioning tasks, and other physiological and educational measures" (Barber 1976:47). At the very least, investigators who do not use a sample of experimenters should, as Barber suggests, state the conclusions from their studies more cautiously, including a proviso that the results may be restricted to the specific kind of experimenter(s) used.

Pfungst, it may be remembered, also underlined that certain questioners were more successful than others in eliciting correct responses from Hans, and that "Hans acquired a reputation for 'Einkennigkeit,' that is, he would accustom himself only to certain persons" (1965:210). "Such a reputation was," as he said, "hard to reconcile with his much praised intelligence" *(ibid.)*. Hans preferred, and performed best for, as Pfungst uncovered, persons who exhibited (1) an "air of quiet authority," (2) intense concentration, (3) a "facility of motor discharge," and (4) the power to "distribute tension economically" *(ibid.)*. In other words, the successful examiner, focusing intently on the horse's tapping or other responses and on the anticipated perception of a correct movement, was able to sustain a tension and to release it at the right moment, in such a way that a detectable movement resulted, a movement used by Hans as a sign to stop performing. In what is still the most thorough examination of minimal, unwitting cues between man and animal, Pfungst succeeded in explaining the bulk of errors made by Hans, not in terms of insufficiencies on the part of the horse, but rather as instances where the questioner was inattentive, tired, unaware of the correct answer, or for some other reason incapable of producing the necessary muscular signal. Pfungst was even able to show the relationship between skepticism on the part of the experimenter and the poorer performance of the animal with such an examiner. A skeptical observer, he noted (1965:144–145), had a lower degree of concentration than one who expected to see Hans perform correctly. The skeptic thus did not relax at the proper moment, resulting most often in Hans tapping with greater frequency than was necessary for a "correct" response to an arithmetical question. The nonbeliever, it would seem, relaxed after his suspicion that the horse would *not* stop at the correct number had been confirmed. Hans, of course, did not know the difference between this and the correctly timed cue of relaxation given off by less skeptical observers.

Despite the fact that the issue is still hotly debated among psychologists (see, e.g., Barber's critique, 1976:64ff.), a recent review of 345 studies of "interpersonal expectancy effects" (or "interpersonal self-fulfilling prophecies") by Rosenthal and Rubin (1978:377) suggests that there is ample reason to believe that

effects of interpersonal expectations are pervasive and of special importance, both scientific and social. . . . Apparently, when behavioral researchers, teachers, or supervisors expect a certain level of performance from their subjects, pupils, or subordinates, they somehow unwittingly treat them in such a way as to increase the probability that they will behave as expected.

One of the findings of Rosenthal and Rubin which should serve as a warning to investigators and experimenters involved in the ape "language" projects is that this interpersonal expectancy effect was especially large for studies of animal learning, a fact which Ellsworth suggests may be due to a combination of three features of such research, namely: (1) the design of the research or the predicted behavior is simple;[18] (2) repeated measures are taken on the same subjects (especially when the experimenter is aware of the subjects' responses); and (3) an experimenter runs subjects in only one condition (Ellsworth 1978:393; see also Carlsmith, Ellsworth, & Aronson 1976). Each of the ape "language" projects suffers from at least one, and usually a combination of, these three factors. This makes it all the more important that, in the future, investigators assign a much greater importance to the exploration of both "procedures for minimizing and calibrating the effects of experimenter expectations" (as far as we know, there has been no controlled attempt to ascertain the impact of experimenter expectations on ape performances) and the fundamental semiotic problem of the "role of nonverbal processes of communication mediating interpersonal expectancy effects" (Rosenthal & Rubin 1978:385).

ANTECEDENTS OF APE "LANGUAGE" RESEARCHES

Alternatively, Ockham allows, dogma—viz., the articles of faith—may impose the benchmark. The prevailing Darwinian tenet that fuels, or constrains, most of the investigations reported throughout this collection postulates (as the subtitle of Griffin's [1976] contribution to cognitive ethology also perfectly captures) the "evolutionary continuity of mental experience." The countervailing older type of psychology defines the animal, following Descartes, as a push–pull automaton, and, as an inevitable consequence of that definition, affirms the superiority of man over the brutes. His slogan, which has influenced the thought of all generations that came after, was formulated in quasisemiotic terms: "instinct and reason, the signs of two natures." The historical roots of the ensuing dialectic are sketched in Limber's paper (this volume; see also Sebeok 1978a), but no doubt take on special evolution-

[18]How complex a task has to be to make Clever Hans cuing unlikely is open to question. Rosenthal (in Pfungst 1965:xxxvii, fn. 4) describes the case of Lady, the talking horse of Richmond, Virginia, who was reputed to be a good finder of lost objects. In one case, when consulted about a missing dog, Lady, operating a special typewriter, spelled out the word "dead."

> In some way, not at all apparent to even a keen observer, the horse's owner, [sensing the dog's owner's belief that her pet had been killed], must have communicated to her the sequence of appropriate keyboard responses. It is one thing to find the cues that start and stop a horse's tapping. It is quite another to find the cues that lead a horse to choose one letter out of 26 and then another and another, especially when the "keys" of the typewriter are quite close together. To learn the unintentional (if it was unintentional) signaling system in this case would have provided Pfungst with another worthy challenge.

On Lady, see further Christopher 1970:40–46.

ary meaning in the context of the great African apes, notably both varieties of *Pan*. On the one hand, as we know since 1975 (King & Wilson), the difference in structural genes between "us" and "them" is astoundingly small, for the average human polypeptide was reported to be more than 99% identical with its counterpart in chimpanzees, although the data, by the same token, also indicated that one quarter of the proteins were genetically different (i.e., while the DNA is very similar, the proteins are substantially unlike; cf. Plomin & Kuse 1978:189). On the other hand, adult chimpanzees do not conspicuously resemble grown-up people, and their respective behavior patterns—most particularly with respect to the semiosic faculty—are separated by a chasm the depth of which remains unplumbed and which therefore persists as a major concern (cf. Ristau & Robbins 1979:295). The fulcrum of our profound phenotypic and adaptive disunion from the chimpanzees and the gorilla must lie in the differential timing of gene expression during brain development. The regulatory changes have a retarding influence upon our unfolding. Fetal growth rates eventuate in hypertrophy of the organ-complex controlling (among other effects) human linguistic competence for speaking and in processing the speech of others, in brief, the language-using animal's species-specific behavior.

As Suzanne Chevalier-Skolnikoff has found (work in progress), baby chimpanzees, gorillas, and orangutans develop behavior patterns during the first 24 months of their lives that closely resemble those of human infants on Piaget's six-stage model of intellectual and motor development. The pivotal difference is that young apes stop advancing in their vocalization at an early age, which then presumably accounts for differential subsequent maturation. While orangutans, too, cry at birth, only human infants go on to more sophisticated vocalizations, to wit, successively, cooing, laughing and babbling, combining phonemes, using words instrumentally, and creating two-morpheme phrases. Alloprimates never coo, never babble, never acquire more elaborate ways of articulating.

Malmi (this volume) has wisely warned against the use of either simple analogies or the design-feature approach to language as arguments for evolutionary continuity in language development. The much more stringent criteria of true homologies, based especially on the examination of neurological substrates, must be met before conclusions about the biological roots of man's semiotic system *par excellence* can be drawn. And "presumed phylogenetic closeness is not sufficient to establish behavioral homology, especially when the behavior does not naturally occur in one of the species but is elicited only through extensive training" (Malmi, this volume:192), as is the case with languagelike systems of communication.[19]

---

[19]On the subject of homology versus analogy, see also, in this volume, Chomsky, Hediger, and Lenneberg. For a recent discussion of the linguistically relevant neurophysiological differences between man and ape, see Popper and Eccles (1977:308–309).

African apes were not always, of course, judged the fashionable, or even the ideal, animal models for simulating man's linguistic behavior. According to Fouts and Rigby (this volume), the chimpanzee series began only in this century; the gorilla then first entered stage center in this decade. As for Asia's sole living species of great ape, Galdikas (1978:291) is of the opinion that "a worse model . . . than wild orangutans could not possibly be imagined."

In the early 1960s, the mystique of the porpoise (or dolphin) began to captivate the imagination of the public. Dolphins were claimed to be capable of producing sounds like those of human speech and to engage in audiomimetic activities. We were told—with undeniable accuracy—by an investigator with medical credentials: "The dolphins may learn English (or Russian) or they may not. . . ." (Lilly 1978:188). Recently, their superiority over primates is contrastively insinuated:

The almost unbelievable ability of the dolphin to match the sounds of human speech can further be appreciated by examining communication with the apes. The human and the dolphin share the capability of communication by means of sounds. The chimpanzee and the gorilla cannot do this with any degree of complexity whatsoever. (*ibid.*:78)

The fact is that marine mammals have never been proven to possess the rudiments of language (cf. Wood 1973:118, specifically with respect to porpoises); in Wilson's judgment (1975:474), among others', the communication system and social organization of delphinids generally "appears to be of a conventional mammalian type." This is not surprising, for the outcome of even the best designed experiment has shown that "dolphin social behavior is much more akin to the example of elephant seal social behavior than it is to human linguistic behavior" (Evans & Bastian 1969:433; cf. Linehan 1979:532; William Langbauer's Porpoise Language Acquisition Project, patterned after Premack's experimental design with Sarah, appears, so far, to have made little progress, *ibid.* 529, 533). Once again, by the stringent application of Ockham's razor, the dolphins' performance is sufficiently explained by the inadvertent conditioning postulated by Bastian, coupled, probably, with a modicum of trial and success. While it is quite safe now to prognosticate that, within the next decade, many computers will be able to talk—if not yet necessarily listen—in "human language," it is equally certain, we think, that no porpoise will ever be able to either generate or understand continuous speech on a remotely comparable level of linguistic and contextual sensitivity. Needless to emphasize, for computers to achieve the implied level of sophisticiation, vast but justifiable funding will be required; money spent on chimerical experimentation with speechless creatures of the deep to be hominified is, however, tantamount to squandering scarce resources.

Apes, Cadillac-like, being also expensive animals to maintain in home or laboratory, it is small wonder that their "linguistic" tutelage has, so far, proceeded exclusively in America, wholly with domestic support. In France, the Renault of animals appears to have been the Greater Spotted Woodpecker *(Dendrocopos major)*, the natural drumming behavior of which was molded by Chauvin-Muckensturm (1974) against an object in association with selective food demands. The woodpecker was taught the use of the following code: one tap for a pistachio nut, two taps for a house cricket, three taps for a mealworm, two plus two taps for a May bug, and two plus two plus two taps for a locust. The investigator's conclusion was that this bird's behavior "seems to represent a phenomenon of man–animal communication analogous to that found in monkeys by the Gardners and Prenack [*sic*]" (*ibid.*:185). This seemingly exorbitant comparison is buttressed by a functional equation of the bird's beak with the simian's hand, an identification which may be less shocking if one is prepared to accept a common characterization of some birds (notably the psittacines) as "the monkeys of the bird world" (Breland & Breland 1966:106).

The code-drumming woodpecker carries us back to code-tapping horses, and thus right to the heart of the prototypal case of Clever Hans, or to similar pseudo-communicative deportment in other domestic species. One of us has dealt with this effect in previous publications, in a semiotic frame (Sebeok 1979c, Chap. 4, this volume; for talking dogs, see *id*. 1979a), and has a book in preparation (Sebeok, forthcoming) on the broader implications and applications of the compelling principle of ideomotor behavior at work and the interacting fallacies such motor automatism nourishes. Here, it may be worth pointing out that, in its crudest form, Clever Hans seems to undergo perpetual reincarnation in one guise or another. Thus, Blake (1977:40) now claims to have "thirty or forty proven [*sic!*] cases" of telepathic communication in horses, while Rowdon (1978:235) copiously expatiates upon the miracles wrought by a certain Mrs. Heilmaier in displaying the linguistic capacities of her dogs; he depicts the powers this lady has over animals in picturesque statements, such as: "I saw [a] chimpanzee leap down from his perch to the bars of the cage as if starved for conversation, to greet her and chatter to her with funny little movements of the mouth." Our credulity is no more strained by such anecdotes than by the tale of one Mucianus about the tame elephant that could write Greek (amusingly pictured in Anon. 1891:291).

EXORCISING HANS'S SPIRIT

From the outset, responsible experimenters with anthropoid apes have been wary of the devastating effects of "illicit communication in the laboratory," as Pilisuk and his collaborators (Pilisuk, Brandes, & Van der Hove 1976) refer to *sub rosa* semiosic activities rampant in research establishments generally. It will be remembered that all the feats of Clever Hans were ultimately uncloaked as having amounted to nothing more than either "go" or "no go" responses to unwitting minimal cues given off by people present in the horse's milieu (for a clear, succinct account, see, e.g., Goldenson 1973:262–279; for an analytic account, see Sebeok

1979c, Chap. 4,this volume). One of the earliest discussions of the possibility of the Clever Hans effect in the ape "language" projects (Ploog & Melnechuk 1971:631–634) suffers from a number of fatal defects. Instead of evidence, heavy reliance was placed upon authority, or the observational and rhetorical powers of prominent scientists:

Brown and others present vouched for [Washoe's] ability to do what was claimed for her. . . . Ploog visited both [the Gardners' and Premack's] laboratories, and he is personally convinced that no answer "leakage" à la "Clever Hans" accounts for the results with either Washoe or Sarah.

The epistemological status of such sincere prose calls for the challenge well posed by Ziman (1978:137–142): "*How* much *can be believed?*" It also readily brings to mind the nuptial pads of Paul Kammerer's midwife toads, apropos of which no less an authority than Professor Stanley Gardiner assured the world that "Kammerer begins where Darwin left off," to which Professor G. H. F. Nuttall added that Kammerer had made "perhaps the greatest biological discovery of the century" (Koestler 1973:91). We must also recollect that the 1911 forgery of "Piltdown Man" was not only maintained, for over 40 years, on scarcely more than the testimony of otherwise competent scientists, such as Sollas, but required a thorough reworking of the ancestral tree of humanity to accommodate "the earliest Englishman," and which was accordingly gnarled by many eminent anthropologists, such as Arthur Keith and Earnest A. Hooton. As Sollas (1915:54) asserted when a question was raised about the possibility that the jaw did not belong with the skull, "The chances against this are . . . so overwhelming that the conjecture may be dismissed as unworthy of serious consideration."

USES OF AUTHORITY AND SELF-CONTROL

There have been two main types of expert witnesses cited by researchers as verification of the accuracy of their results with apes learning ASL: those who have special knowledge of ASL, and those whose areas of expertise do not include ASL but some discipline relevant to the issues raised by this research, such as linguistics, first language acquisition by children, or physical anthropology. In either case, however, the outsider is dependent upon project members for the bulk of the information he receives during his visit. During our stay in Reno, in 1969, for example, we were, as guests, necessarily at the mercy of our gracious hosts, the Gardners, in terms of where and when we could view Washoe, and what additional experiences of the project—in our case, heavily edited films of the chimpanzee, casual testing of her signs in our presence using a box of well-worn objects, and discussions with some of the animal's trainers—we were allowed to have. In every case with which we are familiar, site visitors are carefully chaperoned by members of the local research team, the most important result of which is, unavoidably, that a good deal, if not all, of the former's "first-hand" observations are filtered through the commentary provided by their guides, who often furnish the frames of reference within which the observer is more likely to "perceive" what is consistent with the team's findings and expectations. This semiotic keying (Bouissac 1976, Chap. 10) is especially important when the visitor, although an expert in the field of language acquisition, primatology, or whatever, knows little ASL. Does such an observer actually see the ape's signs, or does he only think that what he saw corresponds to what his guide has interpreted for him, or what his earlier reading of research reports has led him to expect to witness?

Even when the outside observer is fluent in ASL, there may be some question about the reliability of his or her judgments about the ape's behavior.[20] The visitor is, first of all, unlikely to have had much first-hand or possibly even indirect experience with chimpanzees or gorillas, and is thus unable to determine which of the actions he sees performed by such an animal are part of its natural repertoire and which are the result of special training.[21] He will, furthermore, be distracted by the nonsigning behavior he may be witnessing—and attempting to interpret—very likely for the first time. *All* outside observers will be distracted from the signing behavior of the ape to a certain extent by their need to accommodate themselves to the animal's individual personality—not to mention that of each of their hosts.[22] Not only are the observations of such visitors suspect for this reason, but their written reports of their experiences are often anecdotal and/or based upon a reinterpretation of their "first-hand" observations after consideration of the written or film materials made by or about principal investigators and/or their staff.

Hediger (this volume), expressing his misgivings about the legitimacy of claims made regarding the linguistic competence of the anthropoid apes, points to what is a major fallacy underlying all such work, namely, the assumption that scientists, even those unfamiliar with apes, can be counted on to detect social cues and/or other forms of self-deceptive manipulative behavior at play in the interactions

[20]For a single example, see the case cited by Sebeok (this volume:421). One important problem with using deaf persons as outside observers of signing apes is that they will miss any auditory cues given the subject. This is especially relevant in the case of Koko, Nim, and Oklahoma subjects, to whom trainers were allowed to speak and make other sounds, and the flow of auditory signals was not controlled. The same problem would arise, however, in regard to the paralinguistic signs uttered by humans working with the Gardners' chimpanzees.

[21]How many of us, one wonders, are truly aware of the artifice of Chita, the chimpanzee we see in Tarzan movies, the clever chimpanzee on Marlin Perkins's television nature show, and so forth?

[22]The general problems regarding the reliability of outsiders' observations mentioned here are also present in child language research, particularly where the children under study do not speak distinctly. These issues are usually side-stepped by researchers relying upon cross-checking by a number of project personnel, but see our earlier comments on the difficulties inherent in this type of data control.

between man and ape, while those experienced in the nonscientific training of performing animals in circuses, zoos, and similar installations are not called in to give an account of what is going on. This situation somewhat parallels that described by Randi (1975) for the investigation of Uri Geller's claims to teleport objects or persons, read minds, and the like. Geller prefers scientists as witnesses, and will not perform before expert magicians, and for good reason. Scientists, by the very nature of their intellectual and social training, are among the easiest persons for a conjuror to deceive, while a good magician can spot deceptive techniques in a very short time. While what is most likely to be occurring in the ape research is self-deception, in the form of experimental expectancy effects or the "trimming" or "cooking" of data by investigators (Merton 1957), as opposed to outright fraud, the question nevertheless comes to mind whether it would not be wise for the principal investigators of ape "language" projects—not to mention the funding agencies which support them—at least to seek the advice of persons who are practiced in the art of purposefully manipulating animal behavior to create the illusion of humanlike activities.

At the very least, but preferably in addition, those expert in the microanalysis of human communication, in any modality, should be consulted, and not in a casual manner, but under conditions rigidly controlled by the outsiders themselves rather than the research team whose work is being investigated. Someone trained in the analysis of the rhythms of dialogue, interaction ritual, or conversational sequencing, for example, would probably have brought to light much sooner the profoundly important effect on the apes of the signs used by their human interlocutors, a form of discursive cuing only now being uncovered. In the past, it was generally only the chimpanzee's or gorilla's utterances which were recorded and studied, the human's input being either summarized or cut out altogether. This trend is being reversed by Terrace and his collaborators, who have recently given this issue the attention it deserves, and with strikingly illuminating results (Terrace *et al.* n.d.).

Through an analysis of Nim's use of ASL, which related the chimpanzee's utterances to his trainer's earlier signs within the same discourse, these astute researchers found that Nim imitated and interrupted his teacher to a much greater extent than does a language-learning child. The teacher's signing appeared to cue the chimpanzee that it can achieve a reward, a desired object or activity, only if it uses ASL. Since Nim had learned that many of the signs used by the teacher are generally acceptable as responses and therefore useful in obtaining a reward, he imitated some of them together with other generally acceptable signs such as *Nim* or pronouns. Interruptions by the chimpanzee are explained by the authors in terms of the fact that the more rapidly Nim fulfilled the human's requirements, the faster he would reach his goal (*ibid.*:63–66).

Analysis by Terrace and his associates of films of other signing apes revealed a similar dependence on the prior utterances of an animal's teachers, even to the point that, in certain exchanges, most of the animal's signs had already been modeled by the human interlocutor before the animal used them. The authors' comparison of unedited with edited film versions of the same sequence of discourse also shows how this crucial aspect of ape signing is unwittingly masked—and the meaning of the interaction thus remarkably altered—by the usual editing of a film or videotape while preparing it for display.

The startling thing about the study by Terrace and his colleagues is that neither Nim's trainers, nor the many "expert" observers who were fluent in ASL, were aware of the discursive cuing that was occurring. Human attention was focused instead upon the content of the animal's utterances and their nonverbal context. The authors are convinced that the conspicuous differences between the conclusions they drew about chimpanzee linguistic competence on the basis of a traditional distributional analysis of Nim's signs—a method similar to, but much more thorough than, those employed in other projects—and those which resulted from their careful discourse analysis of a complete and painstakingly transcribed film record of ape–human "conversations," casts doubt upon the bulk of previous analyses of the so-called grammaticality of ape sign sequences (*ibid.*:49).

It is not surprising that trainers are by and large unaware of this sort of interactive modeling, since they, not the apes, are the ones doing most of the work during the training and testing sessions. In his discussion of one of his trainers who was clearly giving social cues to Peony, Premack (1976:29) gives us some idea of the considerable number of tasks which the trainer must perform in such situations (e.g., "setting out the test items, following the data sheet—and doing all of this without being bitten!"), while at the same time trying to gain and maintain the attention of the animal. And yet researchers and experts alike continue to believe that they are in control not only of the verbal signs they emit in such a distracting situation, but even the nonverbal signs they give off. Premack, for example, asserts that his trainers were aware of the fact that if a chimpanzee did not know an answer it would look into the trainer's face for clues, but that the trainers controlled this by "refusing" to give clues and by redirecting the chimpanzee's attention to the task (1976:2). The use of self-control as a protection against unintentional cuing is of dubious value. Pfungst (1965), after having decoded the system of minimal cues being given Clever Hans, admits that he was himself unable to keep from cuing the horse, even though he was making a conscious effort not to do so. Rosenthal, in his introduction to Pfungst's account of the investigation of Hans, discusses a number of additional examples of this, in tests of ESP, waterwitching, muscle readers, and several clever animals, where observers were incapable of controlling their cuing of the subject under examination.

Premack claims that, of the 20 trainers he used over the years, only one was found to be clearly giving social cues. "There is no question," he writes, "but that both chimpanzees and children try to use social cues on occasion. However, I have never seen either do so with sufficient stealth to go undetected by an experienced trainer, and especially not by a second trainer who observes the first trainer and the subject" (1976:29). This summary dismissal of social cues as an influence in the interaction between trainers and apes raises more questions than it answers. First, does Premack restrict "social cues" to such obvious, consciously controlled visual behaviors as direct stares and symbolic gestures? If so, he is omitting from consideration a vast array of semiosic processes in both the visual and a host of other channels which have been shown to play an important part in human and animal communication (for numerous references, see Sebeok 1976, 1979c) but which operate largely out of the awareness of interactants. Such an underestimation by Premack of the potential of unwitting cues is curious, in light of his earlier criticism, noted above (p. 6), regarding a general lack of awareness, on the part of scholars, of the power of affective signaling. Marler notes:

Far from being an impediment that somehow blunts the effectiveness of animal signaling behaviors, I would rather view the affective component as a highly sophisticated overlay that supplements the symbolic function of animal signals. Far from being detrimental, it increases the efficiency of rapid unequivocal communication by creating highly redundant signals whose content is . . . richer than we often suppose. (this volume:226)

Since affective, nonlinguistic communication is just as crucial to effective human communication as it is to animal exchanges, it would seem wise to suppose, at least until concrete evidence to the contrary is available, that it *would* come into play in interactions between man and ape.

The second question which arises with regard to Premack's comments is what does he mean by an "experienced" trainer? Nowhere is it specified that Premack himself, or his trainers or outside observers, had received special instruction in any of the techniques of observation and control of subtle nonverbal signals, e.g., control over their own individual facial muscles or training in the recognition of the facial expressions of apes; experience in observing and understanding ape or human body movements and proxemic behavior; training in the apes' use of odors as telltale signs; experience in controlling one's own pupil responses—if indeed this is possible[23]—and in noting those of one's ape partner; tutoring in the use of nonlinguistic acoustic signals, such as throat clearing, breathing patterns, hums, hesitation fillers, and the like, which would make them especially qualified to perceive unwitting nonverbal cues. Without at least some of this sort of sensitivity training, on what grounds can a trainer or outside observer be said to deny conclusively that social cues are not present?

[23]Thomas Mann's fictional character Felix Krull, preparing himself in boyhood for his adult life as a confidence man, reports his success with training himself to control his pupil reactions (in *Confessions of Felix Krull Confidence Man*):

I would stand in front of my mirror, concentrating all my powers in a command to my pupils to contract or expand, banishing every other thought from my mind. My persistent efforts . . . were, in fact, crowned with success. At first as I stood bathed in sweat, my colour coming and going, my pupils would flicker erratically; but later I actually succeeded in contracting them to the merest points and then expanding them to great, round, mirror-like pools.

The very fact that it was only after painstaking analysis of unedited film records that verbal cuing was discovered to be operating between chimpanzees and their trainers, as noted above, not to mention the difficulty with which the Clever Hans and similar cases in the past were "cracked," should put us all on our guard against facile judgments about social cues. Actually being present at a demonstration of an ape's linguistic capacities, plus access to an unedited videotape or film of that same performance, would seem to provide the only type of data by which truly qualified observers could reach a conclusion—maybe—about the presence or absence of subliminal communication between trainer and ape.

However, lacking these, still photographs are themselves enlightening, despite the fact that, in many cases, either the photographer has focused exclusively on the ape or the photograph is cropped in such a way that the human member of the dyad is totally or partially absent from the version the viewer is allowed to see. Nevertheless, when we do catch a glimpse of the trainers in action. they are anything but detached and stone-faced. In conformity with the *National Geographic*'s notorious predilection for the beauty-and-the-beast motif, Francine Patterson has been featured in the majority of photographs illustrating her article (1978a). The richness of the nonverbal communication accompanying the signing of Patterson and Koko leaves little doubt that there is ample opportunity for Clever Hans cuing here. To take but two illustrations from this article, there is a series of four photographs showing Koko signing, according to the caption, "you," "dirty," "bad," "toilet." In each picture, Patterson's head position, facial expression, posture, and gestures change—in response to or in anticipation of Koko's signs, it is impossible to tell. (It is also not possible to tell what took place in the time intervals between the four shots.) Nowhere in the text of the article is it, however, mentioned that Patterson was signaling to Koko. Just what was she doing, then? In another, two-page photograph (pp. 442–443), Koko is shown signing "teeth" by grinning and touching her teeth with an index finger. Patterson is beside her, holding a picture of a grinning chimpanzee, and she is shown touching its teeth with her index finger!

In certain cases, photographs not only furnish important information not specified in the text of the report, but may also contradict assertions which have been made in the text.[24] For example, in Mrs. Premack's book about her husband's project (1976), three photographs (p. 54) reveal that, contrary to the description of working conditions given by Premack in his own book (1976), the chimpanzees—in

---

[24] Barber (1976, Chap. 7) reviews a number of investigations of what he calls the Experimenter Failure to Follow the Procedure Effect. Citing especially Friedman (1967, Chap. 5), he notes that "most experimenters have serious difficulties in following the experimental procedures closely even when the experimental protocol is standardized or is not especially 'loose'" (p. 54). In addition to this, "there appear to be wide deviations in kinesic and vocal behavior on the part of the experimenter" (*ibid.*). Both the "standardized" experiment and the "standardized" experimenter, in other words, are largely an illusion.

this case, Elizabeth—were allowed to position themselves in such a way that they could see their trainer not only before and after being given a problem and solving it, but while they were placing the plastic symbols on their display board. The first photographs make up a sequence of two scenes, in the first of which Elizabeth reads the instruction on the board, and then, in the second, responds by washing an apple in a carton of water. Several things can be said about these two pictures, keeping in mind, of course, that we are not told how much time elapsed between these two scenes, or what behaviors occurred during that interval. First, in the initial photograph, although the caption explains that it is Elizabeth who is "reading" the instruction, *both* the trainer and the ape are shown pointing to one of the tokens at the same time, their hands even appearing to touch. The trainer is, moreover, looking at Elizabeth, head tilted slightly to one side, leaning expectantly toward the board and ape. Elizabeth is looking at the board but is in such a position that she could see (and hear) the movements of the trainer. In the second picture, the trainer relaxes against the back of her chair, her shoulders no longer tense but hunched forward, and she is looking down at the carton of water, grinning. Her whole attitude is one of a person who, tense while waiting to see if the animal could make the right response, now relaxes as it does. Elizabeth, who is either about to place the apple in the water or is removing it, is not only looking directly at the trainer, but returning her grin. At which moment, we might ask, did the trainer begin the process of relaxation, which could have served to inform the chimpanzee that it was on the right path? It was, after all, precisely the conversion of experimenters' anticipatory tension, as Clever Hans approached the correct response to a question, into involuntary cues—such as leaning back and relaxing—which permitted the horse to perform what, to even some of the world's then leading psychologists, appeared to be mathematical and linguistic operations (Pfungst 1965). If, to return to the case of Elizabeth and her experimenter, so many obvious cues are shown in these two photographs alone, how many more were in evidence during the entire session? In light of these observations, the following statement by Premack (1971:820–821) takes on portentous significance:

Was Sarah responding to the plastic language or to nonlinguistic cues arising from the trainer's face and body? In principle, this could be tested by eliminating the nonlinguistic cues. Trainers could wear dark glasses, or, after presenting Sarah with a question, station themselves behind an opaque screen, or simply look away from her. But these measures were practically useless. When the trainer put several questions on Sarah's board and then walked away, leaving her to answer them, Sarah worked erratically or quit altogether. . . . Social contact may be Sarah's primary motivation. In any case, she did not work under these circumstances.

Was it, one wonders, the absence of social reward that caused Sarah to fail, or the lack of social cues?

In the third photograph, a different trainer is seen giving off equally obvious signals as Elizabeth, we are told in the caption, "writes" a message to her on the message board. The trainer is leaning forward expectantly, mouth open (whether to emit sounds or not is, of course, inaudible),[25] her hand poised in the box of tokens (grasping what one would assume is one of the symbols), eyes focused on Elizabeth. Now, if it is the *ape* who is writing the message, why is the trainer handling the tokens? An obvious clue? Possibly, for Elizabeth is looking directly at the trainer's hand which is in the box. Elizabeth is facing the trainer and could also conceivably take in the latter's facial expression, posture, and the like. Why, too, is the trainer staring so intently at the ape? It would seem likely that it would not be necessary for the human to look anywhere but at the display board in order to read Elizabeth's message. Could it be that it is as important for the trainer to be cued by her subject as the other way around? To what extent does she rely upon nonverbal cues given off by the ape to determine what the latter may "mean" by using a given token, or to learn what Elizabeth is likely to do momentarily, thus making it possible for her to intercede—unintentionally, of course—in order to steer the ape in the proper direction through cues of her own?

THE DOUBLE-BLIND STRATEGY

Workers in this field exhibit a touching faith in the so-called double-blind strategy (Ploog & Melnechuk 1971:632–686), borrowed from pharmacology, where it is now all but universally recognized that the method entails a multitude of booby traps on the experimenter's side as well as on the subject's: both "can serve as sources of unintended cues leading to the breakdown of experimenter blindness" (Rosenthal 1976:372; see especially Tuteur 1957–1958).

Researchers have uniformly found it more difficult to test their ape subjects under the more stringent conditions of a double-blind situation than in normal, more sociable circumstances, and both chimpanzees and gorillas alike, whether using ASL or an artificial language, have shown a marked deterioration in their performance in the former, more artificial type of setting. The lack of cooperation on the part of an ape in such situations is usually described in reports in general terms rather than in detail, with frequent allusions to the animal's psychological or emotional attitude toward the "blind" experimenter, reminiscent of Clever Hans being labeled "distrustful" of questioners who were unaware of the correct answers (Pfungst 1965:201).

The casualness of reports makes it extremely difficult in some cases to deter-

---

[25] As Hill (this volume:350) has noted, "even Premack's highly controlled studies" were "polluted" by vocal English, since his trainers were permitted to say the English translation of a token aloud as they put it on the display board. Verbal accompaniments were also an accepted part of the studies by Fouts, Patterson, and Terrace.

mine the significance of the data collected. For example, Patterson devised a test for Koko in which one experimenter baited a small box with "a random selection from a pool of objects representing thirty of the nouns in her vocabulary" (1977:10). A second experimenter, who did not know the contents of the box, stood behind it, and, when Koko started the trial by uncovering the box, asked the gorilla in ASL what she saw. Patterson (*ibid.*) describes Koko's reaction to this test as follows:

This test situation required a fair amount of discipline, and curiously enough, we found that like Washoe, Koko's interest and cooperation could be secured for no more than five trials a day and two sessions per week. Her methods of avoiding the task were varied—she would either respond to all objects with the same sign, refuse to respond at all, or regress to an earlier pattern of asking to have the box opened.

Koko is said to have scored correctly 60% of the time, or above chance, on the series of double-blind tests which were administered to her. It is reasonable to ask, however, how Koko's instances of avoiding the test, e.g., by responding to all objects with the same sign, were scored. We are not told whether or not such inappropriate responses were discounted as avoidance measures or counted as errors. Since these types of responses are not included in any of the four categories of errors listed on page 11 of Patterson's report, we may assume that they are, in fact, not represented in the 60% score noted above. We can only guess whether or not Koko's performance would have been below chance were a less biased accounting to have been made.

The problems with this procedure do not end here, of course. As with so many of the double-blind tests administered to the apes, no allowance has been made for the fact that, when working with such a relatively small corpus of signs, the "blind" experimenter, no doubt a clever human, may be applying certain game strategies which enable him or her to guess at what the input to the chimpanzee or gorilla has been. The "random" selection of objects placed in the test box sounds good, for example, but, if one calculates, on the basis of the various figures and tables provided by Patterson, that, when the double-blind tests were administered to Koko, in September 1975, when she was four, the gorilla probably had only 107 nouns in her vocabulary which could have had objects appropriate for use in the box, each trial, with 30 objects, would constitute 28% of her total inventory of nouns. This would surely allow a good deal of successful guessing to go on by the "blind" experimenter (not to mention Koko), who, as a regular trainer (which we must assume was the case, since Patterson does not say that the person was independent of her project), would surely have a memory sufficient to hold 107 such items.

A second strategy could also have been operating here. It is not specified whether or not different experimenters were used for each of the double-blind tests. If the same individuals were used, then several questions come to mind. First, how was the "random" selection of objects made in each trial? If not done by computer but by human, did the person making the selection choose the objects while blindfolded, perhaps pointing to them with a stick, while a third person picked them up and placed them in a box? If not, that is, if the person is aware of which items he has selected, he would surely be tempted, on subsequent trials, to vary the objects presented. The second, "blind" experimenter could also work according to the assumption that the objects used on one trial would not be likely to recur too frequently in subsequent ones.

There is, furthermore, the possibility that the "blind" experimenter, especially if familiar with Koko's earlier sign performances, could have been cued by the gorilla's nonlinguistic reactions to the objects, which were, after all, familiar to her. If, for example, Koko were accustomed to smacking her lips when presented with a preferred food, the accustomed response during the double-blind test could furnish the experimenter with valuable information about which class of objects— foods Koko likes—the animal has in view. Judging from our own observations of Koko on television, as well as reports of other apes in similar situations, such nonverbal responses are bound to occur even in the double-blind testing context, and it is difficult both to understand why they are so routinely ignored by researchers and to imagine how to avoid them altogether. Even an observer unfamiliar with Koko, for example, might correctly interpret some of the more obvious of the gorilla's reactions, especially when the test objects are repeated on subsequent trials with the same person as witness. In the case of the tests with Koko, the ambiguities are many, since we are not told if the blind observer had prior experience with the project, or if he was engaged in more than a single trial. Furthermore, only the overall score for the whole series of tests was reported, making it impossible to judge the effects of learning by both Koko and her human experimenters.

The Gardners' description (this volume:318ff.) of their double-blind testing procedure administered to Washoe indicates a more carefully controlled situation than the one just discussed. Washoe was seated in front of a booth, before a sliding door which she could open or close, behind which there was a screen on which slide pictures could be projected. One "blind" observer, called O_1 (see the illustration on p. 319 of this volume), stood to one side of the box, and, when Washoe opened the sliding door and a slide was projected onto the screen, asked the chimpanzee in ASL what she saw. O_1 then wrote down Washoe's signed response to this question and deposited the slip of paper into the message slot (MS), where it was retrieved by an experimenter (E_1), who, seated behind the booth and to the left of the compartment where Washoe viewed the slides, operated the slide projector and recorded the observations of both O_1 and O_2. The second "blind" observer, O_2, sat behind the booth, opposite Washoe, whom the former could view through a one-way mirror. In one version of the test, all three human participants were project assistants, while in another, control situation, O_2 was a deaf person without any particular association with the Washoe project, O_1 and E_1 being members of the Gardners' staff.

Despite the added precaution of two "blind" observers, some of the above-mentioned problems concerning experimenter learning, guessing strategies, and the presence of nonlinguistic cues could still be operating during the Gardners' procedure. The research assistants could, through their familiarity with Washoe's relatively small vocabulary, stand a good chance of guessing which object was being shown on the screen, even though the particular exemplars exhibited were changed from those used in training, and the order of presentation randomized (how, we do not know). Non-naive observers could also rely to some extent on familiar nonverbal cues given off by the chimpanzee in response to certain objects. That the two observers may have been learning both the kinds of objects likely to be used in the test and Washoe's nonlinguistic cues is shown by the fact that "their agreement increased from one test session to the next" (this volume:320).

The use of a naive deaf observer in the O_2 position does not necessarily eliminate these problems. For one reason, the Gardners report that this deaf observer was familiarized beforehand with the two vocabulary lists which might be used in the test in which they served as an observer. Why this was thought necessary is not clear, since one would assume that if the chimpanzee's signs are good ASL signs, any native signer would have no difficulty interpreting them at first encounter, without having to study a vocabulary list. This also raises the question of what other types of information were conveyed to the deaf observer during his or her initial training period—e.g., familiar contexts of use of vocabulary items, characteristic nonlinguistic responses of Washoe or E_1 or O_1; (in fact, did either E_1 or O_1 have a hand in training the outside observer?). How "naive," in other words, was this deaf observer when the tests began?

Some of the danger of learning by the deaf observer could have been reduced if the Gardners had used a different deaf person in each trial. As it is, the fact that "the degree of agreement between the deaf observer and the project assistant *during the second test* was well within the range of agreement found between O_1 and O_2, when each was a project assistant thoroughly familiar with Washoe's signing" (this volume:320; emphasis added), may only indicate that the deaf observer was a fast learner.

Outside of such ambiguities, there exists a whole set of difficulties with the double-blind tests administered to "language-learning" apes. Although designed especially to eliminate the possibility of illicit communication in either direction between experimenters and their subjects, a close look at those double-blind tests where enough information about the actual setting and procedures followed is available, reveals that leakage is not only possible in each case, but probable.

Take, for example, the Gardners' own double-blind test. After O_1 writes down Washoe's response, he must pass in front of the sliding door in order to reach the message slot. If, as seems possible, the door is open as O_1 passes by, he would have visual access to the interior of the viewing box. E_1, who operates the slide projector, may have been instructed to leave the slide on the screen until O_1 has delivered a message, in which case the latter will have an opportunity to see the picture and learn whether or not (1) Washoe's response was correct, and (2) the observer's reading of the chimpanzee's signs was accurate. Even assuming that O_1 was not allowed to change his message en route to the message slot, this type of feedback would greatly enhance his ability to predict, in the fashion of a Markov chain (see Cherry 1978), which objects would be projected at a later time, as well as to develop skill in interpreting any idiosyncratic signing behavior by the chimpanzee. What, furthermore, would prevent O_1 from inadvertently indicating to Washoe—and to O_2, who could see O_1 no doubt as he passed behind the chimpanzee—whether or not her response had been correct?[26]

The Gardners report that O_2 communicated his interpretations of Washoe's signs to E_1 verbally, as they sat side by side, in which case how do we know that O_1 could not hear O_2's spoken observations, which might be expected to influence his own, thus leading to an increased degree of agreement between the two? Furthermore, if there is an acoustic channel open between E_1, who knows what the correct responses should be, and O_2, what evidence is there that auditory cues were not given by E_1 to O_2, signals such as "uh-huh," coughs, the sound of shifting in his seat, and the like, which could have provided O_2 with clues as to the accuracy of his own as well as Washoe's performances? Pilisuk *et al.* (1976) have described a host of seemingly innocent sounds which, in laboratory experiments, can and often do function as unwitting cues to subjects. Rosenthal (Pfungst 1965:xxi) also mentions an interesting example of auditory cuing in which the experimenter, using a scratchy pen, unintentionally cued a subject by the recording system he was using (long versus short scratches).

Was the deaf O_2 totally deaf, or could he, too, have taken advantage of such a channel? If totally deaf, to what extent could he control the volume of his verbal report to E_1? And would not E_1, in this case, have a tendency to raise his voice in order to acknowledge O_2's response, thus making it more likely that Washoe and/ or O_1 (both of normal hearing) would hear him? If E_1 could not make use of verbal acknowledgments, did he lean back and sign to O_2? Or could O_2 have leaned back, thus catching a glimpse of the written record being made by E_1?

If there was an acoustic channel open between E_1 and O_2, and between O_2 and O_1 (either through the partition or through the one-way glass while O_1 passed in front of the booth), would not there be reason to believe that O_1 could also hear any auditory cues given by E_1?

Furthermore, since one-way glass does not ordinarily impede the passage of

---

[26]In his article on "Statistical Problems in ESP Research," Diaconis (1978) has demonstrated how even having no more feedback than whether or not he has made a hit or a miss in guessing ESP cards will give an alleged psychic an advantage and thus influence the outcome of the test.

sound, we can assume that Washoe could hear O_2's verbal interpretations of her signs, which might lead her to change her answer, which might be picked up by O_1. We are not told whether Washoe was allowed to do so, or whether only her first response was counted. If there is an acoustic channel open between Washoe and O_2, and between O_2 and E_1, would it not seem likely that Washoe could hear any auditory feedback provided by E_1 as well?

Even if we assume that E_1 could not see or hear Washoe or O_1, he would have to wait until both written and oral interpretations were received before projecting a new slide. This means that whichever observer finished last would have an opportunity to witness a larger set of chimpanzee signs, one of which, but not necessarily the first, being correct.

This test could have been improved by the use of deaf O_1's and E_1's, relying only on written responses.[27] The Gardners seem to have thought of this, but ruled it out on the grounds that Washoe would not have performed as well with a strange O_1. Since the only interaction *required* between O_1 and Washoe was the execution of simple question–answer adjacency pairs, one wonders why this should be the case. Perhaps because this linguistic exchange was *not* the only behavior required by the chimpanzee?

Premack first attempted to eliminate the Clever Hans effect by testing Sarah with a "dumb" trainer, i.e., someone unfamiliar with her token language.[28] As he describes this test (1971, 1976), the "dumb" trainer presented Sarah with a problem on the display board, following only a number code. Sarah's response was communicated by the trainer to a project member, stationed in an adjacent room, by microphone. The project member then told the trainer if Sarah's response was correct or not.

Under these conditions, Premack reports, there was a definite "decrement in her accuracy" and a "deterioration in the form of her behavior" (1976:34):

The most striking aspect of this deterioration was a regression to an earlier form of sentence production that was once her dominant form. Early in training she had not produced sentences in their final order: she put correct words on the board in incorrect orders and then made one or two changes before settling on a final order. Although she had abandoned this mode of production at least ten months earlier, she reverted to it with the 'dumb' trainer. (1971:821)

This is precisely the sort of behavior one might expect the chimpanzee to exhibit if she had been searching for clues from the experimenter. She displays the tokens

---

[27]The double-blind test proposed by Lenneberg (this volume:80), which calls for four experimenters and which could, in principle, overcome some of the deficiencies noted above, has not, as far as we know, been used.

[28]It is not clear precisely what type of experience the new trainer had had with Sarah prior to the tests. We are merely told that Sarah was "adapted" to the person, who "engaged in normal social behavior" with her during a preliminary stage, under conditions which "differed from those of testing" (1976:821).

and then, moving them around, waits for an unintentional sign from the trainer that one arrangement is considered acceptable.

It is not difficult to conjecture how even a "dumb" trainer could inadvertently cue Sarah that a certain sequence of tokens was correct. Since the "dumb" trainer did not understand Sarah's token language, how did he know when she had "settled on" a "final order"? It would seem logical that he would only report her *first* arrangement of tokens—incorrect responses, according to Premack. If the trainer did do so, and, as one must assume from Premack's discussion of this experiment, these responses were not counted as errors, then one may conclude that the project assistant in the adjacent room instructed the experimenter in some way not to go on to the next problem, but to wait for further responses from Sarah. In this way, the chimpanzee could learn—relayed indirectly from the person in the next room— which order was acceptable, since only after that particular attempt would the "dumb" trainer be told to acknowledge her responses.[29]

Another deficiency noted in Sarah's responses was that the verticality of her sentences suffered. "Ordinarily, words were placed more or less below one another, but with the 'dumb' trainer she failed to maintain this orderliness. The sprawling sentence was another characteristic of her early behavior" (Premack 1971:821). Again, if the symbols were "sprawling," how did the "dumb" trainer know in what order to report them to the project member? Did the latter, perhaps, impose an order on the numbers read to him, possibly also conveying to the trainer some information about proper order through the inadvertent signaling of his recognition of a familiar pattern or asking questions about the exact spatial relationship between the tokens on the board?

Premack took into account that the new trainer may have, over a number of trials, learned something of Sarah's code, thereby enabling him to cue her. When he tested the trainer, he found that the latter had in fact guessed some of the rules of the token language, but also that in certain cases his inferences were incorrect, so that in some instances his cues—if responded to by the chimpanzee—would actually have led Sarah to make erroneous responses. Since Sarah did about as well on the first test with the new trainer as on subsequent ones, cuing was not deemed

[29]In addition to this sort of cuing, Sarah may also have been relying on her memory to solve some of the problems in the double-blind test, as Brown (this volume) has suggested. Premack (1976:34) denies that this is the case, saying that, by the time the tests were given, Sarah had experienced 2,600 sentences, and would have had to memorize that many to do as well as she did in the 58 sentences used in the tests. His explanation is, however, not as compelling as it might appear at first blush, for only 14 of the 58 test sentences were new to Sarah, and we are not told whether or not the 44 "old" sentences were taken from recently learned sets. As already noted above, furthermore, the manner in which the test frames were set up was such that Sarah did not actually have to take into account all the signs in a sequence in order to respond correctly, so she might have had to memorize far less than Premack alleges in order to do well. In light of these considerations, Premack's estimate (1976:35) that about 10% of Sarah's accuracy was due to nonlinguistic cues seems unrealistically low.

a significant influence on test results. If, however, the new trainer was serving as a *conduit* for cues coming from the person in the next room, a uniform level of performance is precisely what *would* be expected. While, as Ristau and Robbins (1979) remark, "the possible role of cues in these chimpanzee projects is far more difficult to imagine and cannot be so simple" (288) as those involved in the Clever Hans case, surely there is ample reason to doubt that Premack's "dumb" trainer experiment "is conclusive in that the accurate aspects of her [Sarah's] performance cannot be attributed to subtle cues given by the dumb trainer" (*ibid.*:289).

Premack's more recent attempt (with Woodruff & Kennel 1978) to devise a test for Sarah that is free of social cues involved a paper-marking task utilizing pictures of tokens with which she had been trained, plus pictures of toys, objects, and foods familiar to her. A project assistant placed in front of the chimpanzee trays with pairs of pictures of objects, and pairs of tokens. Sarah was supposed to mark with tape that member of a pair of tokens which correctly described the relationship between the objects (same versus different, similar versus same, and similar versus different). Sarah was later tested on same versus different with letters of the alphabet rather than pictures of objects. The trainer was to leave the room after presenting the trays to the chimpanzee, returning only when Sarah rang a bell.

Although over the five tests given, Sarah's overall score, we are told, was above chance, her accuracy for the concept "different" was only at chance level on two of the tests, as was her performance for the concept "similar" on one of them. The materials, furthermore, as Premack himself admits, tested "only rudimentary cognitive and linguistic capacities" (Premack *et al.* 1978:905). It is, perhaps, significant that as the double-blind controls become more stringent, the tasks required of Sarah become simpler and, from the point of view of linguistic capacity, less interesting.

Even in this experiment, however, the controls were not absolute, and the possibility still exists that social cues were operating in it, despite Premack's claim that "the present results cannot be interpreted in terms of the subject's sensitivity to inadvertent social cues" (Premack *et al.* 1978:905).[30] For instance, it is possible that the trainer unwittingly cued Sarah as he was placing the trays in front of her. He could have inadvertently touched or pointed to a correct response while, perhaps, straightening a paper on a tray, or even merely tensed or in some other way called attention to that hand which was closer to a correct answer. Even a fleeting glimpse at the correct token could serve as a cue, as would the mere positioning of the bell more or less close to a correct response. None of these cues would have been easily visible to the camera, or a second observer, stationed opposite the chim-

---

[30] He does admit, however, that "Sarah's performance may have relied critically on quite different features of her social relationship with the trainer. He was quick to respond to her summons at the end of every trial, and praise or food was soon to follow" (1978:905).

panzee's cage, for the trainer would, for at least part of the time, have had to have his back to both. In addition, it seems unlikely that Sarah would always patiently wait until the trainer had left the room before proceeding to mark the paper, especially since she seemed motivated primarily by a desire for the trainer's presence, praise, or a more tangible reward.[31]

ALONE IN A CROWD

Sarah's paper-marking test raises an additional question concerning the double-blind strategy of the ape "language" projects. The trainer was not the only person capable of cuing Sarah in this context. Judging from the photographs used to illustrate the testing situation (Premack *et al.* 1978:905), there was a photographer present for all or at least some of the test, and when the trainer was not in the room. We are not told in the report if the photographer was familiar with Sarah's token language or not, but, with such a restricted number of tokens as found in this particular test, it would not be difficult for anyone to learn fairly quickly which responses were correct and which ones wrong. Assuming this, then, if that person were in the room when the trainer departed, he would be able to sign to Sarah by optical, acoustic and/or other means, if only by the timing of his camera shots. Even if separated by a one-way mirror, the photographer could inadvertently let Sarah know when she was about to select the correct answer by himself choosing that moment to snap a picture, which Sarah could detect. The same would apply as well to any discontinuous videotaping or filming process. This is based upon the not unreasonable assumption, supported by a study of photographs and films of several of the chimpanzee and gorilla projects, that a photographer or cameraman, if left to select what he will and will not record, is most likely to find photogenic—even dramatic—moments when the animal is doing something significant in terms of the purpose of the occasion, e.g., gesturing distinctly if it is a matter of the use of ASL, or solving a problem with tokens or computer keys, if an artificial symbol system is involved.

One can also speculate—and, since very little information is ever given in reports about the procedures followed for recording events, this is all one can do at present—that there may be some sort of communication which takes place between the photographer or cameraman and the experimenter who is conducting a test. It is certainly not out of the question that a trainer, hearing a camera clicking to a

---

[31] Project reports all too frequently give the impression that ape subjects sit quietly through tests, although this is far from what actually happens. Experimenters must spend a good part of testing time interacting with the animal just to get it under sufficient control to enable them to administer the test, hardly what one would call ideal experimental conditions. If cuing is feasible even when a subject is sitting still and attentive, it is even more so under the sometimes chaotic circumstances created by an ape's natural responses to such man-made rules.

start, could be primed to apply an extra ounce of unintentional pressure on the experimental subject to produce a significant response "for the record." If the cameraman or photographer were a project assistant, and the experimenter an outside observer, the former could actually cue the animal through the latter. Pfungst (1965:210), who seems to have overlooked nothing in his investigation of the Clever Hans case, remarked that, while spectators did not influence Hans, "[t]he effect upon the questioner . . . was unmistakable." Given a calm and confident person, such as Mr. von Osten, "the questioner's zeal was increased and with it the tension of concentration." Less assured questioners, on the other hand, were distracted by the audience so that, rather than improving, the accuracy of their cuing of Hans suffered.

In general, there is a frustrating lack of agreement between assertions made about a project's "usual" procedures and the reporting of any particular training or testing event, leaving the reader with only questions about whether or not a given procedure in fact lived up to the claims made for it. One particularly striking example of this is provided in Rumbaugh (1977:159–160), where Gill and Rumbaugh give the following description of Lana's early training:

Since Lana seemed to thrive on social contact, the behavioral technicians maintained close contact with her and frequently entered her room to "model" the correct behavior, taking her finger and pressing the correct key with it or pointing to the appropriate key or set of keys. Although these supportive techniques were used in the training sessions . . . the experimenters gave no such assistance during the *test* phases in which proficiency levels . . . were assessed. During the tests every precaution was taken to preclude the possibility that any extraneous cues might aid Lana's performance. Blinds were installed to deny her visual access to the experimenter; the sequence of trials for different tasks was randomized; her responses were automatically recorded by the teleprinter; and, when possible, experimenters not involved in training were used to test Lana.

Although it is easy to grasp how cues could have been given during the training sessions (Gill and Rumbaugh mention, for example, that in some cases an experimenter would communicate with Lana by tapping on the walls of her room; in Rumbaugh 1977:172), it is more difficult to assess the degree to which such cues have been eliminated in testing conditions. For example, of the five experiments reported in Rumbaugh 1977 (Part III, Chaps. 9–13), only one actually lived up to the special "blind" conditions noted above, and this consisted merely of a cardboard screen at the window leading into Lana's room.

In another report by Savage-Rumbaugh and her associates (this volume:369), we are told that, in general, "iconic gestures have been repeatedly devised by the animals, and by the experimenters, as an adjunct to the abstract symbols available on the keyboard," such gestures serving "as an intermediate link between symbol and event." But members of the Lana project seldom discuss such gestures in their reports, rarely mentioning the possibility of acoustic, tactile, or olfactory cues, although some sound was able to pass between the animals' cages and the room in which the experimenter was located. In the test reported by Savage-Rumbaugh *et al.* in this volume, for example, the subjects, Austin and Sherman, had to hand each other tools from one room to another; sound surely could pass through the opening needed for this, yet the trainers tested only for visual cues, not acoustic ones.[32] Schubert (1978) questions how "blind" the tests with Austin and Sherman really were. On the one hand, the investigators reported that changes in experimenters resulted in performance decrement during all stages of training, which suggests that some information loss must have occurred when one experimenter replaced another. The possibility of Clever Hans cues being given by the experimenters was not completely eliminated in this experiment, according to Schubert, because, for example, in the naming task, the experimenter stood outside the subject's room and held up a tool so that the chimpanzee could see it through a lexan wall:

> But if the C[himp] could see the tool well enough to distinguish it, he could also see at least part of E[xperimenter]'s hand, and perhaps part of his arm(s) too. How much more information does a chimp need to identify which human (among the small sample of available alternatives) he was dealing with? (*ibid*.:598)

In the majority of photographs of Lana, she is shown alone with her computer console, giving the impression of an interaction completely devoid of human contamination. And yet, as Terrace and Bever (this volume:186) have pointed out, "many of her most striking 'utterances' occur with a trainer present," precisely when, as even the project members themselves agree, conditions are anything but free of social cues. When the trainers do not enter her cage or tap on the walls, they nevertheless:

> can vary the time, rate, and choice of presentation, which leaves open the possibility that Lana's performance is still being shaped by uncontrolled factors (which often appear to be unrecorded), e.g., Lana's cage position, her drive state, the trainer's current assessment of her position and state, and so on. (Terrace & Bever, this volume:186)

Pictures of the more stringently controlled tests raise the question of how "alone" Lana really is even under these less social conditions. Someone, after all, is taking the picture of her, and presumably can be heard by the chimpanzee while doing so. Furthermore, given the large staff associated with this project, one suspects that there might be other detectable personnel performing their varied chores in the wings.

Reports of studies of apes in their natural habitat are similar in their disregard of the presence of photographers and other project staff members. Pictures of Goodall, Fossey, or Galdikas, for example, tend to delineate the romantic image of a brave, young woman alone with the beasts of the jungle. While this may appeal to

[32] See the discussion by Sebeok (this volume) of muscle readers, such as Eugen de Rubini, and their use of such subtle clues as tremors of the floor, faint sounds of feet, movements of arms and clothing, and the like, as cues guiding them to where an object has been hidden.

us much as does the image of Jane in Tarzan stories, or the prototypal beauty in the tale of "Beauty and the Beast," it is hardly an accurate record of the actual, far more complex conditions of research. Designed to bolster the credibility of the findings presented in the narrative part of the report, these illustrations actually obfuscate rather than clarify, at least for the average viewer, who approaches them expecting to have verbal reports confirmed and even amplified.

In this respect, the Lana enterprise resembles the other ape "language" projects insofar as they attempt to communicate a sense of solitary splendor in which experimental animals operate, whereas in actual fact it is usually more a question of the subjects being "alone in a crowd." Consider, as a final example, the experiment conducted by Savage-Rumbaugh and her associates (1978) to test the ability of two chimpanzees, Austin and Sherman, to communicate symbolically between one another. Using the same computer language (Yerkish) designed for Lana, the subjects were trained to identify the symbols for 11 types of food and drink. In the first of a series of tests, the animals shared a keyboard. On alternate trials, one of them was taken into an adjacent room, where it watched the experimenter bait and seal a container with one of the foods. Returning to the first room, this chimpanzee, called the "informer," was asked—by computer, we assume—what was in the container. When the informer had pushed the key on which the proper lexigram was embossed, the second chimpanzee, known as the "observer," who had not been allowed to witness the baiting of the box, but was able to watch the informer's "description" of the contents, was then permitted to request the food via the keyboard. If both chimpanzees responded appropriately, the container was opened and the food given to the subjects. The animals were correct on 33 of 35 trials.

The investigators were aware that, in this test, the experimenter, who knew what was in the box, could cue the animals, which is particularly likely since, as seen in the photograph on page 642 of the report, that person was holding each chimpanzee by a leash.[33] To control for this, in all subsequent tests the experimenter did not accompany the informer into the other room, and the box was baited by another project member. It is not stated, however, that the experimenter was not permitted to see the informer's description of the contents of the container, which means that the moment a correct response was made by the informer—which could be explained adequately on the basis of a simple X–R association—the experimenter, who knew the meaning of the symbols used, was no longer "blind," and would have therefore been in a position to cue the observer during his subsequent performance. This weakness in the experimental design would apply as well to those tests where the animals either were not allowed to see one another pushing the single keyboard they shared, only the lexigram projected above the keyboard,

[33]See Sebeok (1979c:98) regarding how police officers inadvertently communicate their own expectations concerning the whereabouts of suspected criminals to the bloodhounds being used to track them.

or they had separate keyboards in different rooms (separated by glass) on which the lexigrams were arranged in different sequences.

In a final control experiment, the informer was not permitted to use the keyboard, but could communicate in any other way with the observer. The latter was, after 30 to 60 sec, encouraged to use the keyboard to request the contents of the container. As opposed to their high scores on all the earlier tests, Sherman and Austin did only four of 26 trials correctly under these conditions, which the investigators take to show that it was only through symbolic, not affective, signs that the animals had been communicating when more successful. It is possible, however, that the reason the animals' performance was so poor was due to the fact that this was the only test in which the experimenter was truly "blind" and therefore unable to provide cues.

Even precluding the possibility that the experimenter was cuing the chimpanzees, another source of information was available to them which should have been taken into account by the investigators, but was not. Since the container itself was within the reach of at least one of the animals in each test situation, it is possible that its contents could have been made known to the observer by lifting, shaking, or otherwise manipulating it, a trick well known to alleged clairvoyants, who, for example, in order to "see" which of a number of sealed, identical cans is filled with water, may tap their feet, walk around, or in some other unobtrusive way shake the containers, thereby solving the problem by observing the different sounds or other sensations given off by the filled can.[34] In the case of Sherman and Austin, simply picking up the can for a second could tell them whether the food inside was liquid or solid, large or small. Since we are in fact told by the investigators (p. 643) that "attempts to steal the container" were made by the subjects in this experiment, such an alternative explanation of the results does not seem too far-fetched.

Additional doubt is thrown on the investigators' interpretations of Sherman and Ally's performance by close examination of the above-mentioned photograph on page 642 of their report. In the text (p. 643), the authors write that "the chimpanzees were mutually attentive and if one appeared to have difficulty finding a key, the other often tried to assist, though restrained from doing so." In the photograph in question, it is the observer who is seen being restrained by the experimenter, as the informer, according to the caption, "uses the keyboard to declare the container's contents." Why, we may ask, would the observer seek to help the informer, or even make his own request–response, *before* he sees the lexigram touched by the informer? Only, one might answer, if the observer had already learned of the contents of the box by one of the nonverbal means suggested above,

---

[34]See especially Randi's amusing account of Uri Geller's unsuccessful attempts to employ such methods on the "Johnny Carson Show," where Carson, who used to be a stage magician, took special precautions to prevent cuing.

or if he did not have to rely on his own knowledge of the answer for his performance, but rather on that of the experimenter.

THE CLASH OF *UMWELTEN* IN FACT AND FICTION

Critics have, on occasion, raised the question of why the cognitive behaviors alleged to have been elicited in hand-raised apes have in no instance been observed occurring in the wild. This wonder is closely paralleled by a second one: why, if apes have a strong picture-making potential in captivity, have they neither developed nor utilized it in nature (Sebeok 1979b:36)? The usual rejoinders are sanguine: we don't as yet know enough about either the native communication system or artistic tenue of these creatures, but, any day now, someone shall catch them at it. We, however, think yet another fallacy lurks behind this optimism, one that fails to allow for the fact that apes live in a radically different phenomenal world, or what, since Jakob von Uexküll, ethologists call *Umwelt* (Sebeok 1979c, Chap. 10), than we do or than does any other species. An ape is not interested in verbal art, or painting, or the like, but in "apely" objects and relations, in brief, signs that are functionally meaningful to its species preeminently (cf. Hediger, this volume). A narrow segment of their world of signs may, of course, overlap with the modern human *Umwelt*. What the research community is trying to achieve—and, to a limited extent, successfully, by means of *apprentissage* or *dressage*—is to widen this area of overlap; but it by no means follows that the expanded sign repertory will be biologically significant for the trained animal.

That apes can be taught fairly large vocabularies of symbols has been well established by the projects under consideration in this volume. Time and again, however, reports indicate that there is only a faint resemblance between the chimpanzee's or gorilla's application of these newly acquired semiotic tools and that of humans. We have already mentioned that Sarah was primarily motivated, according to Premack, by social needs and food rewards. Such appears to be the case as well for the other animals involved in learning languagelike skills. McNeill (this volume:160), for example, remarked several years ago that Washoe's "reorganization of ASL" to express affect and messages related to social relationships while ignoring the human focus on analysis of objects and relationships between them suggests that chimpanzees are simply not interested in what humans are concerned with, and there is no reason to suppose that they would have evolved, in nature, a communication system at all on a par with human language. Premack's failure to teach Sarah the plastic chip language by the observational method used by language-learning children and their parents was partially due, no doubt, to the fact that the chimpanzee simply "did not focus on those aspects of the situation that were of primary interest to the experimenters" (Ristau & Robbins 1979:275).

Rumbaugh (this volume:249; cf. Rumbaugh *et al.* 1975) has admitted that although his "primary goal was to cultivate in Lana the desire and the skills needed to converse . . . about a wide variety of subjects," Lana, in fact, used her computer language almost exclusively in order to solve practical problems with which she was faced, such as getting a trainer to supply her machine with a favorite food. For Lana, in other words, the symbols she had learned were primarily of instrumental value in achieving goals which could not be obtained otherwise. Similarly, the function of Nim's ASL signs was not to identify objects or convey information about the world, but rather to obtain a reward, whether this was the engagement of a human in some desired activity (such as a game of chase or tickle) or some desired object. Patterson does not specifically mention this aspect of Koko's signing, but a glance at the illustrations of a recent article (1978b) reveals that, like the chimpanzees in other projects, the gorilla signs mainly when in situations where a human is holding out the promise of a reward (e.g., a glass of milk—p. 81; a stereo viewer—p. 82; a game of tickle—pp. 84, 86; a stethoscope—p. 85).

Investigators have sought to narrow the gap between ape and human uses of languagelike symbols by encouraging the intraspecific exchange of messages between the trained animals and between trained and untrained animals, in the latter case with the symbol-wise ape serving first as a teacher.[35] Such attempts have achieved limited success in that the animals do exchange symbolic messages with one another, but there is, as far as we know, no evidence to date that the symbols, in such cases, function in a noninstrumental way (see, e.g., Fouts & Couch 1976; Fouts & Rigby, this volume; Fouts, Mellgren, & Lemmon 1973; Savage-Rumbaugh *et al.* 1978, this volume). Certainly the hope that Washoe would teach ASL to her offspring has not been realized, for Washoe—herself, in effect, raised as a human daughter—had not shown much maternal interest in either of her babies—now dead—which raises the question of how "natural" are the home- or laboratory-raised animals vis-à-vis their wild conspecifics. Irrespective of any training with languagelike symbol systems, the former's rearing in a man-made environment must surely have altered their *Umwelt*. This alone makes difficult any extrapolations from the ways these symbol-using apes utilize their acquired communication skills, even were there to be significant breakthroughs in the future in this regard, to the potential adaptive significance of such symbols for the species as a whole.

What we have at the moment, with respect to the interspecific communication between ape and man in the ape "language" projects, is both accommodation and conflict between *Umwelten*. The chimpanzees and gorillas, placed in a totally man-made environment, whether a private home, experimental laboratory, or primate

---

[35] Such attempts are made difficult by the gradual fading from view of the "star" performers, Lana, Sarah, and Washoe, which is in itself disturbing. "[O]ne begins to wonder," Ristau and Robbins write, "why these performing chimpanzees are no longer the subject of intense study. Haven't they become unmanageable and thus dangerous to their trainers? Have they reached the limits of their abilities and resorting to a variety of learning procedures fails to improve performance?" (1979:294).

research colony, adapt themselves, somewhat reluctantly, by learning a number of arbitrary signifier–signified associations and by utilizing them in situations where trainers will accept no alternative type of response. They will follow certain elementary prescribed rules of play, in other words, but there is no indication that they are playing the same "game." Investigators and experimenters also accommodate themselves to the expectations of their animal subjects, unwittingly entering into subtle nonverbal communication with them while convincing themselves, on the basis of their own human rules of interpretation, that the apes' reactions are more humanlike than direct evidence warrants.

Real breakthroughs in man–ape communication are the stuff of fiction, which usually accompanies, or even anticipates, the stream of scientific research. Ramona and Desmond Morris (1966b, Chap. 2) have surveyed some early science fiction with a simian character, winding up with Pierre Boulle's 1963 satirical novel, best known in this country as *The Planet of the Apes*. The plot of this narrative hinges on the contrast between a language-endowed master race of anthropoids and the human beings who, having regressed into a state of speechlessness, are turned by them into subjects for laboratory training and worse.

As mentioned elsewhere in this book (Sebeok, pp. 426–427), it was Jules Verne, who, in *The Village in the Treetops* (1901), had invented the device of a German savant who undertook a fantastic scientific journey to the central African jungle. Eventually, he is located by two big-game hunters—the American, John Cort, and the Frenchman, Max Huber—who, in the seemingly impenetrable forest, encounter evolution at work. The most fascinating aspect of this late and seldom read book of Verne's about "the so-called language of the monkeys" is that the protagonist and his peregrinations were made up out of bits and pieces suggested by the biographical circumstances and quasi-scientific works of Richard L. Garner, a *bona fide* forerunner of today's primatologists seeking for the roots of language (Sebeok 1979c:268, 291, n. 4). Verne ordered his bricolage into a fictional maneuver, laced with equal amounts of sympathy and raillery, and capped by a moral about cooperation in adversity and its absence when the danger is dissipated.

John Collier's *His Monkey Wife or, Married to a Chimp* (1931), remains possibly the most celebrated spoof in this fictional vein. In the course of this rich and cunningly crafted novel about erotic and racial relations, Emily, the heroine in the title, learns to understand both spoken and written English (she has read, among other classics, the *Origin of Species* and *Murders in the Rue Morgue*), can type quite proficiently, but never masters speech. Even at the end of the book, just prior to the consummation of her miscegenetic marriage, she continues tacitly to gesture "with one or two of those quiet signs by which she managed to express to [her husband's] now subtler understanding almost all that she desired to communicate to him (*my gracious silence* he sometimes laughingly called her). . . ."

An even more thought-provoking novel on this broad theme was published, in English, in 1953, by Jean Bruller (who used the pen name Vercors), under the title,

You Shall Know Them. In a trenchant and suspenseful fashion, it deals with the question: What is man? It does so using by way of a two-faced contrast an invented hybrid species, called *Paranthropus,* an intermediate group of creatures familiarly known as *tropis*. While working out his definition of humanity, the author has much to say about language in ape, man, and the Janus-like ape–man in between. One of the characters, Captain Thropp, who "had read several scientific papers to the Natural History Society on his studies and tests on Great Apes," begins his testimony with a reference to (evidently John B.) "Wolfe's experiments. . . . He gave his chimpanzees a slot machine," he reports, and they "had reinvented money, and even avarice! Not abstract thinking, that?" (cf. Wolfe 1936). He continues: "Sixty years ago [Richard L.] Garner established that there's merely a quantitative difference between our language and theirs: we even have a number of sounds in common with the monkeys." He concludes by obliquely summarizing the highlights of Viki's linguistic tutelage; "[u]nfortunately the young animal died before" her surrogate parents were able to achieve success. The book ends on a hopeful note, where the judge, in his genial summing-up, argues: "Mankind resembles a very exclusive club. What we call human is defined by us alone." In other words, it is up to us to legally admit the tropis to the human community, to share the rights of man. This means searching for a legal basis for agreement to admit new members. But the setting up of such rules and regulations entails a consensus on the definition—or redefinition—of what constitutes language, a task which Chomsky so skillfully and authoritatively undertakes in this volume, but which, Sebeok has argued elsewhere (1978a), may be an inexecutable task, owing to the inherently indeterminate vacuousness of the term.

In a recent publication about "talking dogs" (Sebeok 1979a:4–5), reference is made to Olaf Stapledon's perfervid novel, *Sirius: A Fantasy of Love and Discord* (1944), which deals with the making of a super-sheepdog who develops "true speech," his life and reversion to a feral state, and ultimately his death as an outlaw. What is of interest in this context is that the scientist, Thomas Trelone, who works the remarkable transformation of Sirius, is well acquainted with the Kellogg paradigm (see this volume), but, "[c]ontrary to his original plan," and despite the fact that "apes offered the hope of more spectacular success," opts to use dogs instead. His reasons include that dogs are "capable of much greater freedom of movement in our society," and that he regarded "the dog's temperament on the whole more capable of development to the human level." Thus, à la Gua, the puppy is raised in the familial company of the Trelones' daughter, Plaxy. His dying words were: "Plaxy–Sirius—worth while."

Fanciful fiction featuring apes that learn to attain language capacity reaches its acme in two novels mentioned before (Sebeok, this volume): Peter Dickinson's detective story, *The Poison Oracle* (1974), the entire plot of which is impelled by this very issue of a chimpanzee's putative propensity; and especially John Goulet's *Oh's Profit* (1975), the protagonist of which is a singularly endowed young signing gorilla whose maleficent antagonists belong to a cabal of transformational linguists, thinly disguised but more or less recognizable, we are told, by the insiders who have read this *roman à clef*.

In diverse imaginative ways, the concerns of this literary genre, from Verne to Goulet, are identical with those some of us prefer to struggle with in the mythic world of scientific objectivity: to draw distinctions between man and beast, to identify the one animal endowed with language, separated from but immersed in a sea of speechless creatures, to delineate the nature of language itself and distill the essence of mankind. As Browning's poem, "Bishop Blougram's Apology," professed in an exceptionally nice march of oxymorons:

> Our interest's on the dangerous edge of things.
> The honest thief, the tender murderer,
> The superstitious atheist, demireps
> That love and save their souls in new French books—
> We watch while these in equilibrium keep
> The giddy line midway: one step aside,
> They're classed and done with. I, then, keep the line
> Before your sagest. . . .—just the men to shrink
> From the gross weights, coarse scales, and labels broad
> You offer their refinement. Fool or knave?

CONCLUSION

Throughout this introductory chapter, we have skirted the consequential issue so competently discussed by several contributors to this volume (e.g., Bronowski and Bellugi, Brown, Chomsky, Lenneberg, Limber, and McNeill)—namely, is what is being taught the apes really "language"? We have done so for the simple reason that, at present, of the two related questions posed by Chomsky—*"What is a human language?"* and *"What is a language?"*—neither the first, which is open to scientific, i.e., biological, explanation, nor the second, which is not, can be finally answered. Although the debate over problems such as these is in itself of appreciable value, there seems to be no point in adding further speculative material to the fires of contention.[36]

---

[36]Ristau and Robbins (1979:268) have noted that, even were the question of "what is language" to be resolvable by scientific method, which it is not:

> Just as man's unique and dominant status remained intact when his other 'unique' accomplishments such as tool use and cooperative hunting were observed to occur in other species, so the existence of rudimentary linguistic skills in other species—if demonstrated—will do little to diminish man's radical differences from other species.

In light of the wide attention given this matter in the media and in some popular books (e.g., Hahn 1978), it should, however, be noted that those investigators who accuse critics of the ape "language" projects of being biased in favor of a particular and, they claim, outmoded paradigm which defines *language* too narrowly, may themselves be prone to what Barber (1976:5–6) calls the Investigator Paradigm Effect, through their own special attachment to certain scientific or popular notions, some of which were discussed above (see also Hediger 1974:40; Sebeok, this volume). "If," as one of us has noted elsewhere (Sebeok 1978a:1041), "linguists, such as Chomsky, are to be enjoined from placing what others regard as little more than adroitly presented circus tricks of a handful of captive African apes beyond the pale of language in the technical sense, then, by the same token of a lack of clear definition, the trainers cannot claim a quasi-human language propensity for their charges either."[37]

If the debate over the linguistic status of the signs being taught to apes is unlikely to yield significant novel insights into the phylogenesis of language, the new line of investigation alluded to in this chapter—the critical examination of the interactions between humans and between man and ape in such language-learning situations—promises a rich harvest of information concerning a variety of subjects, including especially interspecific communication between man and ape. Oskar Pfungst, to whom this book is dedicated, can serve as a guide in this undertaking, for his investigation of Clever Hans is still one of the few successful attempts to discover the actual signs which mediated between the expectations of experimenters and the performances of their animal subjects. Pfungst (1965) proceeded from indirect evidence, such as his observations that, as the distance between Hans and his questioners increased, the animal's accuracy decreased, or that his performance suffered if the questioner did not know the correct answer, to direct evidence, from both observations of performance and laboratory experiments, in which a number of elements of the question–answer procedure were systematically altered (e.g., the visual channel between man and animal was blocked). Through the painstaking application of this methodology, Pfungst was able to uncover several types of visual and auditory cues which were being unwittingly given Hans by questioners, his

---

[37]Menzel and Johnson (1978:587), reacting to Griffin (1978a), Premack and Woodruff (1978), and Savage-Rumbaugh *et al.*, this volume, note a final ironic twist to the controversy over anthropomorphism versus anthropocentrism:

The study of "animal language," after the fashion of the target articles, may have, if anything, tended to increase rather than decrease expectations of human chauvinism and presumed "biological superiority," especially in the popular press, where it is more and more often suggested that chimpanzees, gorillas, and perhaps dolphins may deserve special consideration based on the outcome of research projects demonstrating their similarity to humans.

Schubert (1978:597) adds that to "appraise the relative excellence of nonhuman cognitive abilities by measuring the extent to which these conform to those characteristics of our own species" is "a very unbiological approach."

success in this endeavor leading Rosenthal to assert (in Pfungst 1965:xxix) that "it seems clear that neither the strategy nor the tactics of inquiry employed by Pfungst are in any way outmoded or irrelevant to contemporary psychology."

While the basic methodology of Pfungst may still be followed in contemporary assessments of ape linguistic capacities, today's investigator can take advantage of the vast amount of research on nonverbal communication which has been done in recent years. In moving from indirect evidence of social cuing, some of which has been presented here, to direct evidence, the microanalysis of the intraspecific and interspecific communication among man and apes must be performed by persons who have some expertise in one or more relevant areas of nonverbal communication, discourse analysis, *dressage,* and the like, with the support of those especially knowledgeable about experimental design, expectancy effects, and other methodological questions. The examination should be applied to all phases of those undertakings designed to teach languagelike symbols to apes, including the initial familiarization of the subject(s) with trainers and research facilities, all training and testing procedures, and the informal social interactions among project personnel (and outside observers) and the animals.

This work demands extraordinary caution and attention to possible methodological pitfalls, for, as Arthur G. Miller has perceptively commented, "there is a magical or fantasy-like aspect to the idea that one's expectancies . . . can become true merely by entertaining such anticipations" (1978:401), and, when applied to scientific research, is threatening to those involved. To make up for this, "evidence must be sufficiently powerful to counter, as it were, such *a priori* expectations" on the part of scientific investigators *(ibid.).* To be fair, the criteria for acceptance of the work of those whose research is designed to provide direct evidence of nonlinguistic explanations for the apes' use of symbols must be at least as stringent as those applied to work attempting to prove that the ghost of Clever Hans does *not* live on in the performances of today's experimental apes. In fact, both approaches—the creation of ever more carefully controlled double-blind tests, on the one hand, and the observation and experimental manipulation of a full complement of semiosic behaviors, on the other—must go hand in hand, if, as is to be hoped, an accurate appraisal of ape linguistic capacity is to be finally accomplished.[38]

---

[38]Note added in proof: Petitto and Seidenberg (1979) and Seidenberg and Petitto (1979), in two articles which came to our attention too late to be incorporated in this discussion, provide numerous additional examples, taken from the ape sign language projects, of some of the methodological pitfalls mentioned in this chapter. Papers by Fouts, Couch, and O'Neil, and Patterson (the latter constituting the published version of the 1977 manuscript referred to in this chapter), both in Schiefelbusch and Hollis 1979, further illustrate, respectively, the problems of auditory and visual cuing during double-blind tests. See also Terrace 1979 for a summary of his negative findings with respect to ape capacity for syntax, and our article, appearing in the same issue of *Psychology Today,* which represents a drastically abridged form of the present discussion. These and other relevant recent publications will be dealt with at length in the expanded version of this chapter (Sebeok in press).

REFERENCES

ANON. "Wild Animal Training." *The Strand Magazine* 2 (1891): 291–301.

BARBER, THEODORE X. *Pitfalls in Human Research: Ten Pivotal Points.* New York: Pergamon, 1976.

BLAKE, HENRY. *Thinking with Horses.* London: Souvenir, 1977.

BOUISSAC, PAUL. *Circus and Culture: A Semiotic Approach.* Bloomington: Indiana University Press, 1976.

BRELAND, KELLER, and MARIAN BRELAND. *Animal Behavior.* New York: Macmillan, 1966.

BROWN, ROGER. "The Development of Wh Questions in Child Speech." *Journal of Verbal Learning and Verbal Behavior* 7 (1968): 277–290.

CARLSMITH, J. M., PHOEBE C. ELLSWORTH, and E. ARONSON. *Methods of Research in Social Psychology.* Reading, Mass.: Addison-Wesley, 1976.

CHAUVIN-MUCKENSTURM, BERNADETTE. "Y a-t-il utilisation de signaux appris comme moyen de communication chez le pic epeiche?" *Revue du Comportement Animal* 9 (1974): 185–207.

CHERRY, COLIN. *On Human Communication* (3rd ed.). Cambridge: M.I.T. Press, 1978.

CHRISTOPHER, MILBOURNE. *ESP, Seers & Psychics.* New York: Thomas Y. Crowell, 1970.

CLIFFORD, BRIAN R., and RAY BULL. *The Psychology of Person Identification.* London: Routledge & Kegan Paul, 1978.

DIACONIS, PERSI. "Statistical Problems in ESP Research." *Science* 201 (1978): 131–136.

EHRENFELD, DAVID W. *Conserving Life on Earth.* New York: Oxford University Press, 1972.

ELLSWORTH, PHOEBE C. "When Does an Experimenter Bias?" *The Behavioral and Brain Sciences* 1(3) (1978): 329–393.

EVANS, WILLIAM E., and JARVIS BASTIAN. "Marine Mammal Communication: Social and Ecological Factors." In *The Biology of Marine Mammals,* edited by Harald T. Andersen. New York: Academic, 1969.

FOUTS, ROGER S., and J. B. COUCH. "Cultural Evolution of Learned Language in Chimpanzees." In *Communication Behavior and Evolution,* edited by M. E. Hahn and E. C. Simmel. New York: Academic, 1976, pp. 141–161.

FOUTS, ROGER S., JOSEPH B. COUCH, and CHARITY R. O'NEIL. "Strategies for Primate Language Training." In *Language Intervention from Ape to Child,* edited by Richard L. Schiefelbusch and John H. Hollis. Baltimore: University Park, 1979, pp. 295–323.

FOUTS, ROGER S., ROGER L. MELLGREN, and W. LEMMON. "American Sign Language in the Chimpanzee: Chimpanzee-to-Chimpanzee Communication." Paper presented at the Midwestern Psychological Association meeting, Chicago, 1973.

FRIEDMAN, N. *The Social Nature of Psychological Research.* New York: Basic, 1967.

GALDIKAS, BIRUTE. "Orangutans and Hominid Evolution." In *Spectrum: Essays Presented to Sutan Takdir Alisjahbana on His 70th Birthday,* edited by S. Udin. Jakarta: Dian Rakyat, 1978, pp. 287–309.

GARDNER, BEATRICE T., and R. ALLEN GARDNER. "Evidence for Sentence Constituents in the Early Utterances of Child and Chimpanzee." *Journal of Experimental Psychology* 104(3) (1975): 244–267.

GARDNER, MARTIN. *Fads and Fallacies in the Name of Science.* New York: Dover, 1957.

———. "Dermo-Optical Perception: A Peek Down the Nose." *Science* 151 (1966): 654–657.

———. "How to Be a Psychic, Even if You Are a Horse or Some Other Animal." *Scientific American* 240(5) (1979): 18ff.

GILLETT, JOHN D. "The Mosquito: Still Man's Worst Enemy." *American Scientist* 61 (1973): 430–436.

GOLDENSON, ROBERT M. *Mysteries of the Mind: The Drama of Human Behavior.* Garden City: Doubleday, 1973.

GRIFFIN, DONALD R. *The Question of Animal Awareness: Evolutionary Continuity of Mental Experience.* New York: Rockefeller University Press, 1976.

———. "Experimental Cognitive Ethology." *The Behavioral and Brain Sciences* 1(4) (1978): 555. (a)

———. "Prospects for a Cognitive Ethology." *The Behavioral and Brain Sciences* 1(4) (1978): 527–538. (b)

GUTHRIE, R. DALE. *Body Hot Spots: The Anatomy of Human Social Organs and Behavior.* New York: Van Nostrand Reinhold, 1976.

HACHET-SOUPLET, PIERRE. *Le Dressage des Animaux et les Combats de Bêtes, Révélation Procédés Employés par les Professionels pour Dresser le Chien, le Singe, l'éléphant, les Bêtes Féroces, etc.* Paris: Firmin Didot, 1897.

HAHN, EMILY. *Look Who's Talking!* New York: Thomas Y. Crowell, 1978.

HAYES, CATHY. *The Ape in Our House.* New York: Harper, 1951.

HEDIGER, HEINI. *The Psychology and Behaviour of Animals in Zoos and Circuses.* New York: Dover, 1968.

———. *Man and Animal in the Zoo: Zoo Biology.* New York: Delacorte, 1969.

———. "Communication between Man and Animal." *Image Roche* 62 (1974): 27–40.

HEINROTH, OSKAR. "Beiträge zur Biologie, namentlich Ethologie und Psychologie der Anatiden." *Verhandlungen des V. Internationalen Ornithologen-Kongresses* 5 (1910): 589–702.

HESS, ECKHARD H. *The Tell-Tale Eye: How Your Eyes Reveal Hidden Thoughts and Emotions.* New York: Van Nostrand Reinhold, 1975.

JOLLY, ALISON. "The Chimpanzees' Tea-Party." *The Behavioral and Brain Sciences* 1(4) (1978): 579–580.

KATZ, DAVID. *Animals and Men: Studies in Comparative Psychology.* London: Longmans, Green, 1937.

KELLOGG, WINTHROP N., and LOUISE A. KELLOGG. *The Ape and the Child: A Study of Environmental Influence on Early Behavior.* New York: Hafner, 1933.

KING, MARY-CLAIRE, and A. C. WILSON. "Evolution at Two Levels in Humans and Chimpanzees." *Science* 188 (1975): 107–116.

KOESTLER, ARTHUR. *The Case of the Midwife Toad.* New York: Vintage, 1973.

KRESKIN. *The Amazing World of Kreskin.* New York: Random House, 1973.

LENNEBERG, ERIC H. "A Neuropsychological Comparison between Man, Chimpanzee and Monkey." *Neuropsychologica* 13 (1975): 125.

LILLY, JOHN CUNNINGHAM. *The Mind of the Dolphin: A Nonhuman Intelligence.* Garden City: Doubleday, 1967.

———. *Communication between Man and Dolphin: The Possibilities of Talking with Other Species.* New York: Crown, 1978.

LINEHAN, EDWARD J. "The Trouble with Dolphins." *National Geographic* 155(4) (1979): 506–540.

LOCKARD, JOAN S. "Speculations on the Adaptive Significance of Cognition and Consciousness in Nonhuman Species." *The Behavioral and Brain Sciences* 1(4) (1978): 583–584.

MELLGREN, ROGER L., and ROGER S. FOUTS. "Mentalism and Methodology." *The Behavioral and Brain Sciences* 1(4) (1978): 585–586.

MENZEL, E. W., JR., and MARCIA K. JOHNSON. "Should Mentalistic Concepts Be Defended or Assumed?" *The Behavioral and Brain Sciences* 1(4) (1978): 586–587.

MERTON, ROBERT K. "Priorities in Scientific Discovery: A Chapter in the Sociology of Science." *American Sociological Review* 22 (1957): 635–659.

MILLER, ARTHUR G. "And in This Corner, from Cambridge, Massachusetts . . ." *The Behavioral and Brain Sciences* 1(3) (1978): 401–402.

MORRIS, DESMOND. *Manwatching: A Field Guide to Human Behavior.* New York: Abrams, 1977.

MORRIS, RAMONA, and DESMOND MORRIS. *Men and Pandas.* New York: New American Library, 1966. (a)

———. *Men and Apes.* New York: McGraw-Hill, 1966. (b)

MURPHY, GARDNER, and ROBERT O. BALLOU, eds. *William James on Psychical Research.* New York: Viking, 1960.

PATTERSON, FRANCINE G. "Linguistic Capabilities of a Young Lowland Gorilla." Paper presented at a Symposium of the American Association for the Advancement of Science: "An Account of the Visual Mode: Man vs. Ape." Denver, 1977.

———. "Conversations with a Gorilla." *National Geographic* 154(4) (1978): 438–465. (a)

———. "The Gestures of a Gorilla: Sign Language Acquisition in Another Pongid Species." *Brain and Language* 5 (1978): 72–97. (b)

————. "Linguistic Capabilities of a Lowland Gorilla." In *Language Intervention from Ape to Child*, edited by Richard L. Schiefelbusch and John H. Hollis. Baltimore: University Park, 1979, pp. 325–356.

PEIRCE, CHARLES S. *Collected Papers of Charles Sanders Peirce,* edited by Charles Hartshorne, Paul Weiss, and Arthur W. Burks. Cambridge: Harvard University Press, 1935–1966. (References are to volumes and paragraphs, not pages.)

PERELMAN, CHAÏM, and LUCIE OLBRECHTS-TYTECA. *The New Rhetoric. A Treatise on Argument.* Notre Dame, Ind.: University of Notre Dame Press, 1969.

PETITTO, LAURA A., and MARK S. SEIDENBERG. "On the Evidence for Linguistic Abilities in Signing Apes." *Brain and Language* 8(1979): 162–183.

PFUNGST, OSKAR. *Clever Hans (The Horse of Mr. von Osten),* edited by Robert Rosenthal. New York: Holt, Rinehart & Winston, 1965.

PILISUK, MARC, BARBARA BRANDES, and DIDIER VAR DEN HOVE. "Deceptive Sounds: Illicit Communication in the Laboratory." *Behavioral Science* 21 (1976). 515–523.

PLOMIN, ROBERT, and A. R. KUSE. "Genetic Differences between Humans and Chimps and among Humans." *American Psychologist* 34 (1979): 188–190.

PLOOG, DETLEV, and THEODORE MELNECHUK. "Are Apes Capable of Language?" *Neurosciences Research Program Bulletin* 9 (1971): 600–700.

POPPER, KARL R., and JOHN C. ECCLES. *The Self and Its Brain: An Argument for Interactionism.* New York: Springer International, 1977.

PREMACK, ANN J. *Why Chimps Can Read.* New York: Harper & Row, 1976.

PREMACK, DAVID. "Language in Chimpanzee?" *Science* 172 (1971): 808–822.

————. "Teaching Language to an Ape." *Scientific American* 227 (1972): 92–99.

————. *Intelligence in Ape and Man.* Hillsdale, N.J.: Lawrence Erlbaum, 1976.

PREMACK, DAVID, and GUY WOODRUFF. "Does the Chimpanzee Have a Theory of Mind?" *The Behavioral and Brain Sciences* 1(4) (1978): 515–526.

PREMACK, DAVID, GUY WOODRUFF, and KEITH KENNEL. "Paper-Marking Test for Chimpanzee: Simple Control for Social Cues." *Science* 202 (1978): 903–905.

RANDI, JAMES. *The Magic of Uri Geller.* New York: Ballantine, 1975.

————. "Tests and Investigations of Three 'Psychics.'" *The Skeptical Inquirer* 2(2) (1978): 25–39.

RISTAU, CAROLYN A., and DONALD ROBBINS. "Book Review: A Threat to Man's Uniqueness? Language and Communication in the Chimpanzee." *Journal of Psycholinguistic Research* 8(3) (1979): 267–300.

ROSEBURY, THEODOR. *Life on Man.* New York: Viking, 1969.

ROSENTHAL, ROBERT. *Experimenter Effects in Behavioral Research.* New York: Irvington, 1976.

ROSENTHAL, ROBERT, and DONALD B. RUBIN. "Intrapersonal Expectancy Effects: The First 345 Studies." *The Behavioral and Brain Sciences* 3 (1978):377–415.

ROWDON, MAURICE. *Elke & Belam.* New York: Putnam, 1978.

RUMBAUGH, DUANE M., ed. *Language Learning by a Chimpanzee: The Lana Project.* New York: Academic, 1977.

RUMBAUGH, DUANE M., TIMOTHY V. GILL, ERNST VON GLASERSFELD, HARALD WARNER, and PIER PISANI. "Conversations with a Chimpanzee in a Computer-Controlled Environment." *Biological Psychiatry* 10 (1975):627–641.

SAVAGE-RUMBAUGH, E. SUE, DUANE M. RUMBAUGH, and SALLY BOYSEN. "Symbolic Communication between Two Chimpanzees *(Pan troglodytes).*" *Science* 201 (1978):641–644.

SCHUBERT, GLENDON. "Cooperation, Cognition and Communication." *The Behavioral and Brain Sciences* 1(4) (1978):597–600.

SEBEOK, THOMAS A. *Contributions to the Doctrine of Signs.* Bloomington: Research Center for Language and Semiotic Studies/Lisse: Peter de Ridder Press, 1976.

————. ed. *How Animals Communicate.* Bloomington: Indiana University Press, 1977.

————. "Clever Hans & Co." *Times Literary Supplement,* No. 3,990 (1978) p. 1041 (September 22). (a)

————. "'Talking' with Animals: Zoosemiotics Explained." *Animals* 111 (1978): 20–23, 36. (b)

————. "Close Encounters with Canid Communication of the Third Kind." *Zetetic Scholar* 3–4 (1979): 3–20. (a)

———. "Prefigurements of Art." *Semiotica* 26 (1979): 3–74. (b)

———. *The Sign & Its Masters*. Austin: University of Texas Press, 1979. (c)

———. *The Play of Musement*. Bloomington: Indiana Unversity Press, in press.

———. *Clever Hans*. Bloomington: Indiana University Press, in press.

SEBEOK, THOMAS A., and JEAN UMIKER-SEBEOK. "Performing Animals: Secrets of the Trade." *Psychology Today* 13/6 (November 1979): 78–91.

———. "'You Know My Method': A Juxtaposition of Charles S. Peirce and Sherlock Holmes." Bloomington, Ind.: Gaslight, in press.

SEIDENBERG, MARK S., and LAURA A. PETITTO. "Signing Behavior in Apes: A Critical Review." *Cognition* 7 (1979): 177–215.

SOLLAS, WILLIAM J. *Ancient Hunters* (2nd ed.). New York: Macmillan, 1915.

TEMERLIN, MAURICE K. *Lucy: Growing Up Human: A Chimpanzee Daughter in a Psychotherapist's Family*. Palo Alto: Science and Behavior, 1975.

TERRACE, HERBERT S. "How Nim Chimpsky Changed My Mind." *Psychology Today* 13/6 (November 1979): 65–76.

———. *Nim*. New York: Knopf, 1979.

TERRACE, HERBERT S., L. A. PETITTO, R. J. SANDERS, and T. G. BEVER. *Can an Ape Create a Sentence?* New York: Columbia University, unpublished manuscript.

———. "Can an Ape Create a Sentence?" *Science* 206 (1979): 891–902.

TUTEUR, WERNER. "The 'Double-Blind' Method: Its Pitfalls and Fallacies." *American Journal of Psychiatry* 114 (1957–1958): 921–922.

WILSON, EDWARD O. *Sociobiology: The New Synthesis*. Cambridge: Harvard University Press, 1975.

WOLFE, JOHN B. "Effectiveness of Token-Rewards for Chimpanzees." *Comparative Psychology Monograph* (12)5 (1936).

WOOD, FORREST G. *Marine Mammals and Man: The Navy's Porpoises and Sea Lions*. Washington: Robert B. Luce, 1973.

ZEUNER, FREDRICK E. *A History of Domesticated Animals*. New York: Harper & Row, 1963.

ZIMAN, JOHN. *Reliable Knowledge: An Exploration of the Grounds for Belief in Science*. Cambridge: Cambridge University Press, 1978.

COMMUNICATION AND LANGUAGE IN THE HOME-RAISED CHIMPANZEE*

WINTHROP N. KELLOGG

Oral speech develops in the human infant as an outgrowth of his contact with older humans who are continuously using language. A deaf mute fails to speak because he never hears the acoustic patterns which make up words. He has no sound patterns to follow, no models to imitate. If the ear itself is functioning but the child is mentally retarded, he may be able to hear but not to imitate. Again, he does not learn to speak. A normal ear, a normal brain and speech organs, the continuous hearing of spoken language, and a great deal of imitation are necessary for the completion of the process.

The ear, the speech mechanism, and the capacity to imitate are furnished by the child. The linguistic models come from the human environment in which he lives. Also furnished by the child—perhaps as a result of, or in connection with, his imitation—is a long prespeech period in which he produces both vowels and consonants, but not words. This period of prattling and babbling seems to be a necessary forerunner of the words to come. Children who acquire normal speech habits do so as a kind of outgrowth and expansion of this developmental phase (Smith & Miller 1966). In the terminology of the experimental psychologist, it may be thought of as a period of *preconditioning* or *pretraining*.

If special requirements such as these are necessary for speech to occur in a young human, does any other organism below man possess them? The chimpanzee certainly has a good enough ear, as measurements of auditory sensitivity have demonstrated (Elder 1934, 1935). So far as the larynx and speech parts are concerned, the general assumption has been that these also are developed sufficiently well to

*I am indebted to Dr. Keith J. Hayes who has read most of the material in this article in its preparatory stages and has made a number of helpful suggestions. We are also grateful to Dr. R. A. Gardner and Dr. B. T. Gardner for permission to publish information, from one of their research proposals, concerning the progress of the chimpanzee, Washoe, for the first part of their experiment. Further development of the chimpanzee in this remarkable research is anticipated. However, the subject matter as presented here is solely my responsibility.

WINTHROP N. KELLOGG ● Late of the Department of Psychology, Florida State University, Tallahassee, Florida 32306.

permit the articulation of words—although Keleman (1948, 1949) takes exception to this position. The chimpanzee is a great imitator of the movements and activities it sees performed, although it is not as good an imitator as the child (Hayes & Hayes 1952; Kellogg & Kellogg 1967). It does not naturally imitate sounds and noises, like the parrot or the myna bird, which can reproduce human word sounds but are less apt at nonvocal imitation. Also, the development of the chimpanzee brain as compared with that of man remains in doubt.

But has a chimpanzee (or any of the other apes, for that matter) ever been given a really adequate opportunity to learn and to imitate human speech signals as they occur in their natural context? Has a chimpanzee been exposed to the environmental sound models which are necessary—for as long a time and in the same way as human children?

THE APE-REARING EXPERIMENT

The ape-rearing experiment should furnish an answer to such questions. If communication were ever to evolve, it would seem that the environment of a human household would offer the most favorable conditions. To be sure, the keeping of infrahuman primates as pets or playthings is by no means a novel practice and can be traced historically as far back as the ancient Greeks and Egyptians (Morris & Morris 1966). Apes as household pets are not uncommon today and several books by lay authors attest to the problems involved (Harrisson 1962; Hess 1954; Hoyt 1941; Kearton 1925; Lintz 1942). Such ventures have never given any indication of the development of human language. But pet behavior is not child behavior, and pet treatment is not child treatment.

It is quite another story, therefore, for trained and qualified psychobiologists to observe and measure the reactions of a home-raised pongid amid controlled, experimental home surroundings. Such research is difficult, confining, and time-consuming. Too often, unfortunately, its purpose is misunderstood. Since 1932, reports of five such experiments by qualified investigators have been published in the United States and one in Russia. Four of the United States studies were sponsored by the Yerkes Laboratories of Primate Biology at Orange Park, Florida.[1] The animals used in all instances were chimpanzees.

The Russian research and two of those conducted in America had a human child or children as permanent inhouse controls. In the other experiments the chimps were raised in a household with adult humans alone. Table I gives some of the characteristics of the different experiments, including the approximate duration of each, the number of child controls, the ages of the chimpanzees, and the names

[1]Now the Yerkes Regional Primate Research Center of Emory University at Atlanta, Georgia.

Table I. Principal Chimpanzee-Raising Experiments

Publication date	Investigator	Approx. duration	Approx. age of chimp at start	Sex and name of chimp	No. of child controls
1932	Jacobsen, Jacobsen, and Yoshioka[a]	1 year	A few days	F Alpha	0
1932–1967	Kellogg and Kellogg[b]	9 months	7½ months	F Gua	1
1935	Kohts[c]	2½ years	1½ years	M Joni	1
None	Finch	3 years	3 days	M Fin	2
1951–1954	Hayes and Hayes[d]	6½ years	3 days	F Viki	0
1967	Gardner and Gardner[e]	In progress	9–15 months	F Washoe	0

[a]Jacobsen *et al.* 1932.
[b]Kellogg and Kellogg 1932a,b; 1933a,b; 1945, 1967.
[c]Kohts 1935.
[d]Hayes and Hayes 1950, 1951, 1952, 1953, 1954a,b.
[e]Gardner and Gardner personal communication; unpublished proposal and progress report 1967.

of the investigators. In the present article we shall deal only with those aspects of these researches having to do with communication and language. The work of Kohts (1935), Kellogg and Kellogg (1932a,b, 1933a,b, 1945, 1967), C. Hayes (1951), Hayes and Hayes (1950, 1951, 1952, 1953, 1954a,b), and Gardner and Gardner (1967, personal communication, unpublished proposal and progress report, University of Nevada, Reno 1967) is of special importance in this connection. The observations of Jacobsen, Jacobsen, and Yoshioka (1932) do not deal with this topic, and Finch himself never published any of his findings.

THE PRONUNCIATION OF WORDS

The results of such projects show in general that the infant chimp, when properly handled in the home situation, reacts in many ways as a young child does. It adapts rapidly to the physical features of the environment (Kellogg & Kellogg 1933a,b; Hayes & Hayes 1954b), shows a strong attachment for its caretaker or experimental mother, passes a good many of the preschool developmental tests designed for children, and imitates acts performed by adults without special training. Up to the age of perhaps 3 years, its "mental age" is not far behind that of a child. At the same time, its skeletal and muscular development are much more rapid than those of a child.

With regard to the problem of communication, the results at first glance are disappointing. For even in the experimentally controlled environment in which a home-raised chimpanzee is given the same linguistic and social advantages as a human baby, the chimp displays little evidence of vocal imitation. Despite its generally high level of imitative behavior, it never copies or reproduces human word sounds. Yerkes has written with reference to this matter that in neither the studies of Kellogg nor of Finch "were attempts to imitate speech or other indications of learning to use human language observed" (1943: 192). Kohts noted also that her home-raised chimpanzee displayed not the slightest evidence of trying to reproduce any human vocalizations (1935:576).

Moreover, no ape has ever been known to go through the long period of babbling and prattling which, in the human baby, seems to be the necessary prerequisite to the subsequent articulation of word sounds. Vocalized play of this sort was absent in the Kellogg's chimp, who made no sounds "without some definite provocation . . . and in most cases this stimulus was obviously of an emotional character" (1967:281). The Hayeses noted also that their ape was much "less vocal" and was relatively silent as compared to a child (1951:106, 1953).

Despite these observations, the usual chimpanzee noises—such as the food-bark, the "oo oo" cry, and screeching or screaming—were present in all these experiments and were employed vigorously. The use of these and other sounds as natural communicative signals has been examined by Goodall (1965) for chimpanzees in the wild, and by Yerkes and Learned (1925) for captive animals. It is a question whether such sounds can be modified or shaped to fit the human language pattern.

On the positive side belong the remarkable cases of so-called talking apes. A trained chimpanzee studied by Witmer as far back as 1909 was reported to be able to pronounce the word "mama" but only with great difficulty. The "m" of "mama" was well done, but the "ah" was not voiced (1909).

A few years later, Furness (1916), working diligently with a young orangutan, finally succeeded in getting it to say "papa" and "cup." In training the animal to say "papa," Furness found it necessary to place his fingers on the animal's lips and to open and close them in the proper rhythm.

The best known and most successful of these linguistic efforts is that of the Hayeses (1951), who were able to get their chimpanzee Viki to emit recognizable versions of the words "papa," "mama," and "cup." A beginning was also made toward the sound of "up." Viki thereby exceeded the vocabulary level of either of the other apes, although interestingly enough, she pronounced the same words that they had. She had only one vowel for all her word sounds, a hoarse and exaggerated stage whisper.

The first step in Viki's speech training was designed to teach her to produce a sound—any sound—on demand. This was done by reinforcing whatever noises she made during the training session, such as the pleasure barks elicited by showing her food, or the "oo oo" which resulted from withdrawing the food. It was 5 months, however, before the animal could emit a sound promptly on cue, and the noise she made then was a new one; a hoarse "ah," quite unlike the normal chimpanzee vocalizations which had been previously rewarded.

The Hayeses taught Viki to say "mama" by manipulating her lips as she said "ah," then gradually reduced the amount of manipulation as she learned to make the lip movements herself. In this way the animal finally came to say "mama," softly and hoarsely, and without help (although she persisted in putting her own forefinger on her upper lip). Viki's later words were learned more quickly, making use of existing consonant-like mouth sounds which she had often produced in play. Fortunately, her articulation and vocal behavior have been preserved in a sound motion picture film (1950).

These then, "mama," "papa," "cup," and possibly "up," represent the acme of chimpanzee achievement in the production of human speech sounds. But they were learned only with the greatest difficulty. And, even after she could reproduce them, the animal's words were sometimes confused and were used incorrectly. The most important finding of the Hayeses was perhaps not that their chimp could enunciate a few human sounds. It lay rather in the discovery that these sound patterns were extremely hard for the ape to master, that they never came naturally or easily, and that she had trouble afterward in keeping the patterns straight.

COMPREHENSION OF LANGUAGE

The ability of a home-raised chimpanzee to "understand" or react characteristically to spoken words or phrases is perhaps best illustrated by the Kelloggs' ape Gua. These investigators kept a daily record of the language units that both the chimpanzee and her human control were able to discriminate. In the case of the chimpanzee, the words reacted to varied from such relatively simple commands as "No no" and "Come here" to statements like "Close the door," "Blow the horn" (of a car), "Don't put that in your mouth," and "Go to Daddy," "Go to Mama," and "Go to Donald" (as the case might be). In the first 4 months of the study, the chimp was slightly ahead of the child in the total number of spoken phrases to which she could respond correctly. This was no doubt due to her superior locomotor ability since, in the beginning, the human subject was obviously unable to comply with such commands as "Get up on the chair." During the last 5 months of the period of comparison, the child surpassed the ape in comprehension. The total score for the entire 9 months was 68 specific response patterns for the child and 58 for the chimpanzee (1967).

Although the ape was only slightly behind her human control at the end, it is noteworthy that she had earlier scored higher than he. This means that she was overtaken by the child, who accelerated at a more rapid rate. Had the comparison continued for a longer period, all indications are that the human subject would have left the animal far behind in the comprehension of words.

SPONTANEOUS GESTURING

Does an anthropoid ape, maintained in the human household, ever use or develop any system of motions or gestures that carry special significance or meaning? The answer is yes, the amount and type of gesturing depending upon the particular home environment and the particular animal. Regarding this matter, the Hayeses have written about Viki that she "makes relatively little use of gestures of the hand alone" (1954a:299). She would nevertheless take hold of the experimenter's hand and lead him where she wanted to go, an activity earlier observed by Yerkes (1916) in an orangutan with which he worked.

Mrs. Kohts reports that gestures were commonly employed by her chimpanzee Joni and, surprisingly, that many of the chimp gestures were like those used by her son. "Both infants sometimes show a nearly similar gesture language. Thus, 'request' is expressed by extending hand forward, 'rejection of food' by turning face and head aside, 'thirst' by putting hand to mouth, 'desire to draw attention to oneself' by tugging at dress" (1935:544).

The Kelloggs' chimpanzee Gua also employed a kind of language of gesture or of action, but in this instance the gesturing of the ape was generally different from that of the child. Most of Gua's gestures consisted of movement patterns which occurred regularly just before or in advance of some subsequent or final act. In this way they served as preparatory signals for the terminal response to come later. Viewed objectively, these signaling movements can be interpreted as anticipatory reactions that were consistent with and occurred in specific situations. It need not be presumed, therefore, that they necessarily represented conscious or purposeful efforts on the part of the animal to "tell" others what she wanted. Their reliability was confirmed by numerous repetitions. The principal instances of this language of action are given in Table II.

The most significant of the gestures listed in Table II are probably those for "sleep" or "sleepiness," those indicating bladder and bowel needs, and the "help me" signal in drinking a coke. The latter occurred spontaneously during a minor test problem. The animal was seated upon the floor with legs spread apart, and a bottle of Coca Cola with cap removed was placed between her feet. Although she could hold the bottle at the proper angle while drinking, she had not yet learned how to transport it from the floor to her mouth. Unsuccessful attempts consisted of licking or sucking at the opening of the bottle and of overturning it in the crude attempt to pick it up. Finally, after staring at the bottle and looking up at the exper-

Table II. Early Gesture Signals of Chimpanzee Gua[a]

Behavior pattern	Human interpretation
Biting or chewing at clothing or fingers of experimenter	"Hungry"
Climbing into high chair	Same
Protruding lips toward cup	"Drink"
Pushing cup away	"Enough"
Removing bib from her neck	"Finished eating"
Taking hand of experimenter and hanging on it	"Swing me"
Throwing self prone on floor	"Sleepy" or "Tired" (goes to sleep at once when put to bed for nap)
Pulling hand of experimenter to coke bottle	"Help me" or "Lift this for me"
Holding of genitalia	"Need to urinate (or defecate)"

[a]Adapted from Kellogg and Kellogg (1967: 275–278).

imenter, she took his hand in one of her own and drew it gently down to the base of the bottle. This was by no means an isolated instance, since it appeared several times during repetitions of the test. Similar reactions of placing the experimenter's hand on objects to be manipulated were also observed by the Hayeses with their chimpanzee Viki.

TWO-WAY COMMUNICATION BY GESTURE

The spontaneous use of gesture movements by chimpanzees raises the question whether this ability to gesture can be developed into something more. Could an intelligent animal learn a series of regular or standardized signals—as a sort of semaphore system? Even though a chimp may lack the laryngeal structure or neural speech centers of man, it does not necessarily follow that it has deficiencies in general motor activity. Might it therefore be able to communicate back and forth by a series of hand movements, arm signals, and postures? Is two-way communication by gesture possible? This is the question which has recently been asked by the Gardners (1967, personal communication, unpublished proposal and progress report, University of Nevada, Reno 1967) and is now under active investigation by them.

It should be understood, however, that the signs and signals employed by the Gardners constitute a systematic and recognized form of voiceless communication. The alphabet language devised for the deaf, in which each word is spelled out by individual hand and finger movements, would obviously be unsuitable. What the Gardners are using is a series of more general or more encompassing hand and arm

Table III. Some Significant Gesture-Language Signs Used by Chimpanzee Washoe

Meaning of sign	Description	Context
Come–gimme	Beckoning, with wrist or knuckles as pivot.	To persons, dogs, etc.; also for objects out of reach, such as food or toys.
Up	Point up with index finger.	Wants a lift to reach object such as grapes on vine, leaves, or wants to be placed on someone's shoulders.
Hear–listen	Index finger touches ear.	For loud or strange sounds: bells, car horns, sonic booms, footsteps, etc.
Toothbrush	Using index finger as brush, rub front teeth.	At end of meals. Once when Washoe noticed toothbrush in strange bathroom.
Hurt	The extended index fingers are jabbed toward each other. Can be used to indicate location of pain.	To indicate cuts and bruises on herself or on others. Can be elicited by red stains on a person.
Hurry	Shaking open hand at the wrist. [Correct ASL (American sign language) form: use index and second fingers extended side by side.]	Frequently follows signs such as "come–gimme," "out," "open," "go."
Sorry	Rub bent hand across chest. (Correct ASL form; rub fisted hand, circular motion.)	After biting someone, or when someone has been hurt in some other way (not necessarily by Washoe). When told to apologize for mischief.
Please	Rub open hand on chest, then extend in a begging gesture. (Correct ASL form: use fingertips and circular motion.)	Asking for objects and activities. Frequently combined: "Please go," "Out please," "Please drink," etc.

movements (not involving spelling) which serve as substitutes for entire words, phrases, or sentences. The American Sign Language meets these requirements. This is an accepted form of human language and is in active use today in Canada and the United States, principally by the deaf (Stokoe, Casterline, & Croneberg 1965).

The chimpanzee subject of the Gardners' study, a young female named Washoe, has been undergoing training in the understanding and transmitting of sign-language signals since June 1966. The animal lives in a fully furnished house trailer and also has access to children's toys and equipment, as well as to extensive play areas. The human beings who come into contact with Washoe communicate with each other in Washoe's presence only by means of sign language. She hears no human words except those spoken inadvertently by workmen or others not associated with the project. Conditioning methods have been used to establish many of the signs which are employed.

In support of this new approach is the fact that both chimpanzees and gorillas in the wild state are known to use specific gestures and postures (along with noises) for communicating among themselves (Goodall 1965; Schaller 1963). Chimpanzees in laboratory experiments will also adopt characteristic attitudes as a means of communication. An example is the posture of imploring or begging observed by Wolfe (1936). As for the home-raised chimp, the gestures of both Mrs. Kohts's Joni and the Kelloggs's Gua have been noted already (see Table II). There would seem, therefore, to be considerable promise in the gesture method.

After 16 months of training, Washoe was able to use 19 signs reliably. Five more signs were in the developmental stage. A good many of the movements used by the animal are standard American Sign Language signals. Some are variants of the standard and a few are chimpanzee originals. There is evidence that she understands a great many more signs than she can use herself. Some of the gestures employed by Washoe are given in Table III.

The most significant thing about these gesture signals is that they are by no means confined to the names of specific persons or things. (They are not all nouns.) Some of them—for example, "please," "hurry," "sorry"—are verbs and adjectives which apply in varying social contexts and are used effectively in different situations of the same class. As such they are far in advance of all previous chimpanzee efforts to communicate with human beings.

SUMMARY

Although often misunderstood, the scientific rationale for rearing an anthropoid ape in a human household is to find out just how far the ape can go in absorbing the civilizing influences of the environment. To what degree is it capable of responding like a child and to what degree will genetic factors limit its development? At least six comprehensive studies by qualified investigators have been directed wholly or partly to this problem. All these studies employed young chimpanzees as subjects and some also had in-house child controls whose day-to-day development could be compared directly with that of the experimental animal. In general, the results of this sort of research show that the home-raised chimp adapts rapidly to the physical features of the household. It does many things as well as a human child and some of them better (for example, those involving strength and climbing).

By far the greatest deficiency shown by the ape in the human environment is its lack of language ability. This eliminates the verbal communication which humans enjoy, and with it the vast amount of social intercourse and learning which are dependent upon language. Even amid human surroundings, a chimp never prattles or babbles as a young child does when beginning to talk. Although it imitates the behavior of others readily, it seems to lack the ability for vocal imitation. The neural speech centers of the brain are no doubt deficient in this respect, and it is possible also that the larynx and speech organs are incapable of producing the complex sound patterns of human language. One long-time attempt to teach a home-raised chimp to pronounce human words succeeded only in getting the animal to mouth unvoiced whispers of the words "mama," "papa," "cup," and "up."

At the same time, a chimpanzee in the home, as in the wild state, uses gestures or movements as communicating signals. This suggests the possibility of training a home-raised ape to employ a standardized system of gestures as a means of two-way communication. Such an investigation is now under way, using a gesture language devised for the deaf. Considerable progress has already been made in both the receiving and sending of gesture signals by this method. The technique seems to offer a much greater likelihood of success than other methods of intercommunication between chimpanzees and humans.

A WORD BETWEEN US

ERIC H. LENNEBERG

WHAT IS LANGUAGE? (I)

There is nothing obvious about the nature and function of language. The discovery of its nature is as difficult as an attempt to see our own retina or to sense the motion of the planet under our feet. Linguists have been accused by students of animal behavior that they are complicating the picture unduly by the introduction of formalization and by creating an aura of philosophy about language that is—they claim—unnecessary and merely serves to becloud the straightforward and "simple" facts. R. A. and B. T. Gardner, for instance, state (1969):

The theories [on language] that can be constructed are never as interesting as the natural phenomena themselves, and the gathering of data is a self-justifying activity.

They find that "careful scholarship" concerning the extent to which another species might be able to use human language is a less efficient approach toward elucidation of the nature of language than their own "alternative approach [namely] to try to teach a form of human language to an animal."

But suppose we could show that what is being taught to the animal only bears some tenuous and farfetched similarity to language, or that it is actually quite different in the most essential aspects; would that not make the "gathering of data" a rather futile undertaking? As it turns out, it *is* possible to characterize language— to zero in on the question: What is language?—and, having done this, we do, in fact, discover that there is no evidence that the Gardners are teaching another species the use of human language.

It is a pity that the social and biological sciences are so prejudiced against theory. It is actually impossible to gather data without at least an implicit theory; and if such a theory were always made explicit, the data gathering would no longer appear as an end in itself but as something that is secondary to theoretical formulations.

The first step toward an appreciation of the nature of language is an inventory of what is and what is not biologically essential to language communication. I shall

ERIC H. LENNEBERG ● Late of the Department of Psychology, Cornell University, Ithaca, New York 14850.

say there is language communication between two individuals as long as (1) there is substantial agreement between the two concerning the semantic interpretation of most sentences produced by either of them, *and* if (2) these sentences may be judged by an independent speech community to be based upon a natural language such as English, or Turkish, or Navaho.

Condition 1 holds if it can be demonstrated that both individuals assign the same truth value to a sentence that is a proposition (e.g., "the sun is shining today"), or that both individuals would give the same answer to a question concerning an easily verifiable matter of fact (e.g., "is the hat on the table?"), or if both agree on the behavior demanded by an imperative (e.g., "take the shoe from the bed!"). Note that in many instances, although we suspect that language communication between two individuals is taking place, it cannot be demonstrated (e.g, in a prisoner who refuses to talk to his interrogator).

Condition 2 purposely is formulated quite liberally. "Based upon . . ." does not mean that it must be transmitted acoustically or that a given writing system must be adhered to. Finger-spelling, Morse-code transmission, or semaphore may all be based upon English. A production system is based upon a natural language if there is an isomorphism between bona fide utterances in a given language and the products of the system.[1] Where such an isomorphism is lacking, the products of the system usually are degradations (of varying degree) from the natural language.

With these definitions in mind, we may now ask what is the most essential condition that makes language communication possible. We shall start by investigating language communication between human beings because we have so much empirical evidence here, but we shall not prejudge the possibility that language communication between different species is demonstrable. However, we shall presently specify the nature of the evidence necessary to compel us to admit that cross-species language communication has taken place.

When we survey language communication in man, we soon discover that the skills of hearing and seeing are not essential. Congenitally deaf individuals become proficient language communicators through the use of writing or finger-spelling. (In the latter, every letter of a word can be shown by a specific hand signal, although some words are usually abbreviated and represented by a single sign. Thus, isomorphism on a word level is fairly well preserved.) Congenitally blind children learn to speak without undue difficulties; no one has ever doubted a blind person's capacity for language communication.

The study of children with cleft palates and other oral or laryngeal impediments has made it abundantly clear that man's language communication is also not depen-

---

[1] More accurately: if for every element in language L there is a corresponding element in the invented communication system S such that one can be mapped onto the other, and if every operation in L is preserved in S.

dent upon the existence of articulatory skills. It is perfectly possible in individuals who cannot make intelligible sounds (Lenneberg 1962), which may be demonstrated by the methods discussed below.

What, then, are the prerequisites in man for language communication? We cannot demonstrate language communication (as defined above) in babies six months of age or younger. Also, training and exposure to language are definitely less efficient during the first year of life than during the third year. Thus, we may suspect that a certain degree of maturation of the brain is a prerequisite.

Further, we know that the lack of proper exposure to language communication will block language development, and therefore, we may confidently assert that a certain "treatment" of the growing individual is another prerequisite. The treatment consists of *speaking to the child*, where the word *speaking* need not be taken literally. The deaf and blind Helen Keller began to develop a capacity for language communication as soon as a tactual signal system could be developed, and thus, a channel opened through which exposure to language was possible.

I have shown elsewhere (Lenneberg 1967) that there is reasonable evidence to make us suspect that exposure to language must take place during a limited number of years (from age 2 to 12, the period during which the brain goes through its last stages of physical maturation). The observations available so far suggest that under the influence of language exposure, the brain is modified in specific ways during these formative years, making language communication possible. If the appropriate environmental influences are lacking during this period, the child seems to outgrow the time of plasticity, so that a belated exposure may be of little use for the acquisition of language capacities.

There is a wealth of further evidence to indicate that the crucial factor that makes language communication possible is to be found in an as yet difficult-to-specify aspect of brain physiology. Language disturbances consequent upon specific brain lesions have been studied since the middle of the 19th century. The capacity for language communication may be totally or subtotally abolished by destruction of brain tissue while the patient's other mental functions continue to operate fairly satisfactorily. This condition is called aphasia. However, it has not been widely recognized that aphasic symptoms are merely an extreme degree of disturbances that are quite common in their milder form and that occur in a wide spectrum of situations in which the physiological functions of the brain are slightly affected either by drugs, toxic substances, or systemic disease and its pathophysiology. In fact, anything that causes stress may alter brain functions in a mild way, producing slight irregularities that may interfere with language communication. The individual cannot think of the right word at the right time; his articulation may become slurred; stammering may occur; he may begin to speak too much and without finishing anything, and thus become incoherent; or he may show an inability to understand what is being said to him. There are only few symptoms of aphasia that do not have correlates in the language of a person with transient alterations of his normal physiological brain function.[2] (However, many abnormal events in the brain leave language unaffected, and not all failures in language communication are due to abnormal physiology; nor is it possible yet to make accurate predictions about the correlation of clinical speech problems and pathological events.)

All these considerations point to the same conclusion: the most important factor for language communication is located in the brain and has to do with its peculiar function. This may not be a very startling conclusion, but when we put the data together, we find that there is much that is not always recognized and is, in fact, far from obvious.

First of all, we find that there must be some rather specific aspects to human brain function that make language communication possible. This is so because language can be *affected* specifically. The power of learning new facts, for instance, is relatively independent of language capacity. Aphasic adults and speechless children may acquire a great deal of knowledge, even though some aspect of their brain function is insufficient for language communication. On the other hand, some feebleminded children may exhibit some simple forms of language communication but have marked impairment in learning. Chimpanzees give various signs of intelligence and an ability to make associations or to comprehend complex situations, but they do not develop language comprehension from mere exposure to language communication as does a child without a tongue (one such case has been fully documented by Dr. James Bosma of the National Institutes of Health). Language, therefore, does not appear to be simply a huge repertoire of associations (a fact that is also brought into question by the ability of the congenitally blind to acquire language). Nor is it simply a consequence of great intelligence. (The relation that exists is more subtle than can be reported here; see Lenneberg 1967.)

Second, the possibility of localization of certain language functions in the brain attests that there are rather specific specializations for this activity. But one must not conclude from this that these functions are individual skills that reside in given tissues such as Broca's area or the angular gyrus. Rather, the functions underlying language communication are integrated activities involving an intricate net of structures, all functionally interrelated, the activity of one modulating the activity of the other.

Parenthetic Remarks on Cognition

Despite the relative specificity of language functions, we must recognize their intimate relation to cognitive function and to the physiological processes underlying

---

[2]Telegraphic, belabored style is one such exception.

cognition. Therefore, the nature of the claims made here will be better understood if seen in the light of a more general theory of cognition.

We find vastly different types of animals living in essentially the same physical environment. Yet they do not seem to perceive the same things in the same ways. The divergences cannot be explained simply by differences in peripheral mechanisms, especially sensory thresholds. There is substantial overlap in the auditory and visual thresholds of many mammals, but differences in their brain function cause a difference behavioral reaction to what is potentially available to their senses. Ethological research of the last few decades makes it plausible to assume that different brains process physical data in different fashions. One species may treat a set of configurations as essentially similar where another species reacts as if there were irreconcilable differences. These interspecies differences are not very obvious if the stimulus material consists of man-made patterns that do not occur freely in the natural environment (triangles, stripes, geometric forms, etc.). But as soon as different species are confronted with more natural "stimulus textures," the animals' reactions become dramatically different. Little systematic work has been done along these lines, and our impressions are still based on uncontrolled observations made in the field on unrestrained animals. However, the cybernetic studies made by European scholars under the leadership, especially, of E. von Holst are an exception. Even though there is a relative scarcity of experimental data, the conclusion seems safe that different types of brains are associated with different cognitive processes, and that types of cognition are species specific.

Let us focus for a moment on one aspect of cognition: pattern recognition. What are the main features of a device that must recognize some given constellation in the natural environment? For illustrative purposes, we might take an imaginary prosthetic device built for the blind; a portable apparatus that makes one noise when approaching an obstacle and a different noise when brought close to a step-down. The task of this sort of machine is relatively simple—much simpler than most of the recognition tasks solved by animals with quite primitive brains. But even here it becomes obvious that the device cannot be a simple transducer; it must do much more than translate directly, i.e., point by point, its input to some output. It cannot, for instance, transfer light and dark to high and low pitch; we want it to accept as input an infinity of patterns (all the different configurations one sees when walking through streets) but make only three responses (e.g., obstacle, step, safe walking). Thus, the task is one of sorting out or categorizing input. Now, how can the infinitely variable input be processed? The incoming patterns cannot just be passed through a stable bank of templates or filters, because the device would then lack versatility and would make mistakes whenever the patterns in the scanned environment failed to conform to certain rigid specifications. Instead, the machine would have to be sensitive to certain relations that remain constant amid continuously varying factors—it would have to extract invariances from an everchanging world. A machine equipped to behave in this way would, in short, be a kind of computer; the processes intervening between input and output would be computations.

Formally, pattern recognition in animals is no less a computational process than what goes on in artifacts of the sort described. This position is a modern version of a mechanistic point of view. We no longer believe that we can characterize a brain as a mechanical device, but we are still committed to the idea that its operations are based on the same laws of nature that constitute the subject matter of the natural sciences. We are as adamant today as our forerunners were a hundred years ago, that brain function should not be explained by postulation of "vital" forces that are different from those encountered in physics and chemistry. However, the modern version of the mechanistic point of view is more modest in its hopes for actual achievements. We do not expect to be able to give definitive explanations, and we are fully prepared to produce knowledge that continues to be bounded by ignorance. The brain machines that we postulate are "soft" devices, and the computations they perform are the result of biochemical action that sets the stage for neurophysiological processes.

Our attempts to construct formalisms that describe the gross behavior of brains—our attempts to offer mathematics for the biological events—must not be mistaken for theories on the function of the machines, properly speaking. They are no more than descriptions, that is, attempts to systematize scattered observations.

Now let me say a word on the types of computations that may be postulated for brains and their associated cognitive process: recognition. The recognition capacity with respect to a linear stimulus continuum such as light or sound has different degrees of coarseness for different species. This is the same as to assert that animals have different *differential* thresholds. What is a just-noticeable-difference (JND) for the Tasmanian skink, may be a set of colors containing 10 JNDs for the spotted kangaroo. Imagine three animals with the same absolute thresholds— but one has a very coarse perception, another less coarse, the third a very fine one. The animal whose JND interval is the same as the absolute threshold can tell only presence versus absence of stimulation—for instance, a violet color looks the same as a red color. It can perceive all the stimuli (an infinitely large number) within the limits of its threshold, but it can never tell one color apart from any other. The animal whose JND interval is smaller, say a little narrower than the gamut of the animal's absolute threshold, can tell apart only highly contrasting colors. Thus, this animal can tell stimulation versus no-stimulation and low stimuli from high stimuli; it sees two *different* colors but only *two* different ones, even though it is sensitive to an infinity of light stimuli. Finally, there is the animal whose absolute threshold is the same as those of the first two described, but whose JND interval is so small that it may recognize a large number of different stimuli. In no case, however, will there ever be an animal that sees or, in general, perceives *all* the differences that nature provides *objectively*. No animal could possibly be so constructed as to see differences in the number of molecules of two masses, or have a specific taste sensation for every molecular structure. Perception in organisms must always be a many-to-one mapping. Recently, the topologists Zeeman and Buneman (1968) formulated the mathematical concept *tolerance* for the type of mappings involved here. Without going into details, their concept conveys the notion that organisms are tolerant with respect to the stimuli to which they respond; they admit of variations in the stimuli and thus emit a given response to a gamut of conditions.

When an organism responds, it manifests a change in its activity. The most general and, in a sense, simplest type of response is the *perception* of some differences (notice that I think of perception, itself, as a response), i.e., the behavior of neurons in the brain changes from state S to state S_1 at a time when a condition in the organism's environment changes from Y to Z. We may say that the computer, *brain*, performs in this instance an operation: it computes the relationship Y to Z. The organism takes note of the difference between Y and Z, and this manifests itself as the transition from S to S_1.

The relationships that are being computed even by relatively primitive brains are much more complex than ascertaining simple differences. The stimulus patterns surrounding animals are not linear continua, and the simplest vertebrates (and a great many invertebrates) give evidence of having a fairly complex repertoire of operations at their disposal with which to compute intricate relationships.

In short, what is important in this discussion is (1) that cognition is species specific (which, of course, does not claim that the types of cognition of different kinds of animals are unrelated to one another); and (2) that at least recognition and pattern perception must be viewed as a system of relational computations.

WHAT IS LANGUAGE? (II)

Even a superficial examination of language communication suggests that its proper function is predicated upon the proper function of cognition, including recognition. We have seen in Part I that language can hardly be regarded as a collection of associations, a stockpile of individual items where all that is needed for speaking is the acquisition of one item after another. It becomes clearer now that language in all its phases is a process or a mode of dealing with potential information in the environment, a peculiar way of processing data. The learning of a language does not involve the tagging of a particular thing but the selection of a set of principles for doing relational computations upon the environment. The child who was asked what a hole was, answered characteristically, "A hole is to dig." In other words, a hole is not that one unique stimulus configuration in the child's backyard, but is anything upon which a given relational calculus produces an invariant result. The computer that can assign a specific response to all "hole-configurations" must be equipped to carry out a number of operations resulting in fairly specific relational computations. If we admit that brains with their associated cognitive processes are species specific, we must also admit that it is quite likely there will be many types of brains that are not endowed with the requisite capacities to carry out those operations necessary to categorize stimulus configurations into holes versus non-holes. This assertion does not imply that there might not be some stimulus configurations recognizable by means of relational computations that are quite within the capacities of a wide range of animals.

Virtually every aspect of language, including everything in semantics, syntax, and phonology, is a reflection of relational computations performed by our brain upon the objective world. Any one of the natural languages reflects only a small part of the relational computations that the human brain is capable of, and a person who speaks Gururu natively is not restricted to the relational computations that are reflected in his language (as has been claimed occasionally; see Whorf 1956). When a child acquires the language of his surroundings, he learns that certain computations available to him by virtue of having a human brain are labeled either by words or by syntactic means. We say that he understands a word or sentence when he gives evidence that the word can cause the computer—his brain—to go through a certain routine of relational computations. What I am proposing here is that the repertoire of relational computations is the same for all human brains; that a natural language is a system of tags (alas, fairly sloppy ones) for a certain number of modes of relational computations; and that during the acquisition of a natural language, the child learns just which computational processes are being tagged.

Just as there is in sensory perception, to use Zeeman's term, a certain *tolerance* on the set of perceivable points (JND intervals), there appears to be a certain tolerance in natural languages on the exactness of the computational processes that are being tagged by a word. Therefore, it is difficult to say exactly what the necessary and sufficient relations are that categorize a certain stimulus configuration such as *table* or *house* or *red*. Language communication is quite efficient, despite a lack of perfect agreement on just what relations are in question. This is best demonstrated by the process of metaphorizing, which is so characteristic of all natural languages, as well as the use that people make of languages. *Table* is usually the result of relational computations concerning flatness, squareness, proportions, and positions such that one can sit *at* the object (i.e., a certain relationship to the human figure), etc. However, these computations are not rigidly fixed in number and nature. Speakers may extend the meaning of the word, make creative use of it, without danger of being incomprehensible to other speakers of the same language. If I were to speak of a table-mountain, or if I were to state that John shares his table but not his bed with Mary, any speaker of English would have an idea of what I meant, even though I would have taken liberties with common usage. It is exactly this tolerance for groups of relations that are being computed in the semantic pro-

cess that accounts for the fluidity of languages and their relatively quick change in
in the course of history.

THE STUDY OF LANGUAGE COMMUNICATION ACROSS SPECIES

I would now like to address myself to the challenge of the Gardners (1969) and
their claim that they are teaching a form of human language to a chimpanzee
through the medium of what they call the American Sign Language. I assume that
the experiment is being conducted to demonstrate aspects of chimpanzee capacities
that are (a) new to us, and (b) have some relevance to the human communication
via natural languages. In other words, the Gardners are not trying to confirm further
that animals, including chimpanzees, can perform acts upon verbal commands such
as "heel," "sit," "go." Nor do we need experimental data to show that chimpanzees can learn to perform gestures, or that chimpanzees, dogs, or cats will do things
spontaneously to achieve an end, including certain tricks they have been trained to
perform. Thus, the only question on which there is serious disagreement, and a
question that must be settled by careful empirical demonstration, is the extent to
which the animal may be trained to do something that resembles language communication among men. When I say *resemble*, I do not mean some far-fetched analogy
based on utterly superficial similarities or on logical rather than biological similarities (such as, respectively, the speech of the parrot or the language of the honey
bee). In order to show empirically that the chimpanzee trained by the Gardners is
acquiring even a primitive aspect of language, five distinct demonstrations are
necessary:

1. *A test of the language itself (as used from man to man).* We have developed
many communication systems that have very little to do with natural languages.
For instance, the green-red traffic signals are not a simplified language; nor are the
pantomimes that I may invent in a game of charades.

As noted, the Gardners report that they are using for their experiment the
American Sign Language of the deaf. Those congenitally deaf persons in America
who use sign language actually make use of three different systems: first, finger-
spelling, in which every letter of a word is spelled out; second, a system of standardized shorthand signs for entire words; third, nonstandardized, spontaneously
created pantomime. When the deaf communicate, they tend to mix all three, though
the third system serves more as vignetttes to what they are saying (somewhat like
our own use of gestures as we talk). In most casual conversations, the first and
second systems are intermixed freely; when the subject matter is fairly technical,
finger-spelling predominates; but even in light and familiar interchange, finger-spelling continues to be used.

The communication system that is being taught to the chimpanzee is apparently
a mixture of the second and third systems; it differs from the communication of the

deaf in an important way: there is no finger-spelling whatever. Thus, we should first of all discover, what are the capacities of the communication system itself, and what is the proficiency of the trainers in communicating among each other. We would like to know what type of message can be sent, how much detail is preserved in the transmission, what is the mistake rate, etc. To answer these questions, the following experimental paradigm is indicated: two individuals must communicate with each other by signing, one asking questions, the other one answering them. In order to make sure that communication is not helped along by extraneous factors, two additional persons are needed as observers. The first person writes a question on a slip of paper that is handed to the second, who reads it and then translates it into hand signals. The third person sees the hand-signaled question and answers it, again by means of hand signals; the fourth person can only see the third; he does not know what has been asked, but writes on a slip of paper what the answer is. The written questions and answers remain as permanent records and constitute the data from which one may judge the efficiency and versatility of the communication system.

2. *A test of communication efficiency between man and chimpanzee.* If test 1 has been shown to be successful in at least one circumscribed area (cf. also tests 4 and 5, below), we may examine communication across species. The experimental setup must be the same as described for the first test, and would be run in the following way: One man would ask the question in writing; a second would translate it into signs; a third would watch the second and record in writing what he thinks is being signalled to the chimp; and a fourth would watch the chimp and describe his reaction in writing (this man would not know what was actually being said to the animal). It would be a good precaution to send a signal to the fourth man at the conclusion of the communication sent by number two; this would insure that the random behavior of a restless animal (who might blindly go through his acquired list of tricks) is not recorded as a response to a question or command.

3. *A test of the animal's knowledge of language.* A parrot utters words, but all serious zoologists are agreed that the bird does not know any aspect of the language in which these words occur. Mention of the word *know* or *knowledge* need not scare the empiricist. There are objective, behavioral means of demonstrating the existence of knowledge of a language. They consist of limiting our test to questions that may be answered by a yes–no gesture or to instructions that must be followed by appropriate behavior. Now, it is absolutely essential that both questions and instructions be phrased in sentences, however simple in construction and vocabulary, because the sentence (syntax) is the most characteristic aspect of language. There are no natural languages without syntax, and if even the most primitive syntax were unteachable, we could not agree that what has been taught is language-like. In the absence of syntax we would be back to single-word commands, such as "come" or "paw," which we already know animals can follow and which do not

indicate language capacity (just as the jumping over a hurdle upon hearing a buzzer is not an analogue to word acquisition).

Notice that the crucial questions we should like to have answered by the experiment is whether and what the animal can *understand*. The only gestures that the animal need be required to emit are "yes" and "no"—two items of productive vocabulary! Thus, our stipulation is considerably more modest than what the Gardner's ape is at present required to do. All the animal's spontaneous productions should be disregarded in our attempts to assess its *knowledge* of language. This is necessary in order to avoid fooling ourselves in a way similar to that caused by the horse, Clever Hans, and his owner, v. Osten. If we consider the animal's spontaneous productions and assess the appropriateness of these "messages," we are simply testing our own ingenuity to assign interpretations to productions that might, for all we know, have been emitted randomly. It would always be possible to claim that the ape had *intended* to tell us something—the dream it had last night or a thought that crossed its mind (probably concentrated on all sorts of monkey-business).

4. *A demonstration of the existence of simple language—operations*. Naturally, we must use a very simple language, consisting of items that have meaning to the animal in his present environment. On the other hand, we must be sure that at least some of the basic aspects of every language are preserved. To guide us in our construction of the language to be taught, we should include only words or aspects that either are, or potentially could be, present in all languages of the world. At the same time, we should use only items and aspects of language that can be mastered by children in their very early stages of language development. This assures the construction of a truly primitive level of language comprehension. Following is a representative list of items that fulfill these demands. The total number is in fact considerably smaller than the productive vocabulary reported for the Gardners' animal:

> Five object words:
>   hat, shoe, table, window, bed
> Three action words:
>   put, take, point
> Three qualifiers:
>   small, big, black
> Five words indicating position or direction:
>   on–up–high,[3] under–down–low,[3] from–out,[3] to– at,[3] with
> Four function words:
>   no–not,[3] yes, and, is?[4]

[3]I.e., one word with this meaning.
[4]Signal for question.

The total vocabulary of the language need not consist of more than twenty words. The syntax need not consist of more than a single, basic construction used in either the imperative or the interrogative. For instance, "Take [the] big shoe and put [it] under [the] bed!" or "Is [the] big shoe under [the] bed?" This language is extremely primitive, and since we do not require the learner to speak it—only to understand it—even a highly retarded child could manage it.

The number of items and rules is kept here to an absolute minimum, but I consider their nature to be so representative of natural language that any substantial qualitative changes (such as elimination of prepositions or predication) would simply comprise the claim that "a form of human language" is being used. What would remain would be a serious degradation of language communication—not just a primitivization.

5. *A test of productivity*. Even though the subect is not expected to learn to *say* (or sign) more than "yes" or "no," we would have to require that his understanding is *productive*. By this I mean that it is not enough that he learns to answer a small number of stereotyped questions or to execute a few stereotyped commands. The learner must be able to answer essentially any question that is possible within this limited language or to execute any command that can be phrased with the few words. Further, his semantics, too, must be productive. The words "hat" and "table," "big" and "small," "on" and "under" must be understood in essentially all those contexts in which they are understood by the speakers of a natural language, and the mistakes that occur in understanding should be somewhat comparable to those committed by children. (Since the latter are difficult to assess in any objective way, we must be prepared to tolerate fairly broad deviations.)

As soon as the Gardners submit objective evidence on the positive performance of these five tests, I shall admit that the capacity for language acquisition is common to man and to chimpanzee, and that man's superior achievements differ only quantitatively, not qualitatively from those of other primates (or at least the genus *Pan*).

A WORD BETWEEN US

The series of tests above should reveal whether a given subject—human or animal—can discover the intricate net of relationships that underly the meaning of words and whether he can also identify the relationships between these relationships, that is, the syntactic connections between words. The subject will experience great difficulties with this task unless he tends to do with his environment (i.e., unless he cognizes the environment) as members of our species do when recognizing objects around them. Languages merely select from among those computations that the human mind performs when interacting with our surroundings. The word between us is the instrument by which it is possible to bring about accord of computations in different minds; but this can only be achieved if there is a propensity for making just these kinds of computations, of seeing these relationships among the heterogeneities called "the environment." The acquisition of words does not create a propensity for specific computations out of nothing. Teaching a child the meaning of a word does not introduce into his mind a mental operation the way a desk calculator would introduce computing facilities into an empty office. The speech community can do no more than say, "Here is a word *W;* its correct usage requires the application of a given set of relational principles (computations) upon certain aspects of your milieu; find out for yourself how we use the word." In short, the word between us is an incentive to select from among a range of computations, all from the general repertoire of computations for which our brain with its cognitive functions has been built by nature. Whether the brain of a chimpanzee has the same or similar propensities must yet be demonstrated. It is possible, but not probable (Lenneberg 1969).

As an appendix, I wish to draw attention to three common but erroneous views on the essentials of language:

1. It is sometimes thought that an animal approaches man's language capacity to the degree that it acquires a large vocabulary—the more words the closer to man's language. But a language is not a simple store of items; it is a mode of operation. Even if a subject knew the meaning of four thousand isolated words, he would not know the language unless he understood some principles of syntax and some of the subtleties of usage pertaining to each word. The number of items learned is quite irrelevant to language capacity.

2. Those of us who believe that language has a specific biological foundation are frequently misunderstood; our claim is not based on man's capacity to articulate or perceive auditorily but on his capacity to process potentially available information in the environment in highly specific ways. The biological foundation must be sought in the physiological processes underlying man's cognition. From this it should also be obvious that a belief in a biological foundation is *not an anti-environmental* position. Influences from the environment play an indispensable role in the new theory.

3. Finally, it is sometimes claimed that a capacity for generative grammar has been demonstrated simply by showing that the subject is concatenating various items. Actually, this demonstrates very little. The word *generative* in grammer does not refer to the production or making of sentences. It is used as an abstract metaphor and denotes *principles that account for something.* For instance, Peano's five postulates could be said to "generate" arithmetic sentences such as $3 + 2 = 5$; or the concept of the *a priori* generates a rational epistemology. The principle of random selection might generate a table of random numbers. When the word—output of an organism or a machine—is evaluated, we would like to know what the principles are that underlie the production. In the case of a parrot, there *is* a generative principle (replication of some stereotyped phrase); but this principle is different from those that underlie sentence formation of an adult speaker. In short, generativeness in grammar is not synonymous with "combining of words."

THE FIRST SENTENCES OF CHILD AND CHIMPANZEE*

ROGER BROWN

In 1968–1969 several lines of thought about language development came together in an exciting way. Ursula Bellugi and I, in the 1964 paper called "Three Processes in the Acquisition of Syntax" (1964a), had described one sort of common adult response to the telegraphic sentences of very young children, the kind of response called an "expansion." An expansion is essentially a reading of the child's semantic intention. Thus, when Eve said, "Mommy lunch," her mother said, "That's right, Mommy is having her lunch." Dr. Bellugi and I did not commit ourselves as to the accuracy or "veridicality" of these readings. We were not sure whether children really intended, by their telegraphic sentences, what adults thought they intended, and we could not really see how to find out. Our focus was on the expansion as a potential tutorial mechanism. Whether or not children started out intending what adults attributed to them, it seemed to us that expansions would cause them to do so in the end. In 1968, I. M. Schlesinger, working at Hebrew University in Jerusalem, and Lois Bloom, working at Columbia University, independently came to the conclusion that children really did intend certain aspects of the meanings attributed to them by adult expansions (see Bloom 1970; Schlesinger 1971). They did not intend the meanings expressed in the expansions by grammatical morphemes, by inflections, articles, prepositions, and auxiliary verbs. There was nothing in the child's performance to suggest that they had these things in mind. What they did intend were certain fundamental semantic relations such as agent–action, agent–object, action–object, possessor–possessed, and so on. There was, in the child's speech, something to suggest that they intended these relations. This same "something," the aspect of child speech that justified the attribution to them of certain relational meanings, turns out to be missing from the linguistic performance of an important comparison case: the home-raised chimpanzee named Washoe.

Washoe, over a period of more than three years, has been learning the American Sign Language. She now produces many recognizable signs in circumstances

*Some of the research described in this paper was supported by PHS Grant HD–02908 from the National Institute of Child Health and Development.

ROGER BROWN • Department of Psychology, Harvard University, Cambridge, Massachusetts, 02138.

that are semantically appropriate (Table I). What is more, she produces sequences or strings of signs which seem very much like sentences. Does this mean that a chimpanzee has now been shown to have the capacity for linguistic syntax, a capacity we had long thought exclusively human? Perhaps it does. Still there is something missing; the very thing, curiously enough, that justifies the Bloom–Schlesinger semantic approach to the early speech of children.

In this paper the spotlight is on Washoe whose extraordinary feats naturally place her in the center ring. But there are also important events in the other rings. Bloom and Schlesinger are right, I think, and they have increased the power of the analyses we can make of child speech. They also bring to our attention some impressively general and conceivably universal aspects of child speech. As Table II indicates, we begin to have enough analyses of children learning a variety of languages to see what is truly general in language development. And since this paper was written, the range has been expanded by dissertations written at Berkeley on the acquisition of Samoan, of Luo (spoken in Kenya) and of Tzeltal (spoken in a region of Chiapas, Mexico).

Table I. Some of Washoe's Sign Sequences as Classified by the Gardners

A. Two Signs
   1. Using "emphasizers" *(please, come–gimme, hurry, more)*
      Hurry open.
      More sweet.
      More tickle.
      Come–gimme drink.
   2. Using "specifiers"
      Go sweet (to be carried to fruitbushes).
      Listen eat (at sound of supper bell).
      Listen dog (at sound of barking).
   3. Using names or pronouns
      You drink.
      You eat.
      Roger come.

B. Three or More Signs
   1. Using "emphasizers"
      Gimme please food.
      Please tickle more.
      Hurry gimme toothbrush.
   2. Using "specifiers"
      Key open food.
      Open key clean.
      Key open please blanket.
   3. Using names or pronouns
      You me go–there in.
      You out go.
      Roger Washoe tickle.

Table II. The Available Data in Developmental Order

Child	MLU	Age at data	Character of data	Investigator(s)
Eric I	1.10	1; 7	4 h, tape recorded	Bloom
Gia I	1.12	1; 7	7 h, tape recorded	Bloom
Eric II	1.19	1; 9	6 h, tape recorded	Bloom
Gregory		1; 7.5–1; 11.5	Cumulative inventory	Braine
Andrew		1; 7.5–1; 11.5	Cumulative inventory	Braine
Steven		1; 11.5–2; 0.5	12 play sessions, tape recorded	Braine
Christy		2; 0–2; 3	Taped weekly 45 min sessions over 3 months	Miller, Ervin
Susan		1; 9–2; 0	Taped weekly 45 min sessions over 3 months	Miller, Ervin
Kathryn I	1.32	1; 9	7½ h, tape recorded	Bloom
Gia II	1.34	1; 9	7 ½ h, tape recorded	Bloom
Eric III	1.42	1; 10	8½ h, tape recorded	Bloom
Seppo	1.45	1; 11	2 h taped over 1 month	Bowerman
Eve I	1.68	1; 6	3½ h taped over 6 weeks	Brown
Sarah I	1.73	2; 3	3 h taped over 6 weeks	Brown
Seppo II	1.77[a]	2; 2	2 h taped over 1 month	Bowerman
Kathryn II	1.92	1; 11	9 h tape recorded	Bloom
Rina I	1.95[a]	2; 1	2 h taped over 1 month	Bowerman
Adam I	2.06	2; 3	2 h, tape recorded	Brown
Hildegard		1; 0–2; 0	Parental diary, selectively reported	Leopold
Charles		1; 0–2; 0	Parental diary, selectively reported	Grégoire
Edmond		1; 0–2; 0	Parental diary, selectively reported	Grégoire
Zhenya		?	Very selective, interpretative reports	Gvozdev, Slobin
Canta		2; 4	Grandfather's diary, selectively reported	Chao
Washoe		1; 0–3; 0	Foster parental diary, quite fully reported	Gardners

[a]MLU is approximation within ±. 10 due to special problems in calculating for Finnish.

I should like gratefully to acknowledge the courtesy of Dr. R. Allen Gardner and Dr. Beatrice Gardner, who are raising Washoe. They have kindly shown me their films, sent me their diary summaries, and responded to questions I have asked them about this fascinating experiment in comparative psychology.

Once again, and for the third time in this century, psychology has a home-raised chimpanzee who threatens to learn language. Washoe, named for Washoe county in Nevada, has been raised as a child by Allen and Beatrice Gardner of the University of Nevada since June of 1966 when she was slightly under one year old. At this writing the materials available consist of summaries of Washoe's Diary extending to the age of 36 months.

The first of the home-raised chimps was Gua, also a female, raised by the Winthrop Kelloggs nearly 40 years ago (see Kellogg & Kellogg 1967). Gua gave some evidence of understanding English utterances; she responded distinctively and appropriately to about 70, but Gua did not speak at all. Viki, the second chimp to be adopted by a human family and also a female, learned to make 4 sounds that

were recognizable approximations to English words. Viki was given intensive training by her foster parents, Keith and Cathy Hayes, but the 4 word-like sounds
seemed to mark the upper limit of her productive linguistic capacity (see Hayes &
Hayes 1951).

Both Viki and Gua were asked to learn one particular form of language—
speech—which is not the only form. The essential properties of language can be
divorced from articulation. Meaning or "semanticity" and grammatical productivity appear not only in speech but in writing and print and in sign language. There is
good reason to believe that the production of vowels and consonants and the control
of prosodic features is, simply as a motor performance, something to which chimpanzees are not well adapted. The chimpanzee articulatory apparatus is quite different from the human, and chimpanzees do not make many speech-like sounds
either spontaneously or imitatively. It is possible, therefore, that Viki and Gua
failed not because of an incapacity that is essentially linguistic but because of a
motoric ineptitude that is only incidentally linguistic. The Gardners thought the
basic experiment was worth trying again, but with a change that would eliminate
the articulatory problem. They have undertaken to teach Washoe the American
sign language, the language of the deaf in North America. What is required on the
motoric level is manual dexterity, and that is something chimps have in abundance.
They skillfully manipulate so many of man's inventions that one naturally wonders
whether they can also move their fingers in the air—to symbolize ideas.

Why does anyone care? For the same reason, perhaps, that we care about
space travel. It is lonely being the only language-using species in the universe. We
want a chimp to talk so that we can say: "Hello, out there? What's it like, being a
chimpanzee?"

I have always been very credulous about life on other planets and talking animals, and so I have been often disappointed. Remembering the disappointments of
Gua and Viki, I was slow to take an interest in Washoe. From the beginning of their
study the Gardners sent out periodic summaries in diary form to psychologists who
might be expected to take an interest. I glanced over the first 4 of these and noticed
that Washoe seemed to understand quite a large number of signs and that she was
producing a few—in what appeared to be a meaningful way. This much Gua and
Viki, between them, had also done, and it seemed likely that Washoe's linguistic
progress would soon come to an end, but little advanced beyond that of her forerunners. Then, on the first page of the 5th summary, which covers the period when
Washoe was estimated to be between 28 and 32 months old, I read the following:
"Since late April, 1967, Washoe has used her signs—at that time there were six—
in strings of two or more as well as singly. We have kept records of all occurrences
of combinations in the period covered by the previous diary summaries, and found
that Washoe used 29 different two-sign combinations and four different combinations of three signs."

It was rather as if the seismometer left on the moon had started to tap out "S-O-S." I got out the earlier diaries and studied them carefully, and I read with the greatest interest the subsequent diary installments as they came along, and then the Gardners' article "Teaching Sign Language to a Chimpanzee" which appeared in *Science* in 1969. In the spring of 1969, the Gardners themselves paid us a visit at Harvard for two days, showing films of Washoe and discussing her achievements with a group here that studies the development of language in children. We were particularly interested in comparing Washoe's early linguistic development with that of three children whom we have followed for a number of years. In the literature these children are named Adam, Eve, and Sarah.

From an evolutionary point of view the important thing about language is that it makes life experiences cumulative across generations and, within one generation, among individuals. Everyone can know much more than he could possibly learn by direct experience. Knowledge and folly, skills and superstitions, all alike begin to accumulate, and cultural evolution takes off at a rate that leaves biological evolution far behind. Among the various defining features of language there are two that are peculiarly important in making experience cumulative. They are semanticity or meaningfulness and productivity or openness.

Semanticity occurs in some degree in the natural communication systems of many kinds of animal society but productivity does not. Productivity is the capacity to generate novel messages for every sort of novel meaning. Languages have this property because they have grammars which are rules for the compositional expression of meaning, rules which create meanings from words and patterns. Signs in sequence suggest grammar, and so it was a momentous day when Washoe began to produce them. For grammar has, heretofore, been an exclusively human preserve.

WASHOE'S SIGNING PROGRESS

The signs of the American sign language (ASL) are described in Stokoe, Casterline, and Croneberg (1965) and a transformational grammar of the language has been written by McCall (1965).

There are two basic forms of ASL: finger-spelling and signing proper. In finger-spelling there is a distinct sign for each letter of the alphabet, and the signer simply spells in the air. This system, like our alphabetic writing, is entirely dependent on knowledge of the spoken language. In signing proper, as opposed to finger-spelling, the configurations and movements produced refer directly to concepts. Some such signs are iconic, which is to say that the sign suggests its sense. The sign for "flower" in American sign language is created by holding the fingers of one hand extended and joined at the tips, like a closed tulip, and touching the tip first to one nostril and then another—as if sniffing a flower. That is a good icon. Many other

signs are arbitrarily related to their references. Most deaf Americans use some combination of directly semantic signs and finger-spelling. The Gardners only attempted
to teach Washoe the directly semantic signs.

The Gardners are not deaf and did not know sign language at the start of their
experiment. They learned it from books and from a teacher, but do not yet count
themselves really fluent. They and their associates, when with Washoe, and someone is with her all day long, sign, as one would with a child, the names of actions
and things; they sign questions and requests and they just chatter. In addition to
providing this rich opportunity for incidental learning, Washoe's human tutors have
induced her to imitate signs and have used instrumental conditioning (with tickling
as reward) to train her to sign appropriately for objects and pictures in books.

In the first seven months of the project Washoe learned to use four signs with
some degree of appropriate semanticity. The "come-gimme" sign was directed at
persons or animals and also at objects out of reach. The "more" sign, made by
bringing the fingertips together overhead, seemed to ask for continuation or repetition of pleasurable activities and also to ask for second helpings of food. "Up" was
used when Washoe wanted to be lifted and "sweet" was used at the end of a meal
when dessert was in order. In the next seven months 9 more signs were added, and
by the end of 22 months, when Washoe was about three years old, she seemed to
control 34 signs.

In the spring of 1969, the Gardners showed a group of us at Harvard a film of
Washoe looking at a picture book and making appropriate signs as a tutor pointed
and signed "What's this?" On a first showing the performance was rather disappointing. The viewer is not entirely sure that he has seen the signs since there is so
much action going on. However, this changes on a second viewing. The signs of
sign language are not, at first, perceptual segregates for the uninitiated, but even a
single viewing makes them very much more visible. And, probably because so
many of them are iconic, one very rapidly learns about 10–20 of them. I now do not
doubt that Washoe produces the signs.

In the diary reports one can trace the semantic generalization of each sign and
this generalization, much of it spontaneous, is quite astonishingly childlike. To
appreciate the accomplishment it is necessary to recover a certain innocence in
connection with some thoroughly familiar abstractions. Consider the notion connected with the English word "more" when it is used as a request. Washoe started
out signalling "more" with the specific sense of more tickling. Far from adhering to
a particular context, the sign rapidly generalized to hairbrushing and swinging and
other processes involving Washoe which Washoe enjoyed. And then it generalized
further to second helpings of dessert and soda pop and the like. And then to performances of another which only involved Washoe as a spectator—acrobatics and
somersaults. Human children regularly use the word "more" as a request over just
this same range. And when they start to make two-word sentences with "more"

they use nouns to request additional helpings (e.g., "More milk," "More grapefruit juice"), but also verbs to request that processes and exhibitions be repeated (e.g., "More write," "More swing").

The semantic accomplishments are remarkable, but it is the evidence of syntax that most concerns us. Table I sets out some of Washoe's strings or sentences; they are drawn from the Gardners' 5th and 6th summaries which appeared in 1968, and I have selected examples which, in English, look very much like sentences. The classification into combinations using *emphasizers, specifiers,* and *names or pronouns* is the Gardners' own. In Table I we have sign sequences which translate into English as "Hurry open," "Go sweet," "You eat," "Open key clean," "You me go–there in," and so on. How do these multi-sign sequences compare with the first multi-word combinations produced by children learning American English and other languages?

THE AVAILABLE DATA ON CHILD SPEECH

The best index of grammatical development until the age of about 3 years is simply the mean (or average) length of the child's utterances. When speech begins, of course, all utterances are single words, and the mean length of utterance (MLU) has the value 1.0. As soon as word combining begins, the MLU rises above this value. For about 18–24 months almost all grammatical advances have the common effect of increasing the MLU. Because these advances tend to occur in the same order for all children learning American English (and perhaps more generally) when two children have the same MLU values, the internal grammatical detail of their speech is also quite similar.

It is not the case that two children having the same MLU must be of the same chronological age. Children vary greatly in the rapidity with which they progress grammatically and, for that reason, chronological age is a poor index of linguistic level. Figure 1 plots age against MLU for the three children we have studied. Utterance length was counted in morphemes rather than words so as to give credit for inflections like the -s of pluraltiy, the -s of possession, the -ed for past tense, and the -ing of progressive aspect. As can be seen, the utterances of all three children grew steadily longer over this whole period. Eve advanced much more rapidly than Adam and Sarah.

The five straight lines marked with Roman numerals on Figure 1 represent points we have arbitrarily selected for intensive analysis in preparing a 5-stage description of linguistic progress in this period. At our first stage the MLU values for the three children ranged between 1.68 and 2.06. Many utterances were still single words; most were 2 or 3 words long; 5 words was the longest. Washoe seems to have been at about this point when she was 36 months old.

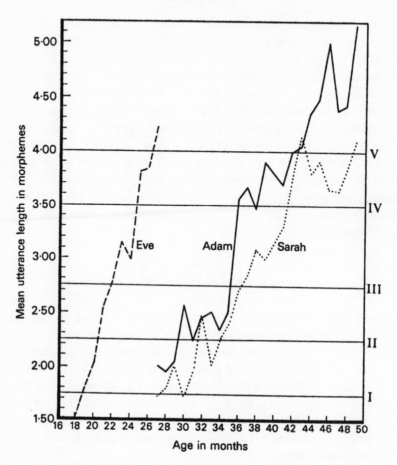

Figure 1. Growth in utterance length with age for Adam, Eve, and Sarah.

Adam, Eve, and Sarah at I, and Washoe at 37 months were not at the very beginning of syntax, and it is desirable to have child–speech data for earlier periods to compare with the earlier data for Washoe. There are, in the literature or in progress, longitudinal grammatical studies which include reports for developmental stages lying between an MLU of 1.0 and the level at which our children were first studied. Combining these reports with ours we have information on an initial period which is bounded by an MLU of 1.0, the threshold of syntax, and an MLU of 2.0, the level of the most advanced child in the set—Adam. For the present purpose we shall call this full interval "Stage I." It seems to correspond fairly exactly with the period for which the Gardners have provided data on Washoe: the age of 12 months through the age of 36 months.

The child–speech data include 18 analyses of 13 children which are fully comparable with one another because they are based on large samples of spontaneous speech, tape recorded in the home. These studies are listed in developmental order above the space in Table II. Adam I, Eve I, and Sarah I belong to the set. So, too, do five analyses of the speech of three American children called Gia, Kathryn, and Eric, which appear in the 1968 doctoral dissertation of Lois Bloom. Then there are single analyses for each of three boys, Gregory, Andrew, and Steven, which were published by Martin Braine (1963) and single analyses for two little girls, Susan and Christy, published by Miller and Ervin (now Ervin-Tripp) in 1964. To this collection of studies of children learning American English, Melissa Bowerman, a doctoral candidate at Harvard, has recently added three analyses of two children learning a non-Indo-European language (Bowerman 1969a,b). The language is Finnish, and the children are Seppo and Rina. All these studies concern Stage I speech; the Roman numerals appearing after the names of some of the children were assigned by the investigators in ordering their own analyses. The developmental order is based on the MLU except in 5 cases for which the MLU was not available. These 5 (Gregory, Andrew, Steven, Christy, and Susan) were placed in the order on the basis of another simple index.

Beyond the studies described above, there is a large literature, mostly American, Russian, French, and German, reporting longitudinal studies of the diary type which are not fully comparable with one another or with the 18 contemporary analyses we have described above. To represent this large literature and broaden the range of languages considered I have selected four studies: Werner Leopold's (1949a) description of the first two years in the grammatical development of his daughter Hildegard who was learning English and German simultaneously; Antoine Grégoire's (1937) account of the first two years of his sons, Charles and Edmond, who were learning French; Dan Slobin's (1966) summary of the detailed account given by the Soviet linguist, A. N. Gvozdev, of the acquisition of Russian by his son, Zhenya; Yuen Ruen Chao's (1951) report on selected aspects of the grammar of his grandson Canta, when Canta, who was learning Mandarin Chinese, was 28 months old. These studies are listed, below the space in Table II, together with the comparison point we never expected to have, the Gardners' study of the first stages in the syntactic development of a chimpanzee, Washoe herself.

MAJOR MEANINGS IN STAGE I

Table III lists 10 kinds of structural meaning that, among them, characterize the majority of two-word sentences in Stage I. In the samples for which complete analysis is possible, those of Seppo, Rina, Adam, Eve, and Sarah, the structural meanings account for about 75% of all multi-word utterances. The characterization by structural meaning is not all there is to say about the first sentences, but it is the

Table III. The First Sentences in Child Speech

I. Operations of reference		
Nominations		*That (or It or There) + book, cat, clown, hot, big,* etc.
Notice		*Hi + Mommy, cat, belt,* etc.
Recurrence		*More (or 'Nother) + milk, cereal, nut, read, swing, green,* etc.
Nonexistence		*Allgone (or No-more) + rattle, juice, dog, green,* etc.
II. Relations		
Attributive	AD + N	(*Big train, Red book,* etc.)
Possessive	N + N	(*Adam checker, Mommy lunch,* etc.)
Locative	N + N	(*Sweater chair, Book table,* etc.)
Locative	N + V	(*Walk street, Go store,* etc.)
Agent–action	N + V	(*Adam put, Eve read,* etc.)
Agent–object	N + N	(*Mommy sock, Mommy lunch,* etc.)
Action–object	V + N	(*Put book, Hit ball,* etc.)

most important information for present purposes. In the contemporary study of child speech, the earliest descriptions published were purely formal and made little reference to meaning. In 1963, Colin Fraser and I described the first sentences as *telegraphic* in the sense that they are almost entirely composed of *contentive* words, of nouns, verbs, and adjectives; the little words or grammatical morphemes which are ordinarily omitted from telegrams are also omitted from early child speech. This is a purely descriptive generalization about surface form. As such, it is correct and has been confirmed in all studies. The characterization of child speech as telegraphic does not provide for *productivity* or the construction of novel combinations. Martin Braine, in 1963, characterized two-word child sentences in terms of a simple but productive grammar involving *pivot* and *open* classes. While Braine's description certainly fits his data, it does not, as both Lois Bloom and Melissa Bowerman have recently shown, fit all children at I, or even all children at the more primitive end of I.

The characterization of two-word sentences in terms of structural meanings was originated in contemporary work by I. M. Schlesinger of the Hebrew University in Jerusalem and by Lois Bloom, both writing in 1968 (see Bloom 1970; Schlesinger 1971). These researchers differ in the grammatical structures they employ, but they agree on one thing: telegraphic sentences looked at in full linguistic and nonlinguistic context can be seen to express certain types of structural meanings. Parents, of course, have always been of this opinion, and they often will gloss a child's telegraphic utterance as a related English simple sentence. When Eve said, "Mommy lunch," her mother said, "That's right, Mommy is having her lunch." When Eve said, "Fraser coffee," her mother said, "Yes, that's Fraser's coffee." The Bloom–Schlesinger approach essentially assumes that such glosses, which the investigator can supply as readily as a parent, are accurate readings insofar as they

interpret the child's words as expressing such relations as agent–object or possessor–possessed. They are not assumed to be accurate insofar as they attribute to the
Stage I child knowledge of inflections, articles, and other grammatical morphemes
for which there is no evidence in his speech. The arguments of Bloom and Schlesinger, as well as the results of our own continuing research, have convinced me
that the early sentences are, in fact, expressions of certain structural meanings.

The classification of Table III is indebted to both Bloom and Schlesinger, but
somewhat different from what either proposes. Set I, *Operations of Reference,* is
made up of utterance-sets, such that each is defined by one or another constant
term appearing in conjunction with various nouns, verbs, and adjectives. Nomina-
tive sentences are always used in the presence of a referent which is pointed at, or
otherwise singled out for attention, and named. In connection with "hi," Bloom
has noticed that children do not use it as a greeting, do not, that is, use it just when
someone hoves in view, but are, rather, likely to light up suddenly and say "hi" to
someone who has been there all along; to someone, to some animal, or to some
thing. Kathryn said "Hi, spoon;" Gregory, "Hi, plane," and Adam, "How are
you, belt." These seem to be expressions of attention or notice. Recurrence of a
referent (Bloom's category) includes the reappearance of something recently seen,
a new instance of the same category, an additional quantity or helping of a substance, a new instance of a quality, and repetition of an action. Nonexistence
(another of Bloom's categories) means that a referent which has been in the referent
field or was expected to appear in it is not now to be seen.

In set II, called *Relations,* we have no repeating words to define each class.
Both initial and final words are varied, and what defines a set is a certain, quite
abstract, semantic relation. The attributive might be said to take some person or
thing (the noun) and *fill in* or *specify* the value of one of its attributes by naming it
with an adjective. The possessive identifies, for a given thing, the person having
special rights with regard to it. The children's possessive typically divides spaces
and objects in the house among family members; it expresses a kind of primitive
notion of territoriality and property. The locative (almost always without the preposition that is obligatory in adult speech) names a locus for a movable thing or a
locus or terminus of an action. Agent–action constructions take a verb and specify
one of its arguments, the argument naming the (usually animate) initiator or performer of the action. Action–object constructions name the other argument for two-
argument or "transitive" verbs: the recipient or target of action. Agent–object constructions seem, in context, to be sentences with verbs omitted. Thus, "Mommy
lunch" means "Mommy is having lunch." The semantic relations, characterized
above, are closely related to the grammatical relations called modifier of a noun-
phrase, subject of a sentence, and object of a predicate. However, the grammatical
relations are defined in purely formal terms, and while they may, in early child
speech, be more or less perfectly coordinated with the semantic rules, the two are
not the same.

In the course of Stage I certain changes occur. At the lower, more primitive end, operations of reference tend to be more prominent than semantic relations. In the middle stretch, operations of reference and two-term relations account for most multi-word utterances. At the end of Stage I, especially in the samples of Adam and Rina, there is a step up in complexity which is manifest in two ways. Several kinds of three-term relations become frequent: agent–action–object; agent–action–locative; agent–object–locative. At the same time one term, the nominal, in two-term relations, which is in the early period always a single noun, begins frequently to be elaborated into a two-word noun phrase. These noun phrases fill all the positions originally occupied by single nouns. The noun phrases are, furthermore, all expressions of possession, recurrence, or attribution. In fact, then, the elaboration of the noun term is accomplished by filling noun positions with just those two-term operations and relations which are noun phrases and which have been long practised as independent utterances. It is quite wonderful to find that these first structural complications take just the same form in Finnish as they do in English. Finally, although I have omitted any discussion of negative, imperative, and interrogative operations from this discussion, all are present in primitive form in Stage I.

The meanings expressed by the sentences of Stage I seem to be extensions of the kind of intelligence that has been called *sensory-motor* by the great developmental psychologist, Jean Piaget. Piaget's studies of the first two years of life (e.g. Piaget 1954) convinced him that the infant does not at first conceive of objects and persons as permanently existing in a single space which also includes the self. Neither does he recognize that objects and persons are all potential *sources of causality* and potential recipients of force. In the first 18–24 months, in Piaget's view, the infant *constructs* the world of enduring objects and immediate space and time. The meanings of the first sentences presuppose the sensorimotor constructions, but they also go beyond them. The aim of sensorimotor intelligence is practical success, not truth; sensorimotor intelligence is *acted* not *thought*. The ability to create propositions that can be expressed in sentences must mature near the end of the sensorimotor period. If the meanings of the first sentences are an extension of sensorimotor intelligence, then they are probably universal in mankind. Universal in mankind but not limited to mankind and not innate. Animals may operate with sensorimotor intelligence, and Piaget's work shows that it develops in the infant, over many months, out of his commerce with the animate and inanimate world.

COMPARISON OF WASHOE AND CHILD

How do Washoe's sign sequences (Table I) compare with the sentences of Table III? "More sweet" and "More tickle" look like expressions of *recurrence*. "Go sweet" (to be carried to the fruit bushes) seems to be an action–locative construction; "You eat" an agent–action construction; and "Gimme please food" an action–object construction. The sentences with *key* appear to express an instrumental relation which also occasionally appears in Stage I child speech. Several of Washoe's three-term sequences look like instances of the three-term relations that appear at the end of Stage I for children; "Roger Washoe tickle" could be an agent–action–object sentence and "You out go" an agent–action–locative. In sum, the strings of Table I look very much the same as a sample of early child speech.

However, there is more to syntax than sequences or strings. The deeper question is whether Washoe was simply making signs distributed in time or whether the signs were *in construction*. What is the difference? As a first approximation, a sequence may simply name a series of ideas which succeed one another in time but do not combine cognitively, whereas a construction puts ideas into one or another structural relation.

In two superficial respects, Washoe's combinations seem to be constructions and not simply sequences. Before one can make a grammatical analysis of child speech, it must be segmented into utterances which mark off just those words that are *in construction* with one another. Segmentation proves to be very easily done, for the reason that children, when they begin to make combinations, already control several of the prosodic patterns that adults use to mark off sentences. One easily hears in child speech the *declarative* pattern, with high pitch and stress near the end and a final fall and fade, as well as the interrogative pattern that ends with a rising pitch. An adult who uses sign language also has devices for marking off sentences. Stokoe *et al.* (1965) say that in the declarative case the hands of the signer return to the position of repose from which they started when he began to sign. In the interrogative case the hands remain, for a perceptible period, in the position of the last sign or even move out toward the person being interrogated. When we talked with the Gardners we asked whether Washoe used such *terminal* or *juncture* signs. Not having been interested in this particular feature they were not quite sure, but since then Allen Gardner has written to me: "Once we started to look for it, it was very clear that Washoe's segmentation (and our own, of course) is very much the same as that described by Stokoe, *et al.* . . . . It is such a natural aspect of signing that we just were not aware that both Washoe and her friends were doing this all along. As in the case of speech contours it is so obvious that you don't notice it until you worry about it."

There is a second surface feature of Washoe's combinations which suggests that they are constructions rather than sequences. In child speech, the very slow rise over time in utterance length seems to represent an increase of information-processing power. The fact that the child at Stage I produces subject and object without any verb surely means that he operates under some kind of complexity limitation. Now it also is the case that Washoe's sequences gradually increase. Two signs are common before three and three precede four. Why should that be so if the sign combinations are not constructions? If they were only signs strung out in time and not interacting semantically and grammatically, then one would think they might be of any length at all, and that there would be no reason for them to start short and become long.

The presence of terminal contours in child speech suggests that certain words are in construction but not what the constructions are; there are no contours to mark off the various relations and operations of Table III. What is there in child speech to suggest that these structural meanings are being expressed and, specifically to the present point, is there anything not also found in Washoe's sign sequences? What there is in child speech, most generally, is the order of the words. The order, generally, is appropriate to the structural meaning suggested by the nonlinguistic situation.

Consider the two drawings of Figure 2. An adult might say of the one on the left, "A dog is biting a cat" and of the one on the right, "A cat is biting a dog." In both pictures just the same creatures and process are involved. The difference is that the arguments of the verb, the agent and object, are coordinated with different nouns in the two cases. It is the structure of the total situation that changes, and in English the changes are expressed by word order, by the order agent–action–object. What would a child of Stage I say in the two cases? Concerning the picture on the left he might say, "Dog bite" (agent–action); "Bite cat" (action–object), or "Dog–cat" (agent–object). In effect, any two of the three terms in correct relational order. Of the picture on the right, he might say, "Cat bite" (agent–action); "Bite dog;" or "Cat–dog." The two sets of pairs are different; there is no overlap. It is this kind of discriminating response, discriminating with respect to the order of elements, that justifies the inference that the child distinguishes structural meanings. What should we say of a child who, in connection with either picture, simply produced all possible combinations and permutations of two content words: "Dog bite;" "Bite dog;" "Cat bite;" "Bite cat;" "Dog cat;" and "Cat dog." We should say that there was no evidence that the structural meanings were understood. This, it turns out, is approximately what Washoe does.

The Gardners have kept careful records of all the occurrences of each combination of signs, and in their 5th and 6th diary summaries, they report that the signs in a combination tend to occur in all possible orders. And that order often changes

Figure 2. Pictures illustrating agent–object relations.

when there is no change in the nonlinguistic circumstances. It appears, then, that we do not yet have evidence that Washoe's sequences are syntactic, because syntax is not just sign-combination but is sign combination employed to express structural meanings. If Washoe does not intend structural meanings, if "Go sweet" and "Sweet go" are not action–object expressions, then what does she intend? What would her stream of ideas be like? It may be that it is a stream of conceptions having no relation beyond order in time. Having thought of "go" she next thinks of "sweet." Washoe's signs may be something like the *leitmotiven* in Richard Wagner's operas. Wagner, especially in the *Ring,* used short musical *motives* with a certain degree of semanticity, enough to enable musicologists to label them with names like *Valhalla, Curse of the Ring, Nibelungen gold, Renunciation of love,* and so on. In given passages the motives succeed one another, and the related ideas may be called to mind in the listener, but they do not enter into relations like agent–object and action–object. They do, of course, enter into musical relationship.

Not every child sentence presents contentives in the appropriate order. There are exceptions such as "Nose blow" (cited by Leopold), "Balloon throw" (cited by Bloom), "Apple eat" (cited by Miller and Ervin) and "Suitcase, go get it" (Adam I). Allen Gardner has written to me that he and Mrs. Gardner have not yet made a frequency comparison for the various orders in which each combination is used and, at this point, the possibility is quite open that Washoe has shown a "preference" for the orders that are correct for each relation. It is going to be interesting to learn the outcome of the Gardners's planned frequency comparisons. It must be said, however, that children show something much stronger than a statistical preference for correct order. In the full data of Table II violations of order are very uncommon—probably fewer than 100 violations in the thousands of utterances quoted. It is definitely not the case that all possible orders of a combination typically occur; they practically never do.

While word order comprises most of the evidence that the child intends structural meanings, there is a certain amount of additional evidence. In 1963, Fraser, Bellugi, and Brown conducted a test of grammatical comprehension using paired pictures for various constructions with 12 children between 37 and 43 months old. Figure 2 is, in fact, taken from that test and is used to inquire into comprehension of the agent and object functions in a sentence. Most of the items in the test are not concerned with the operations and relations of Table III, but there were four test items in all concerning agent and object, and the three-year-olds in the experiment were correct on these 85% of the time. The full report appears as the third chapter in Brown (1970).

Of course, even the child of 37 months is well beyond Stage I, often by a year or more. However, in Britain in 1965, Lovell and Dixon administered the same comprehension test to 20 two-year-olds (average age: 2;6) as well as to older children. The two-year-olds showed a significant ability to decode the contrast.

Finally, we have the results of an action test conducted on one of our own children. Ursula Bellugi asked Adam, when he was 30–31 months old, to act out with toys whatever she said. And what she said involved agent–object contrasts. For example: "Show me, 'the duck pushes the boat'" and, later on, "Show me, 'the boat pushes the duck.'" Adam responded correctly on 11 of 15 such trials, and this result further strengthens the conclusion that Stage I children have the semantic meanings described in Table III.

While I am prepared to conclude that Washoe has not demonstrated that she intends the structural meanings of Table III, I do not conclude that it has been demonstrated that she lacks these meanings. Appropriate word order can be used as evidence for the intention to express structural meanings, but the lack of such order does not establish the absence of such meanings. It does not do so because appropriate word order is not strictly *necessary* for purposes of communication for either the Stage I child or the Stage I chimpanzee. Let us look again at the pictures of Figure 2. If the child uses correct orders for the two pictures, it is likely that he distinguishes the meanings. But, suppose we were parents in the presence of the action pictures on the left, and the child used an inappropriate order: "Cat bite" or "Cat dog" or "Bite dog." We should still understand him and would mentally make the switches to "Dog bite" and "Dog cat" and "Bite cat" which fit the situation. The structure being supplied by our perception of the situation, we can receive the words in any order and understand them as the situation dictates. Even when we are unacquainted with the situation, our knowledge of what is possible in the world enables us to set right some sentences such as "Nose blow" and "Balloon throw" and "Garbage empty." It follows, therefore, that there is little or no communication pressure on either children or Washoe to use the right word order for the meanings they intend. In their world of very simple sentences, which are usually clarified by concurrent circumstances and which often have only one sensible reading in any case, they will be understood whether the order is right or not.

They will be understood, at least, until they begin to want to say things like "I tickle you" and "You tickle me" or "Mommy call Daddy" and "Daddy call Mommy" or "Car hit truck" and "Truck hit car;" and to say these outside of a clarifying action context. In terms of real-world possibilities, the paired propositions are on the same footing. If the propositions do not refer to ongoing actions but to actions at another time, then the listener or viewer, if he is to understand the message correctly, must be given structural signals, of order or of some other kind, to indicate who or what is in each semantic role. In general, as sentences become more complex and more varied and become *displaced* in time from their references, the need to mark attributives, possessors, locatives, agents, objects, and the like grows greater. The capacity for displacement is, like the properties of semanticity and productivity, universal in human languages, and we notice now that experience cannot become truly cumulative until it is possible to report on events not concurrent with the act of communication.

My conclusion, therefore, is that the question of Washoe's syntactic capacity is still quite open. If she fails to mark structures distinctively when a real communication-necessity develops, then we shall conclude that she lacks real syntactic capacity. If, on the other hand, when her sentences become complex and displaced in reference from the immediate context, Washoe begins to mark structure with whatever means is available to her in the sign language—why then. . . . then, there is a man on the moon, after all.

LANGUAGE, NAME, AND CONCEPT

JACOB BRONOWSKI and URSULA BELLUGI

The experiment of teaching a young chimpanzee to use American sign language[1] is an important advance on previous attempts to test the linguistic potential of primates. For the first time, a primate's capacity for a language used by some humans has been clearly separated from his capacity for making the sounds of human speech. In the nature of things, this pioneer study has been made under special conditions, and (like any single study) cannot be assumed to be perfectly representative. Nevertheless, it does offer evidence of a new kind, in the light of which it is timely to reexamine the relation between human language and the signals that animals use or can learn to use.

CHIMPANZEE AND CHILD

It has never been in doubt since the time of Aristotle that language is a characteristically human accomplishment, and that some of the capacities which it demands are either absent in other animals or are present only in the most rudimentary form. Among these is the fundamental capacity to make and interpret the intricately modulated continuum of speech sounds. Lieberman, Klatt, & Wilson (1969) have stressed the differences between the articulatory apparatus of the chimpanzee and that of man. Thus, the Gardners's decision to bypass the articulatory problems of the chimpanzee and undertake instead to teach a gesture language was a good one. They reasoned that the use of the hands is a prominent feature in the behavior

---

[1]See Gardner and Gardner (1969). The Gardners obtained an infant chimpanzee from the wild when she was about 1 year old and began their project in June 1966. The above-mentioned paper describes the first 22 months of the project. The first six diary summaries (unpublished) contain full reports of all new signs and new combinations of signs observed until Washoe was about 3 years old. This article is based primarily on these materials, and we thank the Gardners for supplying them. The interpretations we have made are our own.

JACOB BRONOWSKI • Late of the Salk Institute, San Diego, California 92112. URSULA BELLUGI • The Salk Institute, San Diego, California 92112.

of chimpanzees, who have a rich repertoire of gestures both in the wild and in captivity. By contrast, the futile efforts to teach the chimpanzee Viki to talk (Hayes 1951) had already shown that a vocal language is not appropriate for this species. In 6 years of intensive training, Viki had learned to make only four sounds that grossly approximated English words. The results of the Gardners' efforts with Washoe are spectacular by comparison. By the time Washoe was about 4 years old she had been taught to make reliably more than 80 different signs. This comparative success, therefore, poses a question of substance: What is the true nature of the language performance that has been achieved by a chimpanzee (under these special conditions of training and environment) and how does it differ from that of humans?

We first describe some of the characteristics of the gesture language which Washoe was taught. The Gardners had learned sign language from dictionaries and from a teacher of sign, expressly for their experiment. They used gestures and manual configurations to represent the concepts in sign language and avoided the use of finger spelling as much as possible. All signs are arbitrary to some degree (although some have iconic origins and aspects), and American sign language has many highly arbitrary and conventionalized signs which must be learned. With the addition of finger spelling, it can be used by a literate signer as a direct translation of English in order to communicate with hearing signers; but it generally is not so used among the deaf themselves, whose rules of use may vary in different areas and may not necessarily derive from English. However, the Gardners state that, as far as they can judge, there is no message which cannot be rendered faithfully in translating from English to sign (apart from the usual problems of translating from one language into another). They also report that they tried to follow the word order of English in their signed sequences.

It might be held that, ideally, Washoe's progress should be compared with that of a deaf child of deaf parents who is learning sign as a native language. We cannot yet do this and so must be content to compare Washoe to children learning spoken language. There are grounds for arguing that the Gardners' method of signing makes this an appropriate comparison.

THE CHIMPANZEES' SIGNS AS NAMES

Studies of chimpanzees in their natural environment have indicated that their own communication systems are employed largely to signal motivational and emotional states of the individual. There are few if any calls given by nonhuman primates that convey information about their physical environment. More generally, in communicating among themselves, humans separate the components of their environment and use a great variety of names for them, and animals do not. It has therefore been argued, for example by Washburn (1968) and by Lancaster (1968), that the capacity for language in humans is based on a specific ability to give names to things which is absent in other primates. To quote Lancaster:

An understanding of the emergence of human language rests upon a comprehension of the factors that led to the evolution of a system of names. The ability to use names allows man to refer to the environment and to communicate information about his environment as opposed to the ability to express only his own motivational state. Object-naming is the simplest form of environmental reference. It is an ability that is unique to man.

We now see from the experiment with Washoe, however, that there is convincing evidence that a chimpanzee can be taught to use names for things. Her use of the names she has learned is not more narrow and context-bound than that of a human child. The Gardners report that in general, when introducing new signs, they have used specific referents for the initial training, and that Washoe herself then used signs in ways which extended far beyond the original training. For example, Washoe first learned the sign for "open" with a particular door. This sign she then transferred to "open" for all closed doors, then to closed containers such as the refrigerator, cupboards, drawers, briefcases, boxes, and jars. Eventually, Washoe spontaneously used it to request opening of the water faucet and of a capped bottle of soda pop. Washoe has learned distinct signs for "cat" and "dog" (primarily with pictures of each) and appropriately uses the signs while looking through magazines or books, as well as for real cats and dogs. She also used the sign for "dog" when she heard an unseen dog barking in the distance, and when someone drew a caricature of a dog for her.

There are errors in her spontaneous signing which resemble the overextensions in children's early use of words. Washoe has a sign for "hurt" which she learned first with scratches or bruises. Later she used the sign also for red stains, for a decal on the back of a person's hand, and when she saw a person's navel for the first time. Washoe used the sign for "listen" when an alarm clock rang to signal supper preparation, and then for other bells, and for watches. Washoe also signed "listen" when she found a broken watchband, and when she saw a flashlight that blinks on and off. This is characteristic of the range and extensions of words used by children in the process of first learning a language. There seems little doubt, therefore, that Washoe (and presumably other chimpanzees) can be taught to name in a way which strongly resembles the child's early learning of words.

We must conclude that the prolonged experiment with Washoe proves that the ability to name is not biologically confined to humans. Hence, serious doubt is thrown on any theory of human language which seeks to explain its uniqueness or its origin in a human ability to name.

A CHARACTERIZATION OF LANGUAGE

A searching examination needs to step back instead from the mechanics of human language, and to ask rather what are the global features that characterize it and differentiate it from the sharp and immediate messages that are evoked in animals either by their internal state or by their environment. One such characterization is behavioral—human utterances are more detached or disengaged from the stimuli that provoke them than those of animals—and this is a general feature of human behavior. Another characterization is logical—human language relies on an analysis of the environment into parts which are assembled differently in different sentences. (By contrast, the signals of animals are complete utterances, which are not taken apart and assembled anew to make new messages.) Both the behavioral and the logical component must play a part in any treatment which seeks to relate the way humans shape their utterances to the way that the human brain operates in general.

One of us has formulated such a treatment (Bronowski 1967) in terms which make it possible to see by what steps language might have developed during human evolution. Some of the steps in this sequence are: (1) a delay between the arrival of the stimulus and the utterance of the message that it has provoked or between the receipt of the incoming signal and the sending out of a signal; (2) the separation of affect or emotional charge from the content of instruction which a message carries; (3) the prolongation of reference, namely, the ability to refer backward and forward in time and to exchange messages which propose action in the future; (4) the internalization of language, so that it ceases to be only a means of social communication and becomes also an instrument of reflection and exploration with which the speaker constructs hypothetical messages before he chooses one to utter; and (5) the structural activity of reconstitution, which consists of two linked procedures— namely, a procedure of analysis, by which messages are not treated as inviolate wholes but are broken down into smaller parts, and a procedure of synthesis by which the parts are rearranged to form other messages.

The steps 1 to 4 express the behavioral ability of humans to disengage from the immediate context; without this, it would not be possible to make predicative statements—to give information about the environment in a form which does not imply an instruction to act. Step 5 expresses the logical ability of humans to influence their environment by understanding it; that is, by analyzing it into parts and then making new combinations from the parts.

In this evolutionary characterization of language, the primates can be seen to share in a rudimentary form some of the necessary faculties of the human brain, for example, the delayed response. This may be the case in some small degree also for the separation of affect, the prolongation of reference and even the internalization of language. In these respects we can perhaps find examples in Washoe's use of sign language which resemble some of the earliest stages of a child's language. But the child passes rapidly beyond these precursors into the characteristically human use of language and outstrips the chimpanzee completely. The crucial activity which the child reaches is reconstitution of the language. Human language is highly structured, and as they grow from age about 1½–4 years, children analyze the structure of language in several distinct ways and reconstruct this structure in their own speech. In this ability the nonhuman primates are quite deficient. We have evidence for this defect, for example, in the study by Zhinkin (1963) of the communication system of baboons, and it appears clearly again in the way in which Washoe forms combinations of signs.

We shall compare Washoe's development with that of a child's learning language in terms of the characterizations made above. Since the growth rate of chimpanzees is faster than that of humans, it seems reasonable to compare her development with that of children of the same age.

DISENGAGEMENT FROM CONTEXT

We make the comparison between Washoe and a human child in two parts. One is concerned with the behavioral steps 1 to 4 in which Washoe and the child appear similar at the inception of language, although within a few months the chimpanzee is left far behind. The other is the logical step 5, the reconstitution of language, in which we believe the human capacity is unique:

1. *Delay between stimulus and utterance.* The evidence for this will be found throughout the following sections, and no special discussion of it is required. There is a wealth of research which connects the increase in the delayed response of primates with the development of the frontal lobes of the brain (Warren & Akert 1964).

2. *Separation of affect from content.* The child's learning of language naturally begins in situations heavily loaded with affect. Children's early sentence-words frequently have the force of command or instruction stemming from their immediate desires, discomforts, pleasures, and displeasures ("come here," "give candy"). Washoe's signs are also primarily concerned with immediate situation, her desires, and her emotional states ("hurry open," "gimme drink"). Yet there are some indications of a primitive ability on Washoe's part to separate affect from content of signs; for example, her spontaneous naming of objects around her when there is no indication that this involves the desire for the object or an instruction to someone else.

However, there is a great difference in this regard between the signs produced by Washoe and the sentences of a 3-year-old child. By this age or before, the child is able to make cognitive statements, including those which he may not have heard before. He is able to understand and interpret correctly cognitive sentences without emotional charge. He has mastered the difference between "I want that" and "She fed him," and can separate out the immediate pleasures and emotional components of words from their objective meanings in sentences. There is by now good evidence with children of only 2 years which attests to the ability to understand cognitive statements, including novel ones (Bever, Mehler, & Valian n.d.).

3. *Prolongation of reference.* Children's early one- and two-word sentences (like Washoe's signing) are based almost entirely on the immediately perceptible context. They are often uninterpretable as messages without reference to the situation as context. They are primarily in the present, about objects, persons, or events which are in the here and now. However, they do include rudimentary references to situations in the immediate past ("all gone juice") or demands for something not present ("more cookie"). In this respect, the chimpanzee and young child are not far different. Washoe, for example, signed "listen" when an alarm clock stopped ringing and signed "more food" when there was no food in sight.

At 3 years old, the child comes to present a markedly more advanced picture. He does far more than *name* objects and events not immediately present, as Washoe does. He makes statements which are predicational and cognitive and may refer to events in the more distant past, which have a future sense or intent but are not just demands for action and which involve pretense or possibility. These have been documented for a group of children (Cromer 1968).

4. *Internalization.* There are few indications that gesture language is used as an instrument of reflection by Washoe. She has been seen to name objects while looking through a picture book, and occasionally corrects the signs she makes. Washoe has been observed on several occasions signing spontaneously to herself, in front of a mirror, or in bed at nap time. The Gardners have described these signs as idle chatter.

Weir (1962) collected tape-recorded samples of her 2½-year-old son alone in his room and found that the child clearly uses language as an instrument of exploration. His monologues show a great deal of syntactic play, arrangements and rearrangements, transformations of sentence types, substitution of words in fixed sentence frames, and so forth. It is not just idle chatter, although it has no social function, no content to instruct someone else, and consists in large part of explorations of structure. It is, in fact, the extreme form of that *distancing* from any immediate context which characterizes behavioral modes 1 to 4.

THE CHILD'S SENTENCES

In turning now to the last of the five characterizations of human language, reconstitution, we face a process which is different in kind from the preceding four. In its full meaning it implies an analysis of the sentences the child hears (and indeed of the environments in which the child experiences their meanings) as a condition for the child's formation of his own sentences. In the first place, however, we shall confine ourselves to the child's construction of sentences in a meaningful way from primitive signs or names which are already known. Then, later, we shall ask how the child (and the human mind in general) is able to extract signs or names from their context—is able, in fact, to form concepts by an inner analysis of cognitive sentences.

The most subtle, yet crucial, way in which Washoe's performance falls short of that of a hearing child is in the failure so far to develop any form of sentence structure. The Gardners report that they did not make deliberate attempts to elicit combinations, but almost as soon as Washoe had eight or ten signs in her repertoire, she began to use them in combinations. It is common for her to sign in combinations now, and by June 1968, the Gardners had recorded 330 different strings of two or more signs. A number of these combinations may be spontaneous and original with Washoe; that is, it is unlikely that they are direct imitations of sentences which she has observed. We may compare her combinations of signs with the sequences of words produced by a child of 3. The comparison makes clear both the limitations of the chimpanzee's utterances, and the nature of the capacity and the steps by which a child learns his first language:

1. The child of 3 already gives evidence that he has a concept of a sentence, which includes an understanding of grammatical relations (such as subject of a sentence, predicate of a sentence, object of a verb). These are not only clearly understood but are well marked in the child's own speech. McNeill (1969) suggests that these relations are present before the first combinations of words into utterances in children's speech, and he considers them as a part of children's linguistic predispositions. Our evidence indicates that a child of 3 years expresses the basic sentence relations with great precision in English (where these are often signaled by word order in simple sentences (Brown, Cazden, & Bellugi 1969).

The Gardners in their diary studies report that, for many combinations, all orders of signs have been observed. Various orderings seem to be used indiscriminately by Washoe and do not differentiate the basic grammatical relations. The signs for "me," "you," and "tickle," for example, have occurred in all possible orders in Washoe's signed sequences. These different orders do not seem to refer to different situations in any systematic way. For the same situation (requesting someone to tickle her), Washoe signed "you tickle" and "tickle you." Washoe signed "me tickle" for someone tickling her and again "me tickle" to indicate that she would tickle someone. Washoe's spontaneous signed combinations seem, so far, rather like unordered sequences of names for various aspects of a situation.

2. Children of about 3 years seem to have well-developed means for expressing the full range of basic sentence types. They not only make demands and commands, they also negate propositions and ask innumerable questions. Children seem to have rudimentary ways of asking questions and of negating from the early stages of language development. What develops, along with more complex meanings, are the grammatical rules for expressing those meanings (Bellugi 1967).

The Gardners have concentrated on the question-answer process with Washoe. They write (1969), "We wanted Washoe not only to ask for objects but to answer questions about them and also to ask us questions." They have taught Washoe to respond to questions of several types (for example, "What you want?" "Who that?" "Where Susan?") and in the process Washoe has seen many models. Despite the ample opportunity to learn about questions (and certainly some opportunity to learn negative sentences as well), there is no evidence in the diary summaries that Washoe either asks such questions or negates.

3. The child of 3 organizes his vocabulary into categories and subcategories which resemble in some respects the categories of the adult language. These are combined into sentences not as unordered naming but according to grammatical principles, which include hierarchical organization of the parts of a sentence (Brown & Bellugi 1964b).

We find, in general, that the child forms or extracts rules from the sentences he hears, and resystematizes them in his own speech. The child is not taught and does not need to be taught specifically the underlying rules of grammatical structure, yet careful study of his development shows that he gradually reconstructs the system for himself (often not preciselv the same system as in the adult language, but by stages approaching the complexity of the adult system). Children seem to develop rules of maximum generality, often applying them at first in more instances than required, and only gradually learning the proper domain for their application. For example, 3- and 4-year-old children say things like "He comed yesterday," "It breaked," "I falled;" "two mans," "my foots," "many sheeps." It is clear that these are not phrases that children have heard; they have generalized the past tense and plural forms from regularities like *walked* and *cats*. Children do not need to be taught the rules of grammatical structure because they discover them for themselves, just as they discover and do not need to be taught the rules of correspondence for recognizing the same object under different conditions of light and position. We see that small children whose cognitive powers are limited in many respects show a remarkable ability to reconstruct the language they hear, just as they reconstruct (give structure to) their experience of their physical environment; the process and the capacity are not specifically linguistic, but are expressions of a general human ability to construct general rules by induction. What is involved is not just the capacity to learn names as they are specifically taught by the humans around the child in the early stages. Far more basic and important is the child's ability to analyze out regularities in the language, to segment novel utterances into component parts as they relate to the world, and to understand these parts again in new combinations. It is this total activity, analysis and synthesis together, which is described in the term reconstitution. We conclude by considering this in more philosophical terms.

LANGUAGE AND CONCEPT

It has been proposed (Bronowski 1967) that the human practice of naming parts of the environment presupposes and rests on a more fundamental activity, namely, that of analyzing the environment into distinct parts and treating these as separate objects. That is, there is implied in the structure of cognitive sentences a view of the outside world as separable into things which maintain their identity and which can be manipulated in the mind, so that even actions and properties are reified in words. In this philosophical sense, predication is not merely putting together words in syntactical patterns, or even the manipulation in the mind of ready-made objects and categories. Rather, predication is in the first place a way of analyzing the environment into parts, and only after that can they be regrouped in new arrangements and new sentences.

Thus, a child may first learn the word for *chair* with one particular chair, and may extend it at first to all pieces of furniture without being specifically taught to do so. Through his analysis of sentences about chairs in his parents' speech and his experiences with these sentences ("Please sit in this chair, Mrs. Jones," "John, move your chair around") the child may gradually narrow his use to the range of objects that we might also describe as chairs. It is important to note that there is no way to give a definition of *chair* in terms of size, dimensions, color, material, or other aspects of physical measurements. To recognize another object we have not seen before as a chair, we must ignore many aspects of the differences between chairs, and attend to criteria which include something like the following: A movable seat that is designed to accommodate one person, and usually has four legs and a back. Notice that a chair is a man-made object designed for a specific function or action, and that this is part of its implicit definition. Learning the word for objects like *chair* is considered to be one of the simplest problems of language learning. Yet, for the child to understand his parents' sentence "The chair broke," he must first analyze out the state of being of the chair at the time of the utterance, and then interpret the meaning of the word *broke* (perhaps violently separated into parts, no longer functioning) from this. He can construct the sentence "The toy broke" for himself only after having analyzed out the relevant aspects of the environment in the parts of the sentence above. The new predication can result only after the definitive attributes of *break* and *chair* have been taken apart as independent units, and the activity of predication presupposes this kind of analysis.

What we have been describing in the child is a general characterization of the relation of human thought to the environment. For humans, the environment consists of objects, properties and actions, and we are tempted to assume that these exist ready-made in the outside world, and present themselves simply and directly to the senses. But this is a naive simplification of the complex of interlocking processes by which we are persuaded of the existence and the persistence even of so unitary a natural object as a tree or a bird. Most of what we regard as objects in our environment, however, are far more sophisticated concepts than these. Thus, the logic by which a child unravels the sentences he hears and his experience of the environment together is much more than a capacity for language and expresses in miniature a deeper human capacity for analyzing and manipulating the environment in the mind by subdividing it into units that persist when they are moved from one mental context into another.

What language expresses specifically in this scheme is the reification by the human mind of its experience, that is, an analysis into parts (including actions and properties) which, as concepts, can be manipulated as if they were objects. The meaning that these concepts have derives from their construction (as parts of reality) and cannot be displayed by a direct appeal to the senses, singly or in combination. Very few concepts derive directly from the senses, as the word *cold* does; the great majority are at least as indirect and intellectual as the word *two*. They are constructions of the mind from a variety of contexts, and in making them, the mind acts exactly as the child does who learns to give meaning to a word by analyzing the variety of sentences in which he hears it. Concepts are artifacts extracted by reification from the contexts or sentences in which they occur. Some of them, like *two,* can be taught to animals, but they remain artifacts of the human mind. We may even speculate that the human mind began to reify objects by their function when man began to make tools as functional artifacts for future use.

If the reification of the environment serves to manipulate its parts in the mind, then the laws which distinguish admissible from inadmissible rearrangements round out and complete the same mental process as a necessary part—as the addition *one* and *one* belongs to the concept *two*. That is, we cannot separate the naming of concepts (objects, actions, and properties) from the rules which govern their permissible arrangements—the two form an interlocking whole. Looking for these rules is in essence the search for structural relations in the environment which characterizes the human mind and is the same as the procedure of generalization which in science is called inductive inference (in the widest, nonpartisan sense). For humans, the division of the environment into parts only has meaning if they obey rules of structure, so that permissible arrangements can be distinguished from arrangements which are not permissible. So in human language, words and grammatical structure form an interlocking whole, from which nonsense words and ill-formed sentences are equally excluded. The match between a sentence and the reality that it maps strikes us now, when we know the language, as made by putting the sentence together; but it begins in the first place, in the beginning of language, by taking reality apart. And it is taken apart into words and grammatical rules together (concepts and structural laws)—just as we create a scientific theory of, say, the atomic structure of the physical world by inferring the existence of the elementary particles and the laws of their combination at the same time.

In short, we must not think of sentences as assembled from words which have an independent existence already, separate from any kind of sentence. This puts the matter in linguistic terms; in more philosophical terms, we must not think of the external world as already existing in our consciousness as a previously analyzed assembly of conceptual units, such as things, actions, and qualities. The experience of learning about the world consists of an inner analysis and subsequent synthesis. In this way, human language expresses a specifically human way of analyzing our experience of the external world. This analysis is as much a part of learning language as is the more obvious synthesis of sentences from a vocabulary of words. In short, language expresses not a specific linguistic faculty but a constellation of general faculties of the human mind.

When we watch the way a child learns to speak from his point of view, we become aware of his mental activity in finding for himself inductive rules of usage which constitute both a grammar of language and a philosophy of the structure of reality. The child does not *recapitulate* the evolution of language, of course; instead, he *demonstrates* the logic which binds the development of language to the evolution of the human faculties as a whole. What the example of Washoe shows in a profound way is that it is the process of total reconstitution which is the evolutionary hallmark of the human mind, and for which, so far, we have no evidence in the mind of the nonhuman primate, even when he is given the vocabulary ready-made.

OF LANGUAGE KNOWLEDGE, APES, AND BRAINS*

ERIC H. LENNEBERG

LANGUAGE KNOWLEDGE: FORMAL CONSTANTS OF COGNITIVE ACHIEVEMENT

Every aspect of an organism's behavior bears the indelible imprint of the biological operating principles of its own species. Whatever a cat does, it does in a feline fashion; whatever man does, he does in a human fashion. This may be taken as an axiom (or truism, for that matter, for how could it be otherwise?). An animal cannot change its constitution between behaviors; every movement, every sensation, every insight, every motive is mediated, coordinated, regulated, transformed, integrated, etc., by one and the same nervous system, the same skeleton, the same irritable tissues. Despite the infinite variability in individual acts or skills (many of them dependent upon environmental circumstances), there remain limits to the behavioral variations, and there are physiological constants that are common to literally everything a given animal does. I would like to propose, as a general guidline for research in animal behavior, that one can gain important insights into the biological nature of a species by attempting to discover the physiological constants—by searching for the common denominators underlying the different sorts of behavior in an animal's repertoire.

In this article, I am concerned with man's capacity for language; however, I should like to deal with it in the context of his more general cognitive propensities. All man's intellectual activities must bear the hallmark of the mode of cognition imposed by the operating principles of the human brain. In our present state of knowledge, it is hardly possible to discover the neurophysiological constants of all intellectual activities. But it may be possible to demonstrate that the activities themselves have a common denominator. If we choose an appropriate data language to describe our observations (hopefully a formal language, a logic or an algebra), it should be possible to point to *formal constants* in any two or more intellectual

*This paper has resulted from research carried out under Grant No. 2279 from the Wenner-Gren Foundation.

ERIC H. LENNEBERG • Late of the Department of Psychology, Cornell University, Ithaca, New York 14850.

activities; these formal constants, in turn, may lead the way toward formulating biological hypotheses and eventually even toward neurophysiological discovery.

I shall point to some commonalities in man's capacity for doing arithmetic and his capacity for acquiring a natural language. The formal constant underlying these activities is a peculiar *relational structure*. One could go on to show that man's perceptual activities also share in this formal constant, but space does not permit elaboration of this point here (see Lenneberg 1970).

Once the formal constant has become apparent, I shall try to apply this insight to some common questions that arise in the study of language—for instance, What constitutes primitivity in language? Are language skills a conglomerate of independent traits or habits? How can one decide whether an ape has language? How does one begin to look for brain correlates of language?

The Capacity for Arithmetic

Knowing a language is not the same as knowing arithmetic; nevertheless, the two have much in common. Arithmetic behaves like language in several important ways. All the primitive notions that form the foundation of arithmetic are part and parcel of any natural language, and the operations of grade school arithmetic are derivations of language operations. Arithmetic "sentences" are much more restricted in syntax and vocabulary than are the sentences of a natural language, but arithmetic sentences and language sentences have similar formal properties. It is not difficult to imagine what sorts of tests you would use to examine an organism's knowledge of arithmetic; we shall see that these tests also provide us with a good experimental paradigm for testing the presence of language knowledge.

First, we would test for the presence of arithmetic knowledge in a subject by giving him problems to solve, such as $6 + 4 = ?$ If the subject failed in tasks of this sort, we would try to test for his capacity to understand the basic concepts of arithmetic. To do this, it would be necessary first to analyze arithmetic itself, in order to discover these basic concepts. Let us say that there is a set of elements, called cardinal numbers, and at least one operation, called addition. As a starter, it is tempting to argue that the operation is a more abstract notion than is the concept of numbers and that we should, therefore, treat the capacity for learning numbers separately from the capacity for learning the operation of addition. Furthermore, there is a temptation to explain numbers in a nominalistic vein: a number is "simply the name that we have learned to associate with something we can see"; and counting is "the consequence of another somewhat similar convention, the learning by rote of a sequence" (like the alphabet).

There are three important reasons for not accepting these nominalistic explanations: (1) Although it is possible to see at a glance couplets, triplets, and quadruplets of things, man begins to make mistakes for collections larger than four. The mistakes increase in magnitude and frequency as the collection becomes larger. (2) It is conceivable that a subject could name a trio or a quartet of one set of objects correctly but would fail to see that three marbles have something in common with the three legs of a stool; knowing the *word* "three" does not ensure understanding of the *concept* of "three." Let us call concepts such as *three, one hundred,* etc., classes (following G. Frege or B. Russell). The notion *class* does not imply here the physical existence of a collection, but merely the existence of some principle or criterion by which one may decide whether a given instance is or is not assignable to membership in some group. If I can assert of something that it is *green,* I have used a criterion by which things may be assigned to the class "green things." If I can assert of a collection that it is *three,* I have used a criterion by which any conceivable collection could be evaluated as to *threeness* (the collections may range from fresh eggs to beatitudes). Of course, we have yet to define what the criteria are for the classes called cardinal numbers. But first, back to the reasons for rejecting a nominalistic definition of numbers. (3) If a person knows the correct sequence of number words from one to ten, it does not follow that he can also count correctly seven of his fingers. For instance, many children recite the words and may even turn down one finger at a time, but, when asked how many fingers they have counted after having turned down number seven, will give a random answer. The knowledge of the automatic sequence is clearly not the same as knowledge of how numbers are ordered. To say that counting is a "simple mapping operation" from fingers to objects begs the question of the nature of the number concept. A child is capable of a one-to-one mapping operation without having the number concept; he can, for example, take down the right number of plates from the cupboard for a company of six by giving each plate the name of one of the company. Thus, mapping is a relation that is more primitive than numbers. In short, the concept of numbers cannot be explained nominalistically or as a cultural convention; the number eight is not simply a visual pattern arbitrarily associated with something we can perceive directly.

What, then, is the knowledge of particular cardinal numbers? There is a sizable literature on this problem that need not be reviewed here (see Benacerraf and Putnam 1964). It is possible to define any cardinal number if we allow at least the following three concepts to be intuitively obvious: the concepts of *one,* and *and,* and of *is.*[1] If we don't have to explain what these concepts are or how the subject comes by them, then we can say that we give the arbitrary name *one* to that which

---

[1] I am concerned here with psychological primitives, not with strict axiomatization such as, for instance, Peano's five axioms. Peano introduces the word *successor* axiomatically but makes no mention of the concept *and.* However, the notion *successor* is defined for a natural number system as "any arbitrary number *plus* one." I believe it is correct, therefore, to regard the concept *and* as a psychological primitive.

is perceived at a glance as a single. If we also know what the words *and* and *is* mean, we can say *one and one is* what we call *two; one and one and one is* what we call *three*, etc. Obviously, this isn't much of an explanation. Most disturbing of all is the question, what gives us the wherewithal to count the number of times we have said *and one?* At least we must introduce another notion, namely, that the knowledge of the quantity *one* may be applied to the operation of adding the words *and one* to *one;* thus, if we know intuitively what one is, we can also know intuitively what two is, namely, a single "and-one-operation." Now that we have the general concepts of one and of two, nothing can keep us from applying these numbers to themselves; we can count numbers with numbers, such as one or two singles or one or two couples. If we hear the number two, we immediately understand that it may stand for two couples or two singles. Notice that we had to add a fourth concept to the first three, namely, *iteration*, which makes it possible to extend rapidly our number system *ad infinitum*. This concept answers the question "How often?" or "How many times?"; counting the number of times, we are introducing multiplication, an offshoot of addition.

We still have to face another major difficulty, which should be easy for us to appreciate in our capacity as psychologists. What are the primitive concepts *one, and, is?* What would we have to do, for instance, to demonstrate that an animal knows what is meant by *one?* Clearly, we would have to show that he knows the difference between one and not-one, that is, one and many, singles, and multiples. Harlow's oddity paradigm is an approximation of such a procedure, although in that paradigm it has not yet been demonstrated that the animal has, in fact, learned the property that underlies the entire class of singles, on the one hand, and multiples, on the other. I believe that it should not be difficult to devise an appropriate experiment on this problem, and we need not discuss the details here. I must emphasize, however, that oneness is clearly a *relation;* as is manyness; the classes of ones and not-ones are defined by that relationship. The notions *and* and *is* also imply relations (see the Appendix). Again, it should not be difficult to devise experiments in which the subject's capacity for understanding either of these relations is tested. Imagine an organism that performs five tricks upon five different signals; can it learn to differentiate between the command "trick *A and B*" and the command "trick *A or B*"? An experiment by which the general comprehension of *is* could be tested might follow this procedure: A subject is trained to perform either of two tricks, *A* or *B*. The signals for the performance of the tricks are *x* and *y*, where the only requirement is that $x \neq y$, but either of these variables may take on any value. Thus, one day *x* is a high tone and *y* a low one; the next day *x* is a light and *y* a gong; every day the stimuli change, but throughout the day's session remain constant. If the subject comprehends the notion *is*, he will never make more than one initial mistake, at the beginning of the day's session, because all he will have to find out is what today's *x* "is." The logic here is "if *m* means A and *n* means A, $m = n = x$."

We have made the following points: (1) The operation *addition* expresses a relationship; (2) numbers stand for relationships, and cardinal numbers larger than one express relations between relations; (3) there is no definition in arithmetic that does not itself rely upon a relation, and thus arithmetic is a closed system with respect to relations; (4) the primitive relationships underlying arithmetic with cardinal numbers (1,+, =) can also be expressed in every natural language; (5) the presence or absence of a word denoting a relationship in the vocabulary of an individual is not correlated with that individual's capacity for understanding the given relationship.

It may be well to add here a parenthetical remark that will elucidate my argument in the sections on apes and on brains. Whenever we assert the existence of a relationship, we are implying that something was, is, or will be related to something else; this necessarily implies the existence of a correlated *activity* or *process*. Further, if we say of an organism or artifact that it relates something to something else, we are necessarily talking about some activity in the organism. If an organism can relate things in more than one way, if its output is more than a single relationship, it must be capable of more than a single sort of activity. One might say that such a "device" can be in different activity states, depending on the particular relationship it is computing.

When we teach our children arithmetic, we introduce them at once to the set of natural numbers. If we were trying to teach arithmetic to some other form, say, a chimpanzee, it would, perhaps, be easier to choose a system with a definite number of elements. A good example would be a residue class of, say, only four elements, 0, 1, 2, 3, such as shown in Figure 1. The table included in the figure contains all the addition facts of the residue class (modulo 4). The logic of these additions is illustrated in the dial; it represents a number line turned back onto itself. The origin is zero, and the distance zero to one is unity. From the dial, we see immediately why three and one would be zero in this system, and three and two, one. The subject would have to be trained to point to any one of the four numbers in response to such questions as "What is one and two?" We might use ten of the addition facts

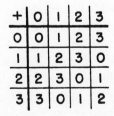
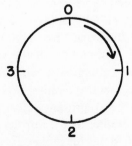

Figure 1. Residue class of four elements: addition table and dial.

for training and demonstration and use the remaining six facts to test his understanding of the underlying principles.

It is always difficult to give a "natural" definition of simplicity or primitivity. Perhaps addition with residue classes is a simpler or more primitive level of arithmetic than adding natural numbers because of the small number of elements. Similarly, the capacity for understanding any one of the relationships that underlie arithmetic *(one, and, is)* may perhaps be regarded as a "precursor" of arithmetic *capacity*. However, if an animal could operate with only one or two of these primitive relationships, one would hardly be tempted to say that it knew "a primitive form of arithmetic," since it could not even learn to count.

The Capacity for Language

Our aim is to show that language behaves like arithmetic in several essential ways. If we succeed, it will be easier to see that certain points that appear obvious enough when we are considering arithmetic knowledge apply equally well to language knowledge.

The study of algebra makes it most obvious that mathematics is concerned with pure relationships. For instance, we allow the symbols a, b to stand for any numbers; then we relate them, such as $a + b$, $a = b$, a/b, etc. As long as we are doing pure mathematics, we are usually not concerned with either giving names to relational constructs or finding out the exact number of a particular relationship that we construct. In natural languages, we have the same situation. We can express pure relationships by letting the symbols a, b stand for any words, and then relate them by saying *a and b, a is b, a to b, a from b,* etc. Evidently the repertoire of relationships in language is more varied than it is in the realm of high school algebra.

In high school algebra, variables such as a and b are intended to represent numbers, where numbers stand for classes or sets. The situation in natural languages is quite analogous. Here the variables a and b also stand for classes, such as nouns, adjectives, verbs, etc. The elements in classes such as nouns, again, stand for classes—not for individuals.

We argued earlier against a nominalistic approach to the meaning of numbers. These arguments also apply to a nominalistic approach to the meaning of words. We can use the same three arguments advanced previously in the section, "The Capacity for Arithmetic." If the variables a and b stand for words, their range is not defined by or limited to what can be perceived at a glance out in nature. Just as numbers represent quantities beyond those that can be perceived directly, words quite clearly can represent constructs that also cannot be observed directly (freedom, truth, brotherhood). Further, the word *shoe,* which seems concrete enough, does, in fact, stand for a class, a point that has been made so often before that it needs no elaboration. Shoehood is a construct, the result of some computation (i.e., operations or relations; I don't care whether it is to be called stimulus generalization or a concept), and the word *shoe* stands for this computation or family of computations.

If someone knows the name of something specific, a proper name, he does not necessarily have the underlying capacity for understanding the principles upon which the classes are constructed to which this individual belongs. If the name of a dish with a certain pattern on it is *three* (or *ding-ding-ding*), it does not follow that the animal who has learned this relationship can construct the class of threeness. Knowing that a certain man is called Pablo Casals is no indication that the knower can understand the class *cellist,* or the class *man,* for that matter.

A natural language can exist without any proper names whatever. And conversely, adding any number of proper names to a language does not change its essential structure; since *name* is an open set, there may be an infinity of names, for all we care. The naming of specifics seems to be quite irrelevant to knowing how a language works. This insight may even be extended to the naming or labeling of specific tricks, such as *heel, beg, show me your nose.* If an infant or animal responds to a verbal command by one stereotyped act, the most important property of language is missing; because no relation is being tagged, there is no generality and no productivity. Formally, the situation is no different from a startle elicited by a bang (or an associated conditioned response), which could hardly be called a primitive form of a natural language.

At one time, it seemed to many of us as if a form of nominalism were introduced into language systems by attaching names to specific sensations; once sensations had been given names, all other words could be defined in terms of these names. There are, however, good reasons now to discredit this idea. First, we don't know what pure sensory qualities are (is *bigness* a sensory quality? is *loudness?*) and second, even in instances where this doubt is minimal, such as in the perception of colors, it may be shown empirically that the words relating to these sensations are always used as relative terms. The word *red* may apply to a narrowly circumscribed range of colors when we refer to a color chart, but when it comes to cows, we are not confused by calling one red, even if we see the difference between that color and the one on the color chart. Every sensory word fits into a system of related sensory words; it is the system that is learned and used, not individual and independent items from it. Color words illustrate this beautifully. As long a child has just one or two words for color, he uses them rather randomly. But when he has four or five color words, his usage suddenly becomes normalized; the relational system has been established, and now the only difference between his usage and that of the adult is that his individual color categories are still somewhat wider than those in the adult system. A semantic study of the words for sensory experience makes it clear that a simple nominalism cannot explain even the use of sensory terms. There is no escape from the necessity of looking toward relations as the basis of meaning.

The third argument used in connection with numbers also applies to language: words by themselves neither define nor delimit nor are evidence for the presence of the underlying relational capacities that appear to be signalled by them. It is common that a child (or a blind person) uses color words without apparently knowing what the words mean. On the other hand, a sighted child may show a preference for red things but be unable to comprehend the meaning of the word red.

We found that in arithmetic the classes labeled by cardinal numbers larger than one cannot be defined without introducing the elementary relation *and;* thus, the construction of the elements in the system of cardinal numbers is intimately related to the addition operation defined for this system. Again, if we look at natural languages as a whole, we find a very similar situation. The elements in the set *natural language L* are words, and the operations are defined by various relationships that may be brought to bear upon the elements. Ideally, we would like to draw a sharp distinction between the lexicon, on the one hand, and syntax, on the other, such that syntax is concerned only with relationships and the lexicon is simply an extensive class of items. However, no such dichotomy is possible. In a highly inflected language, practically all relationships are or at least can be expressed by the words themselves (or by some morpheme). And even in the so-called synthetic languages, such as English, some relationships are signalled by words and some by such syntactic devices as word order. If we look at historical changes in language, it is clear that no relationship whatever is in principle exempt from being taken out of the realm of pure syntax and being incorporated into the lexicon. As a matter of fact, the weight of my argument here has been that every word stands for relational constructs, and therefore the distinction between syntax, on the one hand, and lexicon, on the other, can never be one of relations or operations as against absolute items.

We discussed earlier the bootstrap operation by which the number system is constructed and expanded *ad infinitum.* It consists essentially of iteration of the addition operation and of letting the primitive concepts *one, and, is,* as well as the new concepts generated by means of them, apply to themselves, i.e., letting numbers count the number of times the addition operation has to be repeated. Thus, *one* may be the number of a single count of ten, as in the decimal system. Language also is a system of open sets, and it can be expanded *ad infinitum* by iteration in ways formally similar to the ways in which the number system we have been considering is expanded. Words denote relations, and the relations between words can be denoted by words. Furthermore, any composition of words (a construct of constructs) can be replaced, in principle, by a word, much the way any composition of numbers can be replaced by a number. For instance, anything that has been said or written can be given a name, such as *sentence* or *the complete Shakespeare* (abbreviated to *Shakespeare*), and thus be incorporated into any new composition as a single element. Mappings of this sort have much in common with the mappings implied by an arithmetic sentence. Although there are only certain instances in language in which a given relationship has a unique answer (the author of *Waverley* is Scott) as in certain arithmetic sentences $(2 + 3 = 5)$, we find that any particular ordered or related pair may be equated to any of the elements of a particular set (like the solution set of a given mathematical equation or inequality); for instance, the relationship *young human being* stands in a similarity relationship to *infant, inexperience, issue of fertile parental union,* etc.

The most impressive property of simple arithmetic may be its application to everyday situations, like figuring out our change at the grocer's; it is not clear that language has this particular property. There is no need here to ponder the academic question whether the usefulness of arithmetic is the same as or different from that of language (a somewhat foolish problem anyway). But it should be clear that the *nature* of arithmetic is not its application. Attempts to characterize arithmetic will not be based upon its applications, but upon a description of its peculiar relationships (and, of course, the relationships between them). Just as arithmetic is not "solving equations" or "figuring out percentages," but a system of relationships (the byproducts or benefits of which are the solutions of practical problems), so language is not essentially defined as communication or exchange of information. These are its byproducts, benefits, or applications, but the nature of language is a particular system of relationships, and language knowledge is the capacity to relate in specified ways.

In summary, then, we found (1) that both arithmetic and language are structures of relationships and that there is nothing in either system that can be defined without again making use of some relationship in the definition; (2) that at least the primitive relations of arithmetic $(1, +, =)$ are also marked (either by a word or a syntactic device) in natural languages; (3) that neither of the two formal structures relies upon the use of symbols that stand for unique or specific events or objects; (4) that for both systems we encountered conceptual difficulties in giving mutually exclusive characterizations of elements, on the one hand, and operations, on the other; and (5) that an organism that "has words" does not necessarily have the understanding of the relational constructs that are customarily designated by these words.

Despite these similarities between arithmetic and language, there are, of course, some differences. Some of these are probably less important than others. Let me begin with one that may well be spurious. There seems to be an air of greater concreteness about language than about arithmetic. However, this may be due simply to man's peculiar mode of dealing with the world. For a baboon, language relations may be as abstract as arithmetic ones. And how are we to define *concrete* and *abstract* in objective terms? These notions are as relative as the notions *simple* and *complex*. The appropriate assignment of objects into classes such as *trees, pebbles, mushrooms,* and *none of these* evidently requires the "recognition" or computation of invariances that are as abstract as any notion in geometry or topology.

A more serious difference is that the number of relations in a natural language is clearly larger than in arithmetic, where it is possible to take an exhaustive inventory of all the relations admitted. The primitive relations of arithmetic are, at least potentially, available in language, and we may assume that they come from a universe of relations that is the universal set for both arithmetic and language. Although a natural language is limited by certain conventions, linguists believe (rightly so) that anything may be named in any natural language, and, on this assumption, any natural language is potentially larger in terms of the relations it contains than any mathematical system. It is, therefore, reasonable to believe that arithmetic is a special structure within the more general structure of language.

Although we can say that *one, and, is* are basic, primitive relations in arithmetic, no one has yet succeeded in postulating the minimal set of primitive relations necessary for the construction of a natural language, or even in making a reasonable list of "simple, elementary relations" (no matter whether "necessary" or not for a language). This gives natural languages a certain undefinability and amorphousness.

Closely related to this point is the lack of regimentation in natural languages. Whereas all relationships in arithmetic are clearly defined, particularly those between symbols and concepts, language indulges in frank sloppiness in this respect. Nothing is well defined, not even the relations between symbols and concepts. Even the most elementary relation words in natural languages usually have more than one use, and thus seem to represent a variety of relations. For example, the word "is" may indicate (1) predication-attribution, (2) similarity, (3) equality, (4) identity, and (5) equinumerosity. (One may quibble with this particular list and substitute some other one for it; the point merely is the multiplicity of denoted relationships.) Similarly, the word "and" does not have the same denotation in the sentences (1) *Peter and Paul are hungry;* (2) *One and two are three.* In the case of most nouns, verbs, and adjectives, the situation is even more pronounced. We might, for instance, give a list of relationships that constitute the denotation of the word *shoe* in a particular sentence and context. If we now were to compare this list with similar lists based on different occurrences and usages of the same word *shoe,* we would find that all the lists were somewhat different one from the other, although there would always be an overlap. If we did the same thing in a historical perspective, we would frequently find that the lists have changed with time, so that often there is no overlap between old and new denotations. But even on a synchronous level, we may find that a word is spontaneously being used metaphorically, in which case there may be only a small number of relationships in common with the majority of lists. From this we obtain the following picture: there is a set of operations or relationships (determined by the capacities of the brain) that we can represent as a formal structure called a *space of relations*. Words can be mapped into this space, but points in the relation-space do not stand in a one-to-one or even a fixed relationship to words. Instead, words hover over the space and are capable of being mapped into a region of the space, which I shall call the semantic field of the word. However, the field is ill defined and has a way of sliding around in the space. For example, by changing word usage, the field shifts now in this, now in that direction, and, with the lapse of time (centuries), it may traverse considerable distances in the relation-space.

Because words map only onto ill-defined fields, the interpretation of words between speakers can never be exact. We know with only varying degrees of certainty what our interlocutor means by the words he is using. The efficiency of communication is not vitally affected by this sloppiness as long as the variance in denotation between the two speakers is not too great. In other words, we can *tolerate* a certain degree of shifting and fuzziness of the fields. This notion could be formalized, and the *tolerance upon the fields* could, at least in theory, be defined with some rigor (Zeeman & Buneman 1968).

As in the case of arithmetic, the most general thing we can say about language is that it is a system of relations and that knowing language means relating something to something else in certain ways; oddly enough, what is being related is still relations, a statement that might sound less odd if we could get ourselves to admit that knowledge itself must be understood as a process of relating.[2] Again, we are at a loss to say what constitutes simplicity or single elements in language knowledge. Undoubtedly, whatever it is, it must be certain types of relations, perhaps some of those that are found most widely among the languages of the world—predicate relationships or those embodied in prepositions and conjunctions. But how are we to know?

However, one thing is clear. Elements of language are not simply associations, such as conditioned reflexes. If language means relating in general, then the capac-

---

[2]This idea is suggested by the following train of thought: an organism could know nothing about the world if the world that surrounds it were ideally homogeneous—an ideal Ganzfeld for all senses, never changing over time. In this situation, there would be no contrasts of any kind, not even changes in the organism's own state that it could perceive (for otherwise there would not be total homogeneity for the sensory capacities of the organism). Any deviation from this non-plus-ultra of homogeneity, in either space or time, would at once produce differences that could be assessed by the organism as relations, and the relations would at once introduce notions such as *different from* (and, in man, also *here* and *there*, or *before* and *after*). This way of thinking goes to show that it is *plausible* to regard relations as the irreducible elements of knowledge in animals and man. [However, the thought is not a logical necessity; for instance, an organism could be perfectly tuned to certain types of energy and be responding to these and only these. Thus, if the world surrounding the organism were homogeneous but of property x (homogeneously red, say), the organism would "know" something despite the homogeneity, namely, red-ness.]

ity for language means the capacity for certain relational activities. The activities cannot fail to be of a neurophysiological nature.

APES: HEURISTICS OF CROSS-SPECIES COMPARISON OF COGNITIVE FUNCTION

How are we to know whether an organism, man or animal, has language knowledge? The most commonly used criteria are these: (1) the number of different words the subject produces, (2) the capacity to extend the meaning of the word to more than a single situation or object, (3) the capacity to emit an utterance in an "appropriate" situation, and (4) the propensity for combining words. Unfortunately, none of these criteria is rigorously enough defined to be of much weight in our judgment of language knowledge in a subject. Let us examine each one in detail. (Different criteria are, of course, possible. Hockett [1960a] has used thirteen [for discussion see Lenneberg 1967].)

1. Let us first discuss the size of vocabulary. Recall what we've said about arithmetic. Clearly, we would not judge the degree of the subject's arithmetic knowledge by the size of his number vocabulary. The characteristic relationships inherent in arithmetic may be demonstrated in operations on groups of no more than four elements (or fewer, for that matter). The extension of the number vocabulary might be necessary in order to demonstrate that principles are being applied (relations computed) in producing answers to problems, instead of simple rote memory (if the system allowed only sixteen addition facts, it is conceivable that a subject would simply learn them by heart). But apart from this methodological consideration, nothing would be gained by increasing the group to more than four elements. If we wish to demonstrate a capacity for language, especially in an organism other than man, there is no particular need to insist on a large vocabulary. What matters is a demonstration that the organism has the capacity for performing the basic operations inherent in language, that is, of using words for classes defined by a set of relationships and for relating these classes to one another. Such a capacity can be demonstrated with fewer than 20 words; the size of the vocabulary does not tell us much.

2. The extension of meaning is, of course, an important aspect of the human use of words, but there are difficulties in evaluating this propensity, particularly when we are dealing with animals. Let me illustrate this. When my dog wishes to be let out, he gives a special little bark, while standing by the door. When I took him on a car trip recently, he suddenly gave the little bark. I stopped the car and opened the door, but the dog did not budge. I started to drive again and soon, the little bark. Again, no inclination to get out. At the next little bark I leaned over and opened the window a bit to let fresh air in; the dog immediately stuck out his nose and this was the end of the little barks. How shall I interpret this behavior? Shall I

say "little bark means 'open,'" and the animal has formed an intensive class based on the principle *open?* Or shall I say my dog barks in connection with discomfort? The first formulation would stamp the behavior as "language-like," the second not. There is no way of deciding between these interpretations, because it is impossible to investigate the dog's mind or to know what he intended. Therefore, observations of this kind are essentially useless for judging language capacity. Because of the notion of tolerance with respect to the mapping of words onto relations, it is impossible to say when a subject has made language-like but *novel* (metaphorical) use of a word and when he has made the *wrong* use of a word due to lack of comprehension of the underlying semantic field. As mentioned before, language is different in this respect from arithmetic, where there is no tolerance for the particular mapping of a number word onto the set of relations implied by the general concept of numbers. In arithmetic we can tell at once whether or not a word has been used correctly.

3. Now we come to that will-o'-the-wisp, *appropriate situation.* One of the peculiarities of natural languages (and of the use of natural languages) is that, in one sense, there is no such thing as an appropriate physical situation for an utterance, and, in another sense, any situation is appropriate. If a subject will say *x* only upon presentation of *x,* it is true that the situation is appropriate for *x,* but the use of the word may be quite odd. Characteristically, the sight of an object does not automatically elicit a verbal response in man. On the other hand, man uses language to communicate thoughts, dreams, desires, etc., and this occurs precisely in the absence of the object. But then again, there is nothing wrong with occasionally saying *x* upon seeing it. We see, therefore, that appropriateness of situation may be construed as either saying something in the absence or saying something in the presence of a given situation, and the final criterion for propriety is what we assume to be the intention of the speaker. Sometimes we can make educated guesses about the subject's mental state, but in many other situations this is not possible. Consider a subject who utters such words as "listen" or "one." There will never be a situation that is inappropriate to these words; though *listen* might customarily be uttered in the presence of loud noises, it could either be extended to soft noises or else be interpreted as meaning "lo! no noise." And, whenever the subject says "one," there is one of something around. Suppose the subject has the word "food;" if the word is produced only at precisely the moment that it is "called for," we have no assurance that the word is understood in any sense other than "I am hungry," which is very unlanguage-like. If it is used both at the appropriate moment and at other times, we may assume either that the general meaning of the word is not understood, or that it is understood and that the subject is telling us something about food at odd times. Human communications can take place in entirely appropriate situations but in the complete absence of language knowledge. An American sailor on liberty in Shanghai can pantomime his desire for food or female compan-

ionship by gestures that are appropriate to the situation, but the communication has nothing to do with Chinese, English, or any other natural language. Conversely, it is possible to tell a story that is understood by virtue of shared language knowledge but in the complete absence of an appropriate situation. I think these examples show that appropriateness of situation is not a very satisfactory criterion for estimating the degree of language knowledge in a subject.[3]

4. Finally, let us examine the criterion of spontaneous combination of words. There is a temptation to regard any concatenation of elements as an illustration of the beginning of syntax. But the simple stringing up of meaningful units is not enough. *Tail-wagging* is a meaningful unit—at least we interpret it as meaning ''my state is a happy one.'' *Looking-at* is a meaningful unit—at least we interpret it as meaning ''the target of my gaze produces an internal response in me.'' If my dog looks at me and wags his tail, two units have been combined that might be freely translated as ''Eric, I love you!'' Concatenation, however, may be the result of random generation. Suppose a mathematician wishes to teach an animal a simple arithmetic language consisting of four symbols only. He finds that if he says ''three,'' the animal responds ''one, two.'' He interprets this as a primitive but essentially appropriate and correct *combination of words* by imputing to the animal the intention of wanting to say ''one and two (is three).'' Soon he discovers that when he says ''two'' the animal again says ''one, two,'' which delights him; now he has proved that the animal has learned multiplication. When the animal hears ''one,'' it again says ''one, two,'' which, of course, can mean nothing else but that it has caught on to subtraction. And when it says ''one, two'' upon presentation of ''zero'' his trainer finally has evidence that the animal has spontaneously learned to do addition with the residue class modulo three! The mathematician may, of course, be correct; but then again, the animal may be responding randomly. The combining of words is of interest only if we can demonstrate the particular type of relationship upon which the combination is based.

We have seen that, of these four popular criteria, none is of much help in providing definitive evidence of language knowledge. The difficulty is due largely to the fact that saying something is not evidence of language knowledge. In order to test that knowledge, it is necessary to examine the subject's understanding, either by giving commands that call for the combination of acts in specified ways or by asking questions that call for a yes or no answer. Probably it is easier to begin with commands. We must take care, however, that the commands are not simply the names of a small repertoire of tricks, as argued before (sit, speak, wave bye-bye, etc., will not do, because each elicits a unique response and need not involve any

---

[3]The only sense in which ''appropriate situation'' may be used as an objective criterion is *syntactic;* in the ''sentence,'' *Johnnie of wants a banana,* the word *of* does not occur in an appropriate situation. Appropriateness in this sense should be discussed under the heading of well-formedness in syntax. Whether intentionality may be excluded from such a discussion is still an open question.

relationships; each is just the sign-stimulus for a response). Instead, we must use commands in which we test the comprehension of a small set of relationships. The paradigm is essentially that suggested first for arithmetic: point to the number that equals 2 + 3. In other words, we determine the relationship between the numbers, instead of imputing the relationship to a pair of numbers uttered by the animal.

Let me give a concrete example of some feasible experiments. We shall see, however, that even positive results of this sort fail to prove definitively that chimpanzees can acquire a form of human language. At present, a chimpanzee, Washoe, is said to be learning to use "a form of human language" (the American sign language of the deaf; Gardner & Gardner 1969). So far, it is mostly the animal's achievement in word production that has been described. However, it should be easy to do some controlled experiments in which her understanding is also tested, and thus her comprehension demonstrated. We would continue to use sign language as a vehicle and would confine ourselves essentially to the words that she is reported to have learned. In view of what we have said in the previous section, our experiments would have to concentrate on Washoe's capacity for understanding relationships, but her present vocabulary is rich enough for this.

Suggested procedure. We would use a double-blind test (following, for instance, Pfungst 1911); the interpreter or observer of the animal responses must not know what the preceding commands have been.

Materials. The experiments would be conducted in a room full of objects, including some for which the animal has not yet learned any words. Further, there should be multiple instances of objects for which the animal does have names (several shoes, blankets, etc.).

Language. We would choose nine object words from her present vocabulary—"key," "shoe," "blanket," "hat," "bib," "flower," "toothbrush," "pants," "you"—and three relational words—"out" and "in," plus the additional word "touch" (suggested because Washoe already uses the American sign language symbol for "touch" in the sense of tickle).

The first thing the animal would have to learn is that every command calls for a specific type of action. Examples: To the signal "bib in hat," Washoe would have to respond by putting one of the bibs inside one of the hats. The correct response to "key touch shoe" would be to place a key so that it was in contact with the shoe. The word "out" would, of course, be used as the inverse of "in." If we included commands such as "shoe touch shoe" or "hat in hat," quite a variety of commands could be given. Half of the possible commands would be used for training Washoe; the other half would test her comprehension. (Since all of the experimenter's utterances would be understood as commands, no special sign for "do" would be needed.)

This language is quite simple, although it is impossible to say whether it is primitive in any *natural* sense. If these relationships could not be understood by the animal, perhaps some others could, and we might use our measure of success as a criterion for simplicity. However, unless some comprehension of relationships that exist in language were demonstrated, we could not even speak of *primitive* beginnings of language capacity.

If the animal were successful, we could complicate our first language model in two ways: we could complicate the syntax and we could add to the repertoire of relations. Let us begin with the first. For instance, we could string up relationships, such as in the command "key touch shoe in blanket." Suppose we also complicated the language in the other direction, and added the relationship "and." This would allow us to test for an important feature in syntax and one that is of particular interest because of its perfect parallel in arithmetic. Consider the command, "shoe and key in blanket." This is understood, at least by English speakers, as "shoe in blanket and key in blanket," but not as "shoe and key in shoe and blanket," which would be a possible interpretation, since there would be several instances of shoes and blankets present in the testing room. We can represent this in formal notation as follows:

Let *and* be represented as +, and let *in* be represented by *, then:

$$A + B * C$$

is understood as:

$$(A + B) * C = A * C + B * C$$

but not as:

$$A + (B * C) \; or \; A + B * A + C$$

In other words, it strikes us (as speakers of English) as logically obvious that the *in* relationship applies to both shoe and key, but that the *and* relationship does not apply to both key and blanket.

All languages have some notion of negation and therefore also of affirmation. Suppose we introduced this concept now, so that negation could be used either to cancel the command as a whole or to cancel one of the relationships in a composition containing two relationships, such as *shoe and key in blanket*. The reader may readily convince himself that this new concept makes it necessary to introduce a rather complex set of syntactic rules before it can be used in sentences unambiguously. Without syntactic rules to distinguish between commands such as "shoe and not key in blanket," "not key and shoe in blanket," "key and shoe not in blanket," we would be unable to give unequivocal interpretations to these compositions. Children can answer questions by "yes" or "no," so that the basic concept of affirmation and negation in itself does not appear to be insuperably difficult; nevertheless, they ordinarily have a difficult time in learning the very complex syntax of negative constructions in English (Bellugi 1967). In other words, the syntax of even the most primitive languages quickly becomes complicated by introducing more than just a few common relationships.

If we could introduce hand signs for affirmation and negation, we would have prepared the situation for asking questions that demand a yes or no answer instead of an action. Every utterance could now be prefaced by a sign that tells whether it is a command ("[Put] key in shoe") or whether it is a question requiring a yes or no answer ("[Is] key in shoe?"). This kind of exchange—question and affirmative or negative answer—is entirely universal among all languages of the world, and even small children can quite readily learn to answer simple questions. Could Washoe learn it?

These few examples of syntax should make it obvious that a natural language is more than an accumulation of items. The relations in use interact—relate to each other in peculiar ways—and thus, the complexity of the relational structure as a whole increases exponentially with the addition of every new relation.

In addition to Washoe, another chimpanzee, Sarah, is being trained in language-like tasks by Premack (1969). This animal is reported to be composing sentences and answering questions of an absolutely amazing relational complexity. The vehicle of communication in this case is plastic chips of different shape and color, each standing for a word in English. The chips are taken out of a box and lined up on a board to form sentences. There is some more direct evidence, in the case of this animal, of comprehension of relations; unfortunately, there are no films or slides available of the ape's spectacular success, and no double-blind testing with several independent observers has yet been reported.

While I admire the patience, the skill, the ingenuity, and often the courage behind these training endeavors, I would still have serious reservations with respect to their significance, even if all the experiments reported and proposed could be confirmed and extended. Neither the Gardners nor Premack claim (or expect to show) that there is *no difference* of any kind between the language acquisition capacity of chimpanzees and that of a child without a tongue or one that is congenitally deaf. Thus, the species-specificity of the human form of language interaction and capacity (regardless of its vehicle, i.e., whether audiolingual or visuotactual) remains uncontested. The work on primates can show only that *some* common denominators may exist, a circumstance that is impossible to deny even in the absence of the present training demonstrations. Let us suppose for a moment that the experiments proposed in this article had already been successfully concluded. I would still be hardpressed to say what they go to show. Should we see in these demonstrations a primitive form of man's capacity? the regular onset of human language development but arrested at an early stage? a similar but different form of language capacity? a homologue of our behavior? None of these formulations appears to be entirely justified (Lenneberg 1969).

BRAINS: PRELIMINARIES TO THEORIZING

Let us return to two earlier points. (1) No one can deny that man has, at the very least, an obviously different propensity for language than do other animal forms. (2) Relating something to something else is necessarily associated with a process, and, if language is a set of relational operations, there must exist correlated neurophysiological processes. These two points lead us to ask a number of questions about the brain.

The conjecture that man's language capacity is based on biological species-specific peculiarities does not commit us to any particular hypothesis about anatomical or physiological brain correlates for language, and, since so little is known about brains in general, I have, in the past, refrained from making any specific claims. This caused my critics to accuse me of an "extreme holistic" viewpoint (Whitaker 1969) or of a failure to read the "obvious biological sign posts" (Guttman 1968), which to them apparently spell out a clear story. I think it is much too early to state with confidence what it is in the brain that gives man his peculiar endowment for language. All we can do at present is to infer from the nature of language what general sorts of things we might look for in the brain—what kinds of information about the brain may or may not be relevant to an explanation of language capacity. I shall confine myself here to a critique of some of the more popular assumptions concerning brain correlates of language.

Suppose we could show by careful anatomic dissection of the human brain that the architecture of its connecting fibers is decidedly different from that of other primates and lower mammals. This would perhaps suggest that man's language capacity is due to a unique neuroanatomical specialization; but to lend credence to this, we would also have to make a strong case for the relevance of these peculiar fiber connections to that which we have characterized as the essential nature of language. If we were successful in the latter, we would have fairly well accounted for the mystery of why man has language but chimpanzees do not, and experiments such as those conducted on Washoe and Sarah would be superfluous. However, I think our insights from the first section can be shown to cast serious doubts on the relevance of fiber architecture to language capacities.

In general, we shall see that by combining the question "What brain correlates would we expect for arithmetic knowledge?" with the question "What brain correlates would we expect for language knowledge?" and by trying to answer them jointly, we can at least get a feeling for what is reasonable to expect and what is unreasonable. I shall use this approach to defend the following claims: (1) the architecture of intersensory fiber connections does not explain the capacity for either arithmetic or language operations; and (2) there are no cogent reasons to believe that language or arithmetic knowledge can be further subdivided in such a way that their components or attributes might be correlated with separate, autonomous neural mechanisms, each with a distinct cerebral locus.

Fiber Connections

The concept of a number is independent of the physical shape of its symbol. One hundred may be represented as *100,* or by the spoken word, "satem," or by the instruction, "touch all fingers on both hands as often as there are fingers on both hands." There is no fixed association between any visual, auditory, or tactual pattern and the concept. Consequently, we would not expect the basic capacity for arithmetic to be dependent upon any particular sensory input, and we are not surprised to hear that congenital deafness or blindness (or both) leaves the capacity for doing arithmetic unimpaired. If touch alone (in the deaf and blind) is sufficient to develop arithmetic capacity, it could hardly be argued that arithmetic is possible in man because of some felicitous fiber connection between auditory and visual projection areas.

That the situation for language capacity is quite similar to that for arithmetic in this respect may be seen from the clinical parallels. Neither deafness nor blindness nor the congenital presence of both bars the development of language through the vehicle of touch. There is no question but that an individual like Helen Keller could pass a language test of considerably greater complexity than that outlined in the second section. The argument of the previous paragraph is directly applicable to language capacities. Relations such as *in, out, to, and, not, is* are not dependent on any sensory quality, and the same must be true of those relations that serve to define classes, such as furniture, freedom, grandmother; otherwise, it would be impossible to understand how a person who is both deaf and blind from birth can learn the meaning of these words. (There are a few exceptions: the meaning of some words does require specific patent sensory pathways; examples are words for color, for noise, for touch qualities, etc. However, it is curious to see how relatively unimportant is the incomprehension of such words in the language of the deaf or blind.)

Once a language has been established through whatever sensory avenue, it may continue to function even under conditions of dramatic sensory deprivation. Nor is it necessary for language operations that specific outside stimuli impinge on the speaker in order that a specific utterance may occur. Thus, it becomes difficult to maintain that anatomical connections between cortical projection areas can *explain* man's language capacity, or that a chimpanzee's difficulties in this respect are *due to an absence* of connections from, to, or between the primary visual and auditory projection areas of the cortex.

Furthermore, if we put too much faith in the importance of gross connections, we commit ourselves (by implication) to postulating rather fixed *centers of activities*. There is evidence, however, that *activity centers* are not fixed for life but allow for some degree of shifting during the formative years of the brain. For example, the lateralization of language to the left is dependent upon an intact corpus callosum and the absence of pathology in the left. Patients with congenital absence of the corpus callosum do not seem to lateralize, and an epileptogenic focus on the left may prevent left lateralization altogether. Also, early cortical lesions, both in man and in other mammals, alter the course of normal functional differentiation of cerebral tissues and may, as a consequence, cause a somewhat different distribution of functional loci over the cortex. This relative plasticity in localization of function imposes severe strains on the logic of a strict connectivist explanation of our language capacity. I must stress here that I am not ruling out the possibility that *connectivity* between cells is of a particular kind in man, constituting one of the prerequisites for our characteristic intellectual capacities, including language. But by *connectivity* I mean the fine structure of dendrites and axons. The relations expressed in language are undoubtedly dependent upon nervous activities, and these may well depend on the specific microscopic architecture of neuronal units. This is different from searching for gross subcortical or tangential fiber tracts.

Indivisibility of Language Knowledge

It must be clearly understood that my comments here do not refer to the general skills with which a person communicates verbally. I am concerned only with an alleged possibility of explaining the language capacities of man by gross anatomy. I do not deny that one can correlate clinical speech or language disorders with types and locations of lesions. However, it is not at all clear yet what the cause of this correlation is.

The primary prerequisite for human verbal communication is the capacity for language knowledge; in addition, several other skills are integrated with this capacity. It is possible to knock out some of these other skills by pathology without affecting language knowledge, and the capacity for language knowledge itself may also become disordered independently. In the clinic we see a great number of highly characteristic syndromes, each with its characteristic lesion at specific sites. Thus, from a clinical point of view, I would call myself a "localizer." Since it is certain that there are some brain lesions that leave language entirely intact, we may assert that there is a specific language territory, whatever its topography may be. The following discussion deals only with this territory and, more specifically, with the question whether it is possible to subdivide that territory further in such a way that individual language (knowledge) traits might be assigned to anatomical loci within this territory. The assertion that the latter is possible I call the *strict localization theory of language,* and my aim is to disprove it.

Imagine some mechanism, say a computer; we ask the electrician who assembled the machine to give us a complete list of all its components, and the user of the machine to give us a list of its behavioral traits. Our two consultants work independently of each other; neither knows what the other does. Now, it should be obvious that the items of the two lists will not necessarily stand in a one-to-one relation.

Although there will be some items on the engineer's list that may be related functionally to individual behavioral traits (such that specific dysfunction here will produce specific dysfunction there), there are, logically, four other possibilities, or altogether five ways in which the lists may be related: (1) correspondence is one to one, (2) correspondence is one to many, (3) correspondence is many to one, (4) correspondence is many to many, (5a) correspondence is one to none—in the case of a component, (5b) ditto—in the case of a trait. The first four are too obvious to need elaboration. An example of 5a would be a superfluous or redundant component; for instance, elimination or dysfunction in this case would have no behavioral consequences. In 5b, the list of behavioral traits might include an item actually not attributable to the mechanism of the machine at all, but due to some environmental conditions to which the machine corresponds indirectly; or the compiler of the behavioral traits might have failed to distinguish between behavioral principles (e.g., capacity for multiplication in general) and specific computations (frequent multiplication of $a \times b$).

I assume that a *strict localization theory* of language is one that expects perfect one-to-one correspondences between anatomically identifiable components of the cerebral language territory and specific traits of language. Such a theory makes two untenable assumptions.

The first is that one can compile a list of all the traits that are necessary and sufficient for the characterization of language. I doubt that this is either possible or relevant to the anatomy of the brain. Translate the argument to arithmetic. If we were to consider *any form of arithmetic* in general, our list would be simple: it would contain the single trait *relating*. This is unsatisfactory, because by this definition every organism has a primitive form of arithmetic. If we are to consider a specific arithmetic, such as that taught in grammar school, we would list particular relations (or axioms), and I do not believe anyone could hope to find cerebral loci that bear a unique functional relationship to these components of arithmetic knowledge. In the case of language knowledge, the list of traits would be very long, and the particular entries would probably (at least for the time being) be rather arbitrary because of our difficulty in stating what the universal, common denominators of all possible relationships in languages are.

The second assumption is that traits can be analyzed as if they consisted of autonomous and independent entities. On the behavioral side, it is not clear whether such notions as processing, iteration, phoneme analyzing, and-relations, transformation rules, etc., are autonomous, separable traits. Their individuality may be challenged even on the grounds of logic, let alone on the grounds of psychological and physiological reality. If we cannot make sharp distinctions between these theoretical notions (which I believe can be shown to be true), how much less can we expect to find specific sites for them in the brain!

In summary, I do not deny (in fact, I firmly believe) that specific parts of the brain contribute in their own characteristic ways to physiological functions as a whole. But this does not mean that every aspect of physiological function has an obvious, unique behavioral correlate. Further, there is no question that verbal communication may be interfered with from specific sites and in characteristic ways. It is precisely findings of this sort that constitute strong evidence that our capacity for verbal behavior has a biological foundation. The fact that certain lesions produce characteristic clinical pictures of speech abnormality does not imply, however, an aggregate of independent traits, any one of which could be neatly abolished by a lesion. The advance of a wave of tumbling domino pieces may be arrested by the elimination of any two contiguous pieces; if a couple of particular pieces are particularly prone to elimination (because they happen to be close to a teenager's elbow), we must not attribute to just these pieces the exclusive function of wave advancement. Language operations may be thought of as a family of interrelated activity patterns, which may be deformed by local interference.

DISCUSSION

Man (and probably many other species) appears to react primarily to *relative* differences in energy levels or, in more general terms, to be attuned to relations rather than to absolutes. There is evidence that species differ from one another in the types of relational configurations to which they are most sensitive. It is possible to discover empirically the hierarchy of pattern salience (or stimulus effectiveness) for any given species. If this is done for man, we would go from relations that seem to us "utterly obvious" to circumstances that seem "totally incomprehensible." In the case of man (and probably certain other animals, too), the essential relations that are reacted to are often quite independent of particular sensory qualities. Hence, many patterns and configurations are recognized as similar, even though, for instance, one may be perceived tactually on the skin of the leg and the other auditorily (e.g., the pattern of one long and two short stimuli). Similarly, the relational patterns of language can be extracted from qualitatively different vehicles (speech sounds, Morse code, graphic representation, tactual encoding, etc.). The same is true of arithmetic. Since it is the invariance in relational patterns that counts, while the sensory modality by which the stimulus is received is quite immaterial, one gets the impression that the extraction of particular relations from the environment is an *activity,* not a passive *receptivity*. The same idea is expressed by Piaget's notion of *construction* or by the more recent terms *computations* or *operations*. In each case, the words express the opposite of *simple storing, taking in, replicating*. Man (and, perhaps, chimpanzees) has a propensity for applying, to relations themselves, the relational computations used for processing sensory information; that is, man operates on the products of his own intellectual activities. Thus, man does not only see or hear *two stimuli,* but also can count two *twos;* he does not only name classes of objects, but also can name names. Thus, conceptual systems are built up, of which language and arithmetic are but two examples. All these systems have six important properties: (1) they are self-generating, in the sense that there can be numbers of numbers, names of names, relations of relations, pictures of pictures, thoughts of thoughts; (2) they consist of open sets of intensive classes, because the organism deals only with general characteristics; (3) the constructed systems consist of nothing but relations; (4) the relations constituting the system are readily represented by artifactual symbols; (5) the exact physical configuration of the symbols is of small concern to the symbol maker; (6) there are no other limitations to the quantity or quality of relations that make up a symbolic system than the biologic constraints (the physiological limits of the operating characteristics) of the mechanism that computes the relations, i.e., of the brain.

The argument of this article is best illustrated by means of a metaphor. The patterns of a kaleidoscope are infinitely variable; yet the range of this variability is delimited by the physical properties of the scope. No configuration can escape the limits. Thus, each pattern that is formed has its unique structural characteristics, yet every pattern is related to every other. It is possible to take a single pattern and describe it, for instance, by describing the position and nature of each piece. However, the gestalt that is formed is not a simple aggregate of these descriptive items; removing certain pieces would change the structural characteristics of the gestalt in more complex ways than simple subtraction. In this metaphor, the kaleidoscope corresponds to the brain; each pattern to one conceptual system; the constituents of the patterns to relations. I have pointed out (p. 119) that one can associate with every relation an activity or process (the relations that a computer can compute are dependent upon given activities in which the machine can engage. Thus, the configurations with which we are concerned in this article are configurations of activities or, in other words, the conceptual systems are *activity patterns*. We have shown how relations (as expressed, e.g., in rules) interact with one another, forming a system that is more than the simple sum of its components. In the case of the kaleidoscope, the patterns are end-products; they don't "do anything." In the case of the brain, however, the activity patterns that correspond to conceptual systems have an output: they produce symbols and classes of symbols, such as languages or algebras.

Let us pursue this metaphor a bit further. Suppose we want to find out how kaleidoscopes in general work, but have access to nothing but the patterns they make. Let our first task be to discover which patterns are made by the same kaleidoscope; next we hope to be able to group kaleidoscopes according to the similarity of their constitution. The first task has been achieved if we can at least specify a *class* of patterns by means of a *set* of transformations that can map any one pattern in the class into any other pattern in the same class. In this paper we have looked at two patterns, arithmetic and language, and the thesis has been proposed that one can map one into the other because they have both been generated by the same instrument, the human brain. Other, similar transformations might carry spoken English into the sign language of the deaf, or the logic underlying physics into the logic underlying chemistry. We still know nothing about the nature of the kaleidoscope.

Our next task is much more difficult. How can we arrange or order kaleidoscopes if we know nothing about their internal structure and function? In the case of brains of different animals, we know a bit about their anatomy—enough to know that each species has a species-specific morphology—almost nothing about how they make the patterns we are discussing, and very little about their most intimate, i.e., molecular, structure. We must, therefore, rely largely on the generated patterns for source material. We are aided by the obvious fact that brains have an ontogenetic history. The conceptual systems produced by an immature brain may have no formal equivalence to the systems of mature brains, and, therefore, in terms of our metaphor, the immature and the mature brain each produces a distinct class of patterns with no overlap between them. Consequently, the set of transformations that defines one class is also distinct from the set of transformations that defines the other. Consider now that the set of transformations is actually a formalized description of the kaleidoscope. (It is not a blueprint of how the kaleidoscope works, but the "name" of the kaleidoscope in a formal metalanguage.) Given the fact that immature brains develop into mature brains, there must also exist historical continuity from one set of transformations to the other set. This is to say that one set is in fact *transformed* into the other, giving us a formal criterion for arranging at least certain kaleidoscopes, or brains, into a single order; we might call this order a maturity sequence. (Immaturity is often equated with primitivity or simplicity; I find these words somewhat misleading, because they seem to imply more than is justified.)

Now we must try to arrange all the maturity sequences into a still further, higher order. The motivation for this is, again, biology. We have good reasons to believe that animals have a common origin and that their structural similarity reflects their proximity of descent. Because of this, we are convinced that there must be a yet more general, more abstract level on which one may construct transformations that relate the maturity sequences to one another. Now it is obvious that transformational pathways on this level should be much more involved than those on the previous level, because immature brains grow directly into mature brains, and, therefore, there is also historical continuity between their respective classes of patterns of conceptual systems; but the brains of chimpanzees do not grow directly into those of humans, and, thus, any kind of construct that relates different types of brains to one another is bound to be both more speculative and more arbitrary.

When chimpanzees are taught a form of human language, the implicit hope is to demonstrate phylogenetic continuity between their language capacity and ours. However, to show that some similarities exist tells us very little. If my thesis is correct that language is but one symbolic system out of a whole (intensive) class of such possible systems, all delimited by the instrument that generates the systems (the human brain), then the data from the chimpanzee "language experiments" is simply not rich enough to draw far-reaching conclusions about the proximity of chimpanzee language to human language. Nor are the results sufficient to construct theoretical pathways that relate chimpanzee capacities to human capacities. At the very least, chimpanzee language capacities must be related first to other conceptual achievements by these animals, so that a coherent picture of all the animal's intellectual activities emerges. Only after this has been achieved does it make sense to make comparisons between the chimpanzee's achievements and human achievements, keeping in mind the necessarily speculative nature of this endeavor.

APPENDIX

A short demonstration that *one, and, is* are relations may be in place. In referring to a bunch of mice in front of me, I may say either: there are *six* mice, or there are *three* sets of twins, or there are *two* sets of triplets, or there is *one* litter. It is clear that the concept *one* only makes sense relative to some other concept. Thus, it is intuitively acceptable to say that *one* implies a relation. Take some of the more "obvious" relations, such as *father of* or *bigger than*. These, as any other relations, necessarily imply (1) two sets of classes, S and M, and (2) a mapping, $\rightarrow$, connecting one or more elements in S to one or more elements in M. The act of relating defines some order between elements of the respective sets (in the instance to be considered, one element in each set; therefore, we speak of ordered pairs). For example, Jerry is father of Ben; Roger is bigger than Miriam. Thus, the binary relations considered here "produce" pairs or, in other words, form a new set or class of ordered pairs. In set-theoretical language this set of ordered pairs is a subset of the cartesian product of the two sets ($S \times M$) presupposed by the relation. We see that a relation implies (1) a mapping, (2) the existence of two sets, and (3) the formation of a new set or class.

Notice that the two sets implied by every relation may actually come about by splitting up a single set. If we define the relation R as *is one of*, then R always implies such a splitting-up operation. Let I be the set of integers, then 5 RI, 0RI, etc. The same holds true if I assert of anything that it is *one*.

Let S be defined as $\{1, 2\}$ and M as $\{0, 1, 2, 3\}$. The mapping:

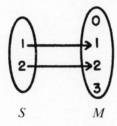

$$S \qquad\qquad M$$

may be taken as the relation *is equal to*. Again, let S be the set of ordered pairs:

and M be defined as $\{0, 1, 2, 3\}$, and consider the following three mappings:

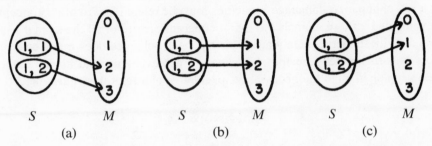

In what sense can these mappings be said to be true? They become true by speci-fying what the relation (or kind of order) is that holds between the elements of the pairs in S. Mapping (a) becomes true by making the relation be *and;* (b), by making it *times;* and (c), by making it *difference between*. This shows that operations (such as addition, multiplication, subtraction, etc.) also imply relations.

CAN CHIMPANZEES LEARN A PHONEMIC LANGUAGE?*

ALICE F. HEALY

The claim that language is a uniquely human accomplishment has been challenged recently by the work of the Gardners (1969) and Premack (1970a), who have obtained quite promising results from their attempts to teach a form of human language to chimpanzees. Earlier attempts (e.g., Hayes 1951) had been largely unsuccessful. One explanation offered frequently for the greater success obtained in the recent studies has been phrased in terms of the difference in the mode of communication employed. The earlier experimenters attempted to teach chimpanzees using the auditory and vocal modes employed in normal human communication. However, both the Gardners (1969) and Premack (1970a) made use of visual and manual modes of communication; a sign language was employed in the former case and a language of plastic tokens in the latter. Thus, it is generally argued that since visual and manual modes of communication are natural to the chimpanzee, a chimpanzee would find it easier to associate ideas or meanings with visual and manual sensations than with auditory and vocal sensations, in accordance with the phenomenon of selective learning (e.g., Dobrzecka, Szwejkowska, & Konorski 1966; Garcia, McGowan, Ervin, & Koelling 1968). In addition, there is evidence (Lieberman 1968; Lieberman, Klatt, & Wilson 1969) that the chimpanzee's vocal apparatus is more limited than the human's so that a chimpanzee is physically unable to produce the full range of human speech.

Undoubtedly, modality of communication was a factor contributing to the chimpanzees' failure to acquire spoken language as well as to their relative success in acquiring a sign or token language. However, communication modality was not the only feature that was switched as the teaching attempts changed from failures to successes. One of the additional features that might be as critical as communication modality in determining whether chimpanzees can learn a given language is the phonemic aspect of language. The earlier studies with negative results employed

*This research was supported in part by PHS Grants No. GM01789 and GM16735 to the Rockefeller University.

ALICE F. HEALY • Department of Psychology, Yale University, New Haven, Connecticut 06520.

a *phonemic language,* by which is meant a language actualized by strings of phonemes; while in the later two more successful studies words rather than phonemes were basic units. Yet the phonemic aspect of language has been shown to be both highly complex and abstract (Liberman, Cooper, Shankweiler, & Studdert-Kennedy 1967; Savin and Bever 1970). In fact, recent research on children's reading problems (Rozin, Poritsky, & Sotsky 1971; Savin 1971) has demonstrated that an inability to perceive phonemes is a likely cause of many children's difficulty with reading.

The question, therefore, arises of whether chimpanzees can in fact learn a phonemic language. Although Premack and Schwartz (1966) and Premack (1971b) describe phonemic systems suitable for the chimpanzee, they have not reported any successful attempts to teach chimpanzees such systems. Similarly, although phonemic sign languages do exist, no evidence is available at present that chimpanzees can learn such languages. However, Gardner and Gardner (1971) have argued that the American sign language (ASL), the language employed in their study with the chimpanzee, is in a sense phonemic. Their evidence is a description of ASL by Stokoe (1960) and Stokoe, Casterline, and Croneberg (1965) which analyzes the signs of ASL into elements called cheremes. Stokoe (1960) contends that the cheremes of ASL correspond to the phonemes of spoken language. However, the cheremes of ASL, which specifiy the place, action, or configuration of the hands used in making a sign, are more closely analogous to distinctive features (Jakobson, Fant, & Halle 1969) of the gestural signs than to phonemes. For example, there are distinctive features that specify the place of articulation of the phonemes just as there are cheremes that specify the place of gesticulation of the signs. Cheremes are not entities themselves, but rather, like distinctive features, are attributes or features of the basic linguistic units which are entities. Therefore, there is still no evidence that chimpanzees can learn a truly phonemic sign language. In a similar way, it would be possible to analyze the tokens employed by Premack (1970a) in terms of elementary attributes or features, such as their size, shape, and color. Such an analysis, however, would in no way provide evidence that Premack's chimpanzee had learned a *phonemic* token language. A phonemic language using tokens would be one, as Premack (1971b) has said, in which each token is made up of certain pieces or units that are shared with other tokens, and that type of token language has not yet been tested.

Consequently, it is still plausible to defend the position that the phonemic aspect of language is unique to humans, and that chimpanzees have not yet been taught and could not learn any phonemic language. This conclusion is consonant with Hockett's (1958:574–575) outline of the design features of human language. Hockett (1960b) declares that "duality of patterning" (patterning at *both* the phonemic and morphemic levels) was the last property to be developed in human communication and the only feature not shared with the communication system of any other hominid and perhaps of any other animal.

ACKNOWLEDGMENT

The author is indebted to Professor George A. Miller for his many contributions to this work.

SENTENCE STRUCTURE IN CHIMPANZEE COMMUNICATION*

DAVID McNEILL

INTRODUCTION

There are now two experiments on teaching chimpanzees a human language that have been in some sense successful (Gardner & Gardner 1971; Premack 1971b). It is an appropriate time to review these experiments in an attempt to determine what they have demonstrated.

What is the purpose of such experiments? The reasons given by the experimenters themselves are not the same, and my purpose, in this paper, is yet again different. It is a good idea, therefore, to describe these differences at the outset. Premack (1970a, 1971b, 1976a) states clearly that his purpose in attempting to teach a chimpanzee language is to validate what he calls a *functional analysis* of language. By this he means a transliteration of linguistic structures into a set of training procedures that one can follow to teach an animal (chimpanzee) these structures. Gardner and Gardner (1969, 1971) are less explicit as to their goals, but seem to be interested in demonstrating that behavior that one would intuitively call *linguistic* can be invoked in some species other than *Homo sapiens*. These are different goals. The difference can be seen from the fact that, in principle, Premack's question could be answered within our own species. For example, except for moral and legal restrictions, Premack could raise a child without any human contact except in controlled experimental settings where the prescribed training procedures are used. Obviously, the Gardners' goal would not be met by such an experiment.

There is a third question that can be raised, which is the question that I will pursue in this chapter. This concerns the possibility that language in our (or any other) species is to some degree a biological specialization. Typically, questions of biological specialization have depended on comparisons of species. However, species comparisons have not been possible with language. But now, for the first time,

*Preparation of this paper was supported by grant from the NSF to David McNeill.

DAVID McNEILL • Department of Behavioral Sciences, University of Chicago, Chicago, Illinois 60637.

the question can actually be investigated with the Gardners' and Premack's experiments. The results, I believe, are more surprising and interesting than many would have anticipated.

Starting from the undeniable fact that only our species has the highly intricate structure we call language, one can ask whether any abilities can be isolated on which this unique biological distribution of language depends; or if it is a result of a general difference in cognitive ability between man and other species. This question is different from Premack's in that one aspect of any biological specialization for human language might well include a high degree of sensitivity to linguistic forms in the developing young. Clearly, it may be possible to find training procedures that instill certain linguistic structures in a chimpanzee even if it lacks such sensitivity. This has been done with many kinds of behavior (Seligman 1970). The question of biological specialization differs from the Gardners' question also because it is conceivable that the chimpanzee and man have abilities in common at some points, but at other points each has certain abilities not possessed by the other. In this case, some aspects of human language could be transmitted easily to a chimpanzee while others would be transmitted only with difficulty and, possibly, would be reconstituted in a form that is more natural for the animal.

In fact, Premack's and the Gardners' experiments taken together, suggest that the chimpanzee resembles man in a number of fundamental respects, but that each species has a distinct set of abilities that are nevertheless linguistic in character. These differences, which may be the result of separate biological specializations, exist alongside a number of conceptual and other linguistic similarities between the chimpanzee and man.

WASHOE

Washoe was estimated to be fourteen months of age at the beginning of the Gardners's experiment. She was kept at home, housed in a trailer, and was in the company of human adults during all waking hours. Apart from simulations of natural chimpanzee vocalizations, the only medium of communication was American sign language (ASL), a system of manual gestures used by deaf people in North America (not to be confused with finger spelling). In ASL typically one sign corresponds to one word, although occasionally signs correspond to groups of words or entire phrases. The Gardners' first report (1969) contains a list of the signs (words) that Washoe acquired in the order in which they appeared. The first twelve of these are as follows: "come–gimme" (one sign), "more," "up," "sweet," "open," "tickle," "go," "out," "hurry," "hear–listen" (one sign), "toothbrush," "drink." Only two signs denote physical objects ("drink," "toothbrush"), the remaining ten referring to relations or actions. Object words were acquired later in

Washoe's development. At first glance this appears to be different from the speech of children at a comparable stage, since with children the earliest words are almost exclusively object words. In an earlier publication (McNeill 1970), I concluded that Washoe differed from children during the one word stage of development. However, it has been noted since that the relatively numerous object words that children acquire are highly transitory, appearing for a few days and then often disappearing completely, to be replaced by other object words (Bloom 1974). Apparently, there is a constant circulation of object words during the single word stage. On the other hand, relational words ("more," etc.) and action words, while much less numerous, are more stable. These new facts change the interpretation of Washoe's single word development. The Gardners adopted an extremely strict criterion for considering a sign to be part of Washoe's vocabulary. A sign had to be used at least once daily for fifteen successive days. If Washoe was like the subject described by Bloom, the Gardners' criterion would have had the effect of eliminating all object words in the early stages. If the same criterion were used with Bloom's data, the resulting vocabulary list would include, like Washoe's, no object-words and only relational and action words. Thus, while we still do not know whether Washoe had an early vocabulary of transitory object words, we do know that the stable part of children's early vocabulary is similar to the stable part of Washoe's vocabulary, both including only relational and action words. There is no evidence that the chimpanzee differs from children at this stage.

WASHOE'S COMBINATIONS

Washoe spontaneously began to produce sequences of signs at an early point in training; 294 combinations had been recorded at the time of the Gardners' (1971) report. Interpretation of this remarkable phenomenon is unfortunately made difficult, but I think not impossible, by the Gardners' practice of disregarding the order in which Washoe made sequences of signs. Whenever the same signs were produced in succession, regardless of word order, the Gardners recorded it as the same combination. Apparently, changes in word order were frequent. This practice discards much of the information essential for determining whether the animal had developed syntactic patterns. The Gardners (1971) claim that Washoe committed no more order "errors" than children do (according to information collected by Brown 1973a), but it is far from clear what this assertion means. The Gardners report no data for the comparison and we do not know what standard was used to determine an "error". In the case of child language, often what seems to be errors from the standpoint of adult grammar turn out to be completely systematic when one takes into account a child's own immature linguistic system (see examples in the various papers in Bellugi and Brown 1964). It does not seem that the Gardners

DAVID MCNEILL

Table I. Sign Sequences From Washoe

come–gimme (one sign) sweet	please up ⎫
come–gimme (one sign) open	more up ⎪
Greg tickle	baby down ⎪
Naomi hug	shoe up ⎬ b
Naomi come	baby up ⎪
Naomi quiet	please more up ⎪
Naomi good	you up ⎭
please out	gimme key ⎫
hurry out	more key ⎪
you me out	gimme key more ⎪
you Roger Washoe out	open key ⎪
you me go out	key open ⎪
you me go out hurry	open more ⎪
Roger you tickle (= Roger, tickle Washoe)	more open ⎪
you Greg peekaboo (= Greg, you play peekaboo)	key in ⎬ c
catch me	open key please ⎪
tickle me	open gimme key ⎪
go in	it open help ⎪
go out	help key in ⎪
in down bed	open key help hurry ⎭
drink red (= red cup)	gimme food
my baby	gimme food gimme
listen food	more tickle
listen drink (= supper bell)	please tickle more
dirty good (= potty)	come Roger tickle
key open food (= open the refrigerator)	out open please hurry
open key clean (= open the soap cupboard)	you me in
key open please blanket (= open the bedding cupboard)	you me out
please song	you me Greg go
song dirty	Roger Washoe out
please song good	you Naomi peekaboo
come hug sorry sorry	you tickle me Washoe
sweet drink (= soda)	you me drink go
please sweet drink ⎫	you me out look
gimme sweet drink ⎪	hug me good (an apology)
hurry sweet drink ⎬ a	Roger Washoe tickle
please hurry sweet drink ⎪	you peekaboo me
please gimme sweet drink ⎭	tickle me Washoe Roger
up Susan ⎫	you me in
Susan up ⎪	you me out
mine please up ⎪	you me go
gimme baby ⎬ b	out you me Dennis (rare)
please shoe ⎪	please me you go (rare)
more mine ⎪	you out me (late)
up please ⎭	

[a] The Gardners add that this request for soda was also made "by variations in the order of these signs". They do not say what the variations are, however. In particular, it would be of interest to know whether the recipient of the action (sweet drink) was always in final position, as in these examples.

[b] Produced in a single situation, where an assistant, Susan, has placed her foot on one of Washoe's dolls.

[c] Produced in a single situation where Washoe obtained a key from an assistant but was unable to open a door with it.

took this precaution into account in determining Washoe's "error" rate and comparing it to that of children.

Table I contains all examples of sign sequences by Washoe that the Gardners (1971) report. Assuming these 91 sequences are representative of Washoe's total output, there is sufficient information here to support certain conclusions about how Washoe put signs together.

In evaluating this evidence, it is important to note that, while ASL does not necessarily follow English syntax, Washoe's trainers tended to transliterate English order into sign, in particular, using the English order of subject–verb–object (Gardner and Gardner 1971:176). Hence, Washoe was presented with linguistic specimens that corresponded to the basic rules of English.

Washoe's meaning is generally clear in these examples. She conveys familiar concepts and seems to organize her perception of the world in ways that are immediately recognizable to adult humans. Specifically, she apparently perceives events in terms of agents and actions ("Greg tickle," "Naomi quiet," "Naomi hug"); actions and recipients of actions ("tickle me," "catch me," "gimme sweet drink"); locations ("in down bed," "go in," "key in"); ownership ("my baby," "more mine"); instrumentality ("key open food," "open key please"); and sequences of action ("you me drink go").

Comparing Washoe to children, or even adults, in terms of the conceptual content of speech, she shows a high degree of similarity. Each of the relations above can be found in the early speech of children. It must be concluded that there is no evidence here of any major species differences.

Washoe has an extensive and mostly appropriate use of several words that refer to conceptual relations. These are the signs that Washoe first acquired as single words; "more," "up," "out," "in," etc. In combinations, they permit her to encode various dynamic and relational interactions, such as leaving a room or locating an object in some particular way. Again, they are used in a way that resembles adult language.

In the context of these similarities between the linguistic processes of Washoe and human children, the differences that follow are particularly striking and significant.

WORD ORDER

Does Washoe follow any rules for arranging signs into sequences? In many respects, her sign sequences are clearly contrary to any use of word order to encode conceptual relationships. For example, each of the following can be found in Table I: a single conceptual relation represented by two different word orders ("up Susan," "Susan up"); two conflicting conceptual relations with the same order in successive strings ("shoe up" where *shoe* probably is agent, "baby up" where baby probably is recipient); non-relational repetitions of referents in a single string ("You tickle me Washoe"); and chaos across several strings ("open gimme key," "in open help," "help key in," "open key help hurry"). These example suggest that the principles for arranging words to express conceptual relationships, if any, are sporadically applied. There is evidence, however, that other forms of regularity exist in Wahoe's sign sequences.

The Gardners (1971) mention two arrangements they have observed in Washoe's signing. One of them is of little interest from a linguistic point of view, but the other is significant. The less interesting arrangment is a tendency for "you" to precede "me" in sequences such as "you me out." The order of words in this case seems to have a purely mechanical explanation and does not reflect a rule of grammar in any sense. The sign for "you" is made by pointing away from the chest toward (and touching?) the addressee, whereas "me" is made by tapping one's own chest, and hence near the region of the chest where the sign for the action will be made next. Given that the sign for the agent precedes that for the action, the preference for the *you–me–action* order can be understood as the result of the greater mechanical ease of signing in this order than in the *me–you–action* order.

The second consistent order reported by the Gardners is the tendency, just noted, to place agents before action words, as in "Naomi hug," and "you me out." In contrast to the placing of "you" before "me," this regularity appears to be genuine. The agent may be either the addressee ("you," "Naomi," etc.) or a non-addressee (always "me," i.e., Washoe herself). The nonaddressee usage is secondary, in the sense that it occurs only in the context of the addressee usage and is limited to self reference ("me"), whereas the addressee usage also occurs alone and includes various referents. Beyond this pattern mentioned by the Gardners, a second apparently real pattern can be found in the examples listed in Table I. Washoe had a strong tendency, 80% in the examples, to place the recipient of an action after the action word, as in "gimme key" and "you tickle me Washoe". In all cases, the recipient is a nonaddressee, who may be either animate (i.e. Washoe) or inanimate (some physical object).

Thus, there are two patterns in Washoe's sequences of signs, one with two versions (addressee and nonaddressee), which may indicate some kind of linguistic structure. The following schemata represent all the discoverable information contained in these patterns (brackets represent single signs, braces indicate alternatives):

1. (Primary) [agent, addressee, animate]—[action]
   (Secondary) [agent, addressee, animate]—
           [agent, nonaddressee, animate]—[action]

2. [Action]—$\left[\text{recipient, nonaddressee,}\left\{\begin{array}{c}\text{animate}\\\text{inanimate}\end{array}\right\}\right]$

The primary version of schema 1 represents such strings as "Naomi hug" and the secondary version represents "you me out." Schema 2 represents such strings as "gimme key" and "(you) tickle me." Comparing only the *primary* version of schema 1 to schema 2, we can see that agent and recipient are redundant with respect to addressee and nonaddressee. That is, agents are always addressees and recipients are always nonaddressees. The converse redundancy also holds, addressees always being agents and nonaddressees being recipients. Agents also are redundant with respect to animateness and conversely, but recipients are not thus redundant since there are both animate and inanimate recipients.

Adding the secondary version of schema 1 simplifies matters considerably. In this version, agent is no longer redundant with respect to addressee, since there is also a nonaddressee agent, the converse redundancy also has disappeared. The redundancy of recipient with respect to nonaddressee and conversely, is similarly interrupted. Only the redundancy between agent and animateness remains. Since there can be animate recipients in schema 2, this redundancy is asymmetrical and we can conclude that in schema 1, both versions combined, it is the agent that is redundant with respect to animateness and not conversely.

Because of the contradictory properties of the primary and secondary versions of schema 1, very little that is positive can be said about what Washoe might have been encoding in her sign sequences. But it is clear that she was not encoding the relation of agent. As noted above, agent is redundant with respect to animateness in both versions of schema 1. There is, therefore, no evidence that Washoe arranged sign sequences so that agents came first. However, we are unable to say on what basis she did arrange sign sequences without further information.

The Gardners report that sequences such as "you me out" occurred early in Washoe's development, and were replaced later by somewhat more exotic sequences such as "you out me." That is, the secondary version of schema 1 disappeared, and was replaced by schema 3:

3. [Agent, addressee, animate]—[action]—[agent, nonaddressee, animate]

The component agent, nonaddressee, animate in schema 3 was always "me" or "Washoe."

This schema coincides very neatly with schema 1 (primary version) and schema 2, and clarifies the significance of the order of signs for Washoe. When schema 3 appeared in Washoe's signing, she consistently followed the pattern: addressee–action–nonaddressee. All other relations were either redundant (recipient), occurred in all positions (agent, animate), or were not used distinctively in any position (animate, inanimate). The exotic string "you out me" has two agents, then, according to schema 3, but this fact did not determine its grammatical structure. Its structure was established by the addressee–nonaddressee relationship. The only exception to this latter pattern is the string, "tickle me Roger Washoe," listed among the Gardners's examples. However, it is not clear that the sign "Roger" was actually a name to Washoe. According to an unpublished summary of Washoe's diary that the Gardners have circulated, it originally meant "please".

The chimpanzee may therefore have imposed her own formula on the sentence structures she observed her handlers using. Washoe's formula does not capture what the handlers themselves encoded (agent, action, recipient), but instead emphasizes a novel relationship as far as grammatical form is concerned, that of an interpersonal or social interaction (addressee–nonaddressee).

NEGATION AND QUESTIONS

Bronowski and Bellugi (this volume) have pointed out that Washoe never incorporated questions or negatives into her combinations of signs. This omission is particularly striking because negatives and questions occurred as single word messages (showing that Washoe had the idea of negating and questioning) and Washoe's handlers used both negative and question forms in combinations, hence, providing Washoe with examples. Children, in contrast, incorporate questions and negatives into even the simplest word combinations, at a stage of development when the longest utterances are only two or three words. "No pocket" and "not truck" are typical negative examples, where "that pocket" and "that truck" are affirmative sentences at this stage. There are surely no reasons of length or complexity that prevent Washoe from producing equally simple negative sentences. The reason she did not must have something to do with the degree of internal organization they require. Early questions in child speech show a similar simplicity. They are formed either by uttering a sentence with a rising "question" intonation, resulting in a yes–no question ("see hole?"), or by combining a Wh-word with another word ("what doing?"). Again, there are no reasons of complexity or length to keep Washoe from producing such forms. This is particularly true of yes–no questions, which in American Sign Language (ASL) are made merely by holding the hands for an extra moment at the completion of a sign gesture. Washoe was able to ask questions this way with single words. Doing the same with a sequence of words would not have been more difficult, if she had known how.

ENCODING PROPOSITIONS ABOUT THE PHYSICAL WORLD

The various characteristics of Washoe's signing discussed above can be unified in terms of a single observation: Washoe apparently does not use word combinations in a way that encodes conceptual relationships. The order of signs in strings, for example, was based on the social relations of addressee and nonaddressee, rather than, as in the case of children, on the physicalistic notions of agent and recipient of an action. This is true despite evidence that the relations of agent, action, and recipient are intellectually within Washoe's grasp. A similar inability is shown in the absence of negatives from Washoe's strings. In particular, nonexistence and denial, two semantic categories of negation that children use at an early age, presuppose some form of an underlying proposition in order to be meaningful. The absence of negative forms from Washoe's sign sequences implies that she was unable to encode such propositions. In the case of nonexistence, there must be a proposition concerning existence, which is negated; "no pocket," meaning that there is no pocket on an article of clothing, presupposes a proposition about the existence of a pocket. In the case of denial, there must be a proposition which makes a claim that is then said to be false; "not truck," meaning that an object is not a truck, presupposes the claim that the object is a truck, which is then denied. Questions likewise depend on an ability to encode propositions by means of word patterns. Yes–no questions are requests for information about the truth or falsity of entire propositions, and Wh-questions are requests for information about parts of propositions. The absence of such questions from Washoe's sign sequences implies, again, that she did not encode propositions when she produced sequences of signs.

Thus, in general, whenever combinations of words serve in the language of adults or children to encode propositions or conceptual relations (components of propositions), it appears that Washoe has failed to use these syntactic patterns in her own linguistic performance. This is true despite the presence of examples of the missing linguistic patterns in the signing of her handlers. Properties of language that do not depend on this use of word combinations, on the other hand, were apparently acquired by Washoe without difficulty. This includes the acquisition of a vocabulary of relation words ("in," "out," etc.), and negation and questions using only one word. The Gardners (1971) repeatedly state that Washoe's signs are semantically associated in strings, but that there is no evidence that her signs are syntactically associated. The semantic association of signs implies a certain similarity of the conceptual processes of the chimpanzee and *Homo sapiens* and suggests that Washoe was able to bring these cognitive processes to bear on her sequences of signs. The lack of syntactic organization implies, however, that she could not use her conception of semantic relations to organize sign sequences. It is indeed the case that, as noted above, Washoe appears to convey information about agents, actions, and recipients. Thus, the grammatical differences between Washoe and human language users are presumably not due to some general cognitive difference between the chimpanzee and *Homo sapiens*. They arise from differences in the ability to use patterns of words to encode conceptual relations, which are themselves more or less shared between the species.

SARAH AND THE PERCEPTIBILITY OF LINGUISTIC STRUCTURE

Premack's experiment with the chimpanzee he has called Sarah (Premack 1970a, 1971b, 1976a) sheds light on the Washoe study by demonstrating a training method that can in some cases overcome the limitation on the chimpanzee's use of combinations, particularly the use of strict word order and negation. This method can be compared to the training "methods" parents successfully employ with children.

The most important result of Premack's experiment is that he was able to instruct Sarah on several of the syntactic forms that Washoe, acquiring linguistic information spontaneously, did not use. For example, Sarah was able to express the idea that two objects were not the same by spontaneously combining the sign for "not" and the sign for "same" (Premack, 1970a). This implies some kind of propositional framework into which a negative operator was placed. Similar examples can be found for quantification ("some," "all") and for the conceptual relations of agent, recipient, and location, each of which is encoded by observing strict word order. As an illustration of what seems to be true grammatical knowledge, we can take the following example (Premack 1976a). Sarah was first taught two alternative ways of encoding attributive statements: "red color apple" and "apple is red"; these and all other utterances being written with bits of magnetized plastic placed vertically on a board, bits with distinct shapes and colors representing different words. These sentences were taught independently and at different times, and no attempt was made to link them until the following test was made. Drawing on previous training, in which Sarah had been taught to use chips meaning "same" and "different" in judging whether objects were the same, Premack now had Sarah judge whether "red color apple" and "apple is red" are the same or different. She was able to do this immediately at her standard level of accuracy, about 80%. A judgment of "same" in this case could be made only on the basis that the two sentences have the same meaning for Sarah—i.e., on the basis of internal relationships that had been decoded from the structured arrays of words. Some sentences that she correctly called "different" are physically more alike than some sentences that she called "same."

These are impressive results. However, we must be cautious before we equate them with the process of acquisition shown by children. Premack has not attempted to simulate language acquisition. His interest has been in finding ways of instilling linguistic information in the chimpanzee, whether or not these are the ways such information is instilled in children. Premack's methods are based on the principle that, at any given time, exactly one element in the training situation should be unfamiliar to the learner, no more. By cascading a series of small steps in the correct order, complex structures can be built up. When learning the word "color" (= "color of"), at first Sarah had only the chip she was to learn to use before her, color. She already knew that whenever the chip representing "?" occurred in a string, it was to be replaced by a word that would make the string into a correct statement. Hence, when she was shown "red ? apple," and "?" could be replaced only with the word "color," her previous experience implied that "red color apple" was true. It remained for Sarah to discover what it was, conceptually, about this statement that made it true. Such inferences evidently are possible for the chimpanzee. The second step of training presented alternative words, hence the need to respond selectively, and further steps required various kinds of generalization.

Adults talking to young children also adjust their speech to the presumed limitations of the child. Remarkably few studies have been made of this adjustment, however. In one of the few studies that exist, Snow (1972) observed considerable simplification of speech directed to two-year-olds, compared to speech to ten-year-olds. Sentences were more often simple declaratives, they were shorter, and they did not include as many pronouns. There was also a tendency to repeat sentences and sentence constitutents in a way that set word groupings off. As an example of the latter, the following seems to be typical: "Put the red truck in the box now. The red truck. No, the red truck. In the box. The red truck in the box." Snow's procedure involved having adults instruct children on how to solve puzzles, the requirements of which would impose the kind of simplification and repetition she observed, so some of these effects may be artifacts of her method. But assuming that similar adjustments occur in normal adult speech to little children, we can see that there are certain similarities between the speech that adults direct to young children and the systematic training procedures used by Premack.

If one thinks of a scale of didactic speech—i.e., speech intended to instruct in language—Premack's "speech" to Sarah would be near (if not at) one extreme and randomly chosen tape recordings of conversations on unrelated subjects would be near the other. On such a scale, the simplified speech of adults described by Snow would obviously be closer to Premack's end. However, the speech of adults to children does not follow the essential principle of Premack's method, that is that only one element should be unfamiliar at any given time. In the example of the repetitions mentioned above, the adult was simultaneously demonstrating declarative sentences, locative phrases, noun phrases, and noun phrases in relation to locative phrases. This discrepancy between Premack's method and the speech of adults to children would be largest at the earliest stages of development, when the child knows least about linguistic structure.

The position of a *training procedure* on the imaginary didactic scale above, influences the extent to which the procedure makes linguistic patterns explicit, and hence, inversely, the extent to which the learner must be able to perceive these patterns on his own. The conclusion to draw from Premack's experiment with Sarah therefore seems to be the following: at Premack's end of the scale, where there is only a single unknown, the word pattern being taught is made fully explicit. The chimpanzee must only be able to infer the significance (meaning) of the pattern. At the child's position on the scale, this inference is also required, but in addition, there must be an ability to perceive the syntactic patterns themselves when they are not made fully explicit. That is, the degree of simplification reported by Snow implies the presence of a perceptual sensitivity to syntactic forms on the part of young children which is not implied by the success of Premack's method. The patterns that children find on their own include precisely those that Washoe, receiving linguistic information from a point that appears to be close to the child's on the scale, is unable to acquire.

Thus, Premack's experiment complements the Gardners' by showing a difference between chimpanzee and children that is correlated with the ability to perceive and presumably use syntactic patterns that encode conceptual relations. The conceptual relations themselves appear to be available to both species, a conclusion that Sarah's performance reinforces quite strongly.

A DIFFERENT GRAMMATICAL SYSTEM?

Attempts to teach chimpanzees English are based on the assumption that if the chimpanzee has any capacity for languagelike communication it is for a language whose structure is basically similar to human language. The assumption implies that the chimpanzee is on the same evolutionary line as man, but has not progressed as far. However, such an anthropomorphic assumption is not necessarily correct. The chimpanzee may already have begun the evolution of a linguistic capacity but not along the line our own evolution has followed. If man's capacity for language itself is in part a biological adaptation to a certain range of life conditions, it is possible that another species has developed an adaptation to accomplish similar communicative effects under different life conditions. I will argue that chimpanzee is such a species. We obtain a glimpse in the Washoe experiment of what may be a fundamentally different linguistic framework, one that has a unique semantic basis and has rules that are unknown in any human language.

Human languages are extremely well adjusted to the problem of describing physical objects and relationships among physical objects. Every utterance, if it is a grammatical sentence, includes at a minimum, a phrase that refers to an object and another phrase that refers to a quality of the object or an action, with the sentence structure indicating the relationship between them. More elaborate sentences add further object phrases and additional relationships. All this is highly familiar. The model of objects is, of course, extended beyond its natural base into realms where real objects and actions do not exist—e.g., to the domain of numerical relationships. Algebraic word problems, for example, are far more difficult to understand than ordinary sentences because in part they cannot be understood as being about objects, even though the sentence containing the problem is geared to describing objects and their relationships. Children, in their earliest speech, however, remain close to the basis of language and confine nearly all sentences to statements about objects and relationships among objects.

Because any human language is so well adapted to describing objects and their relationships, it is difficult for us to see that this facility is in a crucial degree the result of the structure of language, and not solely of the nature of the physical world. The great difficulties that have appeared with automatic pattern recognition by computers show how complex and latent the structure of objects actually is (cf. Neisser 1967). It is not possible to explain the ease with which language describes objects and their relationships by falling back on another process, *the perception of objects,* since this itself turns out to be hidden in obscurity and full of contradictions.

It is clear, however, that in speaking, one does not have to face the epistemological problems that are encountered in automatic pattern recognition. The most successful computer system to date for interpreting sentences from a natural language (Winograd 1972) has built into it a correspondence between the physical world of objects and the grammatical world of noun phrases, predicates, etc. Pattern recognition is not an issue for this program. It seems reasonable to speculate that the human capacity for language evolved under pressures to convey information about objects and their relationships in such a way that a similar correspondence exists in all members of the species, as part of their natural preparation for the use of language. Indeed, this correspondence is so close for human speakers that it is difficult to notice when the correspondence does not hold, as in algebraic word problems, and it is especially difficult to imagine a linguistic system that is based on other principles, not on a model of the physical world. Yet the chimpanzee, apparently with very similar conceptual powers, seems to lack a parallel grammatical structure, and may have in its place a linguistic system based on social and personal interactions.

We have already seen one example of this system in the rule noted before, whereby Washoe put the word for her addressee first and the word for the nonaddressee last, with an action word in between. This use of word order for addressee–nonaddressee is no more arbitrary than the use of word order for action–recipient. The two ordering principles are, however, quite distinct. It is true, of course, that the agent–recipient sentences of English also often encode addressee–nonaddressee. But the latter encoding is not the meaning of the grammatical structure. It is a concomitant meaning arising from the fact that addressees are often agents. For Washoe, the situation seems to be reversed. Addressee–nonaddressee is the meaning of the structure and agent–recipient is the concomitant meaning. This conclusion permits us to see a basis of organization within Washoe's utterances, which otherwise often appear chaotic.

Further examples from Washoe's signing suggest another principle that is grammatical in character but alien to the structural basis of human language. In their first report, the Garnders (1969) describe combinations in Washoe's signing which seemed to function as emphasizers. Words were added to strings not to convey new information but to make the basic message of the string more intense and emphatic. Examples are "please open hurry," "gimme drink please," "please hurry sweet drink," and " open key help hurry." Is there any grammatical principle involved in these strings? According to the unpublished summary of the diary the Gardners kept on Washoe's development, the longest sequences of signs she produced all included this kind of emphasis. Emphasizers were the main source of additional length in strings. We may speculate that she had a rule of the following kind:

$$S \rightarrow P^n$$

where S is a grammatical string, P is the basic meaning of the message and n is the number of occurences of P and depends on Washoe's degree of urgency. An organized string according to this rule, consists of any number of signs in any order which convey a single meaning. This is not as vacuous as it first seems. A string may consist of repetitions ("more more more sweet drink"), different signs ("please out open hurry") or both ("out out please out"). According to the rule above, all have the same grammatical (but not lexical) meaning: something is urgent to a degree that sets $n = 4$.

Beyond the addressee–nonaddressee and emphasis rule, no other grammatical principle stands out. The order of words in Washoe's strings, apart from addressee and nonaddressee, probably depends on the prominence of the things referred to. Since such prominence shifts with time and new situations, word order shifts also. This variability does not require any grammatical structure and does not appear to be part of Washoe's linguistic system. The addressee rule itself might be thought to be the result of a greater prominence of addressees compared to non–addressees for Washoe, except for the fact that, in a large number of cases, the most prominent part of Washoe's message, to judge from the part being emphasized, is the action, and still the addressee is mentioned first. The addressee rule as well as the emphasis rule, may have a different origin, a possibility we turn to next.

ASPECTS OF CHIMPANZEE LIFE

Goodall (1965; van Lawick-Goodall 1971) has written vivid descriptions of the life of free-ranging chimpanzees. From all this material, three observations stand out for our purposes. These have to do with the chimpanzees' lack of interest in physical objects, the importance of social interactions, and the effect of the dominance system.

Chimpanzees are intensely aware of one another. They seem to be in constant social contact through some means or other. This acute sensitivity to other chimpanzees contrasts with what seems to be a remarkable lack of interest in physical objects. Their attention often is not caught even by major new features of the physical environment. Goodall had the rather dangerous experience of chimpanzees almost colliding with her because the animals had not noticed her presence, in spite of her being in clear view. Other chimpanzees did not at first notice a newly arrived photographer, complete with equipment, whereas chimpanzees are ordinarily frightened of humans. Chimpanzee infants, whose play is remarkably like the rough and tumble play enjoyed by human infants, almost never play with physical objects the way human infants do so avidly, even from the earliest months of life.[1]

Chimpanzees are quite alert to selected aspects of the physical environment—the condition of feeding trees, for example, or the presence of the bits of grass they use for catching termites—and young chimpanzees show an interest in these, too. What is lacking in the chimpanzee is the intense human curiosity about objects. There is no exploration of the physical environment unless there is a practical reason to carry it out.

The fact that Washoe did not absorb those linguistic structures that encode relationships among objects thus conincides with an indifference to the physical environment among wild chimpanzees. If the chimpanzee has indeed evolved a capacity for languagelike communication, we should not expect it to resemble human language in structurally encoding physical relationships. In fact, it does not seem to do so. Conversely, we can see in the human's preoccupation with physical objects the basis for the evolution of a specialized communication system that encodes information about objects and their relationships.

The chimpanzees' social structure is complex and important to the animals. As Goodall points out, it is not less important to them than to us, and in some respects may be more important. Every chimpanzee is aware of the social hierarchy in which he finds himself and of his place within it. This dominance hierarchy influences almost everything chimpanzees do—eating, relaxing, mating, even sleeping. If the chimpanzee has evolved a capacity for using grammatical structure based on social interactions, it is possible that the intense awareness of other chimpanzees can be encoded structurally. Corresponding to the human use of word order to encode agent–action, a relationship between object and an event, chimpanzees might use structural means to encode social relationships between individuals. Washoe's use of word order to encode addressee–nonaddressee can be seen in this light. What for her was most salient in situations in which agents and recipients were being related to actions, may have been that there was an addressee and a nonaddressee. It was this fact that she seized upon and made the basis of a grammatical rule.

The impact of the dominance hierarchy on the social interaction of chimpanzees is especially strong on appeasement and begging behavior. A subordinate chimpanzee must perform an elaborate and finely graded ritual of appeasement if he wants to obtain a donation of, say, food from a more dominant chimpanzee.

---

[1]Similar observations have been made of the mountain gorilla by Schaller (1963) and of the Hamadrayas baboon by Kummer (1968) and Omark (1972).

Even when not begging, a display of appeasement may be necessary from a subordinate animal to avoid giving offense and being attacked viciously. In chimpanzee interactions, it is essential for an individual to signal his intentions clearly to other animals, whether he intends to be subservient or aggressive or merely neutral, and to what degree.

Washoe's signing shows many examples of appeasement and begging. Among the latter, "please" combines with "come–gimme," "out," "drink," "open," "go," and many other signs. "Come–gimme" (which is identical to the natural chimpanzee begging gesture) combines with "open," "tickle," "sweet," "listen," "more," and many others. Appeasement was conveyed by "sorry," "hug," "please," and "more" combining with lipsmacking. That is, Washoe used for begging and appeasement most of the signs that played a part in the rule, $S \rightarrow P^n$. This congruence suggests that the rule, with its flexible use of message length, is related to the need for chimpanzees to encode their intentions in social interactions precisely and clearly. Again, assuming that the chimpanzee has evolved a capacity to encode social relationships through structural means, Washoe may have been able to develop a rule for encoding the urgency of messages—begging and entreaties— by arranging signs into strings of definite length and indeterminate order.

Washoe and Sarah may be unique, Sarah because she has been taught, through rigorous means, to use word combinations for encoding propositional relationships; Washoe because she was exposed to ASL. However, Washoe did not encode propositional relationships through word combinations in ASL, although she expressed these in other ways, and she reconstituted the linguistic information she was given along lines completely different from those her handlers presumably had in mind. It is conceivable that her ability to do this would not have been brought to the surface if she had not been given the stimulus of ASL, but it is also conceivable that free-ranging chimpanzees use a native linguistic system organized along the lines suggested by Washoe's reorganization of ASL. This question, obviously, can be answered only by those who can observe free-ranging chimpanzees leading their normal community life.[2]

---

[2]It will not be an easy matter to discover the properties of such a linguistic system, or even whether one exists. The medium may be gestural, but there seems to be no *a priori* basis for guessing what these gestures would look like. The structure of the language will presumably be built around various social interactions, not descriptions of the physical world, and the content of typical messages will probably consist of entreaties, demands, mollifications, and declarations of ownership and location. Linguistic regions may be quite small because of the small areas occupied by isolated reproducing communities. Goodall remarks that occasionally, the chimpanzees would abruptly change their activities and that she could not on these occasions discover what had triggered the change. Conceivably, a coded message had been exchanged, obvious to the chimpanzees but unnoticed by the human observer.

LANGUAGE, COMMUNICATION, CHIMPANZEES*

GEORGES MOUNIN

WASHOE

Following the repeated failures registered between 1930 and 1950 in attempting to set up phonic communication with an ape, Allen and Beatrice Gardner (1969, 1970, 1971) undertook, from 1967 on, to establish communication of a linguistic nature through a totally different medium: the analogical (nondigital) gestural language of the deaf–mute. Their hypothesis was based on the following: First, the phonatory apparatus of the ape differs from that of man, and this element probably accounts for the previous failures observed by authors such as Yerkes. Second, while the young ape very rapidly loses interest in and aptitude for vocal prattle, it retains an abundant gestural prattle, remaining, as an adult, a very imitative and gesticulatory animal.

The learning process took place neither in a laboratory nor in captivity, but in the environment of a home and family. At the outset, Washoe was between eight and fourteen months old. All the human participants in the program (psychologists, advanced psychology students, and others) were obliged to learn American sign language, the code used by American deaf–mutes, and compelled to use only that form of communication in Washoe's presence. The experimenters believed that if the ape heard human language, without being able to acquire it, this would frustrate her psychologically and therefore constrain or disturb her in the apprenticeship of gestural communication. Silence was not prescribed, however; the participants were allowed to make noise, laugh, hum or whistle, etc.

To begin with, the teaching method used was based on the premise that the ape would spontaneously imitate the human transmitters. In order to hasten the imitatory process, however, the Gardners decided to introduce into their sign code certain meaningful gestures employed by Washoe herself,—for instance, her motions expressing "to give," "tickle" (which constituted her chief reward), "to open."

*Translated from the French by Pierre and Mireille Martin.

GEORGES MOUNIN • Université de Provence, Aix-en-Provence, France.

161

Ostensive means were used to speed the acquisition of these signs; the referents of the signs were pointed out, especially where parts of the body were concerned. By so doing, the investigators utilized a behavior pattern common to chimpanzees, who by touching one another seem to establish a kind of phatic and affective communication. Ultimately, Roger Fouts (1972a, b), who participated in the experiment, isolated, described, and systematized the highly efficient technique of *moulding,* in which the experimenter indicated, using both the visual and tactile guidance of his hands, the position, arrangement, and movement required of the ape's hands and arms in order to carry out the sign which she was to learn.

What were the results of this apprenticeship? The learning process took place at the University of Nevada until the end of 1970, and the first phase, that is, the apprenticeship in communication properly speaking, appears to have lasted twenty-two months. (The apprenticeship of Lucy, another ape, at the University of Oklahoma in 1971, under the systematic guidance of Fouts, required only six months.) In the case of Washoe, the first sign, "to come," took her seven months to acquire. After twenty-two months, she was familiar with 34 signs; after forty months, with 92; and by 1973, with 106. In order to be considered acquired knowledge, each sign had to be emitted at least five times. The retention of knowledge was controlled; during the last six months of her stay at the University of Nevada, Washoe made use of between 24 and 30 different signs per day.

Other results should be pointed out. First, the ape produced series of signs, sometimes in order to give indications (thus she had invented "refrigerator" = /to open/ + /to eat/ + /to drink/),[1] sometimes in order to construct types of utterances (/to give/ + /Washoe/ + /tickle/). The signs used may be (temporarily) characterized as corresponding to nouns, verbs, adjectives, adverbs, locatives, and, according to the Gardners, pronouns.

Besides the tests of numerical verification mentioned above, the Gardners made sure that each sign designated a class of referents and not only a single referent. The day Washoe used /toothbrush/ to express brushes other than her own represented an important stage in the experiment. It was also observed that the sign referred to the category of objects as well as to the image of these objects: Washoe was very interested in comic strips and emitted the signs (/cat/, /to drink/, etc.) corresponding to the pictures ("cat," "bottle," etc.). The ape also made use of the signs without incitement from the experimenters, thus without imitative repetition or the stimulation commonly associated with the experimental reward. Finally, the decodability of her gestures was tested by bringing her into contact and communication with deaf-mutes unfamiliar with the experiment but trained in the use of ASL.

---

[1] A word between quotation marks indicates that only the signified (its sense) is being considered; between oblique strokes, only the signifier (its gestural substance); in capital letters, only the referent.

For various reasons, at the end of 1970, Washoe was transferred to the University of Oklahoma, where Fouts had just been hired. There the experiment proceeded, but took on a different orientation based essentially on the study of Washoe's behavior when living with other chimpanzees: domestic home life was replaced by the conditions found in a modern zoo (that is, life spent partially in a cage and partially in semifreedom). Washoe was in periodic contact with five or six other chimpanzees, several of whom had also undergone the learning of ASL after having had previous daily contacts with human phonic language. According to Hewes (1971), an anthropologist who had several times visited Washoe since the experiment began, the changes in the environment and in the conditions of the experiment did not produce any signs of regression in the learning process of the ape. At this stage, however, the main result expected was that Washoe would transmit the learning process to others of her own species, so that communication through ASL could be established among the six or seven chimpanzees. This result was slow in coming; in 1973, Hewes (1973b:6) acknowledged that nothing new had yet evolved on this point.

How should these results be interpreted? In their general conclusions, the Gardners (1969:87) refuse to say whether the set of results obtained authorizes one to speak of a chimpanzee language, in spite of the fact that the terminology of their experiment very often suggests this idea. Hewes (1973a:131) has observed that there have been very few discussions and analyses of the Washoe experiment. The existing ones touch on the psychological problems (comparison with the learning of language by a normal child, by a deaf-mute, by a psychotic child), or are the work of ethologists or anthropologist more concerned, like Hewes himself, with deducing from the experiment a well-founded hypothesis on the (gestural) origin of human language. Paradoxically, even in Hinde's (1972) major work—which in this respect remains very cursive (see Mounin 1974)—Washoe's case did not elicit reactions from semiologists and linguists, though logically their intervention should have preceded all others.

Even in America, psycholinguistic interpretations of the Gardner experiment have been significant but inadequate. Some adopt (implicitly or explicitly) the Chomskyan point of view on language. If syntax, that is, the transformation of deep structures into surface structures, is a language universal, then because language is specifically human, it seems quite unlikely that Washoe could acquire and express the rudiments of syntax, and the series of signs should not be considered syntactic (Bronowski & Bellugi, this volume). If negative transformation is a universal, if wh-questions are a universal, and if imbedded phrases are also a universal, then, because communication with Washoe does not present these characteristic (descriptive) features, there can be no language.

This logical procedure perfectly illustrates Hockett's (1973) warnings: The assertion that universals of language exist is a problem of terminology and therefore is not independent of the theory and methodology chosen (pp. 2, 6). Furthermore, one must not confuse the widespread descriptive features of human language with its defining features, which are differentially relevant with regard to all that is not human language (p. 3). Finally, there is the risk of getting caught in a vicious circle in asserting that a system of communication is not a language if it is not human (p. 5). It is apriorism to state that passive transformation and interrogative transformation are general features of human languages and therefore are among the criterial traits of language (understood as the ensemble of the natural languages of men). This means that the criteria for human language cannot be obtained through mere generalization of its present features, but must be sought through a comparative (descriptive, not a priori) study of all systems of communication, including animal communication. Bronowski and Bellugi, as well as Brown (this volume), confined as they were to Chomskyan psycholinguistics, could not answer the ultimate question: Is there language in Washoe's case? We shall encounter this problem again in the work of the Premacks, who seem to have been strongly influenced by these Chomskyan critics.

Other interpretations have been offered in America, either by psychologists, ethologists, or anthropologists. These, however, are based, indirectly it seems, on the universals of language proposed by Hockett as early as 1960. (Curiously enough, Hockett is rarely quoted, even when evidently utilized, and in spite of the fact that between 1960 and 1968 he had published four successive versions of his views.) Hockett (1963:5), in an empirical comparison between human language and animal communication, lists 16 constitutive features of language (stating clearly, however, that this list is not necessarily made up strictly of defining features). These are the vocal-auditory aspect; directional transmission and reception; rapid fading of (oral) messages; the interchangeability of the transmitter with the receiver; the presence of feedback, enabling continuous control of the emission by the transmitter; specialization (that is, the use of very little [articulatory] energy in comparison with the behavioral results achieved); semanticity; the arbitrariness and discrete nature of the linguistic sign; the possibility of displaced utterances (regardless of the situation of emission, therefore unrelated to the time and space circumstances of that emission); openness (the nonfinite aspect of the number of possible messages); transmissibility through apprenticeship (the noninnate character of the code); duality of patterning (Martinet's *double articulation*); the possibility of false or untrue utterances; the possibility of metalinguistic use (use of the code to speak of the code); and the possibility of learning a second language.

On the basis of these facts, one can generally assert that, in Washoe's case, there is no language properly speaking, since ASL does not comprise double articulation. Thus, the gestural signs are significant units (monemes) incapable of being broken down into distinctive units similar to phonemes; this question has yet to be cleared up, even after Stokoe's (1966) analysis. Moreover, there is no language because Washoe did not use the code to lie or, except for the use of proper names, to convey what she had learned to her conspecifics (Hewes 1973a:141, 1973b:10). Other anthropologists have asked themselves whether the signs of ASL are arbitrary or iconic. (Mostly, they are without a doubt etymologically iconic and have become arbitrary through time, as is the case for most symbols; is the druggist's serpent in France today arbitrary or symbolic?) Still others go so far as to wonder whether a code exists, though without putting forth any arguments (Washburn & Strum 1972).

The Gardners and their readers also use criteria such as the presence of nouns, verbs, adjectives, locatives, and even pronouns to determine whether language exists as far as Washoe is concerned; from a linguistic standpoint, this is Europeocentrism. Languages in which all adjectives can act as verbs or verbs as adjectives, languages in which all nouns can act as verbs, etc., have structures that differ a great deal from those of the Indo-European languages, and yet they are languages. Neither the parts of speech of our Indo-European languages nor their functional classes are necessarily linguistic universals, much less criteria for language.

As far as pronouns are concerned, the lack of linguistic training of those participating in the discussion is worth considering. Hockett does not consider the presence of pronouns as a criterion for language, nor does he regard certain parts of speech as universals or design features (defining features of human language). On this point, the problem was obscured by Benveniste (1966:225–256), since for him *I* and *you* were constitutive universals of the speaker's *own* subjectivity, that is, his consciousness of being a psychological subject. As a matter of fact, children are perhaps conscious of themselves before having acquired language (the mirror phase), but surely through linguistic reference to themselves as third persons ("Peter has a sore" precedes "I have a sore"). By the same token, in the Far East, several nations so rarely use personal pronouns that it has seemed doubtful that the languages contain any. Regardless of how these facts are interpreted, they indicate that human language can or could very well do without pronouns and still not cease to be language, even in its psychological implications. Pronouns are simply a phenomenon of syntagmatic economy whose surprising extension and formal universality (always calling for at least three persons singular and three persons plural) lead us to believe that it could have been invented only once. (This would be a strong presumption for asserting that language has a single origin.) On the existence of Washoe's pronouns, neither the Gardners nor their commentators are very clear. An exhaustive analysis of procedures would be required to get to the bottom of it. From the work published so far, it seems that the functions of proper nouns ("Washoe," "Mary," "Randy," etc.) may have been assimilated to those of the personal pronouns.

The soundest conclusions about the Washoe experiment show that it is a question of quasi-linguistic behavior (Hewes 1973b:130). This formulation, however, is very vague, since it is based on disparate descriptive features of human languages, taken at random, and without previous theoretical justification of the defining criteria assigned to human language.

One can surmise that the methodological failure of this approach, in the Washoe experiment, proceeds from the question "Is there language?" in that it is logically not the first question that should be asked. This question is considered of prime importance only by those researchers whose traditional philosophical backgrounds have accustomed them to consider the words *language* and *communication* as synonyms, so that for them one can speak of language with regard to any system of communication. The deadlock into which this confusion leads the analysis of communication indicates that the problem is not only one of terminological conformity, but also has theoretical and methodological implications.

If we were to ask the priority questions that a semiologist would ask—"Is there communication?" and, if so, "What type of communication?"—we would be confronted with an entirely different and more productive set of suppositions. One must obviously define clearly the concepts to be used. If we were to establish that in order to have communication (in the linguistic sense of the term, specifying human communication), (a) there must be someone transmitting and someone receiving, (b) the transmitter must be aware that his target is the receiver, (c) the receiver must be aware of being the transmitter's target, and (d) the receiver must be capable of becoming the transmitter by using the same channel (as a rule, using the same code or one equivalent to it), then we would know in regard to exactly what criteria we can evaluate the specific relationship between Washoe and her instructors.

As opposed to communication as we have just defined it, training (the other extreme but alternative possibility) also would deserve a strict definition. Neither the Gardners nor their commentators define the term; their notes, however, suggest simple training if (a) there is a one-way relation between A, always the transmitter of a stimulus, and B, always the receiver; (b) the answer–reaction of B to A is always behavioral in nature, without reference to the stimulus code used by A (even though the stimuli of A are linguistic in nature or, more generally, pertain to a code); and (c) there is always a concrete material reward after each correct response to a stimulus.

What, then, becomes of Washoe? The Gardners are correct in asserting that, for the first time, two-way communication through the same channel was established between man and animal. Thus, Washoe responded to the linguistic stimuli of her instructor not only with behavioral answers, but also with linguistic ones. But may we go so far as to say that linguistic communication, in the fully human sense of the word, was established?

This brings us to the second question: "What type of communication?" In spite of the resemblance, what are the differences between communication with

Washoe and communication (linguistic or not) between men? We are quite aware that though she very often answered through the same channel, Washoe was almost never the transmitter, taking the initiative of the communication; statistically speaking, she nearly always remained the receiver, reacting to human communicative stimulation. This confers great semiological importance upon certain rare but well-observed facts: At the University of Oklahoma, Washoe tried to communicate by means of ASL with other chimpanzees and even with dogs. Once, on the man-made island, the presence of a snake had spread panic among the apes, with the exception of one. Washoe stayed by him and, before fleeing, addressed him in ASL: /come/ + /hurry up/ + /dear/. On another occasion, at the University of Nevada, she had spontaneously expressed, without any initial stimulation from her instructor, the utterance /listen/ + /dog/ + /to bark/. The existence of such purely informative messages—even though they are still very rare—is sufficient proof of their feasibility. Such messages are much more important in the eyes of the semiologist than the existence of adjectives or locatives, or even of wh-questions, in Washoe's utterances. Indeed, such messages precisely establish, by means of objective and functional defining features, the kind of communication achieved with Washoe and the actual distance separating it from simple behavioral training in the direction of complete human linguistic communication.

Of course, the Washoe experiment includes many other problems and lessons. For instance, genetically speaking, it would be very interesting to find out what means of communication wild chimpanzees possess; this additional knowledge would be useful in determining a baseline for the Gardners' experiment. Wild chimpanzees have a poor system of vocal calls. Do they have the embryo of a gestural system, or merely, like the animals studied by Konrad Lorenz, behavioral sequences or initiating acts that have indicatory value for conspecifics but cannot be classified as signals or signs explicitly directed towards a receiver? The difficulties encountered in recording in the field and the problems in breaking down the gesture into discrete units of certain relevance have so far prevented even a rough deciphering of these gestural utterances.

Among all the limitations of the experiment and the tentative conclusions that can be drawn from the literature on the matter, it is important that the Gardners and their associates, as they themselves have admitted, were always insufficiently trained as transmitters to be able to "converse fluently" in ASL. This is incidental and can easily be coped with, but it is theoretically important from a semiological point of view in that it considerably weakens social contact. If there were but one chief condition governing communication, it would be that communication originates not from a desire to communicate a pre-existing thought, but, as Karl Marx stressed, from the needs arising from group life. It was wrong to repeat for centuries, from Buffon to Wiener, Colin Cherry, or Chomsky, that the reason animals do not speak is that they do not think. The best explanation is most likely that of an anthropologist like Kortlandt: if communication remains poor in the case of chimpanzees (Washoe's phylogenetic starting point), it is because their type of society (small groups) and social system (for the most part made up of relatively isolated gatherers of fruit and other plant life) does not require a great development; chimpanzees do not speak, says Kortlandt (1973:14), because they "have very little to say to one another." This point of view has deep implications.

One last lesson of the Washoe experiment that should be highlighted is the danger of the inadequate comparisons frequently made by psychologists and psycholinguists; one cannot compare Washoe's performances with those of a normal child, and even less with those of a normal adult, if one is to draw truly significant conclusions. Even if we more reasonably draw a comparison with a deaf child or with a linguistically pathological child, we must not forget that we are trying to juxtapose two primates of which one has a lead of perhaps more than a million years over the other in the area of group life in which social collaboration becomes increasingly complex, the manufacturing of tools, and the development of social communication under such favorable conditions. The other starts almost from scratch: teaching him to communicate means not only training him to handle a code, but also placing him in a pattern of social relations that requires more of him than does the experimental situation.

SARAH

As early as 1966, at the University of California, Santa Barbara, Ann and David Premack (1969; 1970a,b; 1971a,b,c; 1972; Premack & Schwartz 1966) undertook to establish communication of a linguistic nature with a chimpanzee through a totally different channel and code than those used with Washoe. Their ape, also a female, named Sarah, was six years old at the outset of the experiment. She lived in a cage, under the conditions normally found in an animal-psychology laboratory. The code used was solely visual. It consisted of a kind of writing system made up of thin plastic plates in the shapes of abstract hieroglyphics or arbitrary geometric figures, differing not only in shape but also in dimensions and color. (The "word" *apple* was not represented by a plate shaped like an apple or the "word" *red* by a red plate.) Let us call these figures units or signs. The base of each plate, coated with steel, ensured a good grip on a magnetized board. From the very beginning, Sarah seemed to prefer "writing" her messages vertically on the board, a preference which was respected and even codified.

The learning process was thought out according to a rigorous program. It began by appealing to the simplest and most familiar of all social relations: "to give." Each time the experiment was conducted, a fruit was first associated with its signifier (/purple triangular plate/ = "apple," etc.), then placed nearby on a table, and finally given as a reward to Sarah if she correctly placed the plate on the board. This first step, the acquisition of "denominations," was conducted with fruits, common objects, and persons. Next, the experiment proceeded to a second stage dealing with two-unit utterances, such as /Mary/ + /apple/ = "Mary to give apple." The abstract structure "name of giver" + "name of object" was obtained by systematic commutations of the giver paradigm (Ann/, /Randy/, etc.) with that of objects (/banana/, /plate/, etc.). From this point on, Sarah was compelled to learn how to arrange the words in order: /Mary/ + /chocolate/ was deemed grammatical and was rewarded, whereas /chocolate/ + /Mary/ was considered a failure.

The third step was the acquisition of three-unit utterances such as /Mary/ + /to give/ + /apple/ and the fourth step that of four-unit utterances such as /Sarah/ + /to give/ + /apple/ + /Mary/. At this stage, structural commutations bearing upon the paradigm of the beneficiaries produced psychological resistance: Sarah was reluctant to accept beneficiaries other than herself. This obstacle was overcome, however, by substituting a reward better (according to the order of Sarah's preferences, determined by separate tests) than that represented by the object shown in the utterance. New difficulties appeared for the commutation of the action (/ to give/, /to cut/, /to put in/, etc.). Furthermore, using the techniques very clearly described by Premack (esp. 1971a, 1972), Sarah was driven to form the following types of utterances:

1. /apple/ + /similar/ + /apple/, and /apple/ + /different from/ + /banana/.

2. APPLE + /unit of interrogation/ + BANANA + /similar/ + /different/ (= "Does an apple differ from a banana?").

3. An answer to the question by /yes/ or /no/.

4. Acquisition of the metalinguistic use of the code: /plate X/ + /name-of/ + APPLE and /plate X/ + /name-of/ + /apple/.

5. Acquisition of color qualifications (/yellow/, /red/, etc.), of shape qualifications (/round/, /square/), of size qualifications (/small/, /big/).

6. Acquisition of a complex sentence by passage through two elementary utterances: /Sarah/ + /put-in/ + /apple/ + /pail/ + /Sarah/ + /put-in/ + /banana/ + /plate/.

7. Acquisition of the copula (the verb *to be*), from /apple/ + /?/ + /fruit/ to /apple/ + /to be/ + /fruit/.

8. Acquisition of the plural with the following: /red/ + /yellow/ + /to be/ + /plural sign/ + /color/.

9. Acquisition of the complex sentence expressing logical implication (with a single unit, *if . . . then,* to be read as ⊃):/Sarah/ + /to take/ + /apple/ + / ⊃ / + /Mary/ + /give/ + /apple/ + /Sarah/.

This brief description of the procedure already provides us with the essential results. Premack (1972) speaks of a vocabulary totalling 130 signs. The examples show utterances as many as eight units long. The descriptive terminology of the author also suggests that Sarah had learned the equivalents of nouns, verbs, adjectives, pronouns, adverbs, prepositions (*on, in front of*), conjunctives(⊃), and quantifiers (*all, none,* and *several,* which also presented "psychological" problems of acquisition).

The control tests were both numerous at every stage of the experiment and very carefully carried out. For instance, the Premacks went to a lot of trouble to assure themselves that the sign corresponded to a class of referents and not only to a single referent through mere spatial contiguity; correspondence to a single referent is impossible in the negative utterance /purple triangle/ + /negation/ + /name-of/ + /red square/ ("apple" + "no" + "name of" + "banana"). They systematically used the transfer test, which, as I have already mentioned, is none other than the linguistic procedure known as commutation by classes of paradigms. This test establishes that the abstract structure is available without even considering the particular referents. Moreover, they employed several tests to rule out the "clever horse" objection, to which we will return later. Everything was taken into account, from the percentage of failure (20 to 25%) and the "psychological" impediments to the problems arising, in the end, from the uncooperativeness of the now sexually mature subject (1971a:822 n. 11).

Unlike the Gardners, the Premacks based their experiment on the idea that it would allow them above all to "better define the fundamental nature of language" (1972:92) and to answer "the essential question: what is language?" (1971a:808). They purposely assigned themselves "a list . . . of things an organism must be able to do in order to give evidence of language" (1971a:808). Unfortunately, their epistemological rigor ends there. On the one hand, their list contains eight features which seem to comply either with the demands of the Chomskyan universals (there are sentences, there are questions, there is a copula, there are complex sentences), or with everyone's conception of language (there are words, there are quantifiers and other parts of speech, etc.), or with the classical concerns of animal psychology (capacity for abstraction, capacity to function with conceptual classes and not only concrete single referents). On the other hand, they add other features, negative and positive, deriving either (apparently) from Chomsky, from Hockett, or from common sense. Besides, they themselves state (1971a: 809) that "this list [of eight features] is in no sense exhaustive, nor are the items on it of comparable [defining] logical order." Yet, they wish to distinguish between "language as a general system and the particular form this system takes in its use by man" (1971a: 822 n. 1), without realizing that it would be far safer to avoid this a priori polysemy of the word *language,* which has so confused the philosophy of language. It would be better to confine the term *language* to the particular system of man, which is already well defined, and speak of a "system of communication" whenever it has not been established whether a system contains all the defining features of human natural languages. They use the term "communication systems" on one occasion (1972:92), but they do not confine themselves to it. Yet, they are conscious of the risk of trapping themselves in the age-old vicious circle: only men have language, and language is the system of communication proper to man. They also write: "Because man is the only creature with natural language, we tend to assign a definitional weight to every aspect of human language. Yet it is equally reasonable to suppose that only certain features are critical, while others are secondary and should not be given definitional weight" (1971a:822 n. 1). It is regrettable that, having stated this, they lose no time in doing what they have specifically warned us not to do, leaving undone what should be done.

As for Washoe, I have already said that it would be scientifically unwise to consider the noun, the adjective, the pronoun, the verb, and the quantifiers as defining features of language. (The quantifiers are interesting from the point of view of animal psychology, but, rightly, neither Hockett nor Greenberg names them among his universals.) It seems just as unreasonable to refer to them as words, given the great inaccuracy of the term in general linguistics. Like Hockett, I think it advisable to speak of minimal signifier units of a discrete nature, i.e., that can be segmented in a stable manner.

Most Chomskyan universals are questionable if one seeks to establish them as criteria for language. For instance, Hockett does not consider syntax or, in particular, the existence of a functional order in the signs of the utterance as one of his 16 design features, although they constitute a feature of all known languages; one can conceive of languages in which all the functional relations between the units would be ensured by the semantics of those units (cf. François 1968:249). The Premacks went to a great deal of trouble to teach Sarah the usage of the verb *to be,* which is far from a necessary universal: many languages, such as Latin, Russian, and Arabic, do not have a copula for the present tense. The same can be said of the plural, as surprising as this may appear to the common sense of those who speak an Indo-European language. Concerning numerical concordance, the Premacks were uneasy about excluding the following utterance from their procedure: "red" + "green" + "to be" + "plural sign" + "color" + "plural sign." Specifying that "the restriction is temporary" (1971a:816), they did not for a moment suspect that the concordance of certain classes of units is not universal (the average Frenchman, in his philosophy of language, is always amazed at the weakness of the English language, where the adjective is not in concord with the noun). Similarly, because English does not make use of a particle to express the dative case for "Mary gives Sarah chocolate," the Premacks elude the problem which would certainly have worried them (without reason), had they been French, in the utterance "Mary gives apple *to* Sarah."

The absence of the genitive case in Sarah's code also inconveniences them: they regret that certain of their units mean "color-of," "name-of," and not "color" + "of" (with a specific sign). But they are indifferent to the fact that the sign that they translate "insert" in fact means "to put + in" (a good example of a case in which "syntax" is ensured by semantics: the order is unimportant, as one cannot put the pail into the apple). The Premacks also seriously examine the problem of introducing the concepts *animate ~ inanimate*, certainly not defining features of all languages. These examples show at one and the same time the burden resting upon the Premacks from the commonsense presuppositions about languages and the diffuse atmosphere of Chomskyan universals.

Indeed, the learning of the two types of complex sentences is very interesting from the standpoint of animal psychology. The fact that Sarah was capable of carrying out the instruction "Sarah" + "put-in" + "banana" + "pail" + "apple" + "plate" indicates that the abstract structure memorized by the ape was more complex than the linear structure of her sentence. Sarah did not establish an illegitimate functional relation between "pail" and "apple" because they were contiguous. She did, however, establish a correct one between "put-in" and "apple," which were not adjacent in the utterance. It is perhaps more interesting from the point of view of the psychology of intelligence that Sarah should have been capable of correctly interpreting "Sarah" + "to give" + "apple" + "Randy" + "implies" + "Mary" + "to give" + "chocolate" + "Sarah" (i.e., "If Sarah gives Randy an apple, then Mary gives Sarah a chocolate"). Even such complex sentences are not a universal necessary for defining language. The development of the relative sentence, in the families of languages that can be traced back over four or five thousand years, shows that it is a historical acquisition and not a given at the origin of all language. The Premacks are consequently wrong when they also concern themselves with obtaining, after /Mary/ + /to give/ + /apple/ + /banana/ (which is a paratactic structure), a sentence with the conjunction: /Mary/ + /to give/ + /apple/ + /and/ + /banana/. A formally marked conjunction is not a necessary universal.

An appeal to two of Hockett's criteria is very interesting: If one shows that Sarah's code contains all the criteria for language except "rapid fading," then one suggests that that feature is not, as Hockett professes, a defining feature of language (even though it is a physiopsychological feature important for the functioning of language). It is not the same story for another of Hockett's features: "there are no phonemes in the language" (Premack 1971a:809). If one agrees with Martinet that double articulation, in phonemes and monemes, is a main defining feature of language, then one has the scientific means to say how the code used by Sarah differs from a language. (Premack devised and described a code which would contain such duality of patterning, but did not use it [1971a:822 n. 4].)

Thus, generally speaking, whatever the attitude may be regarding the criteria for language, Premack deserves criticism of two kinds. First, he should have systematically tested the communication established by Sarah with every one of Hockett's 16 features, then with every one of the Chomskyan universals (a difficult task, as the list seems to be open), and last with every one of the features that philosophers and common belief assign to language properly speaking. Alternatively, he should have accounted for his own list, specifying the reasons he kept or rejected such-and-such a feature of the three known sets or added such-and-such new features of his choice. By proceeding the way he did, he depended both too much and too little on his predecessors, using them without rigor and as a result not being able to say in the end whether or not there was language. Above all, as was the case with the Gardners and Washoe, he committed a fundamental error of method by asking the question, in the first place, whether or not one could speak of language where Sarah was concerned. Here again, the fundamental questions remain: "Is there communication? What type of communication?"

By provisionally adopting our earlier definition of communication, we rediscover, in another form, the question "Is this communication or training?" Premack was conscious of this problem, but did not give it the main theoretical place he should have in the experiment: he considered it in the classical sense of "the clever horse." This has to do with cases of training carried out at the turn of the century on horses which seemed to obey stimuli of a linguistic nature, i.e., to carry out orders involving linguistic decoding of the utterances emitted ("Show me the letter T," etc.). As a matter of fact, the linguistic stimulus was not decoded as such, but rather associated with nonlinguistic signals revealed by the body or face of the instructor, although too tenuous to be perceived by the spectators. Premack approached this problem by using either an instructor wearing dark glasses (thus eliminating any indications stemming from the gaze), an absent instructor (Sarah would enter the cage where the board was kept only after the instructor had left, having written his message), or an instructor who had not learned the code (thus, transcribing the messages without knowing their meaning, simply by following instructions transmitted by way of numbers: /5/ = "put down the yellow square," etc.). With all these techniques, the score of achievements perceptibly decreased; this is worthy of notice. Another touchy point which does not seem to have preoccupied Premack much is the fairly steady percentage of failure (about 20%). At first glance, this differs a great deal from the human learning process, perhaps because of the instability of attention, the fragility of the motivation, and the superficiality of the social relationships.

The results of the Washoe and Sarah experiments are so fascinating in their technical perfection, and so essential as to the conclusions that can be drawn from them, assuming that they are what they appear to be, that I readily understand the hesitation of ethologists and psychologists in declaring themselves. Many of them, having studied the publications and films on the subject, have expressed the conviction that the experiments once again demonstrate training, but up until now they have been unable logically to formulate their argumentation. As for the linguists and semiologists, there was never any question of underestimating or underevaluating the importance of these two experiments. I believe that they will remain a great moment in 20th-century ethology and animal psychology. Still, as a linguist and semiologist, I am not insensitive to the worries or dissatisfactions that confront the psychologist in explaining them.

For instance, the fact that all the interpretations of the meaning of the experiments are based on linguistic translations of the messages exchanged (all of which are in English) may create the illusion that the systems of communication with Washoe contain nouns, adjectives, adverbs, and even pronouns because English translations contain them. Having tried, in the last three years, to analyse the Sarah experiment, I found most worrisome the following: could the illusion of language come from the fact that the coded messages are always "paraphrased" in English (cf. Premack 1971a:809, 811, where this word is employed)? Could the "meanings" of all Sarah's messages not be reduced to two: (a) "If I put two, three, . . . eight well-defined plates in such an order, I will receive a fruit, a chocolate, or a candy, etc." (b) "If I put two, three, . . . eight well-defined plates in any order other than the first, I will not receive a fruit, a chocolate, or a candy, etc."? Thus, we would be confronted with a very complex case of training, containing a number of exercises—but it would still be a case of training.

David Premack is too good an experimenter and psychologist not to have sensed this objection, although he never phrases it quite in this way. He asks himself, for example, "what the sign /to be/ really means to Sarah" (1971a:815), and he admits that "we cannot know what the connectives [*and, or, if . . . then*] really mean to her" (1971a:822 n. 13). He sets this problem in the light of traditional philosophy: since language is the expression of thought, "was Sarah able to think in the plastic-word language?" (1972:97).

Phrasing the question in this manner—which is neither linguistic nor semiological—enables him to answer in equally traditional philosophical terms, although he admits that "additional research is needed before we shall have definitive answers" (1972:97). The essential part of this answer lies in that Sarah has a capacity for abstraction, i.e., that she can classify different objects from one feature that they have in common (/red/, /small/, /square/, /fruit/, etc.). Animal psychology has been aware of this for a very long time; as early as 1943, Buyssens quoted Verlaine's experiments, which date back to 1928, and those of Koehler on crows, squirrels, parrots, etc., which had confirmed it. The experiment with Sarah contributes the outstanding demonstration that an ape can acquire an abstract structure of utterances filled in by units classified in paradigms. Premack also observed that Sarah could symbolize: she was able to relate the referent APPLE to its sign /purple triangle/, etc., she knew how to use the signs when the referents were not supplied, and she was capable of analysing abstract features (/red/, /round/, etc.) which characterize the referent APPLE by working from the sign /purple triangle/ (1971a:820). In 1972, the only missing element was that there was no proof "that when she is given the word 'apple' and no apple is present, she can think 'apple,' that is, mentally represent the meaning of the word to herself" (1972:97). The results achieved, says Premack, enable him to answer categorically the question of whether Sarah is able to symbolize, and therefore think (1971a:820; 1972:97).

Here again one can discern an error of logic. On the one hand, Premack provides eight defining features of language; on the other hand, he presents another definition of language, along the way: "To think with language requires being able to generate the meaning of words in the absence of their external representation" (1972:97). This is a criterion for language, but language remains traditionally defined as "expression of thought."

Now, Premack's own experiment and the analysis of its results allow us to phrase the question more precisely: Do the ability to abstract and to symbolize, on the one hand, and the ability to communicate (eventually by means of language), on the other, stand for two distinct kinds of learning, most likely different in such a way that the first does not automatically imply the second? The traditional philosophy of language, by defining language as the activity of expressing thought, concentrated on the first kind of learning (to abstract and to symbolize, both considered thinking operations) and did not examine the second (to communicate).

This brings us to the problem of how we are to demonstrate—without first asking if one can speak of language—whether communication with Sarah does exist. What has been said of Washoe thus takes on new importance. That Sarah used a copula is much less important than this other fact which does not retain Premack's attention: following the experimental sessions, Sarah, all alone in her cage (outside any experimental situation), picked up objects or signs and composed utterances on the models of the structures that she had just learned (1971a:810). Can one discern the transition of the main and primary function of her code, social communication, to a secondary use of it, the possibility of developing for oneself the expression of one's own view of the world? Or does this expression only represent play? Likewise, that Sarah apparently knew how to handle the conjunction *and* seems less important than this other fact: Sarah could be driven to describe the instructor's behavior instead of responding to his utterance. For example, she produced /Randy/ + /to put/ + /yellow/ + /on/ + /red/ instead of carrying out the order to place the /yellow/ sign on the /red/ sign. This could mean the beginning of an activity in which Sarah could act as the initial transmitter.

In October 1972, Sarah did not yet ask questions. If she had, it would have been an outstanding fact, not because interrogative transformation is a universal, but because it would have shown that communication with Sarah could really function both ways, with the ape acting as the initial transmitter (without the influence of an experimenter). Premack did not realize the crucial importance of this fact. He merely observed, as a fact devoid of all interest, that "Sarah was never put in a situation that might induce . . . interrogation, because for our purposes it was easier to teach Sarah to answer questions" (1972:95; cf. 1971a:822).

All this leads me to conclude that, in spite of the great interest in this experiment and in its teachings, the question (preliminary to all others) of what defines the type of communication that exists between man and Sarah has not yet been correctly answered, in part because it has not been properly asked. One may still hesitate—for want of being able to prove that Sarah operates as an initial transmitter—to say whether or not it is simply a question of training. Regarding this, certain of Washoe's behavior patterns are more important, and seem to indicate that she is closer to the stage of human communication in its proper sense. If, like Premack, we conclude by saying that "Sarah had managed to learn a code, a simple language that nevertheless included some of the characteristic features of natural language" (1972:99), we conceal the fact that the defining value of these features has not been correctly evaluated, from the theoretical standpoint of what communication is, in the first place.

Regardless of the fact that Premack's experiment may be criticized in such a manner, it nevertheless offers precious theoretical confirmations to the linguist and semiologist. It shows that teaching a language consists simply in outwardly reflecting the articulation of the world of experience, i.e., the organization of knowledge. This justifies a linguistic train of thought extending from Saussure to Whorf (cf. Premack 1971a:810; 1972:95). But to verify that there is no way to acquire a meaning other than differentially—by opposing it to another meaning that can be commuted with it—is a good experimental demonstration in animal psychology of Saussurian semantics: on this point, Premack's considerations on the acquisition of *if . . .then, and,* and the copula are exemplary (1971a:815).

Semiological analysis suggests that the chief limitation of the Premack experiment lies, to begin with, in the *poor social relations* set up by the type of experimentation chosen. Life in captivity and experimental behaviorist conditioning do not stimulate relations that allow true communication, which in turn would initiate the use of and enrich the system of communication taught. The fact that Sarah never saw the humans surrounding her make use of her code can be interpreted in the same way, as Hewes sensed. The same can be said of Sarah's rapid loss of motivation in a situation of conditioning and of her visible loss of interest as soon as the instructor wearing the dark glasses, or the one who did not know her code, entered into the experiment. It is a theoretical error to consider reluctance to communicate as a simple "psychological" and contingent fact, marginal to the interpretation of the experiment. One can equally well say that between Washoe's "clinical" apprenticeship and Sarah's experimental apprenticeship there is the same type of communicational difference—and result—as there is between learning a foreign language in a foreign country, under the conditions of true communication, and learning that language by doing structural exercises in a language laboratory, under the most fictitious and abstract conditions of communication. It is not a philosophical whim to say that the primary and central function of language is to satisfy the need for social communication, but a statement of the phylogenetic and ontogenetic reality. From this point of view, the Washoe experiment, even though less satisfying and less rigorous, is more important than the Sarah experiment, mainly through contrast.

If something more can be learned from these experiments, it is that the ideological presuppositions of an era have an epistemological impact on research. When they state as an obvious fact that "man is the only creature with natural language," the Premacks confine themselves, without knowing it, to a vicious circle that prevents them from accurately stating even the main (preliminary) problems—those of the criteria for communication in general and for human linguistic communication in particular. They conclude by saying (cf. Premack 1971c) that even after their experiments and the meanings drawn from them, *man's uniqueness in matters of language* is preserved, since chimpanzees are unable to teach language to men. This ultimate argument shows that their presupposition was not just a manner of speaking inherited from the traditional philosophical way of posing the problems of language, but rather a preconceived ideological notion: they are not ready to leave the side of those who wish to maintain an ultimate defensive gap in the debate over the radical separation between man and ape. They do not feel free to accept whatever they may find—perhaps the existence of an uninterrupted chain of types of communication, becoming more and more complex, between animal and man.

WHAT MIGHT BE LEARNED FROM STUDYING LANGUAGE IN THE CHIMPANZEE? THE IMPORTANCE OF SYMBOLIZING ONESELF

H. S. TERRACE AND T. G. BEVER

What do you see when you turn out the light?
I can't tell you, but I know it's mine.

—Lennon and McCartney "With a Little Help from My Friends"

Ten years ago, there was little reason to believe that much could be learned from studying language in the chimpanzee. Earlier reports (Hayes 1951; Kellogg & Kellogg 1967) of a chimpanzee's inability to use language seemed to demonstrate that biological factors limited the extent to which the chimpanzee could learn to use abstract symbols. Thanks to the work of the Gardners (1969, 1971), Premack (1970a, 1971b), Fouts (1972b, 1975a,c), and Rumbaugh and Glasersfeld (1973), we now know a chimpanzee has a much larger linguistic potential than was ever imagined. Once allowance was made for the vocal limitations of a chimpanzee (Lieberman 1968; Lieberman, Klatt, & Wilson 1972), it became possible to teach it to use arbitrary gestures, plastic tokens, or the lexigrams of a computer console in a manner that parallels the human use of single symbols. Most provocative have been demonstrations that chimpanzees can use *sequences* of symbols. This suggests a syntactic potential as well as a symbolic one.

In the eagerness to establish whether a chimpanzee can acquire syntactic competence, certain more basic functions of language appear to have been bypassed. Our main purpose in this paper is not only to assess the syntactic accomplishments of Washoe, Sarah, and Lana, but also to delineate other functions of human language that can be studied in the chimpanzee—functions that do not require syntactic competence.

Before considering nonsyntactic functions of human language, it is of interest to digress briefly to compare studies of language in a chimpanzee with studies of

H. S. Terrace and T. G. Bever ● Department of Psychology, Columbia University, New York, New York 10027.

language in another currently popular subject, the computer. At present, there is nothing to suggest that computers can simulate the *acquistion* of language. However, language acquisition by another being may illuminate parallel processes in humans. As compared with the computer, the chimpanzee has many obvious advantages. First, because of its biological and social similarity to man, a chimpanzee is a much more likely source of useful contrastive information on human language. Second, a computer can do nothing more than execute the instructions of its programmer. We cannot be sure *what* a chimpanzee will do with symbols once it has learned their meanings.

Human language has two important functions: (1) it facilitates communication (particularly about events and objects displaced in time and/or space) and (2) it structures how we perceive ourselves and the world. The child's tendency to symbolize interpersonal relations between himself and others is obviously a fundamental ingredient of personality development in all human societies—in particular, the concept of *self* bounded on one side by desires and on the other by social patterns. A child's ability to refer to itself, its desires, and the social pressures of its environment requires little, if any, syntactic ability. Yet this basic function of language has profound effects. We suggest that the mastery of language to express feelings and to encode socially desirable and undesirable behaviors to oneself, may provide sources of motivation for advancing to more elaborate usages of language—usages that do require syntax. A necessary condition of human language may prove to be the ability to symbolize *oneself*. In our own research, we hope to see if this condition is sufficient for an already socialized chimpanzee by teaching it the concept and use of "self."

The assumptions from which our hypothesis derives can be formulated in comparative terms. *All the ingredients for human language are present in other species—they do not become language until an animal learns that it can refer to itself symbolically.* Many animals learn to respond to *symbols,* and to remember them (as in discrimination learning or delayed-response tasks). Many animals also exhibit *hierarchically organized behavior* patterns (e.g., as in Hullian "habit families" or in Tinbergen's analysis of stickleback courtship), and *transformations,* (e.g., shifting from one mode of locomotion to another in maneuvering through a maze, or "displacement reactions" of various animals). Many animals can also perceive and act on *relations* between themselves and other members of their species (e.g., dominance patterns, and courtship interactions). Finally, at least a chimpanzee (and perhaps other species as well) demonstrate a concept of "self" by their ability to deal with their own mirror image (Gallup 1970).

Our hypothesis is that the potential to attach a symbol to *oneself* is an important factor in the recruitment of preexisting potentials for using symbols, hierarchies, and transformations into "language." The motive for his is clear: once we can think about ourselves symbolically, we can think abstractly about our relations to others and to the world. A basic function of syntax is to express the internal relations between different symbols. This, in turn, facilitates the representation and exploration of relations in the outside world. Accordingly, in the present discussion, we consider how language can be used to demonstrate a concept of symbolized self in a chimpanzee before evaluating the extent to which its utterances have a syntactic structure.

One indicator of a symbolic self in a symbol-learning chimpanzee would be specific reference to its own emotions and feelings. As a creature prone to emotional expression of fear, anger, happiness, frustration, like, dislike, sadness, and so on, it may prove possible to teach a chimpanzee to name those states. As far as we can tell, neither Premack nor Rumbaugh has attempted specifically to teach their chimpanzees words describing emotional states. While the Gardners reported that words such as *funny* and *sorry* were part of Washoe's vocabulary, it is not clear that either of these words was used to refer to internal as opposed to external stimuli.

For example, in the Gardners' glossary of Washoe's vocabulary, "sorry" was defined as an "apology, appeasement and comforting; usually the response to *Ask pardon*" (1975b:266). Although Nim, the chimpanzee we are teaching to communicate via sign language, has yet to learn the sign "sorry," it is clear that he has learned to seek reassurance after having done something wrong (such as touching a forbidden object, or biting too hard). On some occasions, Nim will sign "hug" in order to obtain reassurance. From the Gardners' description of Washoe's usage of "sorry," and from our own experience with Nim's attempts to make up by signing *hug*, it appears as if these signs may function as requests for reassurance rather than as descriptions of an emotional state.

As the psychological literature on attribution clearly shows, description of emotional states often entails reference to external and internals states (Bem 1972; Heider 1958; Kelley 1967; Schachter & Singer 1962). It is plausible that chimpanzees can learn to identify emotional expression, as might be portrayed in pictures. It has been shown that monkeys can learn to discriminate between the expression of different emotional states, as shown on video displays (Miller, Banks, & Ogawa 1962). The Gardners (1975) and Fouts (1975c) have demonstrated that a chimpanzee takes well to the task of naming various pictures with appropriate signs. Thus, learning to label emotional displays may not prove difficult for chimpanzees.

Symbolic discrimination of emotional expressions exhibited by others is easier to demonstrate than reference to one's own emotions. Although we have no basis at present for asserting that a chimpanzee would engage in the kind of introspection that is entailed in a description of its own feelings or emotions, we find this possibility intriguing. Introspection of any kind requires a symbolic consciousness of self that has yet to be clearly demonstrated in chimpanzees.

Language plays a vital role in a child's incorporation of the concepts and values of its environment. Although this may seem obvious, its significance does not appear to have been perceived in previous studies of language in a chimpanzee. An unrecognized obstacle in the path of fully simulating the syntactic features of human language in a chimpanzee máy be the absence of the motivation experienced by a child who learns language both to please its parents and to represent its world. Chimpanzees such as Washoe, Sarah, and Lana have obviously been motivated to learn *something*. The question remains, however, whether their motivation and learning have been intrinsically linked to socialization itself.

How human children and chimpanzees learn the names of objects provides an interesting basis for comparing the motivation in each case. Consider, for example, Lana's communication to the computer "what name of this?" when she wanted to learn the name of a box in order to obtain the box, which contained candy. Quite obviously this is an impressive example of a chimpanzee's having learned that things have names. In the case of children, it is commonplace that they ask for the names of objects in their environments. But in the only example we know of in which a chimpanzee asked for the name of something it is clear that learning the name of the box was in the service of obtaining the box and its contents. On the other hand, a child will persist in asking "what that?" simply to learn the names of objects without any obvious extrinsic consequences. The child seems to be acquiring the names of things to use in social discourse, or in the mastery of its world—a motive fundamentally different from that of obtaining the object in question itself.

With these considerations in mind, it is instructive to consider the three major recent studies of language in the chimpanzee. One purpose is to evaluate the extent to which syntactic competence has been demonstrated in the chimpanzee; it is also necessary to evaluate each chimpanzee's socialization and its motivation for using language. We consider first the cage-reared chimpanzees, Sarah and Lana.

PROJECT SARAH

The initial strategy Premack followed in teaching Sarah to use a symbol system, in which words were represented by plastic chips of different shapes and colors, was to utilize behavioral techniques for demonstrating a list of so-called "exemplars" of human language. These included (1) words, (2) sentences, (3) questions, (4) metalinguistics (using symbols to talk about symbols). (5) class concepts such as *color, shape,* and *size,* (6) the copula ("is"), (7) quantifiers such as *all, none, one,* and *several,* and (8) the logical connectors *if* and *then.* As Premack notes, "This list is in no sense exhaustive, nor are the items of comparable logical order" (Premack 1970a:107). In demonstrating a particular exemplar, simple training procedures were used that presumably mapped linguistic knowledge onto discriminations that Sarah had already mastered. For example, the word "apple" was trained by requiring Sarah to offer the word apple (a blue plastic triangle) in exchange for an actual apple. The interrogative was trained by placing the interrogative symbol between objects that are either the same or different. In this case, Sarah is required to replace the interrogative symbol with the appropriate symbol, one that means "same" or one that means "different."

Most of Sarah's behavior appears to have been motivated initially by food reward. Premack (1971b:200) does indicate that "one one occasion, . . . she stole the test material before the lesson, as she has done from time to time, and went on both to produce many of the questions we had taught her, as well as to answer them." Furthermore, after many months of intensive training, Sarah could learn a new symbol–object relationship simply by observing an experimenter pair them together. In these instances, Sarah appears to have been manipulating the materials without the support of an extrinsic incentive. But, as far as can be gleaned from Premack's description of his experimental procedures, initial training was motivated by food reward, as were actual testing sessions.

The solution of the problems presented to Sarah usually required the simple substitution of one symbol for another. Within a particular training set, most of the problems required that Sarah choose only between two alternatives. For example, if she was tested on her ability to identify objects that were the same or different through her use of the symbols "same" and "different," these were her only choices. In other problems, requiring the naming of colors, symbols for numerous colors were provided. However, it was still the case that the only alternatives from which Sarah could choose were symbols designating different colors, i.e., of the same class of answers.

During a typical day, Sarah was given as many as four brief sessions, each containing about 20 problems. All the problems in a given session were of the same type. Thus, one session might concentrate on same-different problems, another on color-of-problems, another on shape-of problems, and so on. Accordingly, it is an open question whether Sarah's usage of these exemplars constitutes a simulation of human language. Training procedures that allow for only one type of problem could establish a *set* for that type of problem and accordingly, may not elicit a *natural* use of language. A related limitation of the results is that during any session the set of symbols that define the answers to that problem (often just two) were all of the same linguistic category. This also is not representative of a human language.

Of course, the most important question is whether Sarah shows evidence of having mastered structural rules governing sequences. The strongest argument that Sarah is sensitive to the structure of sentences comes from the demonstration that she could respond appropriately to a sentence such as "Sarah insert banana pail apple dish." In this situation, Sarah was required to place an apple and a banana in the appropriate container. In order to demonstrate Sarah's understanding of this sentence, she was provided with a choice of fruit and a choice of containers. Premack argues (1971b:215) that the organization of the sentence can be shown to be hierarchical. "Banana and *pail* go together, likewise *apple* and *dish* . . . *insert* not only applies to *banana* and *pail* but to both cases, and finally, . . . it is Sarah who is to carry out the whole actions." Thus, on this interpretation the response could not occur as a result of simple response-chaining.

This interpretation is uncertain for a number of reasons. Typically, Sarah was tested on only one kind of problem for an entire session. Thus the first two symbols (*Sarah* and *insert*) are redundant. Unless *insert* is contrasted with some other verb, it is not clear how one could demonstrate its hierarchical status in the "sentence." Even though Sarah responded appropriately when "withdraw" was contrasted with "insert," one wonders whether Sarah needs to understand the symbol "insert" in order to solve the problem. If the pail and dish were empty before the problem was given, then she need only pay attention to the extralinguistic cue provided by the condition of the container in front of her.

At the very least, we must see many more variations of the sentence "Sarah insert apple pail banana dish," along with appropriate behaviors, before concluding that Sarah understood the hierarchical structure of a sentence. For example: "Sarah withdraw apple banana pail insert peach can apple box." "Sarah wash apple red dish insert banana round dish."

Sarah achieved impressive abilities to deal with symbols as such. She clearly *learned to learn* new symbols—i.e., she discovered that things can have a *name*. Premack has also shown that Sarah has the cognitive ability to solve problems requiring the use of symbols in a metalinguistic manner. For example, Sarah could respond correctly to questions concerning the shape or the color of an apple when all she was shown was the symbol for an apple—a blue triangle. She could *also* answer questions about the plastic symbol itself. However, she did not show unequivocal evidence of the use of sentence structure, nor did her other usages of symbols occur in the varied contexts that characterize human language.

PROJECT LANA

However we interpret Sarah's achievements in relation to human language, they demonstrate convincingly the chimpanzee's symbolic potential. However, a frequent criticism of Project Sarah is that Premack and his trainers were transmitting nonlinguistic cues during the training procedures that shaped her responses (cf. Schachter and Singer 1962). Of course, it is just the personal interchanges required by Premack's technique which gives it a naturalistic relation to human-language learning. In order to test for this so-called "Clever Hans" (or Clever Gretel) phenomenon, Premack employed a naive tester who was unfamiliar with the meaning of the symbols used in Sarah's training. Here was a deficit in the level of Sarah's performance. It was, however, still significantly greater than chance. Furthermore, such a performance drop with a strange tester has many interpretations: we might expect to find a similar decrement in a human child's performance in a similar situation. It is reasonable to argue that personal cues in Sarah's interactions may only have strengthened her understanding of the linguistic situation, as it would a child's. Nobody *tricked* Sarah into learning what a symbol is, which we find to be her most impressive skill.

Of course there might be a trivial Clever Hans effect in which Sarah was responding each time to an unconscious covert personal cue that was functionally the same for each trainer and tester. Unlikely as this seems, Rumbaugh and his colleagues (Rumbaugh and Glasersfeld 1973) are employing a computer as an interactive device to avoid the possibility of linguistically irrelevant cues. This not only minimizes the possibility that the subject could benefit from cues in the physcial proximity of the trainer, it also provides the possibility of a complete record of all symbolic transactions. Instead of plastic symbols, Rumbaugh's experiment with a chimpanzee, Lana, employs lexigrams, each of which appears on the keyboard of a computer console. The functions of these lexigrams are similar to that of the plastic chips used by Premack.

One major difference between Lana's lexigrams and Sarah's plastic chips is that the lexigrams and the computer are available just about all the time. Thus, Lana has continuous access to a much larger set of alternative symbols than did Sarah. At the same time, many of Lana's performances have been trained and practiced in long sessions devoted to just one type of problem. As we indicated earlier, this raises problems of *set* and is a basic departure from what happens in the case of a natural language. Indeed, a general cost of using the computer is the reduction of the social interactions that ordinarily provide the context for the use of symbols.

Rumbaugh's initial report on Lana referred to her ability to "read and write" in "sentence completion" tasks (Rumbaugh and Glasersfeld 1973) that employed the computer's "language." If this was all that Lana had accomplished, it would seem gratuitous to refer to this as "linguistic" behavior. (Indeed, Rumbaugh and his colleagues are appropriately cautious about this—see Glasersfeld 1976.) For example, the sequence they gloss as "Please machine give Lana M & M" or "Please machine give Lana piece of raisin" or "Please machine show Lana movie" can be analyzed as a complex X-R chain. In these sentences, the sentence beginning "Please machine" is completely redundant and obviously requires no understanding of what the term allegedly stands for. What Lana has to learn is simply to associate the symbol "give" with the lexigram (secondary reinforcer?) for such inventives as an M & M, raisin, or apple, and to associate the symbol "show" with the lexigram for such incentives as movie or slide shows. At best, Lana's abilitiy to discriminate between valid and invalid sentence beginnings and to supply appropriate endings for valid sentence beginnings appears to be a weak demonstration of a finite state of grammar, in which possible successive elements of a sentence are determined completely by the immediately prior response. Such two-step chained associations are not hard to demonstrate in many animals. Indeed, this behavior is the formal and pragmatic equivalent of that of a rat in a double T maze.

An important potential of Rumbaugh's approach is inherent to the continuous record it provides of all transactions between Lana and her trainers. It should be possible to have a complete corpus of all Lana's utterances as well as of all the questions that have been put to her. This advantage should, however, be weighed against a certain unnaturalness of the procedures used to train Lana. This includes the fact that the language is artificial, there is minimal socialization, and the fact that all Lana's training requires the constant support of primary reinforcement. It is also not entirely clear that the expensive utilization of the computer has entirely circumvented the Clever Hans possibility, at least when the computer is used as the basis for "conversing" with trainers. The trainers can vary the time, rate, and choice of presentation, which leaves open the possibility that Lana's performance is still being shaped by uncontrolled factors (which often appear to be unrecorded), e.g., Lana's cage position, her drive state, the trainer's current assessment of her position and state, and so on. Since many of her most striking "utterances" occur with a trainer present, this uncertainty is particularly poignant. However, like Sarah, Lana appears to have mastered the notion that objects can be "named," and that she can ask for the name of a new object if she wants it.

As far as we can tell, a complete corpus of Lana's symbolic sequences has yet to be published. Thus, there is no way of evaluating the novelty of particular sequences (which one would assume are drawn from the best examples). Even in the "showcase" examples, it was apparent that Lana can err in generating a sequence. Until one knows the nature and frequency of her errors, the significance of her "sentences," novel or otherwise, remains unknown.

Despite these limitations, Lana's achievements are considerable, and the potential suggested by the computer technique is substantial. Like Sarah, Lana has clearly learned that objects have names. Indeed, Lana appears to have gone one step further in that she has been observed to ask for the name (i.e., lexigram) of new objects (at least when she desires them).

PROJECT WASHOE

Unlike Sarah and Lana, Washoe was reared in a highly socialized environment—an environment that contained many of the features of the social environment of a human child. The same procedures are being used in the Gardners' current study of two infant chimpanzees, Moja and Pili (1975a). As in Project Washoe, the medium of communication in these studies is Ameslan, a language used by the deaf. The obvious social advantage of this natural language is that both the chimpanzee and its human companions can sign to each other spontaneously, without the support of extraneous devices such as plastic chips, or computer lexigrams. The obvious comparative advantage is that sign is a "natural" human language, perhaps with a semantics as rich and a grammar as complex as in any spoken lanaguage (see Stokoe 1960; Bellugi and Klima 1975b). In our opinion, socialization and the use of a natural language make the approach of the Gardners conducive to the development of language skills in a chimpanzee that are most comparable to those of a human child.

For this reason, our project uses procedures similar to those used by the Gardners. In some respects, our procedures for socializing our chimpanzee Nim are more intensive than those used by the Gardners. Unlike Washoe, Moja and Pili, Nim lives together with his caretakers and eats at least one meal a day with them. It is our feeling that a high degree of socialization is not only important in that it provides many opportunities for signing, but also in that it may increase Nim's self-control. Additionally, a high degree of socialization should also prove helpful in establishing a linguistic concept of self, if such is possible.

It is too early to report decisive linguistic consequences of our efforts to socialize Nim as one might a child. At 22 months, Nim has an active vocabulary of about 30 signs; his passive vocabulary is somewhere between 50 and 60 signs. Early this year, Nim made his first combination of two signs, and at the end of the summer he was observed to make combinations of three and four signs. We have yet to perform tests to determine whether Nim's multisign combinations show evidence of syntactic structure. In order to do so, we feel that analyses more stringent than those provided by the Gardners will be necessary.

It is mainly on this point that we part company with the Gardners. It is not sufficient simply to compare the performance of a child to that of a chimpanzee. For example, in a recent paper the Gardners note that "the failure of linguists and psycholinguists to devise a behavioral definition of language is an obstacle that we try to avoid by obtaining observations of the acquisition of sign language by young chimpanzees that can be compared with observations of the acquisition of spoken language and sign languages by human children. Any theoretical criteria that can be applied to the early utterances of children can also be applied to the early utterances of chimpanzees. If children can be said to have acquired language on the basis of their performance, then chimpanzees can be said to have acquired language to the extent that their performance matches that of children" (1975b:244).

To date, the Gardners have shown that semantic interpretations of two-sign utterances (irrespective of order) match those that have been found in children. In a recent publication, they also argue that Washoe responded to wh-questions in a manner similar to that of a Stage III child. In neither of these demonstrations, however, do the descriptions of performance derive from a model of the linguistic structure necessary to capture the complexity of all the children's utterances (cf. Stokoe 1960; Bellugi and Klima 1975b; McNeill 1970)—a complexity that requires structural analysis, not merely behavioral criteria. To use an analogy, the fact that a horse can walk on its hind legs does not prove that it has the walking capacity of a two-year-old child.

CONCLUSION

To summarize, chimpanzees have proved themselves to be an important source of information regarding the development of language. To date, there is no demonstration that the chimpanzees have learned to generate strictly ordered syntactic sequences. Rather, they appear to "speak" by combining semantically related symbols, in most cases, according to preset schemata. The difficulty in proving that they use syntax lies partly in the training procedures used to produce evidence of syntax. If one wants to maximize the chances of producing an approximation of human language, it will not suffice to drill a chimpanzee in separate tasks that represent certain features of human language. It is also necessary to show that these features of language occur under conditions similar to those that obtain for humans. At minimum, linguistic ability should not be specific to particular training situations and the use of language should not be solely in the service of acquiring an external object. We should also like to emphasize that a concept of self whereby a chimpanzee is able to conceptualize its feelings, intentions, and so on, in relation to other individuals in its environment, may be a crucial step in motivating the chimpanzee to acquire the snytactic competence characteristic of human language. As far as we can tell, neither Sarah nor Lana developed a linguistic concept of self; the evidence in Washoe is at best equivocal.

Just as there is no clear answer to when a child "knows" language in a syntactic sense, there is no clear answer to this question in the case of a chimpanzee. Indeed, our willingness to attribute syntactic competence to a young child is often based upon little more than an extrapolation backwards from the anticipated state of adult affairs. Unfortunately for the chimpanzee, this type of extrapolation is not yet possible. Accordingly, evidence that a chimpanzee can generate and understand sentences, as humans do, will necessitate comprehensive demonstrations that a chimpanzee cannot only string symbols together, but that such sequences reveal syntactic structures and that they function as they do in human languages. A single demonstration would fall short of the mark just as it would in the case of human children. A convincing case that a chimpanzee uses language in a human manner will require numerous demonstrations that lengthy sequences are used systematically in order to enable the chimpanzee to describe something about itself in relation to others, to deceive others, to contrast ownership of objects, to describe relations in space and time, and so on. In short, if you want to teach a chimpanzee to use language as humans do, make sure it socializes with humans, and make sure that it uses language to refer systematically to itself in relation to its human companions.

Of course, this leaves open the possibility that any training procedures with chimpanzees will elicit only a "natural" *chimpanzee* language that differs from human language because of intrinsic differences in chimpanzee cognition. To test for this possibility it is also true that a great variety of constructions and symbols will be required to delineate any difference between "human" and such a potential "chimpanzee" language.

It is worth elaborating the notion of "species specific natural language" at this point. In this view, the capacity for symbols cannot be *taught,* but can be trained if the species has the potential for them. It is entirely possible that every primate species has the capacity for some symbolic behavior, but that the effects of learning symbols differ according to the cognitive context of the species. There may be many kinds of language, just as there may be many covert representations and organizations of the world. This makes irrelevant the question of how similar to human language a primate symbol system is. Rather, the question becomes: what is the structure of the primate language itself, and how does it reveal how a primate manipulates symbols? Whereas it is of interest to see to what extent a chimpanzee can use language as do humans, it is of equal interest to see whether they may seek to use language in their own manner.

This view might seem inconsistent with our emphasis on the acquisition of the linguistic concept of "self." After all, if an animal's cognition determines the kind of language it can have, how could an autonomous linguistic development have general effects? In our view, symbols provide an important mode of explicit internal representation of the world, a kind of "private scratchpad." Once such a facility is developed, it may *release* a variety of otherwise latent capacities, even if it does not *cause* them. Our contention is that the symbolization of "self" is such a releaser.

Of course, if we succeed in eliciting a true human language in a chimpanzee, we will face Kafka's question—*Have we taught a chimpanzee to talk or have we released a human being?*

CHIMPANZEES AND LANGUAGE EVOLUTION

WILLIAM A. MALMI

Among the most remarkable developments of recent years has been the opening of a "dialogue" between human and chimpanzee (Fouts 1973; Gardner & Gardner 1969; Premack & Premack 1972; Rumbaugh, Gill, & Glasersfeld 1973). The relevance of chimpanzee performance to human language evolution is controversial, but may be somewhat easier to assess if the debate is framed in explicitly evolutionary terms so that behavioral reconstruction may follow established biological principles. In this pursuit, several lines of research are implied that may refine, if not resolve, some of the questions being asked.

Whether or not *Washoe* (*Washoe* is used herein as representative of all the "linguistic" apes) is using "language" is not particularly useful in this context. The word is used so variously that any consensus is unlikely, a difficulty that is compounded if the dimension of time is introduced: given either a continuum of change or a succession of stages in the transition from chimp-like antecedent to speaking human, where is the linguistic Rubicon to be placed? Moreover, most definitions of language rely on the presence or absence of design features (Altmann 1967; Hockett & Ascher 1964), so that whether a particular output is linguistic is determined by whether or not it involves *syntax, displacement* (Bronowski & Bellugi, this volume), *separation from affect* (Brown, this volume), or some such trait. The result has been that features are proposed which are presumed unique and crucial to human language; these are taught to apes; new features are proposed. But design features are not useful for establishing evolutionary continuity (Lenneberg 1967). Also, it is as unrealistic to expect chimpanzees to emulate all the features of human language as it had been previously to deny that they could learn some of them. The important issue is not whether or not they can learn all the "crucial" aspects of human language, or even what assemblage of traits is to be deemed sufficient criterion of language, but what is the relationship of the performance which has already been demonstrated to language evolution? Does *Washoe* evince language-like behavior because she is employing neural/cognitive systems similar to those underlying human lanaguge as a consequence of common origin, or are the structural similarities superficial and dependent largely on training procedures?

WILLIAM A. MALMI • Department of Anthropology, University of California, Concord, California 94520.

In biological terms, the issue is one of *homology* versus *analogy* (Mayr 1969; Simpson 1961). One of the reasons the design-feature approach is not useful in assessing homology is that traits of common origin may diverge markedly over a short time, whereas traits of different origins may converge so as to appear virtually identical: "homologous structures may be extensively similar or dissimilar. Therefore, similarity is to be considered something quite apart from considerations of homology" (Campbell & Hodos 1970:101).

Before examining ways of assessing homology, it is well to consider briefly some aspects of human and nonhuman cognition and communication in relation to homology. In the first place, presumed phylogenetic closeness is not sufficient to establish behavioral homology, especially when the behavior does not naturally occur in one of the species but is elicited only through extensive training. The point is worth making because, whereas the chimpanzee researchers have, especially more recently, recognized this to be the case (perhaps more so than most of their critics), they sometimes write as though it is not. Fouts, for example, after a good discussion of homology, including the important suggestion that the neural basis of the behavior must ultimately be investigated, goes on to say that "the extreme physiological, behavioral and evolutionary closeness of Man to chimpanzee should actually lead one to conclude that a very similar biological mechanism is operating at the basis of language behavior in the respective species" (Fouts 1974:476). Rumbaugh and Gill (1976a), although careful not to overstate the significance of Lana's performance, cite a recent article by King and Wilson (1975) which indicates 99% genetic similarity between chimpanzee and human, and which to the lay reader may give a false impression of physiological closeness.

The most recent date postulated for the separation of human and ape lines is about 5 million years ago (Sarich 1968). This date, based on biochemical comparisons, is extremely controversial, since most other researchers place the divergence at a minimum of 15–20 million years ago (Poirier 1973). If correct, it would imply a much closer relationship between man and chimpanzee than previously supposed, but it should be remembered that evolutionary rates vary considerably, and that in mosaic evolution some traits are rather conservative, whereas others change rapidly. Human evolution has involved especially dramatic shifts in diet, locomotion, technology, social organization, communication, and the brain. This means that whether apes and humans are remarkably similar depends on what is being compared. In fact, the point of the King–Wilson article is to reconcile the observed genetic similarity with the known morphological and behavioral differences: "Most important for the future study of human evolution would be the demonstration of differences in the timing of gene expression during development, particularly during the development of adaptively crucial systems such as the brain" (1975:114). The 99% similarity in genotype thus reflects recent time of separation, whereas the 1% dissimilarity includes some phenotypically important mutations (the 1%, incidentally, represents at least 40 million base-pair differences).

It is important also to keep in mind that "cognition," like "intelligence," is not just a rather generalized means of coping that improves as we move up the phylogenetic scale; it consists of a number of capacities, some of which humans share with other species, others of which have been sufficiently modified in hominid evolution to be considered species-specific for humans. Language is undoubtedly based on both kinds of cognitive capacity. In emphasizing the similarities between chimpanzee and human linguistic capacities, it must be realized that, unless the nature and origins of the differences can be ascertained, the process of reconstruction can hardly move forward from our chimp-like antecedents. In comparing the formal aspects of simian and human "language" there should be some attempt to distinguish between similarities resulting from training procedures, from features of mammalian cognition, and from capacities more directly relevant to the events of the last million years or so.

Humans are natural language users, but chimpanzees are not. It has, of course, been suggested that chimps do use language in the wild, but that it is too subtle and too different from human speech to be recognized as such. There is, however, no evidence for this, despite close observation of vocalizations, facial expressions, postures, and gestures over thousands of hours by a number of trained observers. It may also be asked why such a system would be so different from our own if it is assumed that chimpanzees and humans are remarkably similar. The natural communication of chimpanzees (Goodall 1965; Reynolds & Reynolds 1965; van Lawick-Goodall 1968a) appears typical of nonhuman primate systems (Marler 1965; Ploog & Melnechuk 1969), and is very different from human language. In fact, it is beginning to appear that nonhuman primate communication is characteristically mammalian (Eisenberg 1973; Moynihan 1970) rather than a novel development, so that the ape, like the monkey, expresses itself in ways that have more in common with a dog or cat than a human language user. The conservatism of mammalian signal systems is remarkable in the face of considerable morphological, socioecological, and cognitive diversity, and serves to underscore the uniqueness of human expression. Comparative neurological work suggests that both animal communication and human nonlinguistic expression are mediated largely by the primitive limbic structures of the brain (Lancaster 1968; Ploog 1967; Ploog & Maurus 1973), whereas language derives largely from reorganization of neocortex (Geschwind 1965; Lancaster 1968; Robinson 1972). Unfortunately, most of this work has been done on humans and a few species of monkey, and there is little information on the substrates of ape communication, although research on the gibbon (Apfelbach 1972) suggests that their call system is organized similarly to monkey vocalization.

On the other hand, it is not surprising that there is evidence (Davenport 1976; Menzel & Johnson 1976) that apes exhibit cognitive/intellectual capacities not shared with monkeys, and that in some respects these are more "humanoid" skills. It would be reasonable to assume that capacities like some of these, in the early hominids, changed in ways that ultimately led to language and speech. Can these aspects of cognition be isolated? What have they evolved for in chimpanzee behavior? After all, preadaptations for a behavior (in this case, language) will not be maintained in a population unless they confer some immediate benefit. If the normal chimpanzee behavior for which such skills have evolved could be demonstrated, it might provide a clue to the origins of human language: greeting? hunting? tool use? Perhaps something very different from what might be expected: "No biological phenomenon is without antecedents. The question is, 'How obvious are antecedents of the human propensity for language?' It is my opinion that they are not in the least obvious." (Lenneberg 1967:234).

To return to *Washoe* and the question of homology, it must be asked, "Are some of the skills underlying *Washoe's* performance particularly relevant to an understanding of human language evolution? If so, what are they and what is their normal employment in free-ranging chimpanzee behavior?" Note that this is a very different question from whether or not *Washoe* is using language as determined by design features of the human and chimpanzee linguistic systems.

Homology is judged on the basis of any of the following:

1. Traits may be traced to their common origin by reference to the *fossil record*. Unfortunately, behavior does not fossilize, so that it must be inferred from morphological landmarks or the archeological record. Compounding the difficulty, the language-relevant soft tissue of vocal tract and brain are not preserved, but must be reconstructed from somewhat unreliable landmarks on the skull. Finally, the fossil record for humans and more particularly for apes is extremely fragmentary.

2. If traits share a *multiplicity of similarities,* homology is likely, especially if such similarities are not functionally related to one another. This approach underlies the practice of amassing a roster of design-feature similarities. Unfortunately, in *Washoe's* case the similarities are largely related to one another not only because some design features are contingent on the presence of other design features, but also because nearly all of them result from training.

3. Behaviors that are similar because of retention of features derived from common origin tend to be *similar in ontogenesis*. Both the schedule and the modes of language acquisition are being compared between chimp and child.

4. Evolution is more conservative in micro- than in macrostructure. The bat is morphologically convergent with birds because of its aerial niche, the porpoise with fish because of its aquatic habitat. In each case, mammalian affinities are revealed by examination of blood, skeleton, eyes, brain, and other features less subject to environmental features than fins, wings, and snouts. Thus, traits should be examined for *minuteness of similarities*. This concern, of course, lies beneath some of the argument over whether, say, *Washoe's* productivity is really very much like linguistic productivity, or whether her acquisition strategies are genuinely similar to those of humans.

In dealing with behavioral as opposed to morphological comparisons, there is a corrollary that may be added. That is, that the biological substrates underlying the performance must be examined. This is because behavior is much more plastic than its physiological basis. Among mammals, and especially the higher primates, learning plays an important role and behavior is very dependent on experience. Humans take advantage of this in training animals, so that apes have been taught to ride bicycles, have tea parties, smoke cigarettes, drink from a glass, and use "language." All but the last can have played no role in chimpanzee evolution, so that success in such activities stems from modification of skills that are present in the chimpanzee repertoire for other purposes. By examining the neural basis of *Washoe's* behavior, it may be possible to determine whether those skills are relevant to human language or whether we are confronted with a situation of clever chimp, clever trainer.

Fouts (1974), in particular, has noted the importance of ascertaining the neural basis of behaviors being compared, but is perhaps unduly pessimistic because it is not presently possible to detail how such neurological organization works to produce the behavior. The neurological investigations of nonhuman primate communication have not explained just how such behavior is produced, but they have suggested a great deal about its organization. The work of Ploog and his associates (1967), and of Robinson (1976) among others, has not only indicated the vast differences between the substrates of language and primate communication, but has suggested why primate signals take the forms they do and express the things they do. Unfortunately, such mapping is difficult, time-consuming, and expensive, and would be difficult to extend to the chimpanzee. Moreover, it requires chronic deep electrode implantation, and, usually, ultimate sacrifice of the subject. Despite the scientific benefits, it becomes increasingly difficult to justify such techniques for chimpanzees due to financial, conservational, and ethical considerations.

There are, fortunately, alternatives that are more feasible, although somewhat less refined than deep electrode implantations. One would be the selective use of drugs. The Wada test, for example, is used to determine language dominance in humans prior to brain surgery. This involves the injection of sodium amytal into the left or right carotid artery, and blocks speech functions in the speech-dominant hemisphere. Would it affect chimpanzee production or comprehension of "language?" If so, is there a lateralized effect? If the behavior is in some sense lateralized, is it consistently dominant on a particular side of the brain for all chimps? Does the Wada test also interfere with other kinds of activities?

Especially intriguing is the suggestion of Raleigh and Ervin (in Harnad, Steklis, & Lancaster 1976) that EEG monitoring be employed. Application of such techniques to language behavior is not new (Lenneberg 1967; McAdam & Whitaker 1971); in fact, Melnechuk has urged that there be electrophysiological investigation of human and nonhuman communication (Ploog & Melnechuk 1971). Analysis need not initially be very refined, but concerned with such parameters as cortical/subcortical, lateralized/nonlateralized, anterior/posterior, and so on. Particularly important is that brain activity be monitored through surface electrodes so that it may be studied in intact, free-ranging animals who have not been subjected to surgery. Assume several possibilities in the case of *Washoe's* performance: similar output—different neural basis (nonhomologous); similar output—similar neural basis (homologous); different output—similar neural basis (homologous).

The first two instances are the most frequently debated, and both are amenable to various kinds of analysis, but it is in the third case that EEG could prove especially useful. Without some kind of ongoing monitoring, it seems unlikely that homology would be suspected in behaviors that appear superficially quite different from one another. In any case, if homologies are suggested they may be telemetered in free-ranging chimpanzees. Moreover, developmental aspects of human and nonhuman acquisition can be compared. Human signers can be compared with chimpanzee signers with speakers with aphasics, and so on: the permutations of comparison are considerable. The potential advantages of EEG analysis outweigh the technical difficulties involved in interpreting the brain-wave patterns, and this avenue should be investigated.

In closing, it should be stressed that, while homologies need to be investigated, preferably by examination of neurological substrates, analogies and the design-feature approach are potentially valuable, if not used to argue evolutionary continuity. By comparing language and nonhuman primate communication in terms of environmental reference, displacement, and duality of patterning, for example, not only can the capacities of the systems be compared, but one may ask, for instance, how duality of patterning might arise, what advantages it would confer, whether it is already present in the cognitive but not expressive capacities of monkeys and apes, and so on. Similarly, there may be analogues in animal behavior that are useful. It has been observed (Nottebohm 1972) that there are remarkable parallels in the anatomy and acquisition of bird song to those of speech. These may be only coincidental, or related to aspects of bird behavior that will reveal little regarding speech, but such similarities are potentially instructive and should not be discarded merely because there is no possibility of homology with human speech. Marler (1976a) has indicated how aspects of speech and bird song acquisition may be related to the development of vocal learning.

Finally, it would be useful to have more extensive investigation of the communication of free-ranging primates. Although there is presently a large *corpus* of data, most of it has been derived either in the course of studying other aspects of socioecology or collected in such a way that signals are separated from the social matrix of which they are an essential part. The relationship of signal form to function, and to social system and ecological context, has been analyzed in only a few extant field studies, none of which involve the great apes.

LANGUAGE IN CHILD AND CHIMP?*

JOHN LIMBER

Today one can scarcely read a daily newspaper or news magazine without encountering a feature extolling the latest linguistic accomplishments of one or another ape. Contemporary introductory psychology textbooks may devote as much space to the achievements of the likes of Sarah and Washoe as they do to the language development of children. With increasing frequency, widely ready journals such as *Science* publish reports of the transmutation of base primates into noble ones. It is no wonder there is a growing belief among students and scientists alike that modern behavioral science has in fact succeeded in teaching human language to apes.

WHAT ACCOUNTS FOR OUR FASCINATION WITH THIS ISSUE?

The answers to this question are varied. Many people have an intrinsic interest in the antics of apes whether in the zoo, laboratory, or circus. The ancients speculated about the possibility of language in animals and the origin of our own. One recent writer (Linden 1975) attributes cosmic ecological significance to the current revival of an old issue. René Descartes, as is well known, argued in the 17th century that the use of language was the critical feature of *Homo sapiens* which distinguished it from the beasts:

For it is a very remarkable thing that there are no men, not even the insane, so dull and stupid that they cannot put words together in a manner to convey their thoughts. On the contrary, there is no other animal however perfect and fortunately situated it may be, that can do the same. And this is not because they lack the organs, for we see that magpies and parrots can pronounce words as well as we can, and nevertheless cannot speak as we do, that is, in showing that they think what they are saying. On the other hand, even those men born deaf and dumb, lacking the organs which others make use of in speaking, and at least as badly off as the animals in this respect, usually invent for themselves some signs by which they make themselves understood. And this proves not merely animals have less reason than

*Portions of this article were presented to the University of New Hampshire Linguistics Group.

JOHN LIMBER • Department of Psychology, University of New Hampshire, Durham, New Hampshire 03824.

men but that they have none at all, for we see that very little is needed to talk. (Descartes 1637/1960:42)

Our exclusive possession of language has always played a primary role in our conception of ourselves in relation to other species.

From another indirectly related perspective, behaviorists in the 20th century see the success in teaching apes human language as a means of vindicating their long-espoused, content-free, species-nonspecific learning principles. Julian La Mettrie, an 18th-century physician, foreshadowed the behaviorist position in his then-infamous *L'Homme Machine,* which was a rejection of Cartesian dualism. First, he argued that language was not the unique feature of man that Descartes claimed. La Mettrie observed that men and animals have at least emotional language in common: "all of the expressions of pain, sadness, aversion, fear, audacity, submission, anger, pleasure, joy, tenderness, etc." Second, he suggested that whatever linguistic deficits animals suffered might be just a matter of impoverished environment, lack of proper training, or both:

Could not the device which opens the Eustachian canal of the deaf, open that of the apes? Might not a happy desire to imitate the master's pronunciation liberate the organs of speech in animals that imitate so many other signs with such skill and intelligence? Not only do I defy anyone to name any really conclusive experiment which proves my view impossible and absurd; but such is the likeness of the structure and function of the ape to ours that I have little doubt that if this animal were properly trained he might at least be taught to pronounce, and consequently to know a language. Then he would no longer be a wild man, but he would be a perfect man, a little gentleman, with as much matter or muscle as we have, for thinking and profiting by his education. (La Mettrie, quoted in Gunderson 1971:30)

La Mettrie clearly anticipated contemporary behaviorists in their stress on the continuity across species and in their efforts to explain the linguistic deficits of apes in terms of experiential factors.

CAN APES ACTUALLY LEARN A HUMAN LANGUAGE?

The final outcome of current efforts to teach language to an ape remains to be seen. However, several firm conclusions can be drawn at this time. The unsuccessful efforts of the Kellogs (Kellogg & Kellogg 1967; Kellogg (this volume) and the Hayes (Hayes & Hayes 1951) demonstrate that a normal human language environment is not sufficient to enable an ape to learn a human language. It is also apparent now that regardless of the ultimate linguistic potential of apes, they are far more adept at learning visual-manual communication than the auditory-vocal processes natural to human language. Finally, and perhaps most important, there is no question but that the methods by which apes must be trained are quite different from the spontaneous, self-organizing acquisition processes employed by children.

Despite these very important differences between human and nonhuman primates, it has become clear from the achievements of Washoe, Sarah, and others (Fouts 1973; Gardner & Gardner 1974b; Premack 1971a; Rumbaugh, Gill, & Glasersfeld 1973) that apes can indeed be taught to engage in an extensive amount of symbolic communication. Of course, not just any symbolic activity can be taken as evidence of human language ability. Some criteria need to be established in order to evaluate the relation between chimpanzee behavior and human language. We can look at these achievements from two somewhat different perspectives. First, these activities may be compared with some of the fundamental attributes of human language; second, the performances of these apes may be compared with linguistic performances of minimally fluent users of a human language—for example, 3-year-old children.

ON THE NATURE OF LANGUAGE

Human language has been analyzed from a variety of perspectives and interests, including the comparative study of languages, the formal structure of language, and the comparison of communication systems across species. Despite the great diversity in languages, there are numerous features common to all of them. From among these I would like to briefly consider three characteristics that seem most relevant in evaluating the linguistic accomplishments of Washoe, Sarah, or any other organism—namely, the separation of language structure from its functions, the use of names, and the concept of syntactic creativity.

The Separation of Structure and Function

Language serves *homo sapiens* in many diverse ways: in communication, as a mode of self-expression, as a means of socialization, and perhaps in aid of thought and creativity. Yet we do not attribute language to humans just because they typically engage in these activities. Fido, drooling and licking his chops as the roast is served, accurately communicates his desire and interest; yet he hardly gets credit for knowing English. The 11-month-old child, whining expressively and reaching for its juice bottle surely conveys its intention; yet we are reluctant to say she is able to talk on the basis of that behavior. Instead, we attribute human language to an individual if and only if we have reason to believe that the individual uses the characteristic structures of a language more or less appropriately. Claims that an organism uses human language or some close approximation must be accompanied by structural evidence that the organism's language is a plausible structural facsimile of a natural human language. To know a language is at least to know the structures of that language.

The Use of Names

A number of the most frequently cited characteristics of human language are inherent in the concept of name. A name is symbolic, meaningful, and arbitrary and enables displacement behavior. Traditional analyses of human language often went so far as to virtually equate the use of language with the use of names. In this "naming paradigm" of language (Fodor, Bever, & Garrett 1974), the meaning of a word is a consequence of its arbitrary association with the thing it stands for, its referent. Learning a language means learning the conventional names for things. Psychologists interested in language were quick to see that this paradigm of language was interpretable within a conditioning model of behavior. A meaningless symbol, a word form, acquires its meaning in a process not unlike the way a conditioned stimulus comes to elicit components of the response previously made to the unconditioned stimulus, the referent. Additional complexities and subtleties of meaning are introduced through the processes of abstraction, higher order conditioning, and generalization. The conditioning model thus suggests an analysis of the difference between proper names and common nouns, a way to generate new meanings from old ("assign learning"; Osgood 1953), and an explanation of the ability to extend symbols appropriately to new instances. Novel linguistic expressions—if considered at all—were presumed to be sequences of responses elicited by novel sequences of stimuli. The expression "big red apple" might be interpreted as the temporally structured responses to *bigness, redness,* and *appleness.* Conditioning was at one time seen as a general phenomenon that could establish meaning of a kind in virtually all species and provide the unifying conceptual apparatus for the learned behavior of all organisms, including language behavior (Mowrer 1954).

Today, the naming paradigm is widely recognized as a very unsatisfactory model of human language. Its primary defects include an inordinate focus on the word rather than the sentence, a failure to distinguish between meaning and reference, and a complete neglect of the syntactic or creative aspects of human language. Naming plays an important, perhaps necessary, role in human language. It does not, however, play a definitive role. An organism's use of names is surely not sufficient evidence to conclude that the organism is using human language. Should one equate human language to the naming paradigm, it follows that any conditionable creature can learn human language.

Syntax and Creativity

Language was for Descartes the one certain indicator of thought, in that language reflected the human ability to produce and understand entirely new expressions of thought. Then, as now, the syntactic creativity or productivity of human language served, more than anything else, to qualitatively distinguish human language from the symbolic activities of all other species.

This notion of productivity or creativity implies much more than an ability to produce and interpret an infinity of sentences. The number of sentences in a human language is only of minor interest. A language whose nonfinite nature was due only to something like recursive intensification, as in "The banana was very, very, . . . very good," would be of relatively little significance to Descartes or ourselves. What is of primary importance is the projective aspect of human language ability. Users of any human language are able to syntactically project novel yet appropriate linguistic expressions onto any of an almost unlimited number and variety of concepts. For the most part, neither the linguistic expression nor the concept itself need have been ever previously experienced by the user; almost never have particular expressions and concepts been experienced contiguously by the user as prescribed in the conditioning model. Indeed, in many cases it is unlikely that *anyone* has ever previously experienced a given combination of concept and corresponding linguistic expression.

Fluent speakers of a human language have available a diverse array of linguistic expressions enabling them to refer to each of a limitless number of persons, propositions, objects, places, desires, beliefs, events, or whatever. For example, a particular place may be syntactically individuated using such expressions as "Urbana," "there," "that city in central Illinois," or "the place we visited last summer." A particular person may be individuated by such expressions as "Otto," "him," "that guy," or "the person that Bill wanted to leave." Each of these nominal expressions may be interchangeably substituted as the noun phrase (NP) constituent in expressions like "[NP] is where she went to school" or "[NP] is my neighbor."

These examples illustrate two universal structural features of human language: hierarchical constituent structure and the use of sentences themselves as constituents within complex nominalizations or syntactically generated "names" (Lees 1960). Such complex constituents as "the place we visited last summer" or "the person that Bill wanted to leave," with their own internal clause structure, fall far outside the scope of the naming paradigm or the conditioning model of language, which concerns itself with the assignment of meaning to previously experienced names or symbols. Learning a human language involves learning to use an integrated set of phonological, syntactic, semantic, and referential rules according to which novel and appropriate linguistic expressions may be constructed. It is this creative aspect, enabling the projection of linguistic expressions onto concepts both old and new, that is the most prominent structural feature of human language.

CHIMP AND CHILD COMPARED

How does the linguistic performance of trained chimpanzees compare with that of those more or less fluent users of language, 3-year-old children? The relevant comparisons may be made along such dimensions as use of names, the syntactic structure inherent in the linguistic behavior displayed, and the demonstrated projective ability.

The Use of Names

Without question, both young children and chimpanzees are capable of an extensive use of arbitrary symbols in referring to objects in their environments. The Kelloggs (1967; this volume), who raised their infant son together with an infant chimp, reported that Gua the chimp comprehended over 100 words during its 2nd year, perhaps more than their son did at the time. The vocabularies of both Washoe and Sarah are roughly comparable in size to those of many 2-year-olds. Sarah, for example, has a vocabulary of about 120 symbols, while Washoe uses about the same number of signs and can understand at least 100 additional signs (Gardner & Gardner 1974b). Whatever limitations there are on the ability of chimpanzees to learn human language, it does not seem that sheer vocabulary size can be much of a causal factor. Even dogs attain vocabularies of similar magnitudes (Warden & Warner 1928). There seem to be no striking quantitative differences among young primates in their ability to use simple names. Whatever differences there are in language ability must go beyond this, even if one ignores for now the matter of how those names are learned.

Syntax in Children

Language development in children follows a reasonably predictable schedule with regard to the features of interest—namely, the use of names, the emergence of hierarchical constituent structure, and projective syntax. Sometime around the 1st year, give or take a few months, children start using single morpheme expressions or holophrases in referring to things. From the beginning of their 2nd year and on into the 3rd, there is a continual increase in the length of a typical utterance from one to several morphemes (Brown 1973). During this time, the characteristic syntactic patterns of the language gradually emerge. Fixing the precise schedule of these syntactic developments depends on a number of things, including the child's articulatory skill—some children's speech is simply unintelligible—and, importantly, the criteria used by the investigator in making inferences about the presence of certain syntactic constructions (Limber 1973, 1976).

A traditional analysis of a child's early utterances using distributional criteria (Chomsky 1957; Hockett 1958) is generally more conservative in attributing structure to a child than is an analysis based on extensive intentional or semantic criteria (Bloom 1970; Brown 1973a). Distributional analysis involves postulating structures—rules and categories—to account for recurring patterns in the actual utterances. Units or constituents which appear in similar environments are taken to belong to similar syntactic categories. It is characteristic of human languages that these units may at one time belong to several hierarchically organized categories. The utterance "Stop!" in English is at once a word, a verb phrase, and a sentence. The logic of the semantic analysis is to postulate a base structure that will represent not only the observable syntactic patterns but the child's presumed intention as determined by the utterance plus the situation. Then the investigator postulates a sequence of transformations that generates the observed surface structure. In mapping the intention onto the actual utterances, these transformations serve primarily as filters which remove most of the structure postulated to represent the intentional and contextual information. With young children, the inevitable result is that more linguistic structure is attributed to the individual on the basis of semantic criteria than the utterances themselves would warrant using distributional criteria alone. One can see that in order for constituent structure of any degree of complexity to manifest itself using distributional criteria, it is necessary that at least some utterances of four or more morphemes in length be produced (Limber 1973). Using semantic criteria, considerable structure may be divined for utterances of even one or two morphemes. Applying semantic criteria, for example, one might well impute much the same grammatical structure to each of the following utterances: "ju," "juice," "want juice," "want drink juice," "me want drink juice." This brings us back to the difference between linguistic function and structure emphasized above. While all of these expressions may be in fact intentionally and communicatively equivalent to one another and indeed to a variety of nonlinguistic expressive behaviors, it is a long and precarious theoretical leap to assume they are linguistically equivalent. Equivalence of function does not imply equivalence of linguistic structure. Getting someone to bring you a drink is surely not equivalent to telling someone that you want a drink. Although inferences as to intent and meaning may prove heuristically useful in the case of children, where it can safely be assumed that the organism is learning a human language, these same inferences may be seriously misleading when applied anthropomorphically to nonhuman organisms, where no such assumption is justified. If we are willing to accept linguistic structures inferred on the basis of meaning or intent, we can anticipate claims of extraordinary linguistic prowess from various regions of the animal kingdom.

By age 3 or 4 at the latest, children regularly produce complex sentences, that is, sentences with hierarchical constituent structure which have embedded sentences or clauses as their internal constituents (Leopold 1949b; Limber 1973). Consider as an example the following utterance spoken by Laura, a girl of 2½, reported in Limber (1973): "I do pull it the way he hafta do that so he doesn't—so the big boy doesn't come out." Notice that "he" and "the big boy" appear in the same linguistic environment, indicating that they are instances of the same constituent for Laura. This is both in accord with our intuitions and any formal analysis of English. The basic structure of that sentence may be informally schematized as follows:

[(NP) (VERB) it the way (SENTENCE) . . .]. It should not be surprising to find instances of similar structure in some of Laura's other utterances, for example, "Do it the way I say" or "I cook it the way mommy makes a cake." A reasonable inference based on such data is that Laura has already learned a close approximation to the English rule for syntactically individuating the numerous "ways" that actions may be carried out, using the process of attaching a complete sentence, more or less, as a constituent to the abstract noun *way*.

Between ages 2½ and 3, this syntactic device emerges in use with other abstract nouns, for example, "Remember the place we went?" or "That's the kind I like," and with certain wh-morphemes, for example, "I think she wants juice," "Baby pretend she eat apple," or "When I was a little girl I could go 'geek-geek' like that, but now I can go 'this is a chair.'" Structures directly analogous to English relativization and nominal complementation thus appear typically in most children toward the end of the 3rd year and early in the 4th year. As in mature English, these complex constructions enable the speaker to individuate entities for which no suitable simple name is available and for which linguistic devices of lesser complexity such as pronouns are awkward or ineffective. It is not a coincidence that the earliest relative clauses produced by children typically are clauses characterizing semantically empty nouns like *one, place, way,* or *kind* (Limber 1973). As in mature English, children use such complex expressions both to refer directly, as in "I know you're there," and indirectly, as in "I know what you did." One can perhaps only come to appreciate the extraordinary referential power of human language by considering how difficult it would be to refer to such things as facts and events within a naming-paradigmlike language in which all possible referring expressions must have been previously experienced by a user—in other words, in a language without a creative, projective capability.

Syntax in Chimps

How do the linguistic performances of Washoe and Sarah compare with those of children in regard to syntactic structure and projective ability? The answer is clear and unambiguous; at present, any normal 3-year-old has far surpassed even the most precocious ape in language structure. Whereas virtually all children use hierarchically structured complex sentences by the beginning of their 4th year at the latest, there is little evidence that any ape ever did. The closest possibility, discussed below, is Sarah's use of her conditional symbol. It is even unclear to what extent the performances of the chimps should be construed as analogous to the use of simple sentences in a human language. A major reason for postulating the sentence as a constituent is that the elements constituting a sentence reoccur as constituents within themselves. In any language without complex sentences, justification for "sentence" must be found in recurring patterns of elements which appear across performances or utterances.

Brown (this volume, 1973a) analyzed the signs of Washoe and found some evidence for internal structure comparable to the structure in the utterances of children at the earliest stages of acquisition. Gardner and Gardner (1974b) reported similar results using Bloom's (1970) procedures of employing semantic criteria to analyze Washoe's signs. On the one hand, there is evidence that Washoe's sign combinations are not totally unstructured or random combinations; on the other hand, it is difficult to justify, even using semantic criteria, that these patterns in Washoe's two to four sign sequences are a result of syntactic rules rather than the semantic constraints inherent in any communication system. Washoe's combinations, like the utterances of children between 1 and 2 years old, are just not of sufficient complexity to suggest anything like the recursive hierarchical constituent structure in the language of 3-year-old children. As Fodor, Bever, and Garrett (1974) concluded in their analysis of Washoe's sign sequences, there is no evidence so far that Washoe represents the semantic relations in those sequences as syntactic relations between parts of a sentence. Washoe, like most children during their 2nd year, has achieved a considerable degree of proficiency in using arbitrary symbols to communicate. This is not to say, however, that Washoe or most 2-year-old children use a human language.

In contrast to Washoe's productive use of sign language, Sarah's training (Premack 1971a) was much more systematic and constrained. In addition, Sarah's training was in the interpretation of symbols rather than in their production. Two aspects of Sarah's performance are of particular significance here: her use of a conditional symbol and her use of the "name of" symbol. Premack's introduction of a conditional symbol, schematized as $[(S_1)$ implies $(S_2)]$, comes closest perhaps to testing a chimpanzee's ability to deal with a structure analogous to an interesting complex sentence. Sarah was trained on this symbol using a number of values for S_1 and S_2. She was then given a series of transfer trials on the symbol with values for S_1 and S_2 that she had not encountered in connection with the conditional symbol but had previously correctly interpreted when presented alone. Example sentences from Premack (1971b) are $[(Red\ is\ on\ green)\ implies\ (Sarah\ take\ apple)]$ and $[(Sarah\ take\ apple)\ implies\ (Mary\ give\ Sarah\ chocolate)]$.

Sarah's performance on this particular symbol does indicate that she is capable of a certain degree of structural analysis. Since Sarah apparently can deal with a fairly large number of potential S_1 or S_2 instances, there is no reason to assume that she has memorized the substitution sentences as a unit and merely learned to substitute unanalyzed units into the conditional symbol itself. Instead, it seems likely that she is able to perform an analysis of each constituent sentence, S_1 and S_2, and integrate the interpretation of those constituents with the significance of the conditional symbol. This in itself does not entail that Sarah treats those S_1 and S_2 constituents as recursively embedded constituents, but asking her or anyone to process sentences of the form [(S) implies (S implies S)] is hardly sensible. It does suggest that at the very least, Sarah has learned a fairly sophisticated substitution in frames technique.

Sarah's use of the conditional symbol and others, including her "compound sentence" and "negation," led Premack (1971b) to conclude that in about 18 months, she had acquired a language competence apparently comparable in many respects to that of a 2- to 2½-year-old child. Premack's conclusion may not be unreasonable if one keeps in mind the respects in which chimps and 2- to 2½-year-old children are not comparable. However, certain features of Sarah's training situation suggest that one should be conservative in estimating her human language ability. Sarah's ability is assessed entirely through comprehension tasks. Premack (1971a) reported her performance on these tasks to be about 80% correct on both training and transfer trials. Although this may seem impressive considering that Premack is defining what is correct, Sarah does have a substantial probability of getting a trial correct simply by guessing from among the limited number of alternatives provided her.

Moreover, forced-choice comprehension tasks with children, and presumably with chimpanzees, are prone to a variety of Clever Hans effects or the possibility that the subjects are using partially valid, nonsyntactic interpretive strategies. For example, young children are known to correctly interpret nonreversible passive sentences but not reversible ones—that is, *The car was kicked by the girl* but not *The boy was kicked by the girl* (Bever 1970). Sinclair-de-Zwart (1973) demonstrated similar phenomena in other tasks. While Premack himself has been sensitive to these problems, secondary reports of Sarah's achievements frequently have not. Often, the claims made on behalf of these linguistic apes in the popular media far exceed the evidence.

Productivity in Chimpanzee Language

All chimpanzees, it seems, quickly learn to use symbols, generalizing them appropriately to new instances and in situations far removed from the original circumstances of training. Washoe reportedly combines previously learned signs into apparently novel sign sequences. However, as noted above, these sequences are of insufficient complexity to infer that they are rule-governed rather than simply patterned by the situation as vaguely prescribed in the naming paradigm. Since Sarah is limited to comprehension tasks, productivity for her is linked by Premack to her use of the "name of" symbol.

Premack's premise, underlying all of Sarah's language training, is that language acquisition is a mapping of preexisting conceptual structure onto linguistic symbols.

If an organism does not have the requisite conceptual structure, that structure must be installed in the organism before it can be mapped onto a linguistic symbol. Therefore, it was only after Sarah mastered the names of a number of concepts that she was given training on the "name of" symbol. This was accomplished by training Sarah to select the "name of" symbol for an appropriate object–name symbol pair and the "not name of" symbol for an inappropriate pair. Training was tested using the well-established interrogative symbol, by asking, for example, "What is the name of the object?" Out of four alternative symbols, Sarah selected the correct one on about 80% of the trials, her typical above-chance but imperfect performance. Premack uses the "name of" symbol to illustrate Sarah's ability to productively generate new names. By this he means that the "name of" symbol can be used to form new symbol–object associations, completely bypassing explicit conditioning procedures. New names are acquired by simply placing the "name of" symbol between a familiar object, previously not named, and a new neutral symbol. Class names including the "color of," "size of," and "shape of" can also be used in this procedure. For example, Premack introduced the symbol for the color "brown" by using the previously known symbols "color of" and "chocolate."

Premack's conception of productivity is quite different from the one discussed above as characteristic of human language. Sarah still requires that the word and the object to be named be experienced contiguously. The "name of" symbol serves to express the appropriate relationship between the symbol and object without further training. Sarah's use of this symbol does not imply that she can generate new names on her own for concepts not explicitly paired with their names, although she can, of course, generalize the names she has learned to new instances of object classes. Premack (1971b) mentions that in Sarah's language she can only *use* words, not make them up. That, however, reflects a very basic qualitative difference between the human use of language and whatever it is that Sarah does. Notice that it is not her symbol system itself that limits Sarah's performance in this regard. A human restricted to Sarah's symbols could indeed productively generate new names for previously unnamed concepts in, for example, the domain of color names. Suppose a human subject knew the various symbols, as does Sarah, for "color of" and a variety of other object names and color names. For any object, O, we might ask our human subject, "What is the color of O?" If our subject in fact knows the color and has a symbol for that color name, we would expect an accurate reply at least 80% of the time. That is what Sarah does. Now, suppose we ask our human the familiar color of an object for which he does *not* have a previously established color name. He will in all likelihood be able to syntactically individuate that color by generating a correct complex "name" in Sarah's language as follows: "The color of O is the color of K," where K is the name of a previously known object having the same color as object O.

A demonstration that Sarah could indeed spontaneously generate a name even for a color of an object would be far more impressive evidence for a humanlike language ability than anything she or any other chimpanzee has accomplished thus far. As far as I can determine, Sarah has not performed in ths fashion; perhaps just as important for an understanding of chimp language, it does not seem that she was ever encouraged to do so.

WHY DON'T APES USE HUMAN LANGUAGE?

The evidence that apes can be taught a human language is hardly compelling. Despite some interesting accomplishments by Washoe, Sarah, and their trainers, nobody as yet is liable to mistake the language of any ape for the language of a normal 3-year-old child. La Mettrie's proposals to eliminate the language gap among primates have not yet proven feasible despite a number of contemporary efforts. What can explain these differences?

Historical Accounts

Most of the potential answers to this question were staked out during the 17th and 18th centuries:

1. *Anatomical differences*. Descartes, of course, knew that differences in behavior were related to differences in anatomical structure. Yet as the previously quoted passage indicates, Descartes ruled out an anatomical explanation for the lack of language in animals on an empirical basis. Specific morphological features of the vocal apparatus appeared neither necessary nor sufficient to explain the vast linguistic differences between humans and animals. Descartes himself may have made anatomical comparisons among various animals to investigate just this point.

2. *Nonmaterial differences*. The anatomical parallels Descartes observed among mammals, including humans, were very important to him, for they seemed to preclude any simple material explanation for the lack of language in animals. These parallels thus supported the nonmaterial explanation of human language inherent in Cartesian dualism. Animals did not use language because they did not have a soul, that is, the power to reason (Gunderson 1971:36). The soul was, for Descartes, a theoretical construct designed to explain obvious and reliable differences in behavior between humans and all other creatures.

3. *Experiential differences*. In *L'Homme Machine,* Julian La Mettrie objected to Descartes' dualistic doctrine that animals but not humans were physiological machines. La Mettrie, we have seen, specifically attacked the empirical foundation for that dualism, arguing that language was not intrinsically unique to humans. This,

however, left La Mettrie without an explanation for the obvious fact that only humans used language. Undoubtedly, La Mettrie would have liked to attribute those differences to anatomical differences, yet he was also compelled to reject anatomical differences as the basis for these language differences. Whereas Descartes knew little, if anything, about primate anatomy, La Mettrie had access to several comparative studies confirming an extremely close resemblance between human and primate vocal apparatus (Vartanian 1960). If these physiological machines were built in the same way, why didn't they function in the same way?

La Mettrie tried to explain the discrepancy by suggesting that animals simply did not have the necessary environmental or educational experience needed to learn a language. Like contemporary psychologists such as Skinner (1957), Osgood (1968), or Piaget (1967), La Mettrie denied the Cartesian contention that human language demanded an explanatory account distinct from that required for other behavior.

Contemporary Accounts of Primate Differences

On the basis of their information, both Descartes and La Mettrie believed that anatomical differences alone were insufficient to explain the lack of language in animals. Today we know that there are indeed differences both in peripheral anatomy and the central nervous system that might be called upon to explain the language differences among primates. In addition, enough is known about the genetic basis for behavior to make it plausible that as yet unidentified genetically determined neural factors differentiate morphologically similar species. Here, in brief, are several widely accepted explanations for the language differences among primates:

Vocal Tract Differences

Keleman (1948), after a careful morphological study, concluded that the chimpanzee vocal tract could not produce the full range of human speech sounds. Several factors led to his conclusion, including the overall tract configuration and relative position of the larynx and his belief that a chimpanzee would be unable to maintain phonation for the duration required for human speech. Similarly, Lieberman, in a recent series of investigations (Lieberman 1968; Lieberman, Klatt & Wilson 1969), proposed that nonhuman primate vocal tracts are inherently limited in the range of vowellike sounds that they can produce. Those primate tracts are said to lack a variable pharyngeal region suitable for producing the full complement of vowel sounds. The implication is that the primates could not produce human speech even if they had something to say.

This explanation of the linguistic deficit of apes on the basis of an articulatory deficit fits nicely into the motor theory of speech perception developed by Liberman and his associates (Liberman, Cooper, Shankweiler, & Studdert-Kennedy 1967). It follows from that theory that any organism with a vocal tract different from a human one would have great difficulty in extracting the relevant acoustic cues from a human speech signal in order to ever break into the system. This would explain why chimpanzees raised in human environments fail to learn even rudimentary human language; they lack the relevant auditory—articulatory transformations. Several investigators (Kellogg this volume; Yerkes & Learned 1925) have remarked on the chimp's lack of mimetic interest in human speech sounds. There is certainly no reason to attribute this to any peripheral auditory deficits in those primates relative to humans (Dewson & Burlingame 1975; Elder 1934; Stebbins 1965).

Can the morphology and position of the larynx in a vocal tract explain the linguistic differences among primates? Despite the simplicity of this peripheralist account of human language, there is reason to be skeptical about it. Both Keleman and Lieberman utilized young primates in their investigations. Keleman himself observed that it might be possible that a mature chimpanzee would have a different laryngeal position than that found in his specimen of age 2, just as a mature human tract differs from that of a young infant. Lieberman's analyses are based on the spontaneous vocalizations of captive 2- and 3-year-old gorillas, a 2-year-old chimpanzee, and 1- to 6-year-old rhesus monkeys. Moreover, Lieberman's argument as to the articulatory deficits of nonhuman primates is based on a seemingly tenuous backward chain of inferences from an incidental sample of vocalizations to conclusions about the articulatory *potential* of the tracts generating those vocalizations. One wonders how well those same inferential techniques would reconstruct the articulatory potential of a myna bird (Klatt & Stefanski, 1974) from its incidental chatter or reconstruct the potential of deformed human tracts from nonspeech vocalizations of those tracts.

Observations on humans with deviant vocal apparatus also militate against a peripheralist explanation of human language ability. Lenneberg (1962) reported a case of an individual with a congenital inability to speak yet who learned to understand English. Even more to the point are the many instances of "vicarious articulation" reviewed in Drachman (1969) and in Luchsinger and Arnold (1965). Here we have cases of individuals overcoming extraordinary deformations of their vocal tracts—for example, loss of the entire larynx or tongue—and still producing intelligible speech. In some cases, this adaptation may occur within weeks after the loss, generally a result of surgery. A remarkable case reported by Drachman (1969) concerns a young girl born without a tongue and having a number of other oral defects. This girl successfully acquired a language with vowels that Drachman (1969:320) describes as astonishingly clear. From an X-ray film taken of the girl while speaking, it appears that she has acquired sufficient control of the facial and mouth floor musculature to provide for a wide range of supraglottal tract shapes. Another case involved a boy who at age 3 lost his larynx and learned to provide a sound source by drawing back the root of his tongue and making it vibrate against the rear of his pharynx.

The obvious implication of such cases is that human language ability, and specifically speech production, involves more than just a standard vocal tract. Engaging in this complex activity not only requires having something to say but also the specific rules for encoding that intention and a vocal apparatus for generating the appropriate signal. As the above cases indicate, humans can recruit an extraordinary range of articulatory gestures in order to produce the desired signal. Drachman (1969) reminds us that this remarkable adaptive capability is not unique to a few freakish cases but is available to all humans whose vocal tracts vary as they grow, eat, smoke, or suck their thumbs. Lindblom and Sundberg (1971) demonstrated this adaptive ability experimentally by systematically disrupting the normal resonant characteristics of the vocal tract by means of a block inserted between the speaker's teeth. To achieve an adequate target vowel, the speaker had to compensate for the abnormal tract shape produced by the presence of the block. Lindblom and Sundberg determined, from examination of the formant frequencies at the moment of the first glottal pulse, that the target-vowel formants were achieved before the speaker had time to utilize his own auditory feedback in adjusting the tract. Apparently, the speaker sensed the potential output of the tract and then made the appropriate compensatory adjustments necessary to attain the desired acoustic target before he began to speak.

All these factors should make anyone wary of claims that the morphology of the human vocal tract is the essence of human language. A normal human vocal tract in itself is neither necessary nor sufficient to account for the linguistic ability of humans. This, of course, was Descartes's conclusion.

Central Nervous System Differences

Today we know of a number of global differences in neuroanatomy among primates. Human cerebral hemispheres may be twice the weight of those in other primates. Furthermore, the human brain is only approximately 40% mature at birth, in comparison to a corresponding figure of about 70% for other primates. This suggests a later evolution of certain portions of the human brain, particularly in the associative areas of the cortex. According to Geschwind (1969), the greatest growth in humans relative to other primates is in the inferior parietal region, the region of the angular gyrus. This region, between the association cortex of the three nonlimbic modalities, is said to be poorly developed in other primates.

Two fairly specific hypotheses concerning the neuroanatomical basis for primate language differences follow from these observations. One is simply that human language ability is a function of brain size, greater size implying greater language capacity. Lenneberg (1967) convincingly disposed of this possibility by showing that nanocephalic dwarfs whose brains were small relative to those of non-human primates did in fact acquire language. Thus, the typically larger size of the human brain is neither necessary nor sufficient to explain language in humans.

The other hypothesis concerns the role of the angular gyrus region and is credited to Geschwind (1969). It is Geschwind's proposal (also see Lancaster, 1968) that the intermodal associations involved in learning and using names are mediated by the structures in the angular gyrus region, which serves as a kind of way station for the formation of intermodal associations without limbic mediation. Thus, naming can in part be understood as a process of rapidly associating a visual or tactile object with a symbolic auditory representation. Moreover, this association and subsequent response can be relatively independent of the state of the limbic system, those portions of the cortex directly implicated as centers of emotion. The essence of Geschwind's proposal is that a unique characteristic of human neuroanatomy enables human language to be used to refer to objects independently of affect. In contrast, non-human primates have communication systems closely linked to limbic functioning.

There is some interesting support for Geschwind's hypothesis. Recall that La Mettrie noted that humans and animals shared a common emotional language: expressions of fear, pain, and pleasure. Descartes, too, was aware of this and sharply distinguished between language reflecting the passions and language reflecting rational thought. Only the latter distinguished man from beast. Contemporary comparative studies (e.g., Marler 1965) have confirmed the fundamentally emotive or affective nature of nonhuman primate communication systems. It is reported that only rarely do messages in these systems directly convey information about the physical environment. Even in those few cases in which such nonaffective information does seem to be transmitted, nonhuman primates show no ability to separate the informational components from the emotional or affective charge (Bronowski 1967).

Geschwind's hypothesis is consistent with a study of evoked vocalization in macaques, in which Robinson (1967) elicited naturalistic vocalizations by electrical stimulation of various limbic sites. There was no evidence that the neocortex participated in any of these vocalizations. Robinson speculated that human language may have evolved in parallel with a common primate affective vocalization system now relegated to a subordinate role in humans. These affective systems, involving facial expression, gesture, and vocalization, are apparently quite similar in form and function across all primates, including humans (Bastian 1965; Lancaster 1968). These similarities across species and the little evidence there is concerning the development of these systems within species (Eibl-Eibesfeldt 1973; Freedman 1975; Sackett 1966) suggest that such systems are far more genetically structured and admit of considerably fewer degrees of freedom than human languages. Finally, this interpretation of the neuroanatomical differences among primates is in accord with the difficulty a number of investigators have found in establishing intermodal transfer in nonhuman primates (Blakemore & Ettlinger 1966; Ettlinger 1967).

Neuroanatomical differences, however, cannot account for all the language differences among primates. Even if Geschwind's hypothesis is correct, it does not take human language much beyond the naming paradigm. Syntax and the creative aspect of human language remain unexplained. Indeed, otherwise-normal blind children acquire the full range of human language structures with no serious difficulty. Moreover, the performances of Washoe and Sarah indicate that the intermodal-association naming issue has been circumvented to a large extent by using reinforcement techniques and by concentrating on visual rather than auditory associations. Sarah's use of the "name of" symbol, in particular, suggests that Premack's training procedure has had the effect of establishing a kind of *learning set* whereby associations between symbol and object may be formed with little need for explicit reinforcement. Davenport and Rogers (1970) reported evidence of intermodal transfer in chimpanzees and orangutans but only after these animals received extensive training. All this suggests that perhaps the differences in brain structure noted by Geschwind are related to the human ability to learn features of their language independently of the limbic system—that is, independently of external reinforcement—rather than being directly involved in the use of language. Speculatively, the angular gyrus region may be involved in the human ability to organize its linguistic environment on its own, whereas the apes' environment must be explicitly structured for them with reinforcement procedures calling attention to normally unanalyzed contingencies.

The evidence relating anatomical structure to linguistic function may be summarized as follows. Although we certainly know much more about the details of primate anatomy today than did either Descartes or La Mettrie, it is apparent that we have little precise knowledge as to how those anatomical structures embody linguistic functions. Differences in vocal anatomy seem insufficient in themselves to explain the linguistic deficits of apes, since humans with gross vocal-tract deficits may learn and use language. Nor is it unusual throughout the animal kingdom to find different anatomical structures realizing the same biological function. Differences in cortical structures exist, but the functions attributed to those structures grossly underdetermine the basic features of human language. It may be no exaggeration to say that the various anatomical differences discovered among primates since the 18th century have little more explanatory force as to the basis for human language than Descartes's explanation involving the soul. The major consequence of recent biological discoveries seems not to have increased our understanding of human language as much as it has served to incorporate materialism into an *a priori* assumption rather than leaving it the open question it was for Descartes.

Cognitive Limitations on Language

Descartes would surely not be surprised at the impoverished level of the apes' language compared to that of humans. Since he believed that language reflected thought, the cognitive or conceptual limitations of a creature would show up as linguistic limitations. Animals say very little because they have very little to say. Versions of this Cartesian view are tacitly held in part by a variety of contemporary psychologists.

Recall Premack's premise that language learning is the mapping of preexisting conceptual structure onto linguistic symbols. The training of Sarah involved finding a suitable symbolic system for her and an effective procedure for associating elements of the symbol system with already existing conceptual elements. From this perspective, the naming paradigm appears as an eminently suitable language model, as it is not the complexity of languages or symbolic systems that varies among species but rather their underlying conceptual structures. Getting an ape to use language like a human would entail establishing a conceptual system comparable to that of a human and then mapping that system onto a set of symbols. If Premack's premise is valid for both humans and apes, then linguistic differences among the primates simply reflect conceptual differences among them once they are all given equal access to an appropriate symbol system.

A variation on the cognitive-limits explanation is implicit in the theory of language development proposed by Piaget and his associates. Within the Piagetian framework (Piaget 1967; Sinclair-de-Zwart 1973), language development is predicated on the development of very general cognitive structures composed of systems of actions established during the first 2 years of life. These action systems underlie future cognitive and linguistic development. Again, the idea is that language development is dependent upon conceptual development. Sinclair-de-Zwart (1973) reports how young children interpret word strings without syntax and argues that the semantic strategies used reflect the underlying conceptual structure upon which language development is based. She proposes that the child first learns to associate the linguistic symbols with elements of the action schema on an intuitive basis and then gradually learns the formal syntactic basis for making those associations through some kind of inductive procedure. The Piagetians, like the behaviorists, stress that no special theory of language development is necessary.

The notion that language ability is constrained by cognitive structure has much to recommend it. It makes little sense to think of learning a human language outside of some particular conceptual and social context. Organisms unfamiliar with base-ball or linguistic theory or language are not likely to use or understand terms like *balk* or *A over A principle* or *name of*. Furthermore, we know that children may not utilize certain linguistic structures that depend on specific cognitive structures until those cognitive structures have developed. For example, a child's use of time-related language structures is related to the development of an adequate conception of time (Clark 1971). Finally, it would certainly simplify the requirements on any theory of human language acquisition if the naming paradigm was even approxi-mately true, that is, if human conceptual structures were more or less isomorphic to the structures of human languages.

A certain level of cognitive structure is surely necessary for learning and using any human language. Yet it is no simple matter to pin the linguistic deficits of apes on any general, independently specifiable, conceptual deficits. The arguments relat-ing linguistic function to conceptual structures are probably less convincing than those relating linguistic functions to anatomical structures. The problem faces in two directions. On the one hand, we do not know what conceptual deficits to attrib-ute to apes to explain their failure to learn human language. The cleverness of chim-panzees is documented in many situations. Moreover, it seems that children come to clearly surpass the intellectual abilities of chimpanzees only after acquiring sub-stantial linguistic ability. Above all, what kind of general conceptual limitations would prevent an ape from learning to use complex linguistic expressions dealing with color names, actions, or locative clauses? From what I make of various accounts of chimpanzee activities (e.g., Hayes & Nissen 1971), there is no question as to apes' ability to deal with and discriminate among colors, actions, and places. They do have the relevant underlying conceptual structures. Why then don't they learn how to syntactically individuate those concepts as do 3-year-old children?

On the other hand, what kind of general conceptual advantages can we attrib-ute to children in order to explain their remarkable linguistic skills? Lenneberg (1967) argued that the development of language bears little relationship to either absolute brain size or variation in intellectual ability. Of course, one frequently used index of human intelligence is vocabulary size, but as mentioned above, chimps and children have relatively comparable vocabularies just at the time when children begin to show their syntactic superiority. Premack (1971b) has remarked that Sarah might easily learn up to 400 symbols if that was his research objective.[1] One might look to advocates of a purely conceptual theory of human language learning for

---

[1] Since completing this paper, I have read a report on the progress of Neam Chimpsky (Nim), a young chimpanzee learning American sign language at Columbia University (Terrace, Petitto, & Bever 1976). Nim has acquired an extensive vocabulary in the range of 200 signs. Nothing as yet in Nim's behavior changes any of my conclusions about language in chimpanzees, although the Terrace *et al.* report is remarkable for its detailed presentation of data and the clear recognition of the importance of using structural or distributional criteria in assessing chimpanzee language.

some suggestion of what specific conceptual structures enable children (but not apes) to learn a language. The Piagetian theory as summarized above involves two basic components: the child's action schemas and its general inductive capacity. As for the schemas, I know of no evidence suggesting that sharp differences exist in action schemas among any of the primate infants. If anything, the manipulative skills of young chimps outpace those of infants. That leaves only inductive ability to differentiate children from apes, that is, to explain why children but not apes are able to go far beyond presyntactic language into the characteristic structures of human language. Induction is the inference of a valid general rule from one or more instances. Now it may well be true that children do exceed apes in inductive ability. But without an account of the origins of potential rules and the means by which one rule is selected over another, induction is very little of an explanatory advance over the Cartesian soul. Virtually, the entire problem of language learning has been packed into an unexplained inductive procedure.

To sum up, cognitive explanations for primate language differences lose much of their intuitive appeal upon close scrutiny. The linguistic differences between young chimpanzees and young humans appear to be far greater than the cognitive differences. Those cognitive differences that emerge along with the development of language are as likely to be consequences of language as antecedents. Moreover, there is little reason to presume that any fine-grained isomorphic relationship exists between conceptual structures and linguistic structures. Very complicated thoughts may be couched in very simple syntax, and very simple thoughts may be propounded with great syntactic complexity. Perhaps most importantly, there does not seem to be any substantive suggestion as to the nature of those general cognitive capacities differentiating the language-using primates from the others.

A Genetic Predisposition for Language?

Biologists recognize that there may be substantial genetically controlled differences in specific abilities even among closely related species (e.g., Manning 1972). Although there is as yet no direct evidence for a genetic explanation for the human use of language, indirect arguments have been presented forcefully by Chomsky (1965, 1972) and Lenneberg (1967). The hypothesis is that human language is a species-specific capacity that is essentially independent of intelligence; children are genetically predisposed to learn any one of the many human languages. Evidence for this hypothesis involves the inadequacies of pure induction to explain any kind of learning, the speed and uniformity of language acquistion by children in diverse environments, and the existence of important similarities across all human languages. Lenneberg's (1967) arguments draw upon Chomsky's hypothesis, certain biologically unique aspects of humans, the uniformity of language development, and certain parallels between human language and other species-specific behaviors. Chomsky's hypothesis is indirect in that he asks what else other than genetically encoded linguistic knowledge could account for the known facts of language development:

We must postulate an innate structure that is rich enough to account for the disparity between experience and knowledge, one that can account for the construction of the empirically justified generative grammars within the given limitations of time and access to data. (Chomsky 1972:79))

Chomsky offers little in the way of specific details as to the nature of this innature structure, but it is not likely to be anything derivable from the peripheral anatomy (Limber 1974), which is of course also genetically controlled. Instead, he suggests that this knowledge involves, *inter alia,* something as to the nature of potential linguistic rules and an evaluation metric for deciding when the child's internalized grammar is reasonably in accord with the grammar behind the utterances of its experience.

Chomsky's "what else" argument has received additional support from the recent efforts to train apes to use language.[2] Gardner and Gardner (1971), in particular, strikingly confirm the difficulties inherent in relying on pure inductive or trial-and-error processes for the acquisition of even overt physical gestures:

It became obvious that most of the signs that we wanted to teach to Washoe could not be taught in a reasonable amount of time if we had to rely on trial-and-error alone. As for shaping, the odds of a reasonable approximation to an appropriate sign occurring in an appropriate situation under conditions that would permit us to administer an appropriate reward were entirely too low. The method of guidance proved to be much more practical for the introduction of new signs than any other method. (p. 133)

The method of guidance involves physically forming the desired sign in Washoe's hands and moving them appropriately (Fouts 1972b). In other words, the Gardners found it necessary to provide a physically structured environment just to enable Washoe to learn her signs. Imitation played an unexpectedly small role in Washoe's training. Despite the chimpanzee's often-mentioned tendency to mimic physical activities, Washoe apparently did not reproduce signs on her own with suitable

---

[2]Some may interpret the achievements of Washoe and other trained apes as disproving Chomsky's hypothesis. Presumably, such individuals also would consider the successful training of a dog to move about on hind legs as evidence against the hypothesis that humans are genetically predisposed to walk. I believe all such achievements must be interpreted within the context of simulation of human behavior. Even a chimp that could successfully pass Turing's test with regard to linguistic ability, although an extraordinary technical achievement, would tell us very little about human linguistic behavior. [See Fodor's (1968) discussion of the "Turing Game."] Making these achievements even less informative than computer simulation is the fact that the program underlying chimpanzee simulation is inaccessible. I discuss the matter of chimpanzee simulation and the related issue of behavioral alchemy in a forthcoming paper.

fidelity or frequency to make imitation a fruitful means of teaching her many new signs (Gardner & Gardner, 1971;136).

What else other than some genetic predisposition or *a priori* knowledge allows children to forego such explicit structuring in learning the appropriate motor control of their articulatory organs? It is hardly surprising that Washoe has not demonstrated acquisition of syntactic rules. Why should we expect Washoe to be any more likely to spontaneously generate an appropriate rule than an appropriate sign? How would one ever use the method of guidance to present an abstract rule to an ape? If it is not possible to convey the rule directly to an organism without language, the trainer ultimately must depend on the appropriate rule arising naturally in the organism's mind. C. S. Peirce (1965), in his important work on induction, long ago anticipated the essence of Chomsky's argument:

It seems incontestable, therefore, that the mind of man is strongly adapted to the comprehension of the world; at least so far as this goes, that certain conceptions, highly important for such comprehension, naturally arise in his mind; and without such a tendency, the mind could never have had any development at all. (p. 348)

CONCLUDING REMARKS ON LANGUAGE, THOUGHT, AND APES

One curious aspect of the literature dealing with teaching human language to infrahumans is a lack of concern for the actual structures of human language. No one, however, expects children to learn an abstract subject like algebra—for which children have relatively little evolutionary preparation—without years of formal instruction from teachers who have mastered the subject themselves. Why then should anyone expect so much more from an ape? A primate egalitarian like La Mettrie might well wonder to what extent the modest language achievements of the apes to date reflect the view of human language inplicitly held by their tutors rather than any intrinsic limits on infrahuman language ability.

This lack of concern for the structural nature of language issues from several sources. These include a failure to distinguish between language function and structure, a belief in content-free universal laws of learning, and uncritical acceptance of the naming paradigm of language. The behaviorist paradigm, with its reliance on bar-press counters and cumulative recorders, has thrived primarily in artificial situations where the structural or topographic nature of a response approaches irrelevance. There is also a widespread intuitive belief that anything which seems easily accomplished—like learning a first language—cannot be much of a problem. This egocentric attitude is common throughout psychology, particularly where consciousness has long been a primary criterion for important psychological phenomena. William James, for example, opened his *Principles* (1890) by asking if such "machine-like" activities as walking, piano playing, or talking, which can be carried out while the mind is absorbed in other things, should be included within psychology. James here is simply echoing Descartes: "The greater part of our motions do not depend on the mind at all . . . respiration while we are asleep . . . and even walking, singing, and similar acts when we are awake, if performed without the mind attending to them" (cited in Vendler 1972:163). Indeed, the neglect of language within psychology (cf. Mowrer 1954) in large part can be traced to both the automatic, unconscious character of human language and the behaviorist belief that all behaviors are pretty much alike.

It should be no surprise that Descartes, with his focus on thought and consciousness, appears little interested in the complexities of human language. Descartes, in fact, had reason to minimize these complexities in order to maximize the importance of thought. Frequently he remarked to the effect that it would be easy for any creature to invent language if indeed it had something to say. If that invention were made out to be a difficult intellectual task, there would inevitably be the suggestion that at least some creatures were able to think but had not yet invented a language. Moreover, if language invention were difficult, how could even the insane, the dull, and the stupid do it? Descartes, to be sure, would not want to conclude that human language ability was akin to those remarkable, automatic, instinctive abilities uniquely possessed by various animal species, such as the web spinning of spiders or the navigation of birds, which were purportedly accomplished by the disposition of their organs rather than through reason.

Descartes, like certain contemporary psychologists, seems to have taken linguistic structure for granted as a relatively insignificant consequence of other factors. There are, of course, several quite diferent interpretations of the place of human language in human behavior. Chomsky's hypothesis has already been discussed. Even if we were to update the Cartesian soul in terms of hypothetical neural programming, Chomsky's hypothesis is quite distinct in that he explicitly postulates a specific *faculté de language* as being responsible for human language ability rather than the Cartesian universal instrument of reason. Chomsky's *faculté de language* is clearly intended to be a different ability than that of general intelligence or reason.

Even more radical in terms of the Cartesian position are those traditions that give language an important causal role in the development of human reason. Chomsky's hypothesis is essentially neutral on this point. No so for the many advocates of linguistic relativity or determinism such as Humboldt (1905), Cassirer (1953), and Whorf (1956), who attribute different cognitive structures to different linguistic structures. Descartes predictably eschews this interpretation: "For who doubts whether a Frenchman and a German are able to arrive at identical conclusions about the same thing though they conceive entirely different words?" (cited in Vendler 1972:180). In yet another tradition, Soviet psychologists (e.g., Vygotsky 1962) have long argued the importance of language as a primary causal factor in human cognitive development. Finally, it has been suggested more than once that human consciousness itself is intimately connected with the evolution of human language (e.g., Jaynes 1973; Wright 1971).

   Evaluation of such propositions has proven extremely difficult. Lenneberg (1967), for example, concluded that linguistic relativity is untestable. Language differences are inevitably confounded with cultural differences. Comparisons made between normal and deaf children are of questionable value. One neither knows what deficits accompany deafness nor how much language a child has internally available despite hearing loss and inability to speak. For such reasons, the empirical evidence bearing on the causal interaction between thought and language remains equivocal. Perhaps the advent of conversational chimpanzees will permit an adequate experimental investigation of this fundamental issue. Experimental comparisons of alternative color coding systems (e.g., Brown & Lenneberg 1954) between animals with and without specific symbols are feasible even within the naming paradigm. Would a chimpanzee that had learned names for items have a symbolic edge in a problem-solving task over an otherwise comparable animal that had not learned names? Such investigations may well advance our understanding of human psychology far more than any research heretofore involving infrahumans.

ACKNOWLEDGMENTS

   The author, a *Homo sapiens,* thanks James R. Davis and William Woodward for their comments on a previous draft of this article and is indebted to the useful comments of an anonymous reviewer.

PRIMATE VOCALIZATION: AFFECTIVE OR SYMBOLIC?*

PETER MARLER

It is a sobering yet ultimately healthy experience to reread the works of great men of science and to discover the extent to which they anticipated ideas and developments that we of our generation had fondly thought to be original. Many biologists have made such discoveries in the writings of Charles Darwin. As a student of animal behavior I have had this experience with the works of some of the great German zoologists such as Heinroth and Lorenz and most recently, in the writings of Robert Yerkes. A quotation from 1925 bears so directly on the occasion of the Yerkes Centennial that I would like to quote it in full. He is speaking about the vocalizations of higher primates with special reference to the chimpanzee:

Everything seems to indicate that their vocalizations do not constitute true language, in the sense in which Boutan uses the term. Apparently the sounds are primarily innate emotional expressions. This is surprising in view of the evidence that they have ideas, and may on occasion act with insight. We may not safely assume that they have nothing but feelings to express, or even that their word-like sounds always lack ideational meaning. Perhaps the chief reason for the apes' failure to develop speech is the absence of a tendency to imitate sounds. Seeing strongly stimulates to imitation; but hearing seems to have no such effect. I'm inclined to conclude from various evidences that the great apes have plenty to talk about, but no gift for the use of sounds to represent individual, as contrasted with racial, feelings or ideas. Perhaps they can be taught to use their fingers, somewhat as does the deaf and dumb person, and thus helped to acquire a simple, nonvocal, "sign language." (Yerkes 1925:179–180)

The anticipation of the Gardners's work is uncanny and a merger of this train of thought with Yerkes's conviction that chimpanzees do indeed have the capacity for ideation and symbolic thought leads directly to Rumbaugh's work at the Yerkes Center. In this paper I dwell on one particular issue that the Yerkes quotation also introduces, that of the relationship between sound production and emotional expression. He represents a view still widespread today, that animal signaling behavior tends to be associated with "affective" processes, in contrast to the

*Research for this paper was supported by NSF grant number BN575–19431.

PETER MARLER ● The Rockefeller University, New York, New York 10021.

221

"symbolic" processes that are typical of our own species. These two alternatives are usually presented as different in kind. This is the issue I want to reexamine.

Chimpanzees have a vocalization known as food grunting (van Lawick-Goodall 1968b) which I prefer to call "rough grunting" (Marler 1976b; Marler & Tenaza 1977) with which Yerkes was undoubtedly familiar and which many of you will have heard. It is rather specifically associated in the wild with the discovery and eating of a highly preferred food such as a bunch of ripe palm nuts or, if the experimentor intrudes, a bunch of bananas. It is well illustrated in the sound film which Jane Goodall and I prepared on vocalizations of the chimpanzee (Marler & van Lawick-Goodall 1971).

This call comes closer than any other in the chimpanzee repertoire to qualifying as a name for a limited class of external referents, in this case, preferred foods. Yet it is clear that the vocalizer is aroused, and many would not hesitate to label this act as a manifestation of emotional arousal: not as "symbolic," but "affective."

In making such a judgment, different people use different criteria. One that recurs focuses on the specificity of the relationship between production of the sound and a particular set of referents—in this case a large class of objects with the abstract property of being good to eat. The larger the class, the more likely it is to be "affective." I want to reexamine some of the criteria used, beginning with this one.

As a result of the work of Fossey (1972) we have begun to understand the vocal repertoire of the gorilla. I was intrigued to learn that this species has a sound with a very similar acoustical morphology to chimpanzee rough grunting—it is known as "belching." However there is a difference in the circumstances of use. While gorilla belching does occur during eating, it also occurs in a wide variety of other situations, associated by Fossey with quiet relaxation, either while resting, or moving slowly through forage and feeding. It seems to have much more of a contagious quality than rough grunting so that choruses of belching may be heard as a group of gorillas moves through dense undergrowth.

Field studies have revealed that the gorilla has a much more coherent social group than the chimpanzee (Fossey 1974; Schaller 1963), which is notorious for the fluid fragmentation and recombination of social units (van Lawick-Goodall 1968b; Wrangham 1975). Gorilla belching seems to serve an important function not only in communicating the discovery of food but also in maintaining group coherence. In a sense the referential designata for gorilla belching seem to be more generalized than those for the rough grunting of the chimpanzee.

Should the gorilla call be viewed as more "affective" than that of the chimpanzee? In an evolutionary sense, one can readily see how a shift of specificity from the relatively proscribed designata of rough grunting to the larger frame of reference of the gorilla call may be viewed as adaptive in light of differences in their social organization. The chimpanzee has no need for a close-range call to mediate coherence of close-knit groups and does not possess one.

How readily can we apply notions of arousal in explaining this difference? In both, vocalizing animals are obviously aroused, although not highly so. Nevertheless one could probably fit them at different levels of arousal along some single dimension. The next question is how many dimensions would be necessary to explain all the signaling behavior of the species? Would a single one going from low to high suffice, or are we compelled to invoke others, perhaps, for example, one involving positive affect and the other negative affect?

Consider another case, Struhsaker (1967) discovered a remarkable array of alarm calls in the vervet monkey (Figure 1). The situations in which the four adult male alarm calls are uttered can be arranged in a series of increasing specificity. The first two, the ''uh'' and the ''nyow'' are evoked by rather generalized stimuli, and seem to function primarily to alert others. The third, the ''chutter'' is associated with a man or a snake, especially the latter. The fourth, the ''threat-alarm-bark'', the only one which is the sole prerogative of the adult male, is typically given on sighting a major predator. It evokes not approach and mobbing as is the case with a snake, but rather precipitant flight to the nearest cover.

This series can be interpreted in terms of an increasing level of arousal, although to provide an effective explanation we have to allow for rather fine fragmentation of the arousal continuum, with qualitatively different calls being given at different levels. Such an interpretation is less straightforward with the alarm calls of the female vervet monkey, including all but one of the vervet series. The ''rraup''

	UH!	NYOW!	CHUTTER	RRAUP	THREAT ALARM-BARK	CHIRP
ADULT ♂	yes	yes	yes (rare)	no	yes	no
ADULT AND YOUNG ♀	yes	yes	yes	yes	no	yes
TYPICAL STIMULUS	Minor mammal predator near	Sudden movement of minor predator	Man or venomous snake — but the chutter is structurally different for man and snake	Initial sighting of eagle	Initially and after sighting of major predator (leopard, lion, serval, eagle)	After initial sighting of major predator (leopard, eagle)
TYPICAL RESPONSE OF TROOP MEMBERS	Become alert, look to predator	Look to predator, sometimes flee	Approach snake and escort at safe distance	Flee from treetops and open areas into thickets	Attention and then flight to appropriate cover	Flee from thickets and open areas to branches and canopy

Figure 1. A diagram of alarm calls of the vervet monkey, stimuli evoking them, and responses to others. After Struhsaker 1967 and personal communication.

and the "chirp", for example, are given in rather specific circumstances and evoke specific and quite contrasting responses, descent from the treetops in one case, and ascent into the treetops in the other. It is not immediately obvious which is associated with a higher level of arousal although careful study might reveal that there are differences. Yet, while continuing to entertain a role for arousal in the programming of these alarm calls, it is here more than anywhere else in vervet vocal behavior, that one is tempted to think of the signals as functioning symbolically, serving a naming function, and designating particular referents.

Even here no alarm call is completely restricted to a single referent. Yet the group of heterogeneous stimulus objects triggering each one also has a set of unifying properties, namely those that constitute a particular class of dangers for which possession of a name, understood by comparisons in such terms, could be of great value to the vervet monkey.

What I am suggesting is that we could think of alarm calls not so much as having large, ill-defined classes of denotations, but rather as representing highly specific classes of dangers, each favoring a particular escape strategy. In this sense they may be viewed as symbolizing such dangers. If they also carry the hallmarks of affective behavior, with simultaneous signs of arousal, perhaps we could think of these as supplementing and enriching the symbolic function rather than excluding it.

There are at least four critical issues of specificity in characterizing symbolic and affective signaling (Figure 2). We assume symbols to have a certain *referential specificity*. Thus, to the extent that the vervet monkey chutter may be given to a man or to a snake, it fails to satisfy one of the implicit conditions for a symbolic signal. However, it might be a mistake to assume that vervet monkeys classify referential events or think about them in the same terms that we do. Such assumptions may be one of the most serious obstacles to overcome if we are to understand the natural communications of primates.

Another aspect of symbolic signaling is *motor* specificity. We require at least the potential ability to perform the motor act of signal production separately from the other behavioral processes associated with responding to the particular refer-

1. Referential specificity (to the class of objects, events or properties represented)
2. Motor specificity (relation to other responses to the referent)
3. Physiological specificity (association with arousal and emotion)
4. Temporal specificity (coupling to immediate referential stimulation)

Figure 2. Features of "specificity" differentiating "symbolic" from "affective" signaling.

ent—an alarm call disassociated from acts of fleeing for example, or a food call separated from eating. This, in turn, relates to the third issue of *physiological* specificity which is what I think we really have in mind when we consider the relationship between the production of a signal and arousal and the emotions.

For purely symbolic signaling, we require disassociation of production from the emotional responses that are normally associated with whatever the animal is signaling about. This relates in turn to what might be called temporal specificity, the binding in time of perception of a referent and signaling about it. Whereas these events are readily disassociated in our own symbolic behavior, they seem to be much more closely tied in animals. Although, as Yerkes reminded us, animals obviously have memories, they do not seem to signal about remembered events, at least not without the provision of some external cue. If rehearsal of some dangerous past experience is in fact impossible for an animal without the physiological correlates and manifestations of terror, one can imagine that out-of-context rehearsal of such events could cause a good deal of confusion. The issue of motor specificity relates to the same fact that the motor acts also reveal to us the animal's emotional state and its degree of arousal.

What we seem to find in so much of the natural signaling of animals is coherence between the production of signal and the other behavioral and physiological events that normally accompany the perception and response to the set of referential objects, whatever they may be. More than any other feature I believe it is this coherence that leads us to separate affective from purely symbolic signaling, where the degrees of specificity in all three of these domains can be so very much greater.

What is the proper position to take on this coherence? Should it be viewed as a primitive trait, such that imperfections in the physiological design blur the distinctions between responding to an event and signaling about the event? My own view of the matter is different. For one thing, such disassocation can occur in animals, as in play behavior, which typically involves a reorganization of the normal temporal relationships of signals and the other behaviors usually accompanying them, and perception of their usual referents. So it is by no means a physiological impossibility for animals, though it may well be something that in most circumstances animals can well do without, other than in play, where youthful experimentation free of affective accompaniments may facilitate ontogenetic experimentation. Far from being an impediment that somehow blunts the effectiveness of animal signaling behaviors, I would rather view the affective component as a highly sophisticated overlay that supplements the symbolic function of animal signals. Far from being detrimental, it increases the efficiency of rapid unequivocal communication by creating highly redundant signals whose content is, I suspect, richer than we often suppose. If at times we ourselves can afford to pare away so much of the affective accompaniment of our own speech this is to serve very special communicative requirements that rarely if ever arise in the natural behavior of animals, even in those as advanced as the chimpanzee.

It is also important not to underestimate the potential richness of the information content of affective signaling. Menzel's recent work clearly points in this direction (Menzel 1974), and one can look further back in time for a similar conviction, as in this quotation from Norbert Wiener's 1948 book on cybernetics:

Suppose I find myself in the woods with an intelligent savage, who cannot speak my language, and whose language I cannot speak. Even without any code of sign language common to the two of us, I can learn a great deal from him. All I need to do is to be alert to those movements when he is showing the signs of emotion or interest. I then cast my eyes around, perhaps paying attention to the direction of his glance, and fix in memory what I see or hear. It will not be long before I discover the things which seem important to him, not because he has communicated them to me by language, but because I myself have observed them.

Thus, with no other signaling elements than signs of arousal and the deictic property of eyes and where they are looking, such behavior has rich communicative potential. This theme, that we have underestimated the potential of affective signaling is echoed by Premack (1975a):

Consider two ways in which you could benefit from my knowledge of the conditions next door. I could return and tell you, "The apples next door are ripe." Alternatively, I could come back from next door chipper and smiling. On still another occasion I could return and tell you, "A tiger is next door." Alternatively, I could return mute with fright, disclosing an ashen face and quaking limbs. The same dichotomy could be arranged on numerous occasions. I could say, "The peaches next door are ripe" or say nothing and manifest an intermediate amount of positive affect since I am only moderately fond of peaches. Likewise, I might report, "A snake is next door," also an intermediate amount of affect since I am less shaken by snakes than by tigers.

Premack goes on to develop the differences between two kinds of signaling, referential (= symbolic) and affective (= excited or aroused), suggesting that information of the first kind consists of explicit properties of the world next door while information of the second kind consists of affective states, that he assumes to be positive or negative and varying in degree. He goes on:

Since changes in the affective states are caused by changes in the conditions next door, the two kinds of information are obviously related. In the simplest case we could arrange that exactly the condition referred to in the symbolic communication be the cause of the affective state.

As Premack indicates, as long as there is some concordance between the preferences and aversions of communicants then a remarkable amount of information can be transmitted by an affective system. While he explicitly restricts himself to "what" rather than"where" one may note, harking back to the Wiener quotation, that incorporation of a deictic component in the signal—pointing or looking—not only indicates where, but also adds a highly specific connotation—one particular apple tree.

While Wiener reflected on the possibilities of an arousal system with a single dimension, Premack has implicated at least two, a dimension of positive affect concerned with attraction, and a dimension of negative affect concerned with apprehension and avoidance. However even two dimensions seem inadequate to explain more than the grossest features of the diversity of signal structure and usage in animals. Again the issue of specificity intrudes. For example there would be considerable advantage, teleologically speaking, if a signaler could give even a general indication of the class of referent being signaled about by positive affect. Otherwise, a hungry animal, for example, might approach a referent indicated by a companion as worthy of positive affect only to discover that it indicated not food but a social companion or a mate. Much time and effort would be saved if, at the very least, the signaler could give some indication as to whether the referent is environmental in nature or social in nature. Environmental designata might include hazards such as predators or bad weather or something positive such as resources, food and drink, or a safe resting place. Social referents. on the other hand, again with either positive or negative connotations, might include a mate or a companion or an infant in need of care, all worthy of approach, or negative social information such as an enraged male of the species, to be avoided.

Intriguingly, I find that some analyses of human emotional states use four dimensions that, with some reshuffling, are reminiscent of those that I have mentioned. Figure 3 is modified from the work of Plutchik (1970), a psychiatrist interested in measuring the behavioral biases of emotionally disturbed patients. His major emotional categories, which can also be viewed as communicative systems, are shown in the center of the diagram. In the outer ring I have added some ethological equivalents to those emotional states at the level of ongoing behavior.

Crude though it is, such a classification encompasses most of the major sets of activities that comprise the ethogram of an animal. It is also compatible with the four affective dimensions of arousal/depression (A) approach/withdrawal (B), object acceptance/rejection (C), and social engagement/disengagement (D) that I have already mentioned (Figure 4).

I am not necessarily wedded to the details of this particular classification. Rather, I am trying in a different way to make the point that in some communicative circumstances there are benefits in an increasing degree of specificity in the multiple relationships between production of a signal, its underlying physiological basis, and the kind of referential events that are contingent to its production. I propose that it may be fruitful to think of symbolic and affective signaling as representing extremes on the same continuum of specificity in these relationships. This would imply that the two signaling modes are more properly thought of as differing in degree rather than in kind. The particular constellation of relationships between referential, motor and physiological specificity that is manifest in so much animal signaling, often at the affective extreme, should be thought of not as something that animals do because they are physiologically incapable of more refined relationships in the

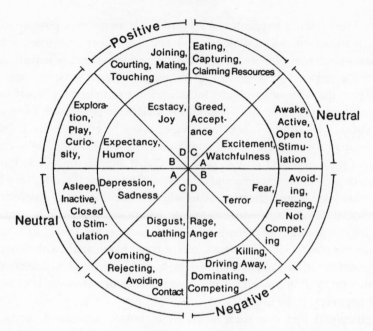

Figure 3. A diagram of human emotional states (inner ring) with some equivalent ethological activities (outer ring) arranged according to whether their connotations are positive, negative, or neutral. A-A = a possible arousal/depression. B-B = a locomotor approach/withdrawal dimension. C-C = an object acceptance/rejection dimension. D-D = a social engagement/disengagement dimension. Modified after Plutchik 1970.

direction of symbolic signaling. Rather, it may be viewed as an adaptive compromise which, even in animals can be seen to move to and fro, in an evolutionary sense, along a continuum of specificity, as with the gorilla belch and the chimpanzee rough grunt, depending on what particular solution solves a given behavioral problem most efficiently.

As Yerkes indicated so long ago, and as we now appreciate even better, it is not that animals are by any means incapable of symbolic thought. The point is rather that purely symbolic signaling is a highly specialized mode of operation which, given all the time in the world, has enormous potential, but used in haste in

(a)	Arousal/depression
(b)	Locomotor approach/withdrawal
(c)	Object acceptance/rejection
(d)	Social engagement/disengagement

Figure 4. Four proposed dimensions of "affective" signaling.

circumstances of dire emergency as is so much more typical of animals, would be hopelessly slow and ambiguous. We see this even in our own behavior, in the intrusion of nonverbal signaling in so many situations that are especially important to us. As everyone ever associated with the name of Yerkes is aware, man has more of the nonhuman primate in him than we often admit.

SUMMARY

An attempt is made to distinguish affective from symbolic signaling in animals and man. In making such a judgment it is common to appeal to the specificity of the multiple relationships between a signal, its external referents, and associated physiological states and behaviors. Referential specificity refers to the class of objects or events represented by the signal. When the class is large we tend to label the signal as affective. Similarly with motor specificity, intermingling of signal production with other responses to the referent is viewed as more typical of affective than of symbolic signaling. When physiological specificity is broad, so that physiological requirements for signal production are accompanied by signs of arousal and emotion, the signal is usually viewed as affective. With regard to temporal specificity, so-called affective signals are usually locked in time to immediate referential stimulation. A review of examples suggests that the degree of specificity exhibited by different animal signals falls on a continuum, so that symbolic and affective signals may be viewed as differing in degree rather than kind. The speculation is offered that in animal signaling, as in human communication, there may be a mingling of symbolic and affective components, the latter having a richer communicative potential than is sometimes supposed.

ACKNOWLEDGMENTS

The author is indebted for thoughtful discussion and criticism of this paper to Steven Green, Donald Griffin, William Mason, David Premack, and Thomas A. Sebeok.

LANGUAGE BEHAVIOR OF APES*

DUANE M. RUMBAUGH

INTRODUCTION

Man's interest in the origins of human attributes has been strong from the time his intelligence first allowed him to reflect upon the nature of things and the relationships between events in his environment. Increasingly, sheer conjecture on the topic has been supplanted by biological and archaeological findings, by formal theory, and by the intrusion of research programs into the study of human attributes and skills once held to be unique, immutable,and/or god-given.

The surge of research interest in the nature of language and the requisites of language acquisition is a case in point. Because only man manifests language with a productive syntax by which novel messages are formulated and publicly transmitted, it is understandable that until now language has been regarded as uniquely human. Language *is* uniquely human in that man and man alone develops this skill with apparent spontaneity and with high probability; however, the results of the last decade's research with chimpanzees that have developed language-type skills have raised a new question: Are the *requisites* to linguistic competence also uniquely human?

The purpose of this chapter is to consider evidence emanating from the various chimpanzee language projects along with selected theoretical perspectives and other related research in an attempt to formulate at least a tentative answer to this question. Sinch language is viewed here as only one of the behaviors which can enhance the probabilty of successful adaptation to challenge from the environment, a general statement regarding behavior is in order.

Behavior and Access to Language Origins

It is through behavior that animal cells, either singly or in concert, effect an interface with the environment. To the degree that the interfacing sustains the life

*Preparation of this manuscript was supported by grants NICHD–06016 and RR–00165 from the National Institutes of Health.

DUANE M. RUMBAUGH ● Department of Psychology, Georgia State University, Atlanta, Georgia 30303.

231

and reproduction of those cells, adaptation is said to have occurred. Single-celled animals or even the single cells of multicellular organisms can and do manifest behavior, for all cells have the general characteristic of being "irritable" or responsive to some type of environmental stimulation, whether external or internal. Clearly, single cells have fewer options for adaptive behavior than *systems* of cells, which can respond in a coordinated manner. Systems of cells are able to adapt through various tactics to a greater variety of environmental challenges and to more difficult ones than are single cells. In general, the more complex the cell system, the greater is the array of behavioral options available for possible adaptation. Man has greatly increased his behavioral options for adaptation to seemingly limitless challenges and niches through the evolution of language. In all probability, the evolution of language has been contingent upon brain evolution and the behavioral plasticity which this evolution provides through open genetic programs (Dingwall 1975; Mayr 1974).

Although this perspective is not new, it should be kept in focus while considering the requisites to language. Although it is possible that genetic alterations or mutations unique to *Homo erectus,* to one or more of his descendants, and/or to *H. sapiens* may have provided totally for man's language skills, it is more probable that language evolved gradually (Toulmin 1971) and that comparative research will yield traces of language origins. Since our immediate precursors have long been extinct, their capabilities for language can only be assessed inferentially through studies of their artifacts (e.g., Marshack 1976) and through their fossil remains. Regrettably, behavior does not fossilize.

As far as direct empirical research on the origins of language is concerned, behavioral studies with extant animal forms offer the only possibility for systematic study. Since apes and monkeys are more closely related to man than are other animal forms, it is reasonable that such research should focus on them, although they are not, of course, the precursors of modern man. Man does share common evolutionary roots with both apes and monkeys, the root common to apes being the more recent one, but since the time of the radiations which gave rise to monkeys, apes, and men as we now know them, all the varieties of primates have been novelly shaped through selective environmental forces. That fact notwithstanding, it remains possible that apes and monkeys (not necessarily to the exclusion of non-primate forms) may have carried forward through their own evolution the behavioral processes and potentials which provided for the evolution of the linguistic competence in modern man. In short, comparative behavioral studies may reveal that the requisites to linguistic competence are not unique to *H. sapiens.*

Language in Comparative Perspective

The pivotal question is: What is linguistic competence, i.e., language? It is easy to define language in such a way as to preclude comparative research. If language is defined in terms of *human speech,* none other than *H. sapiens* can possibly have linguistic competence: no other form is human and even it if could speak, its language would not be human speech. Speech is so salient a characteristic of human language that speech and language have understandably come to be synonyms. Speech is surely the most efficient medium of human language, but to relegate all other linguistic-relevant media such as writing and reading to a secondary or derived status probably hinders efforts to define the psychological substrates of linguistic competence and performance. Dingwall, for example has concluded that apart from speech there are no aspects of language that are unique to man (1975:45).

Rumbaugh and Gill (1976a) have suggested a comparative perspective regarding language. They see it as having private (covert) and public(overt) dimensions and as being only one of many forms of animal communication—one set apart from others by its infinite openness, to use Hockett's (1960a) term. It may or may not be the case that some feral animal form will be found to exhibit openness in its communication, being able to produce and to employ new signals (lexical units) in new combinations (syntax) with other signals to generate and transmit novel messages. Remote as that possibility seems at present, it may still occur in the foreseeable future. That possibility notwithstanding, the comparative language projects of the past decade support the conclusion that chimpanzee subjects benefit rather readily from training in skills which allow for public exchanges of linguistic-type (open) communication. Perhaps they do so because they are naturally equipped with the processes and/or potential for developing the *private* dimensions of language which include symbolizing, the ability to relate different concepts to each other, and the ability to synthesize information and skills for application to novel problems (Mason 1976; Premack 1975). Given the instigation by man and language systems contrived by man which circumvent their inability to speak, they seem able to acquire at least the rudiments of public linguistic skills. The chimpanzee's capacity for such skills is due, possibly, to a brain which bears close similarity to man's (Hill 1972; Jerison 1973; Radinsky 1975; Shantha & Manocha 1969) and to a relatively advanced form of intelligence which is basically, though not exclusively, hominian (e.g., Menzel 1973a, 1974; Menzel & Johnson 1976).

This writer's comparative view of the evolution of the primate brain and attendant abilities for learning, intelligence, and language is as follows: The higher-order psychological abilities of nonhuman primates are a function of the degree to which their brains are hominian. To the degree that they such brains they appear to have sensory, perceptual, and learning processes and intelligence similar to man's (see King & Fobes 1974; Menzel & Halperin 1975; Parker 1974; Warren 1974; Young & Farrer 1971, for representative evidence). To the degree that these parameters are collectively requisite to the covert structures and functions of expressive (public) language, the possibility exists that the psychological foundations of language can be identified. Language in this view is a behavioral domain and not just an attribute of man; other animals, notably the chimpanzee, are capable of behaviors which may form a continuum with human language (Dingwall 1975;Rumbaugh & Gill 1976a).

Primate Intelligence and Learning

Despite the initial interest of comparative psychologists in "animal intelligence" several decades ago, the use of intelligence as a concept came increasingly to be reserved for use in reference to man until very recent years; however, with the development of a more adequate comparative perspective regarding brain development and brain function, it is now reasonable to use the term with certain restrictions in reference to nonhuman primates. If man's intelligence is primarily a function of his advanced brain development, then other nonhuman primates, to the degree that their brains approximate man's, might likewise possess dimensions of intelligence which are requisites to "language." Restrictions in the use of the term intelligence include the following: (1) It is unreasonable to expect nonhuman primate forms to manifest all dimensions of human intelligence. By the same token, it is unreasonable to expect man's intelligence to manifest all dimensions of nonhuman primate intelligence. (2) Major dimensions of intelligence for the various primate forms, including man, have quite possibly been selectively shaped by ecological factors, and this shaping may have occurred through unique interactions of selective pressures across millions of years with the various radiations of primate forms. By implication, wide variation in the dimensions of intelligence can be expected when the diverse primate forms are compared.

By and large, we do not have the data base necessary for laying out the differences between the various profiles of intelligence that this perspective anticipates; however, certain relevant data are already in hand and recent conceptual developments will possibly stimulate the collection of still more. McNeill (this volume), for example, has observed that although chimpanzees *Pan* are intensely aware of one another, engaging in nearly continuous interactions when together, they seem to be much less curious about objects unless those objects relate at a very practical level to such matters as assessing the condition of feeding trees, collecting and preparing materials for "fishing termites" (van Lawick-Goodall 1970), and attacking threatening predators (Kortlandt 1965). McNeill felt that this might help account for the indication that the chimpanzee, Washoe (Gardner & Gardner 1971) was not as adept at encoding relationships among objects as she was at encoding relationships between individuals. Abstract or rule learning is herein considered to be the foundation of the private or covert dimension of intelligence and language referred to earlier. Learning of this type was probably necessary for the accretion of skills that allowed for the elaboration of public language, the type of language which in its decontextualized form becomes "the medium for passing on knowledge" (Bruner

1974: 38). The reader is referred to Rumbaugh (1970, 1971), Warren (1973, 1974), and Mason (1976) for reviews that bear upon comparative learning and representational processes from a comparative perspective.

Cross-Modal Perception

The capacity to equate stimuli on the basis of independent input from diverse sensory modalities may also have become progressively refined as the primate brain evolved.(See Ettlinger 1977 for a review of this topic.) Davenport and Rogers (1970) demonstrated that great-ape subjects are able to select one of two haptically sensed stimuli and match them to a visual sample in tests using novel materials. They have also shown that without special training apes are able to use high-quality photographic representations as the visual stimuli to be matched through haptic choice, and that apes are also able to negotiate at least a 20-sec delay prior to haptic choice. They interpreted these findings to mean that the apes are able to store symbolic information beyond the commonly accepted limits of short-term memory without the benefit of formal verbal competencies to mediate that storage (Davenport, Rogers, & Russell, et al. 1973, 1975).

Attempts by Davenport and others to establish cross-modal competencies with rhesus macaque *(Macaca mulatta)* subjects have succeeded only marginally. Rhesus do have at least circumscribed skills of this type, however, for Cowey and Weiskrantz (1975) have demonstrated their ability to visually select edible from inedible foods with distinctive shapes subsequent to haptic experience with them during feeding sessions in the dark. In this writer's opinion, extant data demonstrating abilities for cross-modal perception more clearly differentiate great apes from rhesus macaques than do any other type of data obtained in formal test situations.

Deficiencies in cross-modal perceptual skills can apparently serve to restrict the perceptual processes necessary to process sensory inputs from diverse environmental sources efficiently. Such deficits would seem to hamper the organism's ability to monitor events in a highly dynamic environment and to automatize patterned behavior (Lieberman 1975b:92), and they would also seem to preclude the decontextualized use of language described by Bruner. Formal language systems may quite possibly serve to integrate environmental events and concurrently enhance the options for adaptive behavior. As Bruner sees it, "though language springs from and aids action, it quickly becomes self-contained and free of the context of action. It is a device, moreover, that frees its possessor from the immediacy of the environment not only by preemption of attention during language use but by its capacity to direct attention toward those aspects of the environment that are singled out by language'' (1974:37).

The Evolution of Human Speech

Animals have relatively sophisticated communication channels which are used in ways that enhance the probabilities of survival. But while some of their species-typical signals might be altered through imitation, stimulation, and learning, "there is no evidence that [animals] continue to recombine qualitatively different vocal units, separately meaningful, into new messages with new meanings. In other words, they have no syntax" (Marler 1975:33). According to Marler, it was the early evolutionary emergence of "templates" into the perception and learning of speech that facilitated the human child's mastery of speech by imitation and the segmentation of highly dynamic speech sounds and prepared as well for "the syntactical openness that made our speech such a powerful force in social evolution" (1975:34). Marler's comparative studies with song sparrows and chimpanzee vocalization patterns and his review of studies dealing with animal communication and human speech perception led him to suggest "that speech arose not just from a general increase in human intellectual capacity, but from a distinctive concatenation of specific physiological abilities on the one hand, and behavioral and ecological opportunities on the other, pressing the exploitation of these abilities to the utmost. Though human in detail, none of the basic physiological mechanisms is in principle unique to man" (1975:34). In Marler's opinion it was probably the discovery and exploitation of the tools used in cooperative hunting that selected for those processes which collectively constitute human language.

The development of competent speech surely followed a gradual evolutionary course, and one which perhaps faltered along the way. The great apes never have evolved the cortical control allowing for the graded repertoire of vocalizations necessary for speech. Various studies (see Kellogg, this volume for a review), notably the one by Hayes and Nissen (1971, reviewed below), have demonstrated that the ape can master only the most rudimentary skills of speech. Thoughts concerning the evolution of human speech are found in theoretical articles and research reports by Liberman, Cooper, Shankweiler, and Studdert-Kennedy (1967), Nottebohm (1972), Lieberman (1975b), Hamilton (1974), Falk (1975), Pribram (1971, 1976), and Dingwall (1975). The work of these investigators suggests that the evolution of human speech involved a number of coordinated evolutionary developments, including the following: refined neocortical control over the movements of the tongue, jaw, lips, larynx, and vocal cords; alterations in the anatomy of the larynx and the relative positions of the epiglottis and velum which facilitate the articulation of a broad range of sounds; cytoarchitectural areas of special control for speech (Broca's area) and word comprehension (Wernicke's area), and the requisite neural tracts for their coordinated function (notably the arcuate fasciculus and the angular and supramarginal gyri); hemispheric asymmetry in sensitivity for auditory speech perception; enhanced neural associations for commerce between the visual, auditory, sensory, and motor cortices; and possibly neurological feature detectors which predispose even the human infant to attend selectively to certain sounds used in speech, to discriminate similar phonemes, and to emulate selectively certain sounds that are germane to natural speech. These and other evolutionary developments are rather reliably associated with hemispheric asymmetry of the planum temporale, which is very pronounced in man, extant but not as extensive in the great apes, and possibly absent in the rhesus monkey (Yeni-Komshian & Benson 1976).

Apart from the above developments and the development of the frontal lobes, there are relatively few characteristics of the human brain which differentiate it from the apes', except for its extraordinary size. For man's average body weight, his brain is approximately 3 times greater than what would be predicted from the brain/body-weight regression line for primates in general (Radinsky 1975).

Summary

The conceptual foundations of ape-language projects are in line with the evolutionary framework of comparative psychology (Church, 1971; Riesen 1974; Savage & Rumbaugh 1977). The projects all rest on the assumption that animals are capable of behaviors which form a continuum with human language functions. Speech has been rejected as a requisite to language, despite its refinement for linguistic communication among humans. Artificial or other "silent" language systems have been devised to circumvent the inability of the apes to master human speech. And apes have been the preferred choice for these projects of animals to date because their behavior and brains bear such close resemblance to ours.

THE LANGUAGE PROJECTS

The origin of the idea that the chimpanzee and the other great apes might be capable of learning a natural human language is perhaps impossible to identify. Hewes (1976, 1977) provides a detailed historical account of the theories of glottogenesis and of speculations regarding the possibility that apes might learn language. This writer is indebted to Hewes for bringing to his attention the conjecture of La Mettrie in 1748 (published in 1912) that it might be possible to teach language to an ape. La Mettrie indicated that he would choose a young ape, one with "the most intelligent face," and one whose mastery of tasks confirmed that it was indeed intelligent. He would pair the ape with an excellent teacher, Amman, whose ability to discover "ears in eyes" of humans born deaf (to teach them language despite their handicap) had left a lasting impression on him. La Mettrie thought that apes might even be taught literally to speak and that if they could learn to do so, they would then know "a language." Unlike the other "true" philosophers of his time, he viewed the transition from animals to man as a continuum, not a "violent" break. And he speculated on the state of man before "the invention of words and the knowledge of language," concluding that man was then but another animal species, though one with less instinct than other animals. La Mettrie viewed language skills as the origin of law, science, and the fine arts and held that they had served to polish "the rough diamond of our mind." Should the ape master language, he too would be gentled; he would no longer be "a wild man, nor a defective man, but he would be a perfect man, a little gentleman. . . ." Nearly two centuries later Yerkes (1927) also speculated that apes might master a gestural, nonvocal language system.

Kellogg's review (this volume) provides the reader with a comprehensive account of early attempts to teach apes to speak. At the turn of the century, efforts by Witmer (1909) and Furness (1916) netted only the slightest suggestion of success—despite ardous effort, their chimpanzee and orangutan subjects mastered only one or two approximations of words. In their own work, the Kelloggs demonstrated that the chimpanzee does have a rather well developed ability to "understand" simple voiced commands; Gua, their subject, was able to respond appropriately to 68 specific commands. In brief, their chimpanzee appeared to have at least a receptive competence for language if not an expressive one. Gua was not subject to extensive, systematic training to develop language, but there was nothing to prevent her from developing an expressive linguistic competence on her own in the enriched home environment in which she was kept had she been so disposed and able to do so. Neither Gua nor any other ape kept as a research subject in a home situation nor any ape kept as a pet has developed a voiced language competence on its own initiative. Although apes do communicate by combinations (not syntactically arranged) of vocalizations, gestures, movements, and facial expressions, they do not develop human language skills simply by living with humans in a linguistically saturated milieu.

Project Viki

The Hayeses (Hayes & Nissen 1971) were the first to conduct a prolonged program with a chimpanzee that included an attempt to teach it to communicate via speech. Their chimpanzee, Viki, was kept in their home and was subject to a wide variety of ingeniously cast projects to tap her "higher mental functions." Tasks entailing the use of instruments; the perception of numbers in a manner that suggests rudimentary counting skills; and the mastery, use, and generation of concepts are lucidly reported by Hayes and Nissen. Viki's limited success in learning sequences of responses in string-pulling tasks was taken to suggest a barrier caused by her lack of language skills. Efforts to teach her speech met with only a modicum of success. Viki chimpanzee mastered four voiceless approximations of "mama," "papa," "cup," and "up." She also used "clicking of the teeth" for a ride in the car, "tsk" to get a cigarette, and another sound, said to be reminiscent of the "clicks" used in certain non-European languages, for requesting to go outdoors (Hayes & Hayes 1950). Ironically, the Hayeses' failure did lead them to consider the use of a gesture language, in part because of a written suggestion by Hewes (see Hewes 1977, for an account of the correspondence); ultimately, however, they did not attempt to teach Viki a gestural language, though their reports of her behavior are replete with instances where she used gestures of her own initiative to attempt communication. For instance, Viki would "ask" for a ride by gesturing toward the car outside, by leading one of the Hayeses by the hand to the drawer which contained the key to the front door, by bringing them a purse always taken along on car rides, and by either showing a picture of a car to the Hayeses or by clicking her teeth together as a "word" for this request. Systematic use of a gestural language with an ape was not implemented until 1966 with the launching of Project Washoe, described below (also see Gardner & Gardner 1971).

The study by the Hayeses was not a failure, however, for it served to underscore the advanced cognitive skills possessed by their subject, Viki. It also served to indicate that if the chimpanzee were capable of language as used by man, its language would not be likely to take the form of speech. Since the time of the study with Viki, a great deal has been learned about the anatomic restrictions of the chimpanzee's vocal tract as far as the production of human phones and phonemes is concerned (Lieberman 1968). Had these limitations been understood in 1947 when the study began, the Hayeses would probably not have used speech as the language medium for training. That probability notwithstanding, we are fortunate that an able pair of researchers did attempt what is now viewed as impossible—the instruction of a chimpanzee in human speech.

Project Washoe

The project undertaken by Beatrice T. and R. Allen Gardner (1971) in June 1966, provided the first strong evidence that the chimpanzee has the potential to master at least the rudiments of a language system devised and used by man. From the Hayeses' project with Viki, the Gardners had concluded that either the chimpanzee did not possess the capacity for significant two-way communication or that, though the capacity was indeed present, human speech was not a feasible medium for such communication.

For them it seemed reasonable to expect that "behavior that is at least continuous with human language can be found in other species" (Gardner & Gardner 1971:118). This perspective, fundamental to the field of comparative psychology, is held in common by all investigators who study the potentials of the chimpanzee for language. More critical, however, are the questions: What is the linguistic domain? and, What is language? For the Gardners, any system is a language if "it is used as a language by a group of human beings" (1971:122).

Speech was ruled out as a feasible choice for cultivating two-way communication skills in their chimpanzee subject, Washoe, a female estimated to be between 8 and 14 months of age when the project began. (The subject was named after Washoe county in which the Gardners' academic home, the University of Nevada, is situated.) The American sign language (Ameslan) was selected as the linguistic medium for the project (1) because it met their definition of a language in that it is used widely by one group of human beings, the deaf in the United States; (2) because the chimpanzee in the field is reported to use gestures in social communications; and (3) because the use of gestures would obviate the many difficulties in the invention of a synthetic language. In view of the lack of a generally accepted set of criteria for determining when a given human being has language competence, the Gardners refused to commit themselves to performance criteria for concluding that their chimpanzee had mastered language; however, they did anticipate that meaningful comparisons might be made between the signing skills of the chimpanzee and those of children.

The diverse aspects of Ameslan and questions regarding its structure, syntax, and production are to be found in Stokoe, Casterline, and Croneberg (1965), Bellugi and Klima (1975a) and Stokoe (1975). For this author, the appropriateness of Ameslan for the Gardners' project, given what was then *not* known about the chimpanzee's propensities for learning language skills, was vindicated not only by their project's results, but by Stokoe's (1975:210) report that, according to their parents' reports, the hearing children of deaf parents have *signed sentences* prior to learning to talk! Such observations seemingly attest to the "naturalness" of communicating by manual gestures and support Hewes's theory (1973b, 1976) that manual gesturing was possibly the original medium of linguistic communication among the precursors of man.

To date, all the ape-language projects have made varying, but nonetheless significant, demands on the subjects' use of hands. The ability to gesture or to sign with facility is contingent, in part, upon a highly mobile hand, the evolution of which is fairly well understood. Through the course of primate evolution the hand has undergone radical modifications (Schultz 1969; Steklis & Harnad 1976) which have made it quite versatile for gesturing. The thumb in the Old World monkeys (*Cercopithecoidea*) evolved to a point of rotation so that between its transverse axis and the transverse axis of the other digits an angle approximating 90° was formed. It also branched more proximally to the second phalangeal joint and acquired increased mobility, thereby allowing for the hand to generate a greater repertoire of finely controlled movements. These trends produced remarkable manual skills in the chimpanzee and the gorilla, many of which are of nearly human quality. The late scientist–philosopher J. Bronowski, in reference to the productions of the human hand, noted that "man alone leaves traces of what he created" (1973:38). For him, "the hand is the cutting edge of the mind," and the ascent of man's civilization "is the refinement of the hand in action." (p. 116)

Washoe's Laboratory

The Gardners were determined to provide, insofar as resources would permit, a rich, stimulating environment within which their subject would live and learn. A trailer home surrounded by a large fenced yard comprised the laboratory context. Within this context, Washoe was treated in many ways as though she were a human child—not with the intention of stimulating a human family-life condition as with the Kelloggs and the Hayeses, but to provide a challenging milieu within which she might learn and apply language skills in an adaptive manner.

The trailer and the yard were equipped with a variety of materials to stimulate Washoe's attention, activity, and language-learning. In addition, trips to various parts of the city, the university, and the country were provided. The goal was to immerse Washoe in a linguistically enriched style of living that would foster whatever language skills her *Pan troglodytes* genetic endowment might support. The linguistic context was restricted to Ameslan, except for the inevitable speech sounds of the neighborhood and the occasional outings. No deliberate attempt was made to keep Washoe's milieu silent, but every effort was made to restrict communications between persons in her presence to Ameslan so that she would have only one linguistic system to imitate.

Teaching Methods

Various methods of teaching signs to Washoe were employed. Initially, attempts were made to shape responses from gestures that in some manner suggested that target sign to be mastered. Efforts to capitalize on Washoe's manual "babbling" were also made initially, but these were abandoned as relatively futile. Guidance was found to be perhaps the best method of all; it entailed holding Washoe's hand, forming her fingers in the desired fashion and moving her hand thus formed through the movement appropriate to the sign to be learned, and then giving her a reward such as food or tickling. Washoe also acquired a few signs through observation (e.g., "sweet:" contact of tongue and lower lip with first and second fingers) of those signing about her and even invented a few signs of her own (e.g., "bib": drawing the outline of one with index fingers on the chest) which, interestingly, proved to be valid Ameslan signs (Gardner & Gardner 1971:139).

Detailed daily records were kept of Washoe's spontaneous signing and her responses to questions such as "What is it?" or "What do you want?" Only when three observers reported Washoe to have signed spontaneously and appropriately to the context was the introduction of a sign recorded. The criterion for determing that a given sign had been *reliably* mastered by Washoe was extremely, and perhaps overly, stringent. It required that Washoe generate at least one observed appropriate and spontaneous use of a sign for 15 *consecutive* days. McNeill (this volume:147) concludes, based on his own work and L. Bloom's, that the object-words that children acquire are highly transitory and that substitutions across days are the rule. By implication, the stringent criterion employed for defining a sign as reliable for Washoe might well have served to exclude object-words in early stages. This risk was perhaps minimized by the Gardners when they began to conduct two formal sessions daily during which time stimuli were introduced for the possible elicitation of those signs not otherwise observed. Although the rationale for the criterion of reliability of Washoe's signs was fundamentally sound, it in no way ensured that the initial corpus was comprehensive; other criteria would have resulted in different definitions of Washoe's lexicon. Consequently, its exact composition may be compared with those of the human child but only with a certain risk of lack of definition.

A self-paced, double-blind testing situation was devised by the Gardners to establish Washoe's signs unequivocally in reference to an array of exemplars. High reliability was reported. Within limits, the semantic value of each sign was inferred as it was or was not given in response to novel exemplars of the referent.

It bears noting, however, that the test procedures did not provide for a way to determine the degree to which Washoe's signs were essentially generalized responses to stimuli similar to those used in prior training. Her competence in the use of her signs syntactically in order to generate and transmit truly novel messages, information, or requests was not tapped by this or any other test. Instances of her using signs in these ways were reported separately; however, they may have been fortuitous productions.

Her extensions of certain word-signs to exemplars quite foreign to those used initially in training are, nonetheless, quite impressive in many instances and should be accepted as evidence that Washoe did conceptualize the meanings of certain signs in a semantic sense. A prime example is in her use of the sign for *open*. Initially she learned this word in reference to a door of the trailer in which she lived. Its use was extended spontaneously (i.e., without special training) to various doors of buildings, to the lids of briefcases and boxes, and most impressively of all, to a water faucet apparently as part of a request that she be given a drink of water (Gardner & Gardner 1969). Extensions of the use of signs in response to other than the exemplars or contexts in which they were initially used were recognized by the Gardners as constituting the best opportunity for inferring the meaning which the words had for Washoe.

Significantly, Washoe not only imitated but did so on a deferred basis. One morning, for example, the Gardners tried unsuccessfully to give Washoe an injection. Later on that same morning, Washoe was observed to be sticking the inside of her thigh with a long nail which she found, her actions not unlike those she saw the Gardners attempting to carry out on her with the hypodermic needle (1971:178). This episode is apparently a remarkable instance of a chimpanzee engaging in an activity that would suggest both a representation of the initial shot-giving episode and the metaphorical use of a nail.

On the question of whether or not Washoe gave evidence of using syntax, the Gardners were properly guarded in their 1971 report. Nonrandom patterns of stringing signs were not necessarily taken to mean that Washoe had some sense or mastery of syntactic principles. After all, no rules for the stringing of signs were required in Washoe's training. In addition, there was always the possibility that generalized imitation of combinations of signs used by the humans might have determined Washoe's patterns. Those views and possibilities notwithstanding, it is impressive and, I believe, highly significant that Washoe did spontaneously string signs together; she did so selectively and in many instances very appropriately (signing at a locked door—"gimme key," "more key," "open more," "open key please," "in open help;" requesting soda pop ["sweet drink"]—"please sweet drink," "gimme sweet," "please hurry sweet drink," "please gimme sweet drink" [Gardner & Gardner 1971:167]). It would also seem that her use of signs was characterized by certain consistencies idiosyncratically determined. In short, Washoe's use of signs in some degree suggests extended linguistic functions, as seen in her propensity for stringing words as though to form phrases, and also an inkling of syntactic sensitivity. At no point, however, was Washoe highly productive in forming and transmitting novel statements in the diverse situations available to her; nonetheless, we cannot assume on the basis of her performance that the mastery of these skills is beyond the competence of the chimpanzee, be it Washoe or any other subject.

At the conclusion of the 1971 report, the Gardners reasserted their reluctance to join the debate on the capacity of an animal (Washoe) to acquire language. They acknowledged that in the final analysis language can be defined so as either to include or exclude a given set of data, and they left it to others to conclude for themselves whether or not Washoe had achieved language.

A later report by the Gardners (1975b) on Project Washoe builds upon the hypothesis that the best evidence of vocabulary use by young children lies in their composition of sentences in response to "wh" (who, what, where, and whose) questions. They predicted that because Washoe was maintained under conditions similar to those of a child, her responses to wh-questions would be similar to those of children. From the 50th and 51st months of the project, Washoe's responses to 50 wh-questions of 10 different question-frame types were examined. The Gardners made a variety of analyses to determine whether "the replies were grammatically controlled by the questions" (1975b:251). The words of Washoe's responses were categorized into proper names, pronouns, possessives, locative verb phrases, etc., for the 10 different question frames; a significant relationship was statistically established, and Washoe was adjudged to be relatively advanced compared to young children (Stage III, Brown 1968). But whether this level of mastery constitutes *grammatical* control is a real question, for grammar subsumes word-order effects. The constituents of Washoe's responses to the wh-questions (e.g., Q. "Who you?" A. "Linn." Q. "Whose that?" A. "Shoes you." Q. "Where we go?" A. "You me out." were by and large appropriate, but the report and its analyses did not address questions regarding the logic inherent in the organization of responses that entailed *more* than one word—a relevant issue in the evaluation of grammatical control.

The Gardners have been sensitive, and quite responsibly so, to the possibility that Washoe's responses might have been inadvertently cued by the questioners. They conclude that their double-blind testing situation for tests of vocabulary "must be the rock-bottom definition of verbal behavior" (1975b:256). They argue further that, in their requirements for *production* by Washoe (i.e., that she form the hand signs for responding) and for other tasks as described above, inherent control against cueing her responses was built in: the cues would have to "contain as much information as the correct replies" (1975b:256). But in instances where her replies consisted of only one or two words, it is possible that cues, *if extant,* might have provided all the information necessary for correct responses: production does not *necessarily* preclude all risk of inadvertant cueing of the subject.

Interestingly, in their most recent article the Gardners refer to Köhler's (1927) early work in which Köhler demonstrates that chimpanzees are able "to combine and recombine their learned responses in meaningful sequences" (Gardner & Gardner, 1975b:245). Since chimpanzees had long been known to possess this type of ability, the Gardners state that they, unlike other researchers, saw no need to determine the chimpanzee's capacity for language. Consequently, they designed their research to determine whether a chimpanzee might use a bona fide human language, i.e., Ameslan. Support for their train of reasoning, that chimpanzees were already *known* to possess the capacity for "language," is generally lacking, for language is surely more than the mere recombination of responses. If it is no more than that, then every animal form that has been conditioned to chain responses in sequences and that modified those sequences might be said to have language. The question of whether or not the chimpanzee is capable of language admittedly revolves around definitional problems just as the Gardners assert—of that there is no doubt. Until those definitional problems are resolved, the question of the chimpanzee's capacity for language should remain open.

Project Sarah

A precursor to Premack's work with Sarah chimpanzee was his earlier work (Premack & Schwartz 1966) designed to determine whether a chimpanzee might

master a joy-stick control so as to modulate the generation of electronically produced sounds for communication purposes. This study, while not successful, was
an admirable attempt to circumvent the chimpanzee's limitations in the production
of vocal sounds for linguistic exchange.

Premack's work with Sarah contrasts sharply with the methods employed by
the Gardners in Project Washoe. In contrast to the Gardners' requirements that
work begin with an infant chimpanzee, that the chimpanzee be maintained in a
linguistically saturated and generally stimulating environment, and that it be signed
to in the course of social exchanges throughout the waking hours, Premack's work
with Sarah (1970a, 1971a, b) was modeled after the general approach and methods
employed by experimental psychologists in laboratory animal research projects.
Sarah, like Washoe, was feral-born and was but an infant (estimated 9 months old)
when received by Premack. For the first year, she was raised in a home where she
could be given the personalized care necessary for normal healthy development.
Beyond that point she was maintained in a laboratory setting and served in various
research projects. Nonetheless, her social contacts with humans continued to be
close and until she was about seven years old she enjoyed the probable benefits of
daily outings, which included romps on the beach. Her linguistic training sessions
took place during specified and limited hours of each day, and the methods
employed in those sessions were more of a discrete-trial type than of the social-
interactional type used with Washoe. All these differences notwithstanding, Sarah
did exceedingly well in her mastery of language skills. How and why she did so
warrants the closest attention, because the results contain clear implications for our
understanding of the roots of language and how it should be defined.

The methods devised for Sarah's training sessions were precise. Each one was
intended to inculcate a specific linguistic function. Premack was interested in the
chimpanzee's intelligence and did not intend that his chimpanzee simulate human
language (as the Gardners had hoped Washoe would do in the mastery of Ameslan).
Rather, he wished to demonstrate experimentally a functional analysis of language
skills.

For Premack, language training was in essence (1) the mapping of existing
knowledge through the acquisition of word meanings, (2) the mastery of a set of
concepts (e.g., name-of, color-of, same-as, different-than, relations between things
in space), and (3) the learning of rules for relating words to one another for the
purpose of constructing and decoding the internal organization of sentences, in
which word order is relevant, (Premack 1970a:113). As important as vocabulary is,
by itself it is not language. Premack never pressed his subject for the mastery of a
large vocabulary; throughout training he stressed the basic functions of productive,
syntactic language.

Sarah's language was a synthetic one, its words consisting of pieces of plastic
of various colors and random shapes. The words could be securely place on a magnetized board, a feature which allowed for the possibility of sustained visual access
to compensate for possible limitations in chimpanzees' short-term memory (see

Rumbaugh 1970, for a review) and also for total control over the linguistic options afforded the subject. Although it is reasonable to conclude that effective social relationships between Sarah and her testers did exist (chimpanzees just don't work well for people they dislike!), they were probably not as pervasive as those between the Gardners and Washoe.

Premack's basic strategy assumed that every complex rule of language can be analyzed into simpler units. By defining these units and devising a procedure for teaching them, one may inculcate language in organisms that do not otherwise have it.

Sarah's first words were taught by pairing agents, objects, and actions with corresponding language elements. Strings of two or more words were then built; by restricting the word options available but nonetheless requiring that Sarah eventually fill in every empty slot in the sentence, word meanings were taught. The negation "no" was introduced as an injunction against certain actions, as in, for example, "No Sarah take banana." Through the presentation of pairs comprised of identical and nonidentical members coupled with the marker for the interrogative and a word to specify either "same" or "different," Sarah learned to respond appropriately by selecting the word for either "yes" or "no," depending both on the members of a pair in a given presentation and on whether the question-word was "same" or "different." Thus, if the two members were identical and Sarah was asked if they were "different," her correct answer was "no."

Through extensions of these procedures, Sarah learned "name-of" as a concept which facilitated her mastery of new names. Language also was used to teach additional language. Dimensional class words for color, shape, and size were likewise used to teach Sarah still other specific words. Quite impressively, the property name "brown color-of chocolate," was used at a time when both "chocolate" and "color-of" were familiar words and "brown" was new. On the basis of that metalinguistic event, Sarah was able to select the "brown" disc (from a set of four distinctively colored ones) and to insert it in a specific dish when commanded, "Sarah insert brown [in] red dish" (Premack 1971b:211).

Sarah also learned to describe whether a card of a given color was on, under, or to the side of a card of another designated color, and she learned to place them correctly in response to requests such as, "red on green." Sentences were built up through the acquisition of unit skills which could be hierarchically organized into logically structured strings of words. Subsequently, mastery of the conditional "if-then" allowed for appropriate performance in response to statements such as, "[If] Sarah no take red [then] Mary give Sarah cracker," (1971b:222; bracketed expressions added by this author).

Premack was and continues to be sensitive to fundamental questions such as, "When does something come to serve as a word?" In addition to demonstrating that pieces of plastic functioned as words in naming tasks and in requests by Sarah for different items, Premack demonstrated that in response to an item identified only by its plastic word piece, the object itself being absent, Sarah could describe its features. For example, in response to the blue piece of plastic which functioned as the word for "apple," Sarah described "apple" as red (not green), round (not square), and as having a stem. Also, she seemingly inferred that a knife had caused an apple to be sectioned and demonstrated an ability to represent or infer the name of the various fruits from which seeds, stems, etc., came (Premack 1975b, 1976b).

Premack's work with Sarah made a powerful impact on other researchers, not only because of the impressive empirical data it produced but also because of the perceptive theory regarding the nature and acquisition of language which it generated. He demonstrated that through use of a functional and analytical approach, novel insights about an exceedingly complex set of phenomena can be gained. At the same time, interestingly, he has retained a willingness to speak both mentalistically and nativistically, as, for example, when he states that "the mind appears to be a device for forming internal representations. If it were not, there could be neither words nor language . . . every response is a potential word. The procedures that train animals will also produce words" (1971b:226). A remarkable blend of apparent mentalism, nativism, and functionalism!

This author agrees with McNeill's conclusion: "Thus, Premack's experiment complements the Gardners' by showing a difference between chimpanzees and children that is correlated with the ability to perceive and presumably use, syntactic patterns that encode conceptual relations. The conceptual relations themselves appear to be available to both species, a conclusion that Sarah's performance reinforces quite strongly" (McNeill this volume:156).

The Lana Project

The germinal idea for the Lana Project came to this author in the fall of 1970 as a direct consequence of the research reports emanating from Projects Washoe and Sarah. The basic motivation behind the Lana Project was to obviate the extraordinary investments of human time and effort required in one-on-one research with single chimpanzee subjects. Alternative methods to those previously employed had to be devised if researchers were to expedite investigation into the nature of language and the capacity for language by life forms other than man. Automation of training procedures seemed the obvious answer. It would serve both to enhance the efficiency of training and to objectify the procedures; it would allow for the automatic recording of all linguistic events and for computerized summary and analysis thereof, and in due course it would provide for a technology with which to support systematic study of the parameters controlling the emergence of initial language skills. Success in this undertaking would also allow for the eventual extension of the training system to research with alinguistic mentally retarded children, who for a variety of reasons find the acquisition of expressive language skills very difficult.

Consultation with a creative bioengineer, Harold Warner, led to the conclusion that such an undertaking was feasible and later to the formation of an interdisciplinary team that worked to define the myriad facets such a system might have. Research resources became available to the project in January 1972, and by February 1973, the system was in operation.

The entire system has been summarized elsewhere (Rumbaugh, Glasersfeld, Warner, Pisani, Gill, Brown, & Bell 1973), and recent articles by Warner, Bell, Rumbaugh, and Gill (1976) and by Glasersfeld (1977) provide technical details of the electronics and of the synthetic language termed "Yerkish" which was devised specifically for the Lana Project. Consequently, only its most salient features will be described here.

The system was designed to provide around-the-clock operations that would not require the presence of a human operator or experimenter. The subject's room, built of plastic and glass, permitted excellent two-way visual access. The subject was provided with a keyboard console equipped with keys, each one serving as the functional equivalent of a word.

The keys were differentiated by unique geometric configurations embossed on their surfaces to designate specific objects, animates, edibles, colors, prepositions, activities, and so forth. Each class of words had a distinctive color for its keys to help the subject learn the various word classes. The keyboard console and the keys themselves were designed to provide for the ready relocation of the keys so that their locations alone would never provide cues as to their function or meaning, and the subject would have to attend to the geometric configurations, termed *lexigrams*, which were to serve as formal, functional words. The keys were back-lighted when in an active state (i.e., when their depression would be sensed by the computer); when depressed, a key became brighter in order to allow the subject to review what she had linguistically produced. When a key was depressed, a facsimile of the lexigram on its surface was produced in an overhead row of projectors. The net result was the production, from left to right in a row of projectors, of a visual portrayal of the sentence or word string.

Outside Lana's room, another keyboard was accessible to the experimenter. Its function was identical to that of her keyboard; the use of either keyboard resulted in the transmission of the visually portrayed sentences to allow for two-way communication and also potentially for conversation. Situated between these keyboards was a PDP–8E computer with the necessary interfacing. It served to monitor all linguistic events, to evaluate their grammar in accordance with the rules of the Yerkish language, and to record via the activation of a teleprinter and a punched paper tape all communications and the time of day they occurred. Syntactically correct requests for a variety of foods and drinks or for a variety of events (such as a movie, projected slides, music, or the opening of a window) were rewarded automatically by the language-training system. Other requests for social interactions (grooming, tickling, swinging, a trip outdoors) were honored if the person named was present and had the time to honor the request.

Initial training began with Lana's learning to use a single key, followed by her mastery of what were termed *stock sentences,* sentences which reliably served to net for the subject drinks or foods of certain types, various types of entertainment, etc. These sentences were taught through essentially standard operant conditioning techniques; at the time of their initial mastery, there was no reason to believe that Lana possessed semantic meanings for any of the keys. Special training procedures instituted to teach her not only the names of things, but also the concept that things do have names (Gill & Rumbaugh 1974) served initially to instruct her, within limits, in the meanings of the various keys. Similarly, Lana was taught eight colors, the names of various items, and several prepositions so as to expand her working vocabulary and to allow her to make reference to things either by name, by attribute, or by location.

On her own initiative, Lana extended her use of stock sentences for other than the originally intended single purposes (Rumbaugh & Gill 1976a, 1977). Furthermore, she extended the use of "no," as taught within the context of answering "yes" or "no" to questions about whether the window or door were shut or open, to include "no" as meaning "don't do that." The initial occasion for her doing so was to protest a technician's drinking of a Coca-Cola when she had none. Still later, she spontaneously used "no" to produce the meaning "it is not true that [food's name] is in the machine for vending" (Rumbaugh & Gill 1976a).

Through the course of other studies (Rumbaugh & Gill 1976a), it was determined that Lana's familiarity with items facilitated her execution of cross-modal judgments of sameness–difference, and that apart from familiarity, names seemingly made their separate contribution to the accuracy of the judgments. But the primary goal was to cultivate in Lana the desire and the skills needed to converse with us about a wide variety of subjects. After all, most everyday language of humans occurs in the context of conversation, in the course of which attempts are made to exchange the proper social amenities, to define topics or problems that need mutual attention, to evaluate alternative courses of action or methods of solving problems, to plan for the future relative to the experiences of the past, and so forth. Significantly, Lana began to enter into conversation without any formal training to do so. Her first conversation was to ask, in essence, whether she might come out and drink some Coca-Cola (Rumbaugh & Gill 1975). Her exact formulation, novelly composed from words and phrases of stock sentences and requests used in the naming training sessions was, "Lana drink this out-of room?" The subsequent question put to Lana, "Drink what?" led to an appropriate refinement, "Lana drink coke out-of room?"

Conversations with Lana have been examined in detail (Gill 1977; Rumbaugh, Gill, Glasersfeld, Warner & Pisani 1975; Rumbaugh & Gill 1975, 1976a, 1977). It seems clear that Lana has been prone to converse whenever she must do so in order to receive something exceptional or whenever something not in accordance with the routine delivery of foods and drinks has occurred—in short, when some practical problem arises for her. She has never *in conversation* commented extensively on this or that as children and adults are inclined when their attention or motivation shifts unpredictably. For Lana, language is an adaptive behavior of considerable instrumental value for achieving specific goals not readily achieved otherwise. To date, at least, she has not used expressive language to expand her horizons except to ask for the name of something which she then requested by the name given (Rumbaugh *et al.* 1975).

Color terms in those natural languages which possess them serve as conceptual codes for describing variations in hue and brightness. Such color-naming systems are not arbitrarily derived; rather, the opponent-process underpinnings of hue perception apparently provide for the evolution of color terms based on physiological contingencies (Bornstein 1973; Bornstein, Kessen & Weiskopf 1976; Hurvich & Jameson 1974). A human, given the necessary visual apparatus, the necessary linguistic environment, and the requisite cognitive capacity, will develop an arbitrary conceptual code which serves reliably to denote nonarbitrary portions of color-space. Essock (1977) asked if Lana, given the fact that she possesses color vision which is extremely similar to that of a normal human (Grether 1940), would do the same.

Subsequent to Lana's acquisition of eight color terms following training with eight arbitrarily selected training colors (red, orange, yellow, green, blue, purple, black, and white), Essock presented Lana with over 350 different Munsell color chips to determine whether she would use her color terms reliably to describe colors perceptually distant from her training colors. Over the course of the experiment, each chip was presented for identification on three widely separated occasions. Lana's response distribution resembled that of human subjects tested under similar conditions. Over 99% of the time only two responses at most competed as possible labels for a given color chip, and when two responses competed, they were always the names for spectrally adjacent hues. Lana's assignment of color names to various areas of color-space did *not* produce areas labeled by a given color term surrounded by increasingly variable responses to colors perceptually distant from the training colors, as should be the case following stimulus generalization. Rather, her responses formed well-defined areas in which a single color name was given to a particular hue despite wide changes in brightness and saturation. This finding is further evidence that Lana does possess sufficient cognitive capabilities to assimilate and adopt arbitrary conceptual codes and to use such codes to describe her perceptual world.

The methods of the Lana Project fall between those of Project Washoe (where the emphasis with an infant chimpanzee was on linguistic saturation, socialization, close human–chimpanzee interactions, and discourse) and those of Project Sarah (where the emphasis was on the design and implementation of rigorous, step-by-step training procedures designed to cultivate well-specified language skills with a juvenile subject). Lana was just two years old when formal work with her began. She had been laboratory-born and reared, and as a result she was probably less advantaged than either Washoe or Sarah, since both were feral-born and home-reared as well. Nonetheless, Lana has both mastered and initiated an impressive array of linguistic skills and performances, as have Washoe and Sarah. Why and how all three have succeeded so well is an interesting question when one juxtaposes the diverse characteristics of the training principles.

Other Ape Language Projects

The projects described above are, at present, all being continued, and since their inception, about a dozen additional projects have begun which extend the original research in various directions. Roger Fouts, initially with Project Washoe, has continued to study Washoe after she was moved to the Institute for Primate Studies of the University of Oklahoma. His studies (Fouts 1974) have included assessments of individual differences in the acquisition of signs (Fouts 1973), assignments of exemplars to conceptual categories, chimpanzee–chimpanzee communication, and the teaching of Ameslan through spoken English.

The last of these studies by Fouts (1974) provides an impressive demonstration of how language mediates the acquisition of new information—in this instance, the meanings of words. Ally, a three-year-old Ameslan-trained chimpanzee, was tested for his understanding of 10 spoken English words used in commands to bring specified items to the experimenter. After he had demonstrated receptive competence with those 10 words, they were divided into two lists of five words each. For one of these lists, Ally then had special training, which consisted of seeing the Ameslan sign for each word for the first time while hearing the word through the modeling and speech of the experimenter. For the second list, no special instruction in Ameslan was provided. Ally was then asked to name via Ameslan each ten objects in a controlled test. He was able to name the five objects for which he had had prior training but not the five objects for which he had had no such training. Fouts's experiment serves as an important demonstration of linguistic instruction with a chimpanzee in which productive competence emerged from initial auditory receptive competence through the contiguous pairings of auditory and visual stimuli.

Additional projects in which Ameslan is being used include the work with Nim chimpanzee by Terrace and Bever (this volume) and Patterson's work (1977) with Koko gorilla (*Gorilla g. gorilla*). Terrace and Bever are concentrating upon the ability of Nim to combine words in accordance with syntactical rules and are hoping to determine the abilities of chimpanzees for linguistic communication and for the report of memories, moods, and other private events. Their study is, at the writing of this chapter, in its early stages.

Patterson's study is unique in that to date it is the only one with a gorilla. Koko gorilla's progress has been as orderly and as rapid as was Washoe's at comparable stages of training. Koko's performance has corroborated essentially all Washoe's proclivities for acquiring new signs, for extending their usage, and for appropriately chaining them together (as suggested by the context). Patterson concludes that both the chimpanzee and the gorilla parallel the human child in the early development of semantic relations.

Chimpanzee Language Skills and Intelligence

In conclusion, a word regarding the chimpanzee and intelligence is in order. Intelligence is admittedly a highly problematic and controversial concept. Although it is beyond the scope of this chapter to review the diverse perspectives regarding intelligence and its worth as a concept, reference to Wechsler's thoughts on the question might assist in the interpretation of data emanating from the chimpanzee language projects. For Wechsler, intelligence is "the capacity of an individual to understand the world about and his resourcefulness to cope with its challenges" (1975:139). In his view, its components are numerous and diverse; its reference systems include (1) awareness of what is being done and why; (2) meaningfulness, in the sense that intelligent behavior is goal-directed and not random; (3) rationality, in that intelligent behavior can be logically deduced and is consistent; and (4) usefulness, in that intelligent behavior is deemed worthwhile by the consensus of the group—i.e., the assessment of intelligence inevitably entails value judgments.

If Weschler's reference system is used to define the intelligent behaviors of the chimpanzees in language-training projects, it is not difficult to select several behaviors which seem to be meaningful and worthwhile adaptations and that are also reflections of awareness, goal-directedness, and rationality. These systems of reference are admittedly elusive, but they do serve to bring attention to the possibility that the operations of intelligence in chimpanzee and human have close resemblance. Despite the fact that the chimpanzee brain is much smaller than the human brain, its basic functions are not, in all likelihood, totally unique.

A COMPARISON OF THE LANGUAGE PROJECTS

The chimpanzee language projects, particularly Projects Washoe, Sarah, and Lana, were characterized by major differences in their subjects' ages, in the languages employed, and in maintenance and training procedures. Nevertheless, all the chimpanzee subjects trained in these projects were able to master significant aspects of linguistic communication. All proved to be relatively adroit at mastering a basic working vocabulary, for example, so that at the end of their training they were able upon the presentation of an exemplar to give a reliably correct response of identification (sign, plastic piece, key depression) through the use of their hands.

Admittedly, a problem exists in defining with any degree of precision the *meanings* of each and every "word" in the respective vocabularies of the various subjects. That problem notwithstanding, each subject did give clear evidence of appropriately extending word usage beyond the specific exemplars of vocabulary training. To review a few of the more important illustrations, Washoe used the word "open" first in connection with one door, then with all doors, then with the lids of briefcases, boxes, and jars, and finally as part of a request that a water faucet be turned on (Gardner & Gardner 1971). Washoe also aptly termed a duck a "water bird," and the brazil nut a "rock berry" (Fouts 1974). Sarah was taught the meanings of new words through instructions and information given her through the use of familiar words; she also was observed both to formulate and then to answer questions in isolation. Lana extended the meaning of "no" from "it's not true," to "don't do that," and to "there is no [food-name] in machine." Also, she has spontaneously asked for things for which she had no formal name by creating metaphorical or descriptive labels. For example, she referred to a bottled Fanta soft drink as "coke which-is orange [colored]"; she asked for an overly ripe banana as "banana which-is black"; and she has variously requested an orange as "apple which-is orange" and "ball in bowl." Fouts (1974) has reported on work with still another chimpanzee, Lucy. In the course of her training, Lucy labeled four citrus fruits "small fruits," radishes "cry hurt food," and watermelon "drink fruit" and "candy drink." Perceptually salient characteristics apparently served as the natural bases for naming.

Both the extended usage of words and the compositions of descriptive new labels should be accepted as evidence of semanticity, both receptive and expressive. Perhaps the most convincing evidence of semanticity, however, is found in the *productive* use of words in syntactic relationships with other words. Evidence that at least rudimentary mastery of syntax was achieved is much stronger for both Sarah and Lana than for Washoe. Sarah's appropriate, differential responses to requests and to if-then conditional communications implied some understanding of syntax, as did Lana's composition of syntactically correct sentences to request that she be allowed to do something new ["Lana drink coke out-of room?" etc., (Rumbaugh & Gill 1975,1976a)], and her ability to interact competently through the course of protracted conversations that were problem-solving in their orientation (Gill 1977). It should also be noted that, apart from such extensions and generative usages, it is problematic to infer semanticity when the sole evidence consists of a correct response of identification to a single exemplar.

All the chimpanzees have shown a readiness either to use words in chains or

to attend to words in sentences. This readiness is possibly significant in that it suggests an informational capacity in the chimpanzee which is a necessary condition
for the syntactic production and interpretation of signal chains, a clear requisite to
the openness of language discussed earlier in this chapter and in the discussion
section which follows.

DISCUSSION

"In cognitive psychology, the instigation to begin peeking behind the curtain
of the abstract organism has begun to arise under the influence of developments in
ethology, cross-cultural research, and linguistics" (Estes 1975:6). The chimpanzee
language projects might well call for a complete opening of that curtain for a variety
of reasons.

Implications for the Definition of Language

Despite the problem that at present no definition of *language* is generally
accepted, there are indications of an increasing acceptance within a variety of disciplines—psychology, anthropology, psycholinguistics, communication—that
Washoe, Sarah, Lana, Lucy, Koko, and other animals have demonstrated communication skills which resemble the language used by man much more than the
essentially closed (as far as we now know), nonsyntactic signal systems of animals
in the field. Continued studies with chimpanzees in language projects should eventually lead to an acceptable definition of language, one cast in nonanthropocentric
terms (e.g., Segal 1977).

As stated earlier, language, in this author's view, is most clearly distinguished
from other forms of communication by a single attribute—its openness, one of several terms used by Hockett (1960a) in his descriptive system of natural languages
(see also Altmann 1967; Hockett & Altmann 1968). Openness here refers to the
introduction of new lexical units (i.e., words) with the possibility that a totally arbitrary relationship might exist between the morphology of the lexical units (sounds,
gestures, geometric patterns, etc.) and their accrued meanings. It also refers to the
generation of new sentences that convey new information as old and new words are
used in interaction (i.e., syntactically) so as to modulate their meanings for the
encoding of information covertly selected for public transmission. The openness of
language allows for all the other characteristics of language described by Hockett,
including displacement, reflexiveness, and lying, which are not otherwise specifically attributable to the physics of propagation and sound conduction (as are, for
example, rapid fading and broadcasting). Because of its openness, language stands

in sharp contrast with the stereotypy of most animal communication, in which there are few lexical elements and no evidence of any recombination of "qualitatively different vocal units, separately meaningful, into new messages with new meanings" (Marler 1975:33).

Implications for Learning Research

The chimpanzee language projects should, in due course, lead to a clarification of a number of problems in the area of learning and of the processes entailed in formulating adaptive behavior patterns such as in the production of sentences). Years ago, Miller (1959:247) speculated that perhaps it is the enhanced capacity of man "to respond selectively to more subtle aspects of the environment as cues" in an abstract manner that makes his mental processes appear to be much "higher" than those of other animals. Another explanation for the same phenomenon, he felt, might lie in man's enhanced "capacity to make a greater variety of distinctive central cue-producing responses, and especially a greater capacity to respond with a number of different cue-producing responses simultaneously so that further responses may be elicited on the basis of a pattern of cues representing several different units of experience" (1959:247).

A concerted effort is now underway to effect a rapprochement between cognitive psychology and the more traditional models of S–R learning theory (e.g., Estes 1975, 1976; Mason 1970; Restle 1975 a,b). Single-unit S–R models, where single responses are assumed to be evoked by single stimuli, cannot account for all behavioral changes that are known to take place within learning experiments (Estes 1975). But if experiments are viewed "as ways of transmitting information about the environment that, if learning occurs, will be incorporated by the animal" (Bolles 1975:272), then one might be able to explain how certain information might not only be used for the determination of an immediate response, but stored for possible reference at some future time.

Bolles considers, for example, that animal subjects might learn at least two types of predictive relationships within conditioning contexts—relationships between the stimuli and the consequences (i.e., light is followed by shock) and between responses and their consequences (i.e., if "such-and-such" responses are made, shock will be attenuated or avoided). He further proposes that animal species differ in terms of their capacities to learn predictive relationships between responses and consequences.

Bolles suggests that the learning of predictive relationships between stimuli and consequences is the more pervasive type and that species are preferentially sensitive to stimuli that are relatively important to them in their natural habitats; in other words biological constraints partially determine what will be readily attended to and learned. By extension, animal species might well differ in terms of their readiness and capacity to integrate sensory events which occur together in time or to integrate motor patterns for responding (Osgood 1957).

The predictive learning abilities of a given species and its mediating capabilities are all important in determining its ability to receive, process, and use information advantageously in formulating novel response patterns, particularly as called for in concept-based behaviors and language. Expressions based on language skills entail the coordination of several complex functions. Problem situations must be defined and analyzed. Their dimensions must receive selective attention. Words must be called up for possible use, and their concatenation must adhere to certain sets of rules. Seemingly, identical problem situations confronted by linguistically trained subjects of various genera might reveal differences in terms of what is perceived or defined as a problem, how it is attended to and analyzed, and how response tactics are deliberated and selected for use. This would be the case, of course, only to the degree that the linguistic expressions would correspond with the psychological processes activated by systematic variations in problem situations—not an unreasonable assumption—but the possibility that language-type training programs and test situations might be uniquely sensitive to species differences in propensities for attending to and for analyzing various classes of problems for the determination of response tactics should not be overlooked. To the degree that species differences in the formulation of linguistic responses would prove relatively insensitive to wide variations in rearing and experience, basic biological differences in proclivity to respond would be implicated.

Language Acquisition and the Formation of Concepts

Language skills are widely believed to facilitate mediating processes, for they provide, among many things, formal labels for organizing related items into conceptual classes. The word-labels of languages are thought to facilitate reference to information units in memory and to facilitate the definition of new relationships between those units as the exigencies of new situations might require. But according to S–S (stimulus–stimulus) theories, as noted by the Kendlers (Kendler & Kendler 1975:236), it is possible for a child, even before he begins to talk, to learn that a given sound pattern refers to a given visual pattern simply by hearing and seeing the two patterns together in time. This type of association requires no labeling response, either overt or implicit. These kinds of associations, the Kendlers held, might well account for the fact that in most children receptive or comprehension language skills antedate production skills.

On the other hand, the acquisition of productive public language does entail both the learning and the appropriate use of relatively specific words and meanings.

How are word concepts formed and how are word meanings agreed upon for public use? Language learning is not a totally arbitrary process. Rosch (1973) has contributed important work and ideas concerning these questions. According to Rosch, perceptual domains exist within which the *core* meaning is not arbitrary but rather is dictated by the nature of the perceptual system. The use of "clearest cases" or "best examples" facilitates the learning of both color and form categories. Rosch obtained evidence to support the hypotheses that (1) it is easier to learn color and form categories in which natural as opposed to distorted prototypes are used as central prototypes in sets of variations, (2) the natural prototype will usually be learned first regardless of whether it is central to the category, and (3) subjects tend to define a given category as a set of variations about the "most typical" or natural prototype. Rosch suggests the possibility that these findings might indicate that even "nonperceptual" semantic categories of languages might adhere to certain general principles of internal structure (with a focal center and nonfocal surrounds) and that these, in turn, might be the key to certain universals of natural language systems. Rosch predicts "that children will be similar to adults in tasks where adults use *internal* structure as the basis for processing . . ., but different from adults in tasks for which adults use *abstract* criterial attributes" (1973:142). Essock's study (1977) of Lana's color classification skills strongly suggests that chimpanzees and man have comparable internal structures for colors.

Extensions of Rosch's ideas and methods might assist us in understanding the degree to which man and apes share the basic processes and structures that are used in learning concepts and categories. To the degree that such research yields positive results, support will be given for the interpretation that apes can master bona fide language skills.

Indirect evidence relative to this point is already in hand. It has been pointed out that Washoe was very young, about one year old, when her language training began [the Gardners now hold that training should start at birth (Gardner & Gardner 1975a)]; Lana was about two years old when her formal training began; and Sarah was about six years old. Despite these major differences in ages, *all* the chimpanzees accomplished basically similar linguistic feats. That they all did so despite substantial age differences might be taken as a reflection of their having similar concept-formation processes and structures which were equivalently functional despite their different rearing conditions prior to language training.

Underlying Processes—Homologous or Analogous?

Extension of Rosch's tactics into developmental-comparative studies of the human child and the chimpanzee would help to answer a key question: Are the processes which support the acquisition of linguistic-type skills in men and in apes

homologous or analogous? To the degree that both ape and child perfect their language-relevant skills in the same basic developmental pattern, to the degree that
training methods have relatively the same supportive or deterring effects upon the
development of those skills, and to the degree that both ape and child have seemingly the same internal structures for their concepts and categories, homologous
processes would be indicated. To this author, it seems quite probable that homologous processes are involved because of the very close biological similarity and
relationship established between man and chimpanzee (King & Wilson 1975).

PROJECTIONS

Animal-Model Studies and Perspectives of Language

Research directed toward accurately assessing the several dimensions of the
capacities of chimpanzees and other great apes for linguistic functions (e.g., Wick-
legren 1974) will undoubtedly continue. Ultimately, if this research is successful,
young apes may come to serve as surrogates for human children in research on the
parameters of language acquisition. Such a development would permit the investigation of certain basic questions which have been precluded to date because of the
ethical constraints which we place upon research with human subjects. But the use
of apes as substitutes for human children in linguistic projects can be justified in the
long term only on the basis of convincing evidence that the chimpanzee's language-
learning processes are homologous to those used by the human child.

Should such evidence be forthcoming, one important topic which would then
be open to research in ways now impossible is the relative efficacy of various methods of cultivating *initial* language functions in human retardates (for suggested prototypes see Brown, this volume, 1973b; Fouts 1973, 1974; Fox & Skolnick 1975).
Other important questions include how one language skill builds on other skills and
how extralinguistic experiences contribute to language learning. These two questions might be investigated by systematically depriving different ape subjects of
various opportunities to learn and experience. Questions regarding apes' abilities to
tell us about their internal states, to teach artificial languages to their offspring and
to translate their natural communication networks to us will be answered only after
many, many years of additional research.

Extended comparative research with apes should also provide us with a better
understanding of (1) the role of language functions in thought and in problem-solving activities, (2) the relationship between intelligence and language-skill acquisition, (3) the contributions of social experience to language functions (Lewis &
Cherry 1977) and to concept of self (Gallup 1970), and (4) the bases for the so-called
universals of natural languages. Regarding the last of these, Chomsky (1968) has

been credited, perhaps inaccurately, with the theory that the normal human child possesses a genetically predicated grammar system which shapes his language-learning efforts. But even "the rankest nativist would not claim that the child is genetically endowed with a complete grammatical system" (Moore 1973:4). The few language universals that can be taken as evidence for the genetic view are seemingly secondary to the diverse networks of rules which facilitate language learning in the broad array of human cultures (Schlesinger 1970). Ape language research should eventually yield a much clearer perspective on the interaction of biological and environmental factors as they influence the acquisition of language.

Language Research and Psychology

In reflecting upon the probable effects which research on language will have on psychology, Slobin concluded that "what is bound to emerge will be a more complex image of the psychological nature of man, involving complex internal mental structures, in part genetically determined, in part determined by the subtlety and richness of the environment provided by human culture, and probably only minimally determined by *traditional* sorts of reinforced stimulus–response connections" (1971:61). Comparative psychological studies will be of critical importance in testing such hypotheses.

MAN–CHIMPANZEE COMMUNICATION

ROGER S. FOUTS AND RANDALL L. RIGBY

This chapter traces the scientific inquiries into two-way communication with chimpanzees from the early attempts to establish vocal communication to ongoing research in gestural and symbolic languages. We shall begin by selectively reviewing the historical speculation concerning the possibility of teaching chimpanzees to speak. This will be followed by a review of early experiments in raising chimpanzees in a human home environment, and by a discussion of the more recent experiments concerning the use of gestural languages and symbols in establishing two-way communication.

The chimpanzee is a nonhuman primate that is very similar to, and at the same time very different from, a human being. From either of these aspects we can obtain a wealth of information on the mental and behavioral capacities of the chimpanzee and, in addition, comparative data to assist in the understanding of human behavior. Since we are interested in two-way communication between man and chimpanzee, we will emphasize the similarities of the communication capabilities of the two species. It is obvious that a physical similarity exists, but it is important to stress that there are also some basic differences. Although man and chimpanzee had a common ancestor, many thousands of years of separate evolution and adaptation have endowed each species with unique physical and behavioral characteristics. However, some remarkable physiological similarities have been found in blood protein and type, chromosomal characteristics, structure, and behavior. The last two were observed by early researchers in man–animal communication.

HISTORICAL DEVELOPMENTS AND SPECULATION CONCERNING COMMUNICATION

The Great Apes

Probably more than any single factor, the physical similarity between man and the great apes aroused the curiosity of those interested in teaching apes to behave

ROGER S. FOUTS • Department of Psychology, University of Oklahoma, Norman, Oklahoma 73019.
RANDALL L. RIGBY • HHB 2/17 FA, APO SF 96251

in ways similar to man. The famous *Diary* of Samuel Pepys reflects this interest in an entry made in August 1661:

By and by we are called to Sir N. Battens to see the strange creature that Captain Jones hath brought with him from Guiny; it is a great baboon, but so much like a man in most things, that (though they say there is a species of them) yet I cannot believe but that it is a monster got of a man and a she-baboon. I do believe it already understands much English; and I am of the mind it might be taught to speak or make signs.

A similar reaction was recorded by Julien Offray de La Mettrie (1709–1751), who, in *L'Homme machine* (1748; see La Mettrie 1912), pondered the varying capacity of animals to learn.[1] La Mettrie, obviously attracted to the striking similarities between man and the apes, proposed teaching sign language to apes in a school for the deaf. His idea was to choose an ape with the most "intelligent face" and send him to school under the teacher Amman (an early writer of books on the education of the deaf). La Mettrie failed to distinguish between monkeys, apes, and orangs (he referred to them interchangeably), but his basic idea was clearly two centuries ahead of its time. It is apparent that he recognized the intellectual capacity of the ape when he wrote:

Why should the education of monkeys be impossible? Why might not the monkey, by dint of great pains, at last imitate after the manner of deaf mutes, the motions necessary for pronunciation? . . . it would surprise me if speech were absolutely impossible in the ape.

Another distinction made by La Mettrie, which was later to become a cornerstone of controversy concerning language acquisition in chimpanzees, was that speech and/or communication with lower primates included the use of gestures. He was obviously influenced by Amman's works, including *Surdus loquens* (1692) and *Dissertatio de loquela* (1700), which contained plans for teaching signs, finger spelling, and lip reading. Training apes to use such communication methods, or at least gestural communication, was to La Mettrie a logical proposal, and he felt that they would be able to master it easily. Unfortunately, La Mettrie was unable to follow up on his idea, and the proposal that apes could learn to communicate with man lay dormant for years. R. M. Yerkes (1925:53), who devoted his entire life to observing and writing about the great apes and other nonhuman primates, echoed La Mettrie's proposal almost two centuries later: "If the imitative tendency of the parrot could be coupled with the quality of intelligence of the chimpanzee, the latter undoubtedly could speak." He predicted future scientific trends when he stated:

I am inclined to conclude from the various evidences that the great apes have plenty to talk about, but no gift for the use of sounds to represent individual, as contrasted to racial, feelings or ideas. Perhaps they can be taught to use their fingers, somewhat as does the deaf and dumb person, and helped to acquire a simple, nonvocal sign language.(1925)

---

[1] The authors are grateful to Gordon Hewes, University of Colorado, for bringing to our attention this work of La Mettrie.

The Influence of the Home-Raising Experiments

With the beginning of the twentieth century and the application of scientific principles to psychological developments and discoveries, an innovative type of animal-training experiment developed. Using an evolutionarily close relative of man, the chimpanzee, the experimenters attempted to duplicate the environment of a human household, the conditions of which are most favorable to language acquisition in man. Because the requirements for human language were obviously met in such a setting, i.e., adequate social atmosphere, sufficient periods in which to practice babbling, and an appropriate model for vocal imitation, it was assumed that the chimpanzee would acquire vocal language. Lightner Witmer (1909) summarized the rationale for this type of experiment:

While my tests of Peter give no positive assurance that he can acquire language, on the other hand they yield no proof that he cannot. If Peter had a human face and were brought to me as a backward child and this child responded to my tests as credibly as Peter did, I should unhesitatingly say that I could teach him to speak, to write and to read within a year's time. Peter has not a human form, and what limitations his ape's brain may disclose after a persistent effort to educate him, it is impossible to foretell. His behavior, however, is sufficiently intelligent to make this educational experiment well worth the expenditure of time and effort.

In general, the rationale for the home-raising experiments that followed were based on the above assumptions. If the necessary language eliciting environment were provided, perhaps the puzzle of nonhuman primate language could be solved.

As pointed out by Kellogg (this volume), keeping nonhuman primates as pets in the home is certainly not a novel idea and can be traced back several centuries. Such practices frequently occur today, but no instances of intellectual language use by such pets have been reported (Kellogg this volume; Yerkes 1925). A major problem is that the great majority of those who have nonhuman primates in the home are not familiar with language training and are ill equipped to observe and record the animals' reactions. However, as Kellogg (this volume) observes, "It is quite another story for trained and qualified psychobiologists to observe and measure the reactions of a home-raised pongid amid controlled experimental home surroundings." We will describe several home-raising experiments, all of which were designed to determine the extent to which a nonhuman primate could acquire a vocal language capability.

Peter

The first chimpanzee to be mentioned was named Peter and was observed by L. Witmer (1909). Although this study was not a home-raising experiment, the rationale and desire to undertake such a task with an ape was inspired by this and similar reports found in the literature during this period.

Peter was a 4- to 6-year-old chimpanzee owned by a man named McArdles, who had trained him for 2½ years. Employed by Keiths Theatre in Philadelphia as an example of "a monkey who made a man of himself," the entertaining, humanlike chimpanzee aroused the curosity of Witmer, who made arrangements to test Peter to determine the extent of his intelligence. The tests were conducted in the fall of 1909 at the Psychological Clinic in Philadelphia. Most of them involved motor coordination and simple reasoning tasks—opening a box to obtain articles inside, unlocking locks with keys, and driving nails into a board with a hammer—all of which Peter was able to do with relative ease. It was observed that Peter showed only imitative writing movements and possessed no special writing ability. He was able to articulate the word "mama," however; it is noted that he did so with considerable effort and with apparent unwillingness. The articulation of the sound "m" was said to be perfect, but the second "ma" in "mama" sounded more like "ah" and was inaudible more often than not. Witmer noted that the sound was a hoarse whisper rather than an articulated word and that Peter always tried to speak with the inspired rather than the expired breath. Peter was trained to articulate the sound "p" in only a few minutes, leading Witmer to conclude:

This experiment was enough to convince me that Peter can be taught to articulate a number of consonantal sounds and probably to voice correctly some of the vowels. . . . If a child without language were brought to me and on the first trial had learned to articulate the sound "p" as readily as Peter did, I should express the opinion that he could be taught most of the elements of articulate language within six months' time.

Witmer also observed that although Peter was unable to speak, he was nonetheless able to understand spoken words. Using the analogy of Helen Keller, who first comprehended the use of symbols in the place of objects, Witmer proposed that in a similar manner Peter could be made to comprehend symbols as representing objects; with further training to articulate these symbols, he would be able to communicate. Recognizing that early language training would be crucial in the event that vocal language could be taught to the chimpanzee, Witmer predicted that "within a few years, chimpanzees will be taken early in life and subjected for purposes of scientific investigation to a course of procedure more closely resembling that which is accorded the human child." Clearly a precursor for things to come, Witmer's prediction has been attempted on several occasions.

Joni

Joni was a male chimpanzee raised and observed by N. Kohts and her family from 1½–4 years of age (Kohts 1935). The period of observation was from 1913 to 1916, and the daily events pertaining to Joni's behavior were recorded and later compared with those of Kohts's own human child, Roody, during the period 1925–1929. Kohts's manuscript was prepared in the early 1930s, 15 years after the chimpanzee had been observed and tested. The comparison of the developmental sequences of the home-raised chimpanzee and of the child probably reflects the influence of Yerkes. Only a small part of the report is devoted to language capacity in the chimpanzee, and this is in direct comparison to that of the human child. No special attempts were made to train Joni to use articulate language, the author's purpose being to record the language capability emitted in the home environment without any special training. Kohts reported that Joni was able to produce at least twenty-five sounds elicited by various stimuli within his environment, but they were clearly his own natural sounds to express emotions and desires. Our own observations of chimpanzees are to a very great extent in agreement with Kohts's, in that vocalizations in chimpanzees seem to be elicited by the environment (Fouts 1973).

During the period of observation the chimpanzee never attempted to imitate the human voice or to express himself by other than his own characteristic utterances. The lack of vocal language communication was interpreted by Kohts to mean that the mental capability of the chimpanzee was qualitatively different from that of the human child. This was possibly in reaction to Yerkes, who had published his book *Almost Human* in 1927. Kohts responded by saying: "Not only is it impossible to say that he is 'almost human'; we must go even further and state quite definitely that he is 'by no means human,'" This conclusion was based mostly on language capability, for Kohts was able to observe many instances of human–chimpanzee commonality in play behavior, emotional expression, conditioned reflexes, and "a few" intellectual processes, including curiosity, recognition, identification, and "sounds of an undifferentiated nature." It was concluded that the more biologically important the function the greater the capability of the ape to approach or surpass that of man. Conversely, the higher the intellectual process involved (Kohts considered language the highest) the more dominant man became over the apes.

Kohts's observation that there is a qualitative difference between humans and chimpanzees stems from her emphasis on vocal language capability rather than language capability regardless of mode. To view intellectual processes solely from the position of vocal language ability, in our opinion, limits the possible avenues for two-way communication with nonhuman primates. We conclude from reading Kohts's account that the study was undertaken with the expectation of establishing two-way vocal communication with the chimpanzee within a matter of months. When this expectation was not met but was later achieved with Roody, Kohts's interpretation was that the chimpanzee was qualitatively different from humans in mental capabilities. A valuable lesson learned here is that physical similarity does not necessarily mean total similarity. The chimpanzee may look and act like man but its mode of communication is not necessarily the same. Kohts's conclusion that a qualitative difference in intelligence exists between humans and chimpanzees because of differing communication modes is a common, prejudicial misjudgment.

Gua

The experiment involving the infant chimpanzee Gua evolved as a direct result of the influence of Witmer (1909) and Yerkes (1925) and Yerkes and Yerkes (1929), who had expressed opinions concerning the feasibility of raising an infant chimpanzee in the home. Gua was 7½ months old when she was obtained by W. N. Kellogg and L. A. Kellogg (1967) from the Yerkes Experimental Station in Orange Park, Florida. Gua lived in the Kelloggs' home for nine months and was afforded the same surroundings and treatments as their similarly aged son, Donald. During this time the Kelloggs were able to distinguish four naturally occurring sounds made by Gua. All of them seemed either to be elicited by the external environment or to be the result of the emotional state of the chimpanzee. The sounds—barking, food bark, screech or scream, and the "oo-oo" cry—are similar to those reported by other observers of chimpanzees (Yerkes 1925; Yerkes & Yerkes 1929; Kellogg this volume). In attempting to teach the two syllable word "pa-pa," the Kelloggs noted that Gua showed considerable curiosity in the facial movements although she never tried to imitate the sounds. Upon being encouraged to do so by manipulation of the lips, Gua made occasional lip reactions but did not attempt to produce the sound. The Kelloggs proposed that if the chimpanzee ever progressed to actual articulation of human sounds it would be under training circumstances similar to those experienced by Gua. At the same time, the Kelloggs were attempting to teach their son to articulate spoken words, also without success. Although Donald was clearly making gurgling and babbling noises, Gua was not able to make such sounds.

Viki

One of the most successful attempts to train a chimpanzee to speak was conducted at the Yerkes Laboratories of Primate Biology in Florida by Keith Hayes and Catherine Hayes (1952). They obtained a female chimpanzee only a few days after birth and tried to provide a background of experience resembling that of a human infant as closely as possible. The chimpanzee, Viki, lived in the Hayeses' home for six years and learned to produce four words, recognizable as "mama," "papa," "cup," and "up" (Hayes & Hayes 1951). In agreement with an earlier report (Witmer 1909), the Hayeses found the chimpanzee's vocal expression hoarse and seemingly difficult for the animal to produce. Viki's language training, which extended over several years, consisted of manipulating her mouth and lips and subsequently rewarding approximations to desired sounds. As training progressed, fewer manipulations were required to obtain the desired syllables. In this manner Viki was able to produce the words "mama" and "papa" on demand, although her difficulty in speaking was apparent. The words "cup" and "up" were more readily added to her vocabulary as they closely resemble sounds naturally produced by a chimpanzee. In his review of home-rearing experiments, Kellogg (this volume) indicates that this project represents the acme of chimpanzee vocal achievement in human sounds. But it must be noted that even after these words were learned they were often inaudible and used incorrectly to identify objects.

Other Experiments

It is generally recognized that of all the animals, the chimpanzee most closely resembles man in physiology, intelligence, and imitative ability (Yerkes & Yerkes 1929). Attempts at home-raising experimental animals other than chimpanzees have been reported, but none have produced any notable differences in vocal language capability from that of the chimpanzee. Furness (1916) attempted to train a young orangutan in a home-type experiment but after extended training was able to report only limited success. The orangutan was able to say "papa" and "cup" but these words were hoarse and were produced with considerable difficulty. These findings are consistent with later ones concerning vocalizations in chimpanzees (Hayes and Hayes 1952). For the orangutan to produce the sound "papa," Furness found it necessary to manipulate the animal's lips with his fingers. "Cup" was pronounced with relatively greater ease, probably because it is a more naturally occurring sound for such primates, as Hayes and Hayes (1952) noted.

Toto, a gorilla, was raised in the home of A. M. Hoyt (1941) in Havana, Cuba. Notable comparisons between Toto and other home-raised pongids were recorded, but if the gorilla produced any humanlike sounds they were not reported by Hoyt.

Conclusions about Home-Raising Experiments

The relative lack of success in home-raising experiments designed to teach vocal speech to nonhuman primates has led a number of people to conclude that there exist qualitative differences between man and apes that constitute a definition of man. According to this line of reasoning, Noam Chomsky (1968) and other linguists imply that man is unique because he uses language, which provides a linguistic reservoir from which he can structure thought. By rearranging the linguistic symbols he can alter his thought; he can use the symbols in relation to tense; he can abstract thought; and, ultimately, by modifying his concepts he can produce novel combinations of symbols, which he can relate to other humans by the use of language. According to Chomsky this capability makes man unique, the only animal capable of creative thinking.

Chomsky's argument is to a great extent based on a traditional point of view. His assumption is that if an event cannot be observed it does not exist. We consider this to be reasoning based on "negative evidence." In the present instance it is erroneous to assume that language *per se* does not exist because vocal language is not exhibited to any great extent by chimpanzees. If all the researchers had assumed that language ability did not exist in chimpanzees, scientific attempts to investigate communication with nonhuman primates might well have ended with the home-raising experiments. Instead, the rationale for such experiments was changed because it was felt that attempts to communicate vocally were leading nowhere. Kellogg and Kellogg (1967) summed up, "We feel safe in predicting . . .

it is unlikely any anthropoid ape will ever be taught to say more than a half-dozen words, if indeed it should accomplish this remarkable feat."

The results of the home-raising experiments indicated two possibilities. The first was that vocal communication with lower primates on a significant level is impossible because it is beyond the capability of animals other than humans. The Hayes and Hayes (1951) study, which involved the most highly controlled and systematic attempt to teach vocal communication to a chimpanzee, has often been cited by those who take this position. The negative-evidence rule applies here: Those who hold this position see no communication; therefore the capacity for communication does not exist.

The second possibility (based on anatomical evidence recently confirmed by Lieberman 1968) is that speech as commonly used by humans is not a suitable medium of communication for chimpanzees. Similarly, Gardner and Gardner (1971) have reported that the reason chimpanzees do not learn to speak is behavioral as well as anatomical. Some specific portions of an animal's behavioral repertoire are highly, and perhaps completely, resistant to modification; attempts to teach vocal language to chimpanzees have apparently failed because vocalization is resistant to modification (Gardner & Gardner 1971).

Although attempts at vocal training were largely unsuccessful, chimpanzees use their limbs, particularly their hands, in a highly efficient and well-coordinated way (Hayes & Hayes 1952; Kohts 1935; Riesen & Kinder 1952; Witmer 1909). Having noted that a species-specific characteristic of using hand gestures to communicate either greeting or threat had been observed in both wild (van Lawick-Goodall 1968b) and captive chimpanzees (van Hooff 1971), Gardner and Gardner (1971) proposed a method of communication using a form of gestural language. They selected the American Sign Language (Ameslan), which does not totally satisfy the linguistic criteria of language capability (Chomsky 1968; Gardner & Gardner 1971), but nonetheless provided an excellent means of determining linguistic abilities of chimpanzees. Ameslan was chosen because it is actually used as a language by a group of human beings (Gardner & Gardner 1971). Widely employed by the deaf, it is composed of gestures, mainly executed by the fingers, hands, and arms, and it has specific movements and places where the signs begin and end in relation to the signer's body. The signs are analogous to words in a spoken language (Stokoe et al. 1965).

The results obtained by Gardner and Gardner with their chimpanzee, Washoe, indicated that their choice of language medium was a good one. By using the chimpanzee's natural ability to use gestures and her imitative and intellectual capacities, they were successful in teaching Washoe to use Ameslan. The Gardners provided evidence that a true two-way, nonvocal communication channel between man and chimpanzee can be established. Recently, using Ameslan, experimenters have begun to explore the intellectual capacities of chimpanzees (Fouts, Chown, &

Goodin 1973; Fouts, Mellgren, & Lemmon 1973; Mellgren, Fouts, & Lemmon 1973). This method of communication has begun to reveal conceptual processes in chimpanzees that were heretofore impossible to determine, and what has been accomplished shows that the nature of animal intelligence may at last be studied to an extent never before attempted.

PROJECT WASHOE

Project Washoe was the first successful attempt to teach a nonhuman primate in human language. The project was begun at the University of Nevada in Reno by R. A. Gardner and B. T. Gardner (1969, 1971) in June 1966 and was terminated in October 1970, by which time Washoe had acquired a vocabulary of over 130 signs. Although the number of signs is relatively small, Washoe was able to use her vocabulary very well. She could readily produce spontaneous combinations of signs that demonstrated her syntactic ability, and her combinations were contextually correct. She was also able to transfer her signs and combinations to novel situations with ease and with a high degree of reliability.

Washoe, an infant female chimpanzee estimated to be 8–14 months of age in June 1966, was obtained by the Gardners from a trader in the United States. It is assumed that she was born in the wild and was raised for several months by her natural mother before being captured. Washoe was raised in the Gardners' back yard, an area of 5,000 square feet. She lived in a completely self-contained house trailer (8 × 24 ft), which provided for her toilet, kitchen, and sleeping needs. Throughout the project her researchers used only Ameslan to communicate with Washoe, and, when in her presence, they used only Ameslan to communicate with one another.

McCarthy (1954) has indicated that the barren social and physical environments once common to institutions for the mentally retarded are not favorable conditions for the development of language in humans. Gardner and Gardner (1971) extended this hypothesis to the chimpanzee and claimed that the same could be said for the raising of chimpanzees in cages. Since one would expect less linguistic capability from chimpanzees raised under these conditions, Washoe was provided with an environment that was kept as interesting as possible. Her teaching program was made part of her environment and was maintained throughout her daily routine.

A member of the research team was with Washoe during all her waking hours. At the change of shift two researchers overlapped for an hour. She was often afforded additional companionship when visitors were present and during frequent outings in the nearby community. She played and climbed trees and playground equipment in the Gardners' backyard. The function of the researchers on the team was to keep Washoe totally immersed in Ameslan. Since their principal job was to be a conversing companion and a model for Washoe, they used Ameslan in normal daily routines, games, and general activities. They chatted with her in Ameslan while cooking meals, cleaning, brushing her teeth, playing with her in her sandbox, and correcting her lapses in toilet training.

Methods of Acquisition

Manual Babbling

A number of methods of acquiring the signs were examined. Manual babbling, which is considered analogous to vocal babbling in human infants, was infrequently observed early in the experiment, but as the project progressed, the occurrence of babbling increased. The increase continued until the end of the second year, after which time manual babbling was rarely observed. This decline is attributed to Washoe's progress in acquiring a vocabulary during this period, and it was concluded that the acceleration of signing may have replaced the babbling, just as babbling in humans decreases as vocal speech develops. Gardner and Gardner (1971) reported that the only sign attributable to this method of acquisition was the "funny" sign, which is produced by touching the nose with the index finger.

Shaping

Shaping procedures similar to those used in operant conditioning were used in teaching Washoe new signs. She was rewarded whenever she made an approximation to a sign in order to encourage her repeating it. Successively closer approximations were then rewarded, and in this manner several signs were acquired. For example, originally Washoe would bang on doors with her fists when she wanted the door to be opened. This natural response was shaped using rewards for successive approximations to the correct sign. The sign for "open" consists of placing the two open palms against the object to be opened and then moving them up and apart. Washoe quickly acquired the "open" sign and soon generalized it spontaneously from doors to other objects such as books, briefcases, boxes, and drawers. Although shaping procedure was used a great deal during the first year to introduce new signs, it soon became apparent to the Gardners that this technique was not as efficient as other methods.

Guidance

The Gardners considered guidance the most effective method of teaching a new sign to Washoe. It consists of physically molding the hands and arms in the appropriate position for the sign, usually in the presence of an object or action that represents the sign. An example is given by Gardner and Gardner (1971):

The sixth sign that Washoe acquired was the sign *tickle*. It is made by holding one hand open with fingers together, palm down, and drawing the extended index finger of the other hand across the back of the first hand. We introduced this sign by holding Washoe's hand in ours, forming her hands with ours, putting her hands through the required movement, and then tickling her.

Molding

A more recent investigation into the optimal training method (Fouts 1972b) utilized a design intended to test three procedures: molding, which involved physically guiding Washoe's hands into the correct position and movement for the sign; imitation, which consisted of the experimenter's making the sign and Washoe's being required to imitate his example; and free style, which was a combination of molding and imitation. Under the conditions of the experiment it was found that the molding procedure produced the most rapid acquisition of signs, followed by free style (which is not significantly different from molding), and lastly by imitation. Although imitation showed the poorest performance in the experiment, it nonetheless resulted in the acquisition of signs.

Observational Learning

Gardner and Gardner have reported that during the project Washoe apparently learned to comprehend Ameslan as practiced by her researchers although no overt efforts were made to teach her the signs. Since it was the researchers' practice to converse in Ameslan in Washoe's presence, the Gardners called this process of learning *observational learning*. It is analogous to Fouts's (1972b) use of imitation, although in the Fouts experiment a deliberate effort was made to train the chimpanzee to produce a particular sign. For example, Washoe learned the sign for "sweet" (which is produced by touching the lower lip or the tongue with the extended index and second fingers of one hand while the remaining fingers are pressed into the palm) merely by observing her trainer. "Flower," which Washoe at first signed incorrectly, was later corrected by observing how the researchers made the sign. Gardner and Gardner (1971) indicate that because Washoe was totally immersed in an environment of sign language, she often acquired signs after several months' exposure to them without any concerted effort on the part of the researchers to teach them to her. It would appear that she spontaneously acquired the signs.

Recording Washoe's Signs

The various signs given by Washoe were recorded each day. A major portion of the recording procedure was concerned with whether the signs were spontaneous or prompted. Spontaneous signs were noted when Washoe made a correct signing response in answer to a question or made one entirely on her own, such as "open." Prompted signs were those for which a correct response required assistance from one of the members of the research team. Additionally, information concerning the correctness of the form of the response was recorded. The criterion for reliability of responding was whether Washoe could produce the response spontaneously and correctly on 15 consecutive days. In addition to keeping a signing record, the researchers kept a daily diary of the uses Washoe made of various signs and combinations and the contexts in which she used them. Later in the project a daily tape recording was added to the recording procedures. One of the researchers observed the team in a training session and verbally recorded the training procedures and Washoe's responses using a whisper microphone.

The first 36 months of the project yielded notable results in Washoe's ability to acquire signs. By the end of this period she was using 85 signs reliably. By June 1974, she was using over 160 signs reliably. A cumbersome recording process is required in a project such as this; therefore it should be stressed that the reported size of Washoe's vocabulary is limited more by the researchers' ability to handle the recording and testing of the vocabulary than by Washoe's ability to acquire signs.

Testing Procedures

Because the approach in this project was at variance with any previous attempts, new ways of testing had to be designed in order to quantify the accumulated data accurately. Several methods were tried before a test was found that satisfied the requirements of both the researchers and the chimpanzee. A double-blind procedure (a procedure in which the observer evaluating the behavior of the subject does not know what treatment the subject has received until after the evaluation is completed) was used to control for the possibility that the experimenter might cue the subject as to the correct answer.

The first test used flash cards. Washoe was shown pictures on large cards and then questioned about the cards. This test had some drawbacks, the main one being that the test was experimenter-paced and so required an excessive amount of discipline. The box test was then devised. Washoe was to identify three-dimensional objects placed in a box by the researcher. Although this test was a great improvement over the flash-card technique, it was logistically difficult to conduct. It was replaced by the slide test, which was similar to the box test except the 35mm color transparencies were used as exemplars. The slide test was efficient and easy to administer, and it differed from the other two tests in that it could be paced by Washoe.

The Gardners report that in the initial slide test Washoe correctly identified 53 items out of a possible 99. This performance was considerably above the chance level, which was 3 correct responses out of a possible 99. Although the correct responses were obviously encouraging to the researchers, Washoe's errors produced equally interesting data in that the errors fit into meaningful conceptual categories. Conceptual categories are such things as animals, foods, or grooming articles. Instances of responding to conceptual categories occurred when Washoe signed "dog" in response to a picture of a cat, "brush" for a picture of a comb, and "food " for a picture of meat. Also it was reported that when pictures of three-dimensional replicas of objects (e.g., toy cats, toy dogs) were used in the box test, the "baby" sign occurred frequently among Washoe's errors. When the slides were of real items, the "baby" error did not occur; however, the "baby" error continued to occur when a picture of a replica was used. For example, when a slide of a doll resembling a cat was shown, Washoe made four errors out of ten trials, and all four were the "baby" sign. However, when a photograph of a real cat was shown, she correctly identified and signed seven out of eight trials, and the single error was not the "baby" sign.

One important criterion of language use is that words must be used with others to form phrases, or more appropriately, combinations. Washoe first signed the phrase "gimme sweet" and "come open" during the tenth month of the project. At this time she was approximately 18 to 24 months old (Gardner & Gardner 1971). It is interesting to note that similar instances of language use appear in humans at approximately the same age. Fouts (1973) has recently found that the chimpanzee is capable of forming combinations at a much earlier age. The discrepancy between the new findings and those reported by Gardner and Gardner (1971) is most likely due to the comparatively late start in training Washoe to use Ameslan (8–14 months), whereas in Fouts's research the teaching of Ameslan was begun at a much earlier age. In a more recent project by Gardner and Gardner (personal communication), two young chimpanzees acquired their first signs at age 3 months. Schlesinger and Meadow (1972) have found a correspondingly early acquisition of signs in humans, in which deaf children exhibited the use of signs at 5 months of age. The two-month difference is probably not due to qualitative differences in intelligence but to comparatively faster motor-coordination development in chimpanzees (Kellogg & Kellogg 1967).

The segmentation of combinations by Washoe was done in much the same manner as executed by human signers. Washoe would keep her hands raised in the signing space until she had completed the combination. She would terminate the combination by making contact with some object or surface, comparable to the human signer's hands in repose.

The contextual relevance in which Washoe used her signs was found to be very good. She would correctly sign intended destinations with phrases such as "go in," "go out," or "in down bed." When playing with members of the research team she did not generalize but referred to each one by name, signing "Roger you tickle," "you Greg peekaboo" (Gardner and Gardner 1971). Washoe signed at a locked door on thirteen separate occasions, and each time her contextual relevance was appropriate and correct: "gimme key," "more key," "gimme key more," "open key," "key open," "open more," "more open," "key in," "open key please," "open gimme key," "in open help," "help key in," and "open key help hurry" (1971:167).

The Gardners analyzed Washoe's two-sign combinations according to a method proposed by Brown (this volume) for use with children. It was found that her earlier combinations were comparable to the earliest two-word combinations of children in terms of expressed meanings and semantic classes. Longer combinations were often formed by Washoe by adding appeal signs, such as "please" and "come," to shorter combinations. Between April 1967 and June 1969, 245 different combinations of three or more signs were recorded in the researchers' diary. About half were formed by the addition of an appeal sign. The remaining ones were introduced by new information and relationships among signs, such as pronouns or proper names.

In analyzing Washoe's combinations the Gardners found evidence that possibly indicates that Washoe had specific preferences for word order. The combination "you me" was preferred in over 90% of the samples taken, with "me you" used in the remaining 10%. Earlier instances of this combination showed a preference for "you-me"-action, but later this changed to a preference for "you"-action-"me" order. The Gardners were reluctant to accept this order as an indication of syntax in Washoe's manual language. They point out that it may merely be her imitation of the preferred order of the members of the research team. From a behavioral viewpoint, however, there appears to be little difference between Washoe's preferences in word order and language behavior in human children in learning syntax.

Fouts (personal observation) has also noted the use of sign order to express meaning in signing chimpanzees. One of his chimpanzees, Lucy, has a definite preference for the sign order "tickle Lucy" when asking someone to tickle her; and when the order is reversed to "Lucy tickle," she correspondingly tickles her companion.

WASHOE AND THE OKLAHOMA CHIMPANZEES

Project Washoe was terminated in October 1970, when several members of the research team were receiving their degrees and leaving the project. Washoe was then brought to the Institute of Primate Studies in Norman, Oklahoma, directed by W. B. Lemmon. The Institute has many chimpanzees in its main colony, and since Washoe's arrival, it has been directing research with chimpanzees reared in private homes around the area. In both the colony proper and the private homes, the animals have been taught to use Ameslan.

One of the first experiments using Ameslan conducted in Oklahoma (Fouts 1973) was designed to determine the relative ease or difficulty of acquiring signs by four young chimpanzees, two males and two females. An interesting finding was that just as in humans, chimpanzees have different rates of acquiring signs. By means of a molding procedure (Fouts 1972), a total of ten signs were taught to the chimpanzees in daily 30-min training sessions. The acquisition rate for each sign was compared on the basis of the number of minutes of training necessary to reach five consecutive unprompted responses. After the chimpanzees had acquired the ten signs, they were tested on nine of them (all nouns), with the sign for "more" (an adjective) excluded from the test. Testing was conducted using the double-blind box test procedure, similar to the test described by Gardner and Gardner (1971).

The results indicated that some of the signs were consistently easy or difficult for the chimpanzees to acquire. The mean times for acquiring the signs ranged from 9.75 min to 316 min. The mean times to reach criteria for each chimpanzee across signs were: 54.3, 79.7, 136.4, and 159.1 min. These differences may be due partially to the individual chimpanzees' behavior in the training sessions.

All the chimpanzees performed above the chance level during testing. The correct responses in the double-blind box were: 26.4%, 58.3%, 57.7% and 90.3% correct. The low score of 26.4% obtained by one of the chimpanzees may have been a result of the difference between acquisition and testing. This particular chimpanzee seemed to require much praise and positive feedback from the experimenter for her correct responses when acquiring the signs. However, in the double-blind testing situation, the observers who recorded her scores were unable to give her any positive feedback, and, as a result, her performance would begin to deteriorate noticeably after the initial trials in the test were completed. Another important finding was that Washoe was not the only chimpanzee that had the capacity to use Ameslan.

Mellgren *et al.* (1973) examined the conceptual ability of a chimpanzee in regard to the category of items by studying the relationship between generic and specific signs in Ameslan. The subject in this experiment was Lucy, a 7-year-old chimpanzee that had been raised in species isolation (by humans without ever seeing another chimpanzee) in a human home since she was 2 days of age. She had been taught Ameslan for 2 years and had a vocabulary of 75 signs. Our objective in the experiment was to determine if a new sign would become generic or specific relative to a category of items. Lucy had previously learned 5 food-related signs: "food," "fruit," and "drink," which she used in a generic manner and "candy" and "banana," which she used specifically. The sign we chose to teach her was "berry," and the category of items consisted of 24 different fruits and vegetables, ranging from a quarter of a watermelon and a grapefruit to small berries and berry-like items such as blueberries, cherry tomatoes, and radishes. The exemplar for the "berry" sign was a cherry.

The 24 different fruits and vegetables were presented to Lucy in a vocabulary drill, and her responses were recorded. As each item was presented she was asked: "what that?" She was allowed to handle the various foods and eat them if she wished. The order of presentation was varied from day to day. The food-related items were presented along with at least 2 other items that were not in the fruit and vegetable category but were items for which she had a sign in her vocabulary. For example, she would be asked to identify a shoe, a string, and then a fruit or vegetable. Following her response to the food items, she was questioned about such things as a book or a doll. For the first 4 days of training, data were collected to obtain a systematic baseline of her responding to the 24 items to determine her usual response to them. On the 5th day she was taught the "berry" sign, using the cherry as the exemplar, and for the second 4 days the "berry" sign remained entirely specific to cherries. On the 9th day she was taught the "berry" sign again, but this time blueberries were used as the exemplar. On the 9th and 10th days she called the blueberries "berry," but on the 11th and 12th days she switched back to what she had previously called them; but throughout these 4 days she persisted in calling cherries the sign she had originally been taught for them, "berry." By her responses it was apparent that she preferred to use the "berry" sign in a specific sense.

Lucy's conceptualizations of fruits and vegetables were also examined. She showed a preference for labeling the fruit items with the "fruit" sign in 85% of the trials and for using the "food" sign in 15% of the trials. A dichotomy of responding to the two categories was indicated because she preferred to refer to vegetables as "food" 65% of the time and as "fruit" 35% of the time.

In a very revealing finding, Lucy created novel combinations to describe her perception of the stimulus item. She preferred to call a watermelon "candy drink" or "drink fruit," whereas the experimenters referred to it with entirely different signs, that Lucy did not have in her vocabulary ("water" and "melon"). Another, more striking example occurred with radishes. For the first three days of the experiment Lucy labeled them "fruit food" or "drink." On the 4th day she bit into a radish, spat it out, and called it "cry hurt food." She continued to use "cry" and "hurt" to describe the radish for the next eight days. Sixty-five percent of the "smell" signs were used to describe the four citrus fruits by labeling them "smell fruit," probably referring to the odor released when one bites into the skin of citrus fruits. The spontaneous generation of novel combinations not only demonstrated Lucy's ability to form new combinations but also indicated her ability to use her existing vocabulary of signs to map various concepts she had about the categories of the fruits and vegetables she was presented with.

A good understanding of vocal English appears to exist in a number of the home-reared chimpanzees near the institute. Some of them were exposed to vocal language before being taught Ameslan. To determine the relationship between their English vocabularies and their Ameslan vocabularies, a study was undertaken by Fouts, Chown, and Goodin (1973) using Ally, a young male chimpanzee that was being home-reared. A pretraining test was administered to determine his understanding of ten vocal English words. He was given such vocal commands as "Give me the spoon," "Pick up the spoon," "Find the spoon." Ally had to obey the command by choosing the correct item from a group containing several other objects. Only after correctly obeying the command five consecutive times was he considered to have an understanding of the vocal English word. Training was begun by dividing the ten signs into two lists of five signs each. One experimenter attempted to teach Ally a sign using only the vocal English word as the exemplar. Following this, a second experimenter, who did not know which words had been taught or if any had been acquired, would test Ally on all five, using the objects corresponding to the vocal English words. For the second list of five words the experimenters exchanged roles. Ally was able to transfer the sign he was taught to use for the vocal English word to the object representing that word. This finding is very similar to the acquisition of a second language in humans. A second possible implication is that the learning occurred via cross-modal means.

The initial research in gestural language ability in chimpanzees indicates that chimpanzees can produce novel combinations of signs in their existing vocabularies. Humans have this capacity also, but in addition, they are able to understand novel combinations produced by someone else. A recently completed experiment (Chown, Fouts, and Goodin 1974) seems to indicate that chimpanzees are able to understand novel combinations when they are used as a command. In the first phase of the experiment Ally was taught to pick out 1 of 5 objects in a box and place it in 1 of 3 places. For example, a command might be "Put baby in purse." When Ally was sufficiently adept at this, the second phase of the experiment began. New items that had not been used in training were placed in the box and a new place to put the object was added. A screen between the experimenter and Ally prevented the experimenter from giving helping cues. With 5 items to choose from and 3 places to put them, chance would produce 1 correct response in 15 trials. But Ally's performance was far above the chance level. During one test session he responded correctly to 22 of 36 commands.

Chimpanzee-to-Chimpanzee Communication Using Ameslan

Because a major criterion of language is that it be used by members of the same species, we have been interested in intraspecific communication using Ameslan in chimpanzees. Fouts, Mellgren, and Lemmon (1973) explored various conditions under which such communication might be observed. The experimental subjects were Booee and Bruno, two young male chimpanzees who had already acquired a vocabulary of 36 signs each (see Figures 1–4 for examples of Bruno signing). The experimenters intended to keep the chimpanzees' vocabularies small so that they could examine the animals' acquisition of new signs when Washoe was introduced into the signing dyad, but their vocabulary has already increased to over

Figure 1. Bruno signing "hat."

Figure 2. Bruno signing "look."

40 signs. Activities such as tickling, play, mutual comforting, and mutual sharing are most conducive to communication in Ameslan between the chimpanzees. Booee and Bruno, however, seem to prefer their own natural communication over Ameslan, perhaps because of their relatively greater exposure to other chimpanzees. Washoe and Ally are quite different, and prefer to use Ameslan when communicating with humans or other chimpanzees. When Booee and Bruno reach a

Figure 3. Bruno demonstrating his ability to sign ''drink'' with either hand.

preset criterion of reliability in chimpanzee-to-chimpanzee communication, Washoe will be introduced into the dyad. In April 1974, Ally was introduced to Bruno, and a good deal of communication from Ally to Bruno has been observed and recorded, mostly signs referring to food or play.

Manny, a young chimpanzee in the colony, has acquired from Washoe the ''come hug'' sign, which is used correctly when the chimpanzees greet one another or are engaged in mutual comforting. Kiko, a 3 year-old chimpanzee acquired the ''food'' and ''drink'' signs from Booee and Bruno and displayed them correctly,

Figure 4. Bruno signing "baby."

but he was stricken with pneumonia and died. Another mature chimpanzee, housed in close proximity to Washoe, has often shown the "food," "drink," and "fruit" signs, but when offered a drink or a piece of fruit he has failed to show the signs again. Apparently, he has failed to make the important connecting link between the sign and the object it represents.

One thing that makes research with chimpanzees so interesting is their ability to use Ameslan in situations other than experiments and at times when it is not expected. Both Lucy and Washoe have spontaneously invented signs for objects that were not in their vocabulary. Washoe invented a sign for "bib" by making the outline of a bib on her chest (Gardner & Gardner 1971), and Lucy invented a sign for "leash" by making a hook with her index finger on her neck. Washoe was observed making a new combination, "gimme rock berry." When the experimenter approached to question her about the seemingly incorrect sign, he found that Washoe was pointing to a box of brazil nuts on the other side of the room. On another occasion, Washoe referred to a rhesus monkey (with whom she had earlier had a fight) as a "dirty monkey." Until this time she had used the sign "dirty" to refer to feces or soiled items, usually as a noun. She now uses it regularly as an adjective to describe people who refuse her requests.

SARAH AND COMMUNICATION VIA PLASTIC OBJECTS

Premack (1970a, 1971a, b) and Premack and Premack (1972) were the first to devise an artificial system for two-way communication between two species by using variously colored and shaped pieces of plastic to represent words. Sarah, the subject used in their research, is a wild-born female chimpanzee estimated to be six years of age when the project was begun. Since the Premacks' artificial language is visual and written they decided to use the chimpanzee because of the similarities between its visual system and man's.

Pieces of plastic that vary in size, shape, texture, and color are used to represent words. The plastic pieces are backed with metal and can be arranged in lines on a magnetized board. In this manner, the pieces can be displaced in space but not in time on the board, thus avoiding the necessity of relying on memory (which is an integral part of human languages, gestural or vocal). The use of pieces of plastic allows the human experimenter to determine which pieces of plastic will be available to the subject at any time. Also, the design of the experiment controls for the individual difficulty of any problem, but at the expense of the spontaneity of usage typically found in human language. It should also be pointed out that when an experiment is so over-controlled, it limits the possible findings to those preconceived by the human experimenter rather than exploring the full mental capacities of the subject.

The Premacks' approach emphasizes the functional aspects of language by breaking into behavioral constituents and providing environmental contingencies for each constituent they selected for training Sarah. They briefly summarize their work as follows: "We have been teaching Sarah to read and write with various shaped and colored pieces of plastic, each representing a word; Sarah has a vocabulary of about 130 terms that she uses with a reliability of between 75 and 80 percent" (Premack & Premack 1972:92).

The Premacks appear to work from the premise that the relational and logical functions of language are derived from operant procedures, and therefore they use training procedures based on operant methodology. The procedure is of the same type as that used by psychologists to train pigeons or rats to peck keys or press bars. They reduce the constituents to very simple steps and then use standard operant techniques to train the subjects. For example, training may be started by placing some fruit on a board and allowing Sarah to eat it. Next, Sarah is required to place a piece of plastic representing the same fruit on a magnetized board before being allowed to eat the fruit.

As training continues in this manner, new pieces of plastic are added simultaneously with new aspects of the situation, e.g., a different kind of fruit or a new person. For example, a new piece of plastic may be added when Sarah's reward is changed from bananas to apples; or she may be required to place a piece of plastic representing the new trainer's name (e.g., Mary or Jim) on the board. In the next step a piece of plastic representing the word "give" is introduced, and Sarah is required to place it in between the trainer's name and the name of the fruit. Finally, a piece of plastic representing Sarah's name is introduced, and it has to be placed at the end of the sequence. Sarah appears to have little difficulty making these conditional discriminations. She may also be induced to respond to a sequence like "Sarah give apple Mary" by offering her a piece of chocolate if she gives up her apple. By using a more preferred item (chocolate) they are able to induce Sarah to place the pieces representing "Mary give apple Jim."

Premack and Premack report that Sarah is able to use and understand the negative article, the interrogative "wh" (who, what, why), the concept of name, dimensional classes, prepositions, hierarchically organized sentences, and the conditional:

Sarah has managed to learn a code, a simple language that nevertheless included some of the characteristic features of natural language. Each step of the training program was made as simple as possible. The objective was to reduce complex notions to a series of simple and highly learnable steps . . . compared with a two-year-old child Sarah holds her own in language ability. (Premack & Premack 1972:99)

LANA AND THE COMPUTER

Rumbaugh, Gill, and Glasersfeld (1973) devised a computer-controlled training situation that objectively examined some of the language capacities in a chimpanzee. They used Lana, a 2½ year-old female chimpanzee, as their subject. After 6 months of training, Lana was able to read projected word characters, complete an incomplete sentence based on its meaning and serial order, and reject an incomplete sentence that was grammatically incorrect.

Rumbaugh et al. (1973) are using a PDP–8 computer with two consoles containing 25 keys each. On each key is a lexigram in "Yerkish," an artificial language developed by the experimenters. The symbols are white geometric figures, created from 9 stimulus elements, used singly or in combination. The keys, on which the symbols are displayed, have colored backgrounds made up of 3 colors used singly or in combination. When the key is available for use by Lana it is softly backlit. When Lana presses a key, it becomes brightly lit. When a key is not available for use, it has no backlighting. When Lana depresses a key, a facsimile of the lexigram appears in serial order on one of 7 projectors above the console. The computer also dispenses appropriate incentives to Lana when she depresses the keys in the correct serial order in accordance with the grammar of Yerkish. She may ask for such things as food, liquids, music, movies, toys, to have a window opened, to have a trainer come in, and so on, when they are available (that is, when the appropriate keys are softly backlit). There is also a console available only to the human experimenters so that the computer can mediate conversations between the experimenters and Lana.

Lana's training was begun by requiring her to press a single key in order to receive an incentive. Next, she was required to begin each request with a "please" and end it with a "period." The depression of the "period" key instructed the computer to evaluate the phrase for correctness of serial order. If it was correct a tone sounded and Lana would receive what she had asked for; if not the computer would erase the projector display and reset the keys on the console. Later, Lana was required to depress holophrases (e.g., "machine gives M & M") in between a "please" and a "period." Following this, she was taught to depress each key represented in the original holophrase (e.g., "Please/Machine/give/M & M/period"). Then the keys were randomized on the console and she had to select and press them in the correct serial order.

One very interesting finding was that Lana soon learned to attend to the lexigrams on the projectors without training. She would erase sentences in which she had made an error, by pressing the "period" key, rather than finish them. On the basis of Lana's spontaneously learning to attend to the projected lexigrams and their order, Rumbaugh *et al.* (1973) were able to examine her ability to read sentence beginnings, to discriminate between valid and invalid beginnings, and to complete sentences. In their first experiment, Lana was presented with one valid sentence beginning ("Please machine give") and six invalid beginnings. She could either erase them or complete them. If she completed them she had to choose from correct lexigrams (e.g., "juice," "M & M," "or piece of banana") and incorrect lexigrams ("make," "machine," "music," "Tim," "movie," "Lana"). "Music" and "movies" were incorrect since the computer was programmed to accept these with "make" rather than with "give." Lana's performance on the various aspects of this test ranged from 88% correct ("please machine give," the valid beginning). The second experiment was the same except that "make" was substituted for "give" in the sentences with the valid beginning. Lana's performance was 86% or more correct in this experiment. The third experiment used only valid beginnings with varying numbers of words in them: e.g.: (1) "please," (2) "please machine," (3) "please machine give," (4) "please machine give piece," (5) "please machine give piece of," and so on. Lana ranged from 70 to 100% correct on the various beginnings.

Rumbaugh *et al.* (1973) concluded that Lana accurately read and perceived the serial order in Yerkish and was able to discriminate between valid and invalid beginnings of incomplete sentences in order to receive an incentive.

COMPARISON OF THE METHODS

Using artificial languages or Ameslan avoids the problem of vocal communication. In one project using an artificial language (Rumbaugh *et al.* 1973) the human element has been removed by using a computer as an intermediary, and because a computer is used the experimenters are able to keep an exact record of the chimpanzees' communication. However, this refinement is also expensive in that the chimpanzee is forced into a strict and rigid paradigm that allows only that behavior to appear that will fit into the experimental situation. For example, the computer is not programmed to accept novel or innovative uses of the language. Although this method has managed to find and confirm conclusively such behavior as responsiveness to word order (syntax) in exploring the mental capacities of the chimpanzee, the artificial language approaches are limited to examining only behavior conceived of by the experimenters, and not responses created by the chimpanzees.

It is our contention that when conducting initial experimental research, such as we are doing with chimpanzee language acquisition, the situation must be structured only to the extent that control of the experiment remains in the hands of the experimenter. This point was made by Köhler (1971):

Lack of ambiguity in the experimental setup in the sense of an either-or has, to be sure, unfavorable as well as favorable consequences. The decisive explanations for the understanding of apes frequently arise from quite unforeseen kinds of behavior for example, use of tools by the animals in ways very different from human beings. If we arrange all conditions in such a way that, so far as possible, the ape can only show the kinds of behavior in which we are interested in advance, or else nothing essential at all, then it will become less likely that the animal does the unexpected and thus teaches the observer something.

We feel the gestural language approach more closely fits the idea expressed by Köhler. Using this method we can simultaneously conduct highly controlled experiments and allow the chimpanzee to show, spontaneously, many of its capabilities. This is possible because the chimpanzee, not the experimenter, has the control of its language use. The chimpanzee can from within itself make statements or do what it wishes without having to rely on an arbitrary symbol to do so. It is not bound to a computer program or by the limits of the experimenter in making vocabulary available.

For example, in the study done with Lucy on the 24 different fruit and vegetables (Mellgren *et al.* 1973), had we limited her possible responses to the five food-related signs in her vocabulary, we would not have discovered her conceptualizations of the items. Nor would Washoe be able to insult people by calling them "dirty" if she had not been allowed to change a noun into an adjective. Similarly, Washoe would not be able to refer to brazil nuts as "rock berry."

Each approach has its advantages, and it is up to the scientists to decide whether they wish to examine only those things they are capable of conceiving of, or if they are willing to accept some help from the chimpanzee in examining the animal's mental capacities.

CONCLUSIONS

The language skills of the chimpanzee are similar to those displayed by humans, although many definitions of language have attempted to exclude the chimpanzee from the realm of language as used by humans. An often-quoted, popular paper, by Bronowski and Bellugi (this volume) lists five characteristics of language: delay between stimulus and utterance, separation of affect from content, prolongation of reference, internalization, and reconstitution. Bronowski and Bellugi contended that Washoe probably met the first four, but failed to demonstrate the last. They define the structural activity of reconstitution as consisting "of two linked procedures—namely, a procedure of analysis, by which messages are not treated as inviolate wholes but are broken down into smaller parts, and a procedure of synthesis by which the parts are rearranged to form other messages" (this volume: 106).

Bronowski and Bellugi conclude: "What the example of Washoe shows in a profound way is that it is the process of total reconstitution which is the evolutionary hallmark of the human mind, and for which so far we have no evidence in the mind of the nonhuman primate, even when he is given the vocabulary ready made" (this volume: 113). We disagree with their contention of "no evidence." Most certainly the empirical evidence presented earlier in this chapter demonstrates that chimpanzees have the capacity for reconstitution, particularly Lucy's reference to a radish as "cry hurt food" and Washoe's calling a brazil nut "rock berry." We contend that these gestural utterances more than meet the most restrictive definitions of reconstitution and represent a remarkable intellectual and linguistic accomplishment, given the limited vocabulary we have allowed the chimpanzees to learn. And if Bronowski and Bellugi's contention that reconstitution is the "evolutionary hallmark of the human mind" is correct, then we must assume that the capacity for language was in the repertoire of the species before the great apes split off from hominoid evolution. Another alternative may be that the basis of language is not unique to language *per se*, but is actually the basis of other behaviors of which language is just one product.

COMPARATIVE PSYCHOLOGY AND LANGUAGE ACQUISITION*

R. ALLEN GARDNER AND BEATRICE T. GARDNER

A science of psychology must assume that behavior is lawful. We recognize diversity, but we assume that both similarity and difference are products of the same fundamental laws that combine and recombine in unique ways to yield the rich diversity of behavior that we observe between and within species. The proper analysis of behavior is not in terms of simpler behavior and more complex behavior or in terms of simpler organisms and more complex organisms, but rather in terms of general functions such as perception and learning that are found in all forms of behavior. We avoid the invention of new laws of behavior for each newly discovered level of complexity in favor of the formulation of more powerful generalizations.

The history of other sciences favors the movement toward fewer and more powerful generalizations. For some psychologists, however, the laws of human behavior must be different from the laws of animal behavior, and certain kinds of human behavior are immune to the laws that govern other kinds of behavior. Chomsky has won a wide following by statements of the following kind:

When we ask what human language is, we find no striking similarity to animal communication systems . . . human language, it appears, is based on entirely different principles. This, I think, is an important point, often overlooked by those who approach human language as a natural, biological phenomenon. . . . As far as we know, possession of human language is associated with a specific type of mental organization, not simply a higher degree of intelligence. (1972)

This view argues from surface dissimilarities to the conclusion that there are fundamentally different sets of laws for different types of behavior.

Our research on teaching sign language to chimpanzees proceeds from the

*Research supported by NSF grants GB–35586 and BNS–75–17290, by NIMH research development grant MH–34953 (to B. T. Gardner), and by The Grant Foundation, New York, New York.

R. ALLEN GARDNER AND BEATRICE T. GARDNER • Department of Psychology, University of Nevada, Reno, Nevada 89557.

assumption that any form of behavior, human or animal, if it exists at all, exists as a natural, biological phenomenon. If a form of behavior such as human language appears to be different in character from other forms of human and animal behavior, then we do not abandon the search for general laws; instead, we question the adequacy of existing observations. Before Project Washoe, there were two general courses that might be pursued to obtain more relevant data. A linguistically sophisticated attempt could be made to study animal communication, or a means could be found to teach a human language to an animal. With respect to the first alternative, there has been little progress so far.[1] With respect to the second alternative, much progress has been made, and this is the subject of the present report.

PROJECT WASHOE

Although there have been several, unsuccessful efforts to teach chimpanzees to speak, Project Washoe was the first attempt to teach sign language to a chimpanzee. Within 51 months of training, Washoe (who was about 11 months old when training began) had acquired 132 signs of American sign language (Ameslan), as determined by strict criteria of usage established in the course of the project. These signs were used for classes of referents rather than for specific objects or events. Thus, the sign "dog" was used to refer to live dogs and pictures of dogs of many breeds, sizes, and colors, and for the sound of barking by an unseen dog as well; while the sign "open" was used to ask for the opening of many doors (e.g., to houses, rooms, cupboards) or containers (e.g., boxes, bottles, jars) and even—an invention of Washoe's—the turning on of a water faucet. In addition, there were several hundred signs of Ameslan that Washoe learned to understand during this period.

We were able to integrate signing into Washoe's repertoire of behavior so that spontaneous communication by means of Ameslan was a common feature of her interaction with other beings, both human and animal, both friends and strangers. Throughout the course of the project, a systematic program of testing was administered to Washoe (1971, 1974b, 1975b) but the resulting record represents only a fraction of Washoe's verbal behavior, because so much of her use of Ameslan was spontaneous. She learned to ask for goods and services, and she also answered questions with verbal descriptions and commentary about the world of objects and events that surrounded her. Washoe's descriptions and comments were not limited

---

[1]This is largely because the study of animal communication continues to employ techniques that defy meaningful comparison with observations of human linguistic communication. This point has been discussed in detail elsewhere (Sarles 1969, 1974). Essentially, the problem is that if human speech were also studied by the correlation of sound spectrograph recordings with gross motor activity, we would also be forced to conclude that human speech has no linguistic significance.

to replies to our questions; she initiated many of the conversations with questions and opening statements of her own.

As soon as Washoe had about 8 signs in her vocabulary, she began to use them in combinations of 2 or more, and soon combinations such as "You me hide," and "You me go out there hurry," became very common. In our report of the first 36 months of Project Washoe (Gardner & Gardner 1971), we were able to draw several close parallels between Washoe's early combinations and the early combinations of the young children reported by Roger Brown (this volume).

One example of the parallelism is shown in Table I. Commenting on these early results, Roger Brown 1973a concluded that "the evidence (from Gardner & Gardner 1971) that Washoe has Stage I language is about the same as it is for children." Project Washoe continued for another 15 months, and during the last third of the project, there was no indication that we were approaching any asymptote of chimpanzee capacity. On the contrary, progress continued to accelerate. In the further development of her combinations, in her replies to wh-questions, in her use of negatives, prepositions, and locatives, Washoe compared favorably with children at Brown's Stage III and beyond (Gardner & Gardner 1974b, 1975b).

Table I. Parallel Descriptive Schemes for the Early Combinations of Children and Washoe

Brown's scheme for children[a]		The scheme for Washoe	
Types	Examples	Types	Examples
Attributive: (Ad + N)	*Big train; Red book*	Entity and attribute:	*Drink red; Comb black*
		Animate and trait:	*Washoe sorry; Naomi good*
Possessive (N + N[b])	*Adam checker; Mommy lunch*	Entity and possessor:	*Clothes Mrs. G.; You hat*
		Entity and possessive:	*Baby mine; Clothes yours*
Locative: (N + V)	*Walk street; Go store*	Action and locative:	*Go in; Look out*
Locative: (N + N)	*Sweater chair; Book table*	Action and location:	*Go flower; Pants tickle[c]*
		Entity and locative:	*Baby down; in hat[d]*
Agent and action: (N + V)	*Adam put; Eve read*	Agent and action:	*Roger tickle; You drink*
Action and object: (V + N)	*Put book; Hit ball*	Action and object:	*Tickle Washoe; Open blanket*
Agent and object: (N + N[b])	*Mommy sock; Mommy lunch*	(Not applicable)[b]	
(Not applicable, see text)		Appeal and action:	*Please tickle; Hug hurry*
		Appeal and entity:	*Gimme flower; More fruit*

[a]Brown, this volume
[b]Indicates types classified in more than one way in Brown's scheme and only one way in our scheme.
[c]Answer to question, "Where tickle?"
[d]Answer to question, "Where brush?"

This article is a survey of the new line of research that began with Project Washoe. It includes a description of the methods of Project Washoe and its current sequel in our laboratory, the rationale for these methods, the current findings and the future goals.

APPROPRIATE LABORATORY CONDITIONS

In some areas of research it may be reasonable to compare caged chimpanzees with caged members of other species, but in the case of language acquisition, human beings are the only standard of comparison. It would be unreasonable to compare chimpanzees that were maintained in the caged environments of a primate facility with human children that were maintained in a middle-class home. Only the most devout Chomskian nativist would expect language to unfold normally in human children who were kept in cages with infrequent, sporadic exposure to their native language. Most psychologists are convinced that there is an intimate relationship between language acquisition and every other aspect of behavioral development, and many of us believe that language acquisition may be especially sensitive to the quality of the behavioral environment.

Our procedure can be distinguished from the procedures of other ape projects that have sprung up since Project Washoe on the basis of our emphasis on the quality of the behavioral environment throughout the waking day. This aspect of our procedure has a twofold purpose: the best language acquisition is to be expected under the best conditions, and the best comparisons between child and chimpanzee are those that are made under the most comparable conditions. It is feasible to provide comparable conditions because of the high degree of similarity between human and chimpanzee childhood.

Morphological and behavioral similarities between chimpanzees and humans have long been recognized, but recent research indicates that there is an even closer relationship between chimpanzees and men than most authorities would have conceded twenty or even ten years ago. In blood chemistry, for example, relationships between species have been quantified, and the chimpanzee is not only the closest animal to man by this measure, but the chimpanzee is also closer to man than the chimpanzee itself is to any other species (Goodman 1976, Sarich & Cronin 1976). Where reliable information is available, the sensory systems of the chimpanzee and human are found to be highly similar in structure and function (Prestrude 1970). Under natural conditions, infant chimpanzees are weaned at about five and continue to live with their mothers until they are seven or eight—often ten years old. The youngest chimpanzee mother that Jane Goodall observed at the Gombe Stream was 12 years old (1974). All these discoveries, but most particularly the discovery of the prolonged childhood and adolescence of the chimpanzee, indicate a close relationship between child and chimpanzee.

Figure 1. Pili, at 1½ years, looking through his picture book. In this laboratory, the chimpanzees see numerous exemplars of the signs they acquire.

Because of this close relationship, we can maintain our chimpanzee subjects in an environment that is very similar to that of a human child. Their waking hours follow a schedule of meals, naps, baths, play, and schooling, much like that of a young child; their living quarters are well stocked with furniture, tools, and toys of all kinds, and there are frequent excursions to other interesting places—a pond or a meadow, the home of a human or chimpanzee friend. Whenever a subject is awake, one or more human companions are present. The companions use Ameslan to communicate with the subjects, and with each other, so that linguistic training is an integral part of daily life, and not an activity restricted to special training sessions. These human companions see that the environment is maximally stimulating and maximally associated with Ameslan. They demonstrate the uses and extol the virtues of the many interesting objects. They anticipate the routine activities of the day and describe these with appropriate signs. They invent games, introduce novel objects, show pictures in books and magazines, and make special scrapbooks of favorite pictures, all to demonstrate the use of Ameslan (Figure 1).

A Natural Language

We call certain hand movements made by our chimpanzee subjects signs because their form and their referents are practically the same as the form and referents of signs made by human beings. The similarity is so close that the earliest signing shows immature variations and developmental changes in form that are the same as those of young children. Our descriptions of the signs have been confirmed by dozens of fluent signers who have observed our subjects in person, and by thousands of fluent signers who have viewed films and video tapes of our subjects using signs. In addition, we have tested the ability of deaf observers to identify the signs made by Washoe when the observers were under double-blind conditions.

Ameslan itself existed in the United States for more than 100 years before Project Washoe. Its roots in European sign languages can be traced for several hundred years. An extensive and rapidly growing body of literature reveals that Ameslan is a natural language as close to any of the spoken languages as modern linguistic analysis can determine.[2] In addition, research on the development of Ameslan in deaf children indicates that the patterns of development of Ameslan are similar to the patterns of development of spoken languages (Bonvillian, Nelson, & Charrow 1976; Hoffmeister, Moores, & Ellenberger 1975; Klima & Bellugi 1972; Schlesinger & Meadow 1972).

[2]A sign language consists of a system of manually-produced visual symbols, called signs, that are analogous to words in a spoken language. Stokoe (1960) has shown that just as words can be analyzed into phonemes, so signs can be analyzed into cheremes, a small closed set of distinctive features, meaningless in themselves but which combine to form morphemes, or signs that denote meaning. In commenting on this analysis, Carroll points out:

> to isolate cheremes . . . (Stokoe) . . . proceeded as any linguist would in identifying the phonemes of a new language—by noticing recurrent and discriminating features of utterances—or the analogue of utterances—in sign language.

The sign morphemes, like syllables, were seen to be composed of cheremes. Three varieties of cheremes are distinguished, corresponding to three aspects of the formation of a handsign; its position (i.e., location relative to other parts of the body), its configuration, and its motion. The total number of cheremes is roughly comparable to the number of phonemes in natural language—somewhere between 20 and 60 in most languages (Carroll 1961). Recent psycholinguistic studies, based on the errors that occur in short term memory tests, have shown that signs are perceived and processed not as holistic gestures but in terms of this cheremic structure (Bellugi, Klima, & Siple 1975).

Because of the cheremic and morphemic construction of signs, a very large number of signs are possible. New signs are continually introduced into Ameslan and old signs drop out, just as is the case in spoken languages (Fant 1972). Some signs are highly iconic, while others are quite arbitrary, but all are arbitrary to some degree. Even for a highly iconic sign, it is often the case that different iconic attributes of the same referent are arbitrarily selected for representation by the conventions of different sign languages (Schlesinger, Presser, Cohen, & Peled 1970). Consequently, there are regional and national differences in sign languages as in spoken languages. Deaf persons report that they experience considerable difficulty in conversing with the deaf of another country (Battison & Jordan 1976), and an experiment on referential communication with deaf users of Ameslan and of five foreign sign languages confirmed this difficulty in understanding foreign signers (Jordon & Battison 1976). Within Ameslan itself, signs have undergone historical changes which enhance the ease of articulation. These changes have reduced iconicity, and produced signs that are more arbitrary and conventional (Battison 1974; Frishberg 1975; Woodward 1976). Thus, only a small percentage of the signs in Ameslan have a meaning that is transparent to persons who are unfamiliar with the language. For many signs, the outward appearance of the gesture is misleading: there is a consensus among nonsigners in the meaning attributed to such signs, but it is a wrong meaning (Hoemann 1975).

Thus, we do not assert that our chimpanzee subjects use the signs of Ameslan. We do not argue from analogy. We do not base our conclusions on the assumptions of any linguistic theory. Instead, we can marshal a very large body of empirical evidence and independent linguistic analysis, all agreeing with the conclusion that the communicative gestures of our chimpanzees would be called signs if they were used by human children and that the chimpanzees use signs in a rudimentary, childish form of Ameslan.

Fluent Signers as Adult Models

Perhaps the greatest discrepancy between Project Washoe as we originally conceived it and as we were able to execute it lies in the weakness of the adult models that we could provide. One high-school age deaf boy worked with us for two summers. Two recent graduates of Gallaudet[3] worked with us during a part of the last summer of the project. Occasionally there were visitors who were fluent in Ameslan. For the most part, Washoe's human companions and adult models for signing could be described as hearing people who had acquired Ameslan very recently and whose fluency, diction, and grammar were limited because they had very little practice except with each other and with Washoe.

In the current project we have been able to recruit deaf people and people whose parents were deaf to serve on our research staff. News of the success of Project Washoe was warmly received in the deaf community, and this has helped in the recruitment. Many articles about our research have appeared in popular publications for the deaf, and our research films have been captioned and shown to deaf audiences on television. We have presented invited lectures and film demonstrations at Gallaudet College, the National Technical Institute for the Deaf, and other colleges that have special programs for the deaf. With many educated deaf research assistants, including married couples, we may be able to reproduce the ideal version of the Kellogg experiment: an infant chimpanzee raised in a family consisting of deaf parents and their children.

With the recent increase in technically sophisticated analyses of the structure of Ameslan by linguists who are also fluent signers, the parallels between the formal structure and communicative functions of Ameslan and spoken languages are becoming more evident (Stokoe 1972; Woodward 1972). As in the case of English, there are order constraints in Ameslan that affect the intelligibility of sign sequences (Tweeney, Heiman, & Hoemann 1977), but a language in the visual mode uses many unique devices to convey grammatical information, and the recent analyses offer a detailed description of some of these visual devices (Bellugi & Fisher 1972; Covington 1973; Friedman 1975; Woodward 1973, 1974).

There is no known limit to the size of a legitimate Ameslan vocabulary or the length of a legitimate Ameslan message. In transmitting information, pairs of fluent Ameslan signers are as effective as pairs of fluent English speakers (Jordan 1975).

[3]Gallaudet College, founded over 100 years ago, is in Washington, D.C. and is the national liberal arts college for deaf persons.

Newborn Chimpanzees as Subjects

Washoe's exposure to Ameslan did not begin until she was nearly one year old. The effective beginning of Washoe's exposure to sign language was further delayed (1) by the fact that most of the human participants in the project were themselves only beginning to learn sign language, and (2) by the need during the first months to spend a great deal of time and effort establishing laboratory routines for a project of this type.

The exposure of human children to their native language begins at birth, and most theories of language acquisition assume that the exposure which takes place during the first year is particularly important. Evidence from the study of institutionalized children indicates that an impoverished linguistic environment within the first two years of life results in long-term deficits in language development (Dennis 1973). In a project that aims to provide the most favorable conditions for language acquisition and to compare the development of chimpanzee infants with human infants, the best way to proceed is to start with newborn subjects.

For the four subjects of the new project, exposure to sign language began within a few days of birth. Chimpanzee Moja, a female, was born at the Laboratory for Experimental Medicine and Surgery in Primates, New York, on November 18, 1972 and arrived in our laboratory in Reno on the following day. Chimpanzee Pili, a male, was born at Yerkes Regional Primate Research Center, Georgia, on October 30, 1973 and arrived in our laboratory on November 1, 1973. Chimpanzee Tatu, a female, was born at the Institute for Primate Studies, Oklahoma, on December 30, 1975 and arrived here on January 2, 1976. And our newest subject, chimpanzee

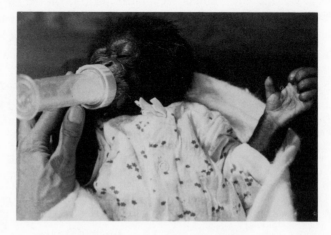

Figure 2. Pili at one week. For the subjects of the new project, exposure to sign language begins within a few days of birth.

Dar, a male, was born at Albany Medical College, Holloman AFB, New Mexico, on August 2, 1976 and arrived here on August 6, 1976 (Figure 2).

Washoe participated in our sign language project just over four years; the next four subjects were born in 1972, 1973, 1975, and 1976, and participated from birth. Thus, statements about developments during the first year in this laboratory are based on the data for five young chimpanzees (Washoe, Moja, Pili, Tatu, and Dar); Pili died of leukemia on October 20, 1975, just before he reached his second birthday (Muchmore & Socha 1976), so that statements about the third and fourth year of training are based on the Washoe and Moja data alone.

Intellectual Maturity

Project Washoe ended when Washoe was only five years old. At that point she was still very immature. For example, she had only five of her adult teeth and had lost only six of her milk teeth. Her progress in sign language was still accelerating and there was no indication that her intellectual development was leveling off. Acceleration had become so great that we could no longer keep track of new developments with the stringent daily reassessment procedure that we had used early in the project.

Common laboratory lore has it that as they reach adolescence, captive chimpanzees become ill-tempered, aggressive, dangerous, and generally unmanageable. Most captive chimpanzees that are more than a few years old are kept closely confined in bare cages nearly all the time. When they are very small, they may be taken out and petted from time to time, but as they grow larger they are rarely handled or allowed out of their cages (except with squeeze cages or after tranquilization). What happens to young chimpanzees under these conditions is what you would expect to happen to human children under comparable conditions (and this expectation is confirmed in understaffed though otherwise well-equipped institutions for the mentally retarded). Their behavior becomes increasingly pathological with each year, until they are dangerous and unmanageable.

Most of the laboratory lore refers to that kind of chimpanzee. But consider the following excerpt from a report by Keith Hayes to the Board of Directors of Yerkes Laboratories that is dated March, 1954, when Viki was six years and eight months old:

The indications are that it will be well worthwhile to continue our study for some years to come. We expect this to be facilitated considerably by Viki's new house on the laboratory grounds, which we have been preparing during the past six months. Although we will not actually be living in this house with her, one or another of us will be there with her during all her waking hours. We expect to be able to provide her with at least as broad a background of experience as heretofore, and we will continue to study her behavior as intensively as in the past.

It should be noted that this change in housing has not been necessitated by any recent increase in Viki's destructiveness or aggressiveness. Rather, the change has been desirable for some years past, since she has always been inclined to dismantle the house and its furnishings. Although she continues to be as friendly and moderately well-disciplined as ever, her increasing size and developing skills make her potentially, if not actually, more troublesome, as time goes by. (Viki died of encephalitis just before her seventh birthday.)

According to Premack, when Sarah reached this age she was too dangerous to handle. The difference, of course, is that at this time Sarah was being kept almost exclusively in solitary confinement in a 3 × 6 meter cage (Ploog & Melnechuk 1971).

As Washoe became older, we found her more manageable and more easy to deal with. If we could have continued the project in its original form, there is every reason to believe that her progress in sign language would have continued to accelerate for years. From Jane Goodall's observations it takes from 12 to 16 years for a chimpanzee to become an adult. In order to determine the highest level of two-way communication that can be achieved by chimpanzees, it is necessary to continue our special laboratory conditions from birth to intellectual maturity.

Multiple-Subject Project

It was appropriate to have only one experimental subject in Project Washoe because so much new ground had to be broken. Now that the general method has been shown to be feasible and productive and critical aspects of procedure have been worked out, a great deal of replication is in order. Ideally, replication should be carried out with several chimpanzees in the same laboratory at the same time. As in most types of research, such a procedure yields greater precision of replication and many practical advantages. In a Washoe-type project there are additional advantages: signing between chimpanzees can be studied, the social lives of the subjects are greatly enriched, and it becomes more feasible to work with older subjects.

Having demonstrated a significant degree of two-way communication between man and chimpanzee, the first objective that springs to mind is to determine the highest level of communication that could be reached. The next is to study signing between chimpanzees. It is not hard to establish cooperative laboratory tasks in which chimpanzees communicate with each other; they do not need Ameslan for that (Menzel 1973c). Our objective has been to observe and record linguistic communication between chimpanzees. The most favorable subjects for such research are those for whom Ameslan has been an integral part of all their daily lives. Only subjects who have been prepared in this way can be expected to use sign language as an integral part of their communication with each other (Figure 3).

Now, newborn subjects enter the project at intervals of one or two years, so that at any given time the subjects are at different ages and different levels of devel-

Figure 3. From a tree, 2½-year-old Moja signs "come" to 1½-year-old Pili. The chimpanzees use sign language to communicate with each other, especially during play and when changing location.

opment. The reason for this procedure is that after studying the extensive records at the Gombe Stream Reserve and after observing wild chimpanzees there for ourselves, it became clear that the relationships that form among young chimpanzees of different ages are more appropriate for our purposes than the relationships that form among evenly matched youngsters. Especially in the early years of the new project, when the subjects are very young we expect more two-way communication for subjects that differ in size, sophistication, and skills. In addition, the big brother/sister relationship seems much more likely to lead to one chimpanzee learning things from another, as indeed has been the case. Ways of greeting, playing, and signing have been transmitted from an older subject to a younger one.

Starting with the subjects at intervals has very important practical consequences as well. Our worst practical problem in Project Washoe was the selection and training of new assistants. In a project of the Washoe type, the chimpanzees must be free-living and must have almost unlimited contact with human beings as well as access to human artifacts. They must also submit to the drill and testing of the laboratory routine. All research assistants must achieve a high degree of behavioral control. We found that those assistants who could achieve behavioral control over Washoe were also able to maintain good control as she grew older; for them, Washoe became easier to deal with as she grew older. Nevertheless, as our subjects grow stronger and more active and more sophisticated in the ways of hazing the new recruits, the task of achieving behavioral control in the first place becomes a formidable one for the inexperienced. This is a serious problem, because there is an inevitable rate of turnover in student assistants and they must be replaced rapidly in order to maintain the pace of the project. By starting newborn subjects at intervals we can continually train new assistants with very young chimpanzees. This procedure not only lowers the rate of failure but also permits us to select those assistants who work with the older subjects from a pool of experienced chimpanzee workers.

The multiple-subject version of Project Washoe became a fairly extensive enterprise by the time that there were three chimpanzee subjects. At that time, we moved with all the chimpanzees and several of the research assistants to a secluded site that was formerly a guest ranch. With some remodeling of existing buildings, the ranch became a research facility: there are quarters to house the chimpanzees and, nearby, quarters for the human companions who monitor them throughout the night; there is a unit for work with groups of subjects and units for individual testing; and there are buildings and rooms to serve communal needs such as meetings, shop, kitchen, office, and so on. The centralized facility provides the most favorable conditions for the study of communication between signing chimpanzees. In line with the long-term plan of the research, the facility assures the maintenance of laboratory conditions and procedures from birth until adulthood. This would permit us to work with the valuable adult subjects whose sign language training has been in progress for the longest time and who might participate in the teaching of their offspring.

LINGUISTIC DEVELOPMENT

Vocabulary

In two earlier articles (1969, 1971), we described in detail the procedures that we used to teach Washoe her first signs. These procedures consisted of conventional training techniques—shaping, molding, and modeling—suitably adapted for use with our relatively unusual subject, conditions, and task. Working in our laboratory, Fouts (1972b) was able to measure the relative effectiveness of three major techniques, and confirmed the hypothesis that a mixed method in which the teacher varied his technique in response to the subject was much more effective than any pure method of teaching. Fouts also found that the relative effectiveness of different methods depended upon the developmental level of the subject. In particular, the relative effectiveness of modeling increased as Washoe grew older. This agrees with our basic program of teaching, in which sign-by-sign procedures are prominent in the early stages of vocabulary building only. Gradually, these procedures give way to imitation, particularly spontaneous imitation of signs that have been used in conversation by the adult models.[4]

Signs that Washoe had acquired by observing her adult models without direct training began to appear early (Gardner & Gardner 1969) and appeared with increasing frequency at later stages of development (Gardner & Gardner 1971). In the long run, all sign-by-sign procedures place severe restrictions on the rate of vocabulary acquisition. Because observational learning is effective in our laboratory, our subjects are continuously exposed to the full range of the language of their adult models.

Criterion of Reliability

Before our first publication of the Washoe data (1969) we established a very strict and conservative criterion that had to be met before a sign could be placed on

---

[4]Fouts's hypothesis grew out of his insight that the mixed method was the closest to the method that we were actually using in our laboratory, and that the source of its advantage was in the sensitivity of a good teacher to the responses of the subjects. At the time of our early reports of Project Washoe much more than today, psycholinguistics was dominated by Chomsky's doctrines of innateness and antibehaviorism. A common response to our detailed descriptions of teaching procedures was the assertion that no teaching procedures are involved in the language acquisition of human children. This assertion always seemed to us to contradict our informal observations of parents and young children. The Chomskian child absorbing language from the air seemed more remote from normal human children than did our chimpanzee subjects exposed to a continual patter of simple declaratives interspersed with probing questions.

On this point as on many others, empirical studies of human children have provided observations that agree with ours. A large body of recent research indicates that parents throughout the world speak to their children as if they had very similar notions of the best way to teach language to a young primate. In a volume devoted to this new evidence Snow remarks:

> The first descriptions of mother's speech to young children were undertaken in the late sixties in order to refute the prevailing view that language acquisition was largely innate and occurred almost independently of the language environment. The results of those mothers' speech studies may have contributed to the widespread abandonment of this hypothesis about language acquisition, but a general shift from syntactic to semantic-cognitive aspects of language acquisition would probably have caused it to lose its central place as a tenet of research in any case. All language learning children have access to this simplified speech register. No one has to learn to talk from a confused, error-ridden garble of opaque structure. Many of the characteristics of mother's speech have been seen as ways of making grammatical structure transparent, and others have been seen as attention-getters and probes to the effectiveness of the communication (1977: 31, 38).

the vocabulary list. Each time that we have published a vocabulary list (1969, 1971, 1972, 1975b) we have restated the criterion and included detailed descriptions of the form and usage of each item on the list. Early in Project Washoe, we settled on the following procedure. When a sign had been reported on three independent occasions by three different observers, the day of the third report was entered as the date of introduction of that sign into Washoe's vocabulary. The sign was not listed as a reliable item of vocabulary, however, until it had been reported to occur spontaneously and appropriately at least once on each of 15 consecutive days. Various forms of prompting figured prominently in our teaching methods (1969, 1971), but an occurrence was reported as spontaneous only if there had been no prompting other than a question such as "What that?" or "What you want?" Thus, if Washoe noticed a passing cat through the window and signed "cat" to call our attention to it, this was recorded as a spontaneous sign, and if she signed "cat" when a picture of a cat was shown to her in a drill session, this was also recorded as a spontaneous occurrence.

The 15-day criterion is strictly observed; if no observation is recorded on any day, the count is restarted. It is a very difficult criterion; although some signs reach criterion within a month, most require more than a month, and many require several months. Consequently, each list of reliable signs that we publish underestimates the observed vocabulary at the time the list was compiled (1969, 1971, 1972, 1975b). Because our subjects are free-living and no more deprived of basic needs than most middle-class human children, we cannot force them to make any signs. This means that a sign that has met our criterion of reliability was fairly easy to elicit during at least one 15-day period. In Project Washoe, we found that such signs remained easy to elicit, indefinitely. We continued to use the same stringent criterion through Project Washoe and have continued to use it in the current project.

Reassessment

No matter how stringent the criterion, a cumulative list of vocabulary items that have met the criterion once is of limited significance unless we demonstrate that each item remains a permanent part of the effective vocabulary; that is, that old signs are not forgotten as new signs are acquired. Early in the development of our subjects, nearly all of the verbal output of each day can be recorded together with context notes because the amount of output is small and contexts, including the verbal contexts, are fairly simple. With increases in vocabulary and increases in the complexity of contexts and interchanges with their human companions and each other, it becomes impractical to obtain complete records of each day's verbal behavior. Moreover, as the vocabulary gets larger, it includes items that may not occur daily. Thus, live cows are more common in Reno than in New York City, but there are many days in which there would be no opportunity to name a cow unless a deliberate attempt was made to present a live cow or a picture of a cow to our chimpanzee subjects.

When there were 13 signs on Washoe's list of reliable signs, we conducted two formal sessions each day in which we deliberately introduced the appropriate context for each sign that had not yet been reported during that half of the day. After there were 57 signs on the list, the procedure during the second half of the day was altered so that in the second formal session of the day we made extra tries only for the signs that had not yet been reported previously in the day. After there were 108 signs on the list, we divided the list into six groups of signs; one group was composed of special signs that had their contexts deliberately introduced each day, while each of the other five groups were selected for special attention in rotation on successive days. As a result of these efforts, we were able to document the permanence of all the signs that were listed on Washoe's reliable list. For example, during the last month of the project there was only one sign on the entire list that failed to appear spontaneously and appropriately at least once.

These procedures of reassessment, together with the special attention given to eliciting new signs that had not yet met criterion, were carried out in addition to the teaching of new signs, the double-blind testing (1974b) (described below), the systematic sampling of spontaneous utterances (1971, 1974b, d) (described below), the systematic tests for comparison with child data (1975b), and the extensive motion picture film and video tape recording. This heavy program of documentation placed severe limits on the rate at which the vocabulary could be enlarged.

Looking back over the Washoe data we saw how adequate documentation of the vocabulary list could be maintained with less frequent testing, and we instituted a more efficient procedure in our current project. When the size of the reliable vocabulary reaches ten, one day each week an attempt is made to elicit every sign on the list by deliberately introducing appropriate contexts when necessary. We can continue this procedure until the list for a given subject reaches 100. At that point (which Moja reached in her 40th month), we divide the list as before and continue until the list becomes so long that the only practical procedure is sampling.

In addition to the weekly test of total vocabulary, each item in the vocabulary is assigned as Sign of the Day, according to a system of regular rotation. All persons working with the subject attempt to elicit the Sign of the Day, and every occurrence of this sign is recorded. While vocabulary size is small, each sign receives this special attention at least once a month, but as vocabulary size increases, the period between reassessments becomes as long as three months. For every occurrence of the Sign of the Day, both form and context are described as completely as when the sign was first being acquired. In this way we can trace changes in form and developments in semantic range for a given sign, as new signs enter the vocabulary.

Semantic Range

Whether the context is judged to be appropriate or inappropriate, careful notes are kept on all usage, particularly on unusual or unexpected usage. We are able to trace the development of reference for each sign. Washoe's "more" sign was initially taught her as a way of asking for more tickling. From tickling, the "more" sign transferred spontaneously to many other contexts. First, she used "more" to ask for more of other services such as brushing and swinging; next, for second helpings of food and drink; and eventually, for her friends to repeat performances such as acrobatics or animal imitations (1969). Brown (this volume) has remarked on the similarity to the development of reference in human children.

Inappropriate usage often reveals a pattern, and some of our most valuable observations concern inappropriate usage. Thus, Washoe's "flower" sign was introduced into her vocabulary by delayed imitation and she used it to ask for or to name flowers. Soon we also noticed that it was used in several inappropriate contexts that all seemed to involve odors, as when opening a tobacco pouch or entering a kitchen filled with cooking odors. Taking our cue from these observations, we introduced the sign for "smell," by molding and prompting. Gradually, Washoe made the distinction between "flower" contexts and "smell" contexts by using the two separate signs, but "flower" continued to occur as an error for "smell" contexts (1969).

First Signs of Child and Chimpanzee

Washoe was approximately one year old when her training in sign language began. After seven months of exposure to the conditions of the project, her vocabulary consisted of the signs, "come-gimme," "more," "up," and "sweet." By contrast, both Moja and Pili started to make recognizable signs when they were about three months old (1975a). The date of appearance of a new sign is taken as the day that the third of three independent observers reported an appropriate and spontaneous occurrence. By this criterion, Moja's first four signs ("come-gimme," "more," "go," and "drink") appeared during her 13th week of life. Pili's first sign appeared during his 14th week, and he had a four-sign vocabulary ("drink," "come-gimme," "more," and "tickle") by his 15th week. These earliest signs met the three observer criterion within a few days of each other, both for Moja and for Pili. Similarly, by 13 weeks Tatu's vocabulary included five signs ("go," "drink," "more," "up," and "come-gimme") and this was also true for Dar, whose vocabulary at 13 weeks consisted of "more," "tickle," "come-gimme," "up," and "drink."

The age at which these infant chimpanzees produce their first signs may seem early when compared with the age at which hearing children produce their first words, but it is not so different from the age at which deaf children produce their first signs. There are parental reports of first signs appearing between the fifth and sixth month for children who were exposed to sign language from birth. Possibly, signs are easier than words for the human infants. Possibly also, it is easier for an adult to recognize and then encourage the infant's first approximation at a sign than it is to recognize the earliest attempts at a word.

Two measures of vocabulary growth are of particular interest because comparative data for children are available (Nelson 1973). Nelson used maternal reports for the appearance of new words and presented age norms for a sample of 18 English-speaking children to reach a 10-word and a 50-word vocabulary. The age for a 10-sign vocabulary was 5 months for Moja, Pili, and Tatu, and 6 months for Dar, but 25 months for Washoe; in the case of children, the age at 10 words ranged from 13 to 19 months, with a mean of 15 months. Pili was 21 months and Moja 23 months when each attained a 50-sign vocabulary, whereas Washoe was 37 months old. For children, the age at 50 words ranged from 14 to 24 months, with a mean of 20 months.

The new subjects have progressed faster than Washoe, if progress is measured from the start of training; since Washoe did not arrive in our laboratory until she was nearly a year old, Moja, Pili, Tatu, and Dar are even farther ahead in terms of the age at which a particular size of vocabulary is reached. Even though they attained a vocabulary that was comparable in size, the older Washoe and the younger subjects of the current project showed interesting differences in the sophistication of their usage of signs (Goyer 1977). For example, Pili acquired the sign, "toothbrush" when he was one year old, and his early usage of the sign was to demand his toothbrush when it was shown to him at the end of meals. Washoe's first use of the sign "toothbrush" occurred when she was about 2 years old, and quite outside the mealtime context in which we used her toothbrush and modeled the sign. She was visiting the bathroom of one of her friends, and looking at a mug full of their toothbrushes, when she first signed "toothbrush." Again, Moja, when she was 3 years and 4 months old, began to use the sign "time" for the immediate future. If we followed her "time" by questions, such as "time for what?," Moja would sign again, and add "brush" or "eat" or "chase," which clarified her utterance. Washoe began to use the sign "time" when she was 4 years and 3 months old, but, from the very start, she almost always combined "time" with a sign designating action: "Time drink," "Time out," "Time tickle."

A great deal of the early output of our subjects is more like baby talk than like adult signing. Again, Ameslan parallels spoken languages in that babyish styles of signing are generally recognized. In their early development our chimpanzee subjects exhibit babyish versions of signs that are strikingly similar to the early signing of human children. Prominent examples are: enlarged signing area, as in the overhead version of "more"; orienting signs toward the self instead of toward the receiver, as in the baby version of "bird"; and reducing a hand configuration to one pointing finger, as in the baby version of "bug" (Gardner & Gardner 1973).

The most striking parallelism between the early signs of the chimpanzees and the early words of the children is in the content of the first vocabulary. They are parallel both in the proportions of items falling into the major functional types and in the referents as well. Following Nelson's methods (1973), we assigned class descriptions to vocabulary items and identified general nominals ("bib," "flower," "toothbrush"), specific nominals ("Moja," "Pili," "Susan"), action words ("go," "hug," "up"), modifiers ("good," "hot," "mine"), and personal-social terms ("no," "please," "won't") that appeared in the chimpanzees' vocabulary before they were two years old (Table II). General nominal terms were the most prominent functional type in the early or 50-item vocabulary. For children, as well as for Moja, Pili, and Washoe, such terms comprised about half the 50-item vocabulary. Among these general nominals, the words for milk and apples, for dogs and cats and birds, for shoes and hats, for watches and lights and keys were common in the early vocabulary of the young children studied by Nelson, and the signs for these objects were in the early vocabulary of at least two of the three chimpanzee subjects.

For chimpanzees, as for children, generalization of names to categories takes place from the very beginning of language learning. Thus, with the Sign of the Day reassessment procedure, we found that Moja, by age two, had used her sign "drink" to name liquids of different kinds—milk, water, broth, coffee, and orange juice; in different containers—her own infant bottles and plastic cups, her companions' mugs, and soda-pop cans; and in different forms—a dripping kitchen faucet,

Table II. Parallel Grammatical Categorizations for the Early Vocabulary of Children and Chimpanzees

Categories used by Nelson[a]	Nelson's data for children (N = 18)		The data for chimpanzees (N = 3)	
	Examples	Mean percentage of 50-item vocabulary	Examples	Mean percentage of 50-item vocabulary
I. Nominals				
specific	*Mommy, Daddy, Dizzy*	14	*Betty, Pili, Susan*	7
general (includes pronouns)	*doggie, light, milk*	51	*dog, light, milk*	49
II. Action words	*go, out, up*	13	*go, out, up*	24
III. Modifiers	*big, hot, mine*	9	*good, hot, mine*	8
IV. Personal-Social	*no, please, want*	8	*no, please, refusal*	11
V. Function words	*for That?, Where*	4	—	0[b]

[a]Nelson (1973).
[b]Before reaching a 50-item vocabulary, our fourth subject, Tatu, was using *That?*, which is identical to the name-asking questions categorized as Function Words for children.

raindrops on the outside of the window. As in the case of children, there were also overgeneralizations to food, and to nonedibles, as, for example, when requesting a balloon. For action signs, such as "cry," "go," or "up," the course of generalization showed extensions beyond self-reference, and thus, the development of an action sign into a verb. The sign "up" appeared very early in the vocabulary of all our chimpanzee infants, and its earliest usage was a request to be picked up. Our sampling procedure showed that Moja persisted in this usage for a year before also signing "up" for the second person (e.g., "up" to a seated human companion, after tugging at her hand to get her to stand), and the third person (e.g., the phrase, "up light" after Moja had succeeded in switching on a ceiling light). A more advanced development in the extension of reference was recorded for Moja at age 42 months. She signed "orange" for a bright spot of orange light, reflected from an automobile tail light. Before that, she had only used this sign for the fruit or pieces of the fruit.

As would be expected, there was considerable overlap in the early vocabulary of Washoe, Moja, and Pili: 30 of the first 50 signs were common to all three of the subjects. In a few cases, the same referent was named with a different sign by Washoe and by the new subjects, as when Washoe signed "bed," but Pili signed "sleep"; or the new subjects acquired two separate signs, "dirty" and "potty," for the semantic range that Washoe covered by the single sign "dirty." Further development of these differences in vocabulary will be of considerable interest, since the signs involved are distinct as constituents of sentences (Gardner & Gardner 1975b).

So far, the most notable exception to the overlapping vocabularies has been the repertoire of negative signs ("can't," "no," "won't") which Moja, Pili, and Tatu started to use within the first 15 months. Washoe did not use negative signs

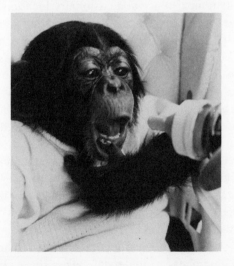

Figure 4. Moja, at four months, signs "drink" for her bottle. The age at which the infant chimpanzees produce their first signs is about three months, and it is about five months for human children.

Figure 5. Moja, at three years, replies "tree" to the question "What that?" of a birch tree.

until the 30th month of her training. Phrases containing negatives were produced by Moja, Pili, and Tatu within their first 24 months: "hug no," "gum no," "potty can't," "drink can't." Negation in phrases is significant because it is a relatively advanced development in children (Brown this volume) and because such phrases effectively double the number of statements that the subjects can make, as in the request "hug" and the refusal "hug no," or indicating "gum there" and commenting "gum no" about its nonexistence at a customary location.

As the size of vocabulary increases beyond 50 signs, and particularly from the end of the second year, when signs for modifiers begin to appear (color adjectives, "large" and "small," "mine" and "yours," etc.), it becomes increasingly difficult and indeed ambiguous to introduce new signs by displaying exemplars of the referent. The chimpanzees themselves have used different signs for the same object, depending on its use, as when Moja called a bandana a "handkerchief" and also a "peekaboo," or requested her favorite cup, which was red and transparent, by signing "drink," "cup," "see," or "red glass" at different times. Now we have to use the verbal context provided by signs, often very abstract signs, to specify the referent for a new sign. Thus, Moja and Pili, as well as Washoe, used the sign "listen" for wristwatches, which seemed appropriate to their early interest in watches as noisemakers. When Moja learned the sign "wristwatch," we could explain "That name wristwatch" and "That make listen," and used signs such as "make" and "name" when questioning her about wristwatches, eyeglasses, and so on. We can verify the chimpanzees' understanding of these abstract signs by adapting procedures described for Washoe (Gardner & Gardner 1975b) and asking such questions as "What name that?" "What color that?" "What size that?" and "What for that?" about the same object (Figures 4 and 5).

STRUCTURE

Combination, Inflection, and Prosodic Features

In our laboratory, we have avoided formal teaching of structural aspects of Ameslan such as the rules of combination, inflection, and prosodic pattern. In this way we can observe spontaneous developments. Of course, the human signers in the laboratory attempt to model good Ameslan, and they are more likely to respond appropriately to complete and well-formed utterances. This is similar to the conditions under which human children develop the structural features of their native languages. It is developments that are spontaneous in this sense that are most prominent in theories of human language acquisition.

In the speech of very young children, there is evidence of the basic prosodic patterns such as juncture and the contrast between statement and question (Brown 1973a). Similarly, Washoe used the manual activities and facial expressions that signal the prosodic features of Ameslan (Covington 1973; Stokoe, Casterline, & Croneberg 1965). Although our early records of Washoe's use of prosodic features are incomplete (because of our lack of sophistication in Ameslan at that time) we have good records of the earliest use of prosodic features by our new subjects. These new subjects began to use the questioning junction by the tenth month, and used three major contrasts—statement, question, and emphasis—regularly, before the end of their first year.

Word order, inflection, and grammatical markers are the major syntactical devices of human languages, but each natural language has its own distinctive mix of these devices. To some extent, these devices are used redundantly, but presumably, a balance must be struck for a language to function with efficiency. For example, an increase in the use of word-order makes possible a reduction in the need for inflection. Up to the present time, however, no linguistic theory has been proposed that could describe the balance that must be struck in any given language. This gap in linguistic theory is sufficient to make us doubt that an adequate artificial language could be invented in the near future. So far, the artificial languages invented by Premack (1971b) and by Rumbaugh et al. (1973) use symbol-sequence (as an analog of English word-order) to the exclusion of any other structural device. It is impossible to say whether or not such a single-device language could ever be functionally equivalent to any natural language. It is certain that such a language presents problems of acquisition that are grossly different from the problems faced by the learner of any naturally occurring human language.

Ameslan uses all the syntactical devices of spoken languages with its own distinctive combination of sign-order, inflection, and markers. Because it is quite different from the combination found in English, good Ameslan is usually difficult or impossible to render into good English without resorting to paraphrase (Fant 1972; Stokoe *et al.* 1965). To add to the difficulty, the human experimenters, themselves, were only slowly mastering the inflectional peculiarities of Ameslan during the course of Project Washoe. Washoe began to use some of the characteristic inflections of Ameslan before the end of her first year in Reno, and was making considerable use of Ameslan inflections during her last year in Reno. Now, in the current project, many of the experimenters are expert at the inflections and markers of Ameslan, and all are fairly fluent. Not only are we presenting better models for our new subjects, but we are also better able to recognize and record the earliest use of structural devices and to trace their development as Brown (1973a) has for human children.

Ameslan Only

Because the structure of Ameslan is so different from the structure of English, it is difficult, perhaps impossible, to speak good English and sign good Ameslan at the same time (Bonvillian *et al.* 1976). Attempts to do so are much like attempts to speak English and write German simultaneously: the fluency and intelligibility of one or both of the languages must suffer. Those who have acquired Ameslan recently as a second language soon find that they are speaking English sentences and producing the signs for a few key substantives. The only way we can present good models of structured language to our chimpanzee subjects is by restricting verbal communication to Ameslan. It is true that apes can be taught a modest vocabulary of signs with much less rigorous procedures than we maintain, and without the rule of "Ameslan Only." But if the development of linguistic structure is a primary objective of the research, then the subjects must be presented with the best possible model of the language that they are to acquire. This is the crucial reason for our rule of Ameslan Only in the presence of the chimpanzee subjects.

Word Order and Sign Order

Two-word, telegraphic combinations such as those in Table I form the bulk of the utterances that have been observed in the early speech of young children. Brown and Herrnstein (1975) state that about 75% of the children's combinations across various languages fit the categories of Table I. By the third year of Project

Washoe, 78% of Washoe's two-sign combinations fit the categories of Table I. For children, the two-word stage is seen as a great advance over the single-word or holophrastic stage of language acquisition, but utterances are still highly telegraphic; that is, functors and inflectional morphemes are virtually absent. The obvious absence of other grammatical devices in these "early sentences" forced many writers to overstate the case for word order as a grammatical device in early speech. Some were ready to claim that two- and three-year-old children are nearly perfect in their grammatical use of word order (Brown this volume), and sometimes these extravagant claims have been contrasted with our rather conservative statements about Washoe's use of sign order.

As more child data have accumulated, however; it has become clear that the evidence for grammatical use of word order does not fit the simple models that were first proposed. When the evidence is compared, rather than the claims, the parallels between child and chimpanzee are much stronger than the contrasts (see Table II). Neither the children nor the chimpanzees are rigorously precise or even narrowly consistent in their use of word order or any other grammatical device.

Children usually place names and personal pronouns in sentence initial position. Thus, Bloom (1970) showed that, for a sample identified as Kathryn I, "Mommy" combined with many different formatives but was placed in sentence initial position in 29 out of the 32 combinations that contained the name "Mommy." For a larger sample that we obtained during the fourth year of Project Washoe, the different names of Washoe's human companions and the pronoun "you" were placed in sentence initial position in 87 out of the 96 combinations that contained names or the pronoun "you." The name "Washoe" and the pronoun "me" (that can serve either as a subject pronoun or an object pronoun in Ameslan) present a markedly different picture. These signs appeared in sentence initial position in only 53 out of the 158 combinations in which Washoe referred to herself. This agrees with the contextual evidence that Washoe, like human children, was more likely to refer to herself as the object and to her companion as the agent of her sentences.

The presence of different sign orders in different contexts is stronger evidence for grammatical use of order than the preponderance of one order or another in spontaneous samples. For example, Brown states that in naming constructions, "there" serves as a demonstrative, and the correct word order for children is Demonstrative + entity as in "there cat"; but in locative constructions, "there" serves as a locative and the correct order is Entity + locative as in "piggie there." Perhaps, because they are infrequent in spontaneous samples, no child data have been tabulated for Entity+ locative, but in Kathryn I, Bloom (1970) reported 17 constructions in which nouns were combined with demonstrative pronouns, "indicating a particular instance of the referent which she (Kathryn) names." In every one of the 17 cases, the demonstrative pronoun was found in the sentence initial position that is considered correct for Demonstrative + entity. As in the case of the child samples, Washoe's typical samples mainly contained naming constructions of the demonstrative pronoun, "there" when combined with nouns. Thus, in samples taken during the fourth year of Project Washoe, there were 40 constructions that included nouns combined with the demonstrative pronoun "there." In 35 of these 40 cases, Washoe was spontaneously naming pictures in a book or objects in view or she had just been asked a naming question such as "What that?" In 34 of these 35 cases of naming constructions, "there" was used in sentence initial positions, as in Kathryn I. We also provided controlled conditions in which a specified set of "where" questions were asked. This gave us obligatory contexts for locative replies, and Washoe used "there" in sentence final position in 20 out of 26 constructions when replying to locative questions, e.g., Q. "Where cow?" A. "Cow there," and Q. "Where tickle?" A. "Tickle there."

In his more recent writing, when Brown refers to evidence for early use of word order he specifically limits his conclusions to English-speaking children. This is a major retreat from the early position with respect to word order, the position that was taken by most psycholinguists at the time of the first reports of Project Washoe. Now, reviewing the evidence from around the world, Brown (1973a) comments on "the extravagant variation in the data," saying, "This is a set of outcomes that offers something to disconfirm almost any hypothesis." He concludes that "What is clearly not the case, is that the human child finds it necessary to settle on particular orders to be used consistently to express particular semantic relations." This dethronement of word order as a "universal" feature of child language across the human languages of the world demands a considerable revision of the early commentaries on Project Washoe. In fact, the chimpanzee data are like the child data in showing a certain degree of consistent word order, a certain degree of varied word order correlated with varied contexts, and a residue of free variation easily attributable to the immaturity of the subjects and the ambiguity of some of the context notes.

First Combinations of Child and Chimpanzee

Our chimpanzee subjects commonly use signs in combination, and they begin to do so before they are one year old. During this early period, we record all combinations or phrase tokens, although we report development in terms of the number of different combinations, or phrase types, that have been recorded. The age at which the first ten different phrases had been recorded was 6 months for Tatu, and 7 months for Moja, Pili, and Dar. On this measure of linguistic achievement, also, the current subjects were ahead of Washoe, whose age at ten phrases was 24 months, and of English-speaking children, where the age at ten phrases ranges from

16 to 24 months (Nelson 1973). Here again, children who sign are more comparable to our subjects, for recent information indicates that they start to form phrases at 9 to 10 months (McIntire 1977).

During the first two years of development, it was feasible to make complete inventories of the subject's phrases, together with notes on context. With this method of recording, we found that, in their first 12 months, Moja, Pili, and Tatu produced 75 different phrases or phrase types. In almost every case, substantives were appropriate to the context; that is, "Tickle more" was signed between bouts of tickling, "There milk" when the subject indicated a bottle of milk, and so on.

The early phrases can be grouped according to the signs that they contain to reveal productive construction patterns. Thus, during her first 12 months, Tatu followed the sign "there" with object names to produce the set of phrases, "There milk," "There drink," "There eat," "There diaper," and "There hat," all of which expressed the same basic relations. For all three infants, Moja, Pili, and Tatu, the phrases recorded in the first 12 months included the basic sentence relations that children express in their early two-word utterances: Agent and action ("Susan brush," "Pili potty"), Action and object ("Up me," "Gimme drink"), and Demonstrative and entity ("There drink," "There diaper") (Brown 1973a). Elementary negative sentences occurred when negative particles were incorporated in phrases. As with children (Bloom 1970; Bowerman 1973), notes about context showed that in these elementary negative constructions, the negative particle was sometimes combined with an element being negated (as in "Hug no," "Potty can't," and other examples cited earlier) and sometimes with a positive alternative, as in "No, drink" when refusing solids and requesting milk, or "No, go" in reply to the request "Please kiss" from a departing companion.

As the subjects develop, and the frequency of signing increases, maintaining a complete inventory of phrases and context descriptions becomes impractical. Often, the brisk pace of an extended conversational interchange precludes verbatim recording. From the end of the subject's second year, it becomes necessary to rely on a different method of recording. During a specified period of time, usually 20 minutes, we record everything signed by the chimpanzee and a description of context, which includes all signing addressed to the chimpanzee. These comprehensive signing records require a team of two persons, both of whom remain near the subject throughout the recording period. One member of the team whispers an immediate spoken transcription of signing, together with notes about context, into the microphone of a miniature cassette recorder. The other person performs the usual role of teacher, caretaker, playmate, and interlocutor. The technique is based on that used by Brown and his co-workers for the study of language development in Adam, Eve, and Sarah (Brown 1973a), and by Bloom in her extensive study with three other English-speaking children (1970). Table III shows that in terms of quan-

Table III. Description of Language Samples for Washoe and Young Children

	Signing chimpanzee Washoe	English-Speaking children ($N = 6$)[a]		Signing children ($N = 6$)[b]	
		Median	Range	Median	Range
Age at start of sample (months)	59	20	(18–27)	38.5	(25–75)
Duration of individual samples (hours)	3.4	6.5	(2–9)	3.8	(2–13)
Total number of utterances	346	713	(207–2760)	642	(271–1865)
Utterances per hour	103	184	(52–356)	176	(116–207)
Utterances longer than one morpheme	37%	22%	(9%–54%)	—	
Type/token ration	72%	43%	(37%–71%)	—	

[a] Brown 1973a; Bloom 1970.
[b] Hoffmeister, Moores, & Ellenberger 1975.

titative characteristics such as rate of utterances and proportion of phrases, the technique yields material for young chimpanzees that is well within the range of language samples for these young English-speaking children, and of language samples obtained by videotaping young children who sign (Hoffmeister *et al.* 1975).

Earlier in this report, we stated that a visual language such as Ameslan employs many unique devices to convey grammatical information. Thus, in our analysis of Washoe's combinations (1971) we described the visual devices she used that enabled us to consider two or more signs as a unit. Washoe punctuated her signing by keeping her hands within a limited area in front of the chest until the end of an utterance; then she allowed her hands to fall from this signing space and rest on a nearby surface or on her lower body. This device for punctuation is used by the subjects of the current project also, and is quite similar to the visual patterns that mark off sentences for human signers (Stokoe *et al.* 1965).

In the current project we have profited from recent advances in the linguistic analysis of Ameslan, and have recorded further variations in forming signs that signal grammatical relations. Orienting or moving signs for actions along with the line of sight between signer and addressee has been recognized as an inflection of the sign, used by human signers to imply that "I" or "you" is the agent of the action (Bellugi *et al.* 1975; Fant 1972). We noticed that the infant chimpanzees regularly placed certain signs on the addressee's body as well as on their own. The chimpanzee's lips or the addressee's lips have been used as the place for "be-quiet," the back of the chimpanzee's hand, or the back of the addressee's hand as the place for "tickle," and so on. By recording the contexts in which this contrast in placement of signs was used, we obtained evidence that Moja, Pili, and Tatu used this device as an inflection for agent of action for the signs "be-quiet" and "tickle," as well as for such signs as "chase" and "finish." The infants expressed contrasts such as "You be-quiet" and "I be-quiet" in this way within the first 12 months.

A later development for Moja was the incorporation of a common movement into a set of signs that differ in form to produce the negative inflections "don't want" and "don't-know." This is known as negative incorporation, and has been noted by linguists as a characteristic of native signers. Those who acquire Ameslan as a second language commonly use two separate signs: "not" and "know," "not" and "want," "not" and "good" (Woodward 1974).

PLASTIC TOKENS AND LIGHTED PANELS

Before Project Washoe, there was a long history of laboratory investigations of the intelligence of nonhuman species by means of forced-choice tests. During the last three or four decades in particular, a great deal has been learned about the intelligence of nonhuman primates largely as a result of the work of Harry Harlow and his students. Along with the substantive results of this research has come a great deal of sophistication about the pitfalls of investigating the intelligence of nonhuman beings. Following Project Washoe, two investigators, David Premack (1971b) and Duane Rumbaugh (Gill and Rumbaugh 1974; Rumbaugh 1977; Rumbaugh, Glasersfeld, Warner, Pisani, Gill, Brown, & Bell 1973), have attempted to demonstrate that chimpanzees have the intelligence to solve certain highly contrived forced-choice problems and that solution to these problems is evidence of language. Unfortunately, both Premack and Rumbaugh *et al.* seem to have forgotten all the hard-won technical expertise of the primate laboratory. Thus, despite the paraphernalia of the conventional laboratory such as caged subjects, elaborate hardware and elaborate software, and strict schedules of reinforcement, they have fallen into some of the worst of the classical pitfalls.

Rote Memory

In the course of experiments that required intensive training of individual subjects, experimenters have discovered a great deal about the ability of nonhuman primates to memorize the correct choice, when given many trials on a large number of unrelated forced-choice problems. A typical example is illustrated by an experiment of Farrer (1967) in which 24 arrays of four choices were presented to chim-

PROBLEM NO.	PICTURES				CORRECT POSITION
	Lever - 1	Lever - 2	Lever - 3	Lever - 4	
1	+	⑨	l	×	4
2	▲	—	▣	◎	2
3	⑨	ⓦ	+	×	3
4	ⓑ	—	ⓡ	▣	1
5	⑨	▲	—	×	2
6	+	×	ⓡ	ⓑ	4
7	l	⑨	ⓦ	▲	3
8	l	⑨	—	◎	1
9	+	◎	ⓧ	×	3
10	▲	▣	l	×	2
11	ⓦ	◎	ⓑ	l	4
12	◎	ⓧ	▣	—	1
13	ⓧ	ⓦ	—	⑨	1
14	×	▣	○	ⓡ	2
15	ⓡ	—	+	▣	3
16	▲	⑨	l	ⓡ	4
17	×	▣	◎	ⓑ	3
18	ⓑ	l	×	—	4
19	ⓑ	+	▣	—	1
20	+	⑨	l	ⓑ	2
21	×	ⓦ	—	ⓑ	2
22	▲	×	⑨	ⓡ	1
23	▣	ⓑ	×	l	3
24	▲	—	▣	ⓡ	4

Figure 6. The problems used in Farrer's picture memory test. Farrer's chimpanzees memorized these 24 discrete stimulus arrays, and were 90–100% correct in choosing a particular member of each array, with the arrays presented in random order.

panzees on a set of lighted panels very similar to those used by Rumbaugh *et al*. The arrays, which were presented randomly, are shown in Figure 6. Farrer found that his laboratory chimpanzees had little difficulty in memorizing the correct choices in these arrays.

Learning Sets

Certainly, one of the most important discoveries of this century was Harlow's discovery of learning sets in monkeys. It is now well-established that a monkey or an ape that has the benefit of experience with a long series of forced-choice problems can solve an entirely new and unrelated problem in very few trials. Quite often monkeys can achieve 100% correct performance on the second trial after 50% performance on the first trial. What this means is that when transfer tests are used to determine whether a sophisticated monkey or ape has mastered a general concept, such as shape or number, only first-trial performance can be counted. After trial one we are measuring new learning. The fact that the new learning is faster than the original learning is only relevant when we can compare the speed of new learning in experimental and control groups of subjects.

Plastic Tokens (Premack 1971b)

Premack has presented a large number of two-choice discrimination problems to a chimpanzee named Sarah. With rare exceptions, all the problems that Premack presented to Sarah were of a two-choice type illustrated in Table IV. Most of Premack's interpretation of Sarah's performance depends upon our agreement that his plastic tokens represented a code in which one-for-one substitutions could be made between the tokens and English words. The first column of Table IV lists Premack's English translations of all the pairs of questions and two-choice answers that could be presented to demonstrate the correct usage of the token that represented the noun "apple" when used with the token for the operation "name of."

In order to give the reader of this article a better idea of what Sarah saw on a given trial, we have translated the English back into a set of typographical symbols that represent the different plastic tokens. In this code:

  i = "apple"
  j = "banana" (or any other token for any other fruit)
  x = an apple (a real exemplar of any other fruit)
  $\hat{\phi}$ = "name of"
  $\tilde{\phi}$ = "not name of"
  A = "Yes"
  B = "No"
  Y = a banana (or any other real exemplar of any other fruit)
  ? = a token that Sarah learned to replace with one of the answer tokens
        available. Premack gives this token many English glosses that
        serve to make his translation read like a proper English sentence.

The second column of Table IV contains a literal representation of all the questions and two-choice answer alternatives as Sarah saw them, with the only difference being that we have used typographical symbols rather than plastic tokens. The third column of Table IV presents all the "answers" that Sarah gave, if the array that remained after she had replaced the ? token with one of the two answer tokens can be said to be an "answer" produced by Sarah.

It is clear from the answer column that Sarah could have produced four of the six possible "answers" by memorizing the array $i\phi x$. Premack, of course, presented Sarah with other sets of questions, such as a set about bananas that could be "answered" by memorizing the array, $j\phi y$, but none of Sarah's "linguistic operations" required more than eight such arrays. This is well within the rote-memory

Table IV. Premack's Analysis of "Name of"

Premack's English gloss	How the display looked to Sarah before and after her answer	
	Before	After
Q: What is the relationship between "apple" and an apple?	i ? x	i ϕ x
A: name of/not name of	ϕ $\bar{\phi}$	
Q: What is the name of apple?	? ϕ x	i ϕ x
A: "apple"/"banana"	i j	
Q: What is "apple" the name of?	i ϕ ?	i ϕ x
A: apple/banana	x y	
Q: Is "apple" the name of apple?	? i ϕ x	A i ϕ x
A: yes/no	A B	
Q: What is the relationship between "apple" and a banana?	i ? y	i $\bar{\phi}$ y
A: name of/not name of	ϕ $\bar{\phi}$	i ϕ y
Q: Is "apple" the name of a banana?	? i ϕ y	B i ϕ y
A: yes/no	A B	

capacity of chimpanzees demonstrated by Farrer and others. But what about the last two lines of Table IV that required Sarah to use the negative tokens? At first glance the need to use the negative tokens seems to multiply the number of combinations that Sarah would have to memorize, because each new name token can enter into many new invalid combinations, such as $i\phi y$, $i\phi z$, and so on. Here again, the constraints of the forced-choice technique come to the rescue. Given the alternatives presented, all Sarah had to remember was that she would be rewarded if she used the ϕ or the B token when Premack presented any array that was not on her short list of memorized arrays. [Throughout his book (1976a) Premack cites numerous examples of tests in which all or most of Sarah's errors were made on trials that required her to use a negative token.] Since Sarah's program of training and testing concentrated on each "linguistic concept" for days and weeks at a time, she could have solved Premack's entire battery of problems by rote memory alone.

Premack attempted to show that Sarah had learned general principles by presenting her with transfer problems at the end of each phase, such as the "name of" phase:

The subject has never failed a transfer test. Though often trained on as few as two positive and two negative instances of the concept, she has invariably been able to apply the words to cases not used in training. (1971b)

But the smallest number of trials that Premack has reported for any transfer test was ten trials, and his criterion for passing a transfer test has always been 80% correct. For 20 or 30 years it has been recognized that the only acceptable evidence for concept formation in a learning-set sophisticated primate is the first trial of a transfer test (Thomas & Kerr 1976). After that we are no longer dealing with transfer except for the transfer of a win-stay lose-shift strategy, without any concept formation whatsoever. A learning-set sophisticated rhesus monkey could have passed all Premack's transfer tests at the average level he required.

Lighted Panels (Gill & Rumbaugh 1977; Rumbaugh 1977; Rumbaugh et al. 1973)

Rumbaugh, Gill, and Glasersfeld replaced Premack's plastic tokens with a keyboard consisting of plastic panels that could be backlighted to present different patterns. When the lighted panels were depressed by their chimpanzee subject, Lana, the patterns also appeared on a display at the top of the keyboard. As in the case of Premack, there is supposed to be a one-to-one correspondence between the patterns and English words, except that usage is more restricted. For example, a pattern can be used only as one of the parts of speech, even though most English words can be used as nouns, adjectives, or verbs. Thus, the pattern that is supposed to correspond to the English word "drink" can be used only as a verb, never as a noun. Strings of patterns are said to be grammatically correct, but all she has to do to get a reward is to produce one of a very small number of correct color sequences. There is no reason to suppose that Lana's productions have any semantic content. In fact, there is some internal evidence that her earlier productions had none. After more than six months and thousands of trials in which Lana was supposed to have asked for M & M's and bananas, Gill and Rumbaugh report that it took two more weeks and 1,600 trials before Lana could reliably light up the sequence $A - B - x$ when the experimenter showed her an M & M, and the sequence $A - B - y$ when the experimenter showed her a banana (English translations: $A - B - x$ = "name of this M & M"/$A - B - y$ = "name of this banana.")

Further tests of naming reported by Rumbaugh et al. followed hundreds, sometimes thousands, of trials of drill on the same small set of keys and stimulus arrays, and none of these tests can have placed any great strain on the rote memory of a chimpanzee. In their so-called "naming task," the same 36 objects (six exemplars of six objects, the different exemplars of the same object differing only in color) were presented over and over again for hundreds and hundreds of trials. It is instructive to compare this with the procedure that we describe in the next section of this article.

RIGOROUS TESTING UNDER NATURALISTIC CONDITIONS

With caged subjects and forced-choice tests, the results that Premack and Rumbaugh et al. have presented thus far are more parsimoniously interpreted in terms of such classic factors as Clever Hans cues, rote memory, and learning sets. With our free-living subjects and naturalistic conditions, we developed testing procedures that rule out these alternative interpretations.

A Test of Vocabulary

Our first tests required Washoe to communicate information about objects that she alone could see. In describing the testing procedures, it will be helpful if we define a vocabulary item as a class of referents that could be correctly named by the use of a given sign, and an exemplar as a member of one of the item classes. The essential features of the procedure were (1) that each exemplar be presented on one trial only, so that every trial was a transfer trial; (2) that vocabulary items be presented trial by trial in a sequence that could not be predicted either by the subject or by the observers; and (3) that the sign that Washoe made would be the only information available to the observers.

Our most versatile and successful testing procedure made use of 35-mm color transparencies to present exemplars. With photographic slides rather than actual objects, a wide range of exemplars can be presented. It became possible to produce exemplars for most of the nouns in Washoe's vocabulary, and it became easy to produce many different exemplars for a given item, so that the tests could consist of exemplars that were all new. Thus, we could photograph the entomological collection at the biology department and produce exemplars of the item "bug," or go down to the parking lot and produce abundant exemplars of the item "car."

The apparatus used in the slide test is shown in Figure 7. For this test, we used a cabinet built into the wall between two rooms (R_1, R_2 of the laboratory). The slides were back-projected on a screen (PS) that was mounted inside the cabinet. The cabinet and the screen were flush with the floor, and O_1 standing beside the cabinet could observe Washoe's signing without seeing the pictures. The test was self-paced: Washoe began a trial herself by unlatching the sliding door (SD) of the cabinet. When she opened the door, O_1 asked her (in signs) what she saw, and wrote down her reply on a slip of paper that was passed to E_1 through a message slit (MS) in the wall. In the other room, E_1 operated a carousel projector, presenting the slides in a prearranged random order. A one-way vision screen (1-wG) was set into the top of the cabinet and permitted a second person, O_2, to observe Washoe without being seen by Washoe and without being able to see the projection screen. Thus, O_2 could confirm the signs reported by O_1.

The slide test is a very sensitive test of vocabulary. In a two-choice test such as the WGTA, the subject's expected chance score is 50% correct. Such tests require very high scores or a very large number of trials to differentiate the subject's performance from chance responding. In our tests the subject could make many

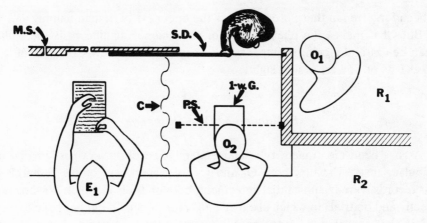

Figure 7. Double-blind testing apparatus used in Project Washoe.

different responses, and hence the chance expectation of obtaining the correct response—that is, the correct sign—for a given exemplar was extremely low. For a test involving N items, Washoe's expected chance score was $1/N$ items correct, if her responses were confined to the items of the test. Since other signs (extralist intrusions) were also reported, the expected chance score was actually lower than $1/N$. For example, a typical test that we administered to Washoe involved 32 different items, with four exemplars of each. This 128-trial test was administered in four sessions. Washoe's expected chance score was 1 out of 32 or four trials correct. In fact, according to O_1, Washoe named 92 of the 128 exemplars correctly, and according to the second blind observer, O_2 Washoe's score was 91 out of 128.

The increased sensitivity of our testing procedure made it possible to learn a great deal from Washoe's errors, as well as from her correct naming. A particular form of error could occur infrequently and still be detectably different from chance responding. The analysis of errors showed that incorrect responses reflected conceptual relations among signs: the common error for a photograph of a dog would be a sign for another animal such as "cat" or "cow," while the common error for pictures of fruit would be another type of food, such as "meat" or "cheese" (Gardner & Gardner 1971, 1974b). It is instructive to contrast our procedure with the procedures that Rumbaugh and Gill present as evidence for "object and color naming" (Gill & Rumbaugh 1974). In their most advanced test there were precisely six vocabulary items, ball, box, can, cup, shoe, and bowl, and six exemplars of each, the different exemplars of the same vocabulary item differing only in color. Thus, there was a grand total of 36 exemplars altogether, and these were presented over and over again, for hundreds of trials day after day. Premack's procedures were even narrower and more repetitive. For example, the token for apple referred to objects that were always red, always round, and always had stems on top. What has been forgotten in these procedures is the openness of human naming and labeling. But what makes the phenomenon of class names significant is the ability of human beings and signing chimpanzees to name a brand new exemplar of a class such as cats or flowers at first sight.

Evaluation by Deaf Observers

As has been mentioned, two independent blind observers participated in the vocabulary tests. Usually, both O_1 and O_2 were members of the research staff. Their records were in substantial agreement, both on trials in which Washoe replied correctly and on trials in which she made an error. Since O_2 did not have to interact with Washoe in any way, persons who were fluent in Ameslan but strangers to Washoe could serve as observers of her signing. In this important control condition, there was also substantial agreement between the reports of the two blind observers, and their agreement increased from one test session to the next. Two observers, whose previous contact with Washoe had been minimal, participated in the tests. For one, agreement with O_1 was 67% during the first test session that he observed, and rose to 89% during the second test. For the other, agreement with O_1 was 71% during the first test, and also rose to 89% during the second test. The degree of agreement between the deaf observer and the project assistant during the second test was well within the range of agreement found between O_1 and O_2, when each was a project assistant thoroughly familiar with Washoe's signing. This implies that Washoe's signs were standard items in Ameslan, and that her accent could readily be learned by persons fluent in Ameslan.[5]

Test of Combinations

In the current project, we can expand the test material so that subjects are required to tell us more about the exemplars than just their names. For example, combinations of names with color or numerical modifiers can be illustrated and tested. The use of locative constructions can be tested with a set of photographs illustrating spatial relations between pairs of objects, such as "X behind Y," "X in Y," "X under Y," and so on, for which O_1 would ask the subjects, "Where X?" or "Where Y?" The use of order to communicate meaning also can be tested with our procedures. This especially important extension of the test requires the use of short

---

[5]We controlled for the possibility of improvement by increaced familiarity with a particular list of vocabulary items by using two different (though partially overlapping) lists and by familiarizing the deaf observers with each list before the start of each test session.

loops of motion picture film. By showing reversible relations between pairs of animate nouns, such as "A chases B," "A kisses B," "A feeds B," and so on, we can measure the ability to use rules of order in the replies describing the scenes.

Grammatical inflections characteristic of Ameslan, such as reduplication to represent plurality, inflectional endings for the agentive suffix, and directional modifications of signs to represent agent and object roles, can also be illustrated and tested. The pictorial exemplars can include a singular book, tree, or house and a plurality of books, trees, and houses; and pairs calling for the agentive suffix such as boat and sailor, kitchen and cook, and so on.

REPLIES TO WH-QUESTIONS

Sentence Constituents

Most of the work in child psycholinguistics has been concerned with the evidence for grammatical competence that can be found in the spontaneous speech of children (Brown 1973a). One source of evidence consists of the relation between question and reply when the question is of the type referred to as a "wh-question" (Brown 1968, Ervin-Tripp 1970):

In fact, each interrogative word is a kind of dummy element, an algebraic 'x,' standing in the place of a particular constituent of the sentence. . . .The dummy word asks for specification of that constituent. It marks the spot where information is to be poured into the sentence, and the form of the dummy . . . whether *who, what, where, when or why*—indicates the sort of information required?. . . . in general, each kind of Wh question calls for an answer which is an instance of particular major sentence constituent. (Brown 1968: 280–281)

The production of well-formed wh-questions is a rather late development in the speech of children. According to Brown, the first evidence that they have the relevant grammatical competence is found in their ability to produce grammatically correct replies to wh-questions.

In a sense, the ability to restrict replies to the correct grammatical category is more significant than the ability to give sematically correct answers. For example, when someone asked Washoe, "Who that?" while indicating Roger S. Fouts, all semantically correct replies had to include the name-sign "Roger," either alone or in a phrase such as "That Roger." Nevertheless, incorrect name-signs such as "Susan" or "Greg" were still correct in a way that replies such as "hat" or "black" or "Tickle me" were not. The fact that Washoe usually answered "Who that" questions with name-signs is in some respects more significant than her ability to associate particular name-signs with particular individuals in her acquaintance. The restriction of replies to a correct grammatical category is also a more general characteristic of replies to wh-questions, because there is an important group of questions such as "Who good?" and "What you want?" that refer to matters of opinion and preference rather than to matters of fact. Strictly speaking, the only objective judgment that an observer can make about the correctness of a reply to such a question is a judgment about the correctness of its grammatical category.

Question Frames

From the many questions found in everyday conversations with Washoe, a list of ten question frames was selected that was judged to be representative. These are listed in Table V. A question frame in this investigation was constructed by combining one interrogative sign with one or more additional signs in such a way that a category of replies was specified. The category that was specified for each question frame is listed in Table V under the heading Target Categories. Also included in Table V are typical examples of questions and replies that were recorded during this investigation. This procedure for constructing question frames and specifying target categories in closely parallel to that of Brown (1968) and Ervin-Tripp (1970) in their studies of this aspect of the spontaneous speech of young children. Unlike the child studies that depend on adventitious occurrences of relevant questions and

Table V. Relations Between Questions and Replies

Question type	Target category	Examples of questions and replies	
Who pronoun	Proper name	*Who you?*	*Me Washoe*
		Who me?	*Linn*
Who action	Proper name or pronoun	*Who smoke?*	*You smoke*
		Who go out?	*You me*
Who trait	Proper name or pronoun	*Who good?*	*Good me*
		Who pretty?	*Washoe*
Whose demonstrative	Possessive	*Whose that?*	*Mine*
		Whose those?	*Shoes yours*
What color	Color	*What color?*	*Bird white*
		What color that?	*Green*
What demonstrative	Common noun	*What that?*	*Book*
		What that?	*Food fruit*
What now	Verb	*What now?*	*Tickle*
		Now what?	*Time drink*
What want	Common noun or verb	*What want?*	*Want berry*
		What you want?	*You me out*
Where action	Locative	*Where we go?*	*Out*
		Where shall Susan play-bite you?	*Susan bite there*
Where object	Locative	*Where shoes?*	*There shoe*
		Where baby?	*(fetches doll)*

answers, we have presented questions belonging to specified question frames in a systematic and balanced fashion, while at the same time embedding the questions in the normal course of daily conversations.

Comparison with Human Children

Brown (1968) made extensive tape-recordings of three children in conversation with a parent at home, and then sampled replies from

"all child answers to all Wh questions produced by parents . . . centering on [Stages] I, II, III, and V" (pp. 281–282).

He reports that when

the three children attained [Stage] III they were producing numerous declaratives with noun-phrase subjects, main verbs, noun-phrase objects and locative adverbials. By [Stage] III also, they were correctly answering questions calling for just these constituents about half the time . . .the children were able to take a Wh word supplied by a parent as a signal to supply an appropriate constituent member. (pp. 283–284)

Thus, the wh-questions that the children were answering correctly about half the time were those containing the interrogative *who, what do, what,* and *where.*

On the basis of similar text-eliciting procedures with children from 1.9 to 2.5 years old, Susan Ervin-Tripp (1970: 7) reports that

all five had clearly mastered the locative features of *where,* and the nominal, non-animate marker for *what.* Four of the five controlled the possessive, + animate, + NP marking for *whose* by 2.3, and four controlled + animate as a feature of responses to *who,* at the outset of text-collections.

In terms of the actual performance that indicates such mastery and control, the data for this group of children appears comparable to the data reported by Brown. Ervin-Tripp presents the complete set of replies to who questions for one child at 2.6 years. This set contains replies to 22 Who questions, and the child answered appropriately just half the time. Specifically, there were 11 nonimitative replies that contained animates ("Carol," "Paul," "Mommy," "Mother," and "baby"). The remaining replies contained nouns, as in the reply "Meat" to "Who's eating?", actions, as in replying "Shaving" to "Who's watching Daddy?"; and imitations of parts of the preceding questions, as in the interchange, Q. "Who took Suzy to bed?" A. "Take Suzy bed."

The data of both studies are too scanty and too uncontrolled to permit distributional analyses of the type that we have presented for our subject (Gardner & Gardner 1975b). The appearance of appropriate words in about half the replies were sufficient for Brown and Ervin-Tripp to claim that children have grammatical "mastery" or "control" of a particular question type. Both agree that Stage III children should be credited with that much mastery of who, what, and where questions. They disagree about the ability of Stage III children to respond appropriately to whose and what do questions. Neither found evidence for mastery of why, how, or when.

The questions used with Washoe contained the interrogatives who, what, where, and whose, and Washoe's replies contained appropriate sentence constituents 84% of the time, even when appropriateness was much more narrowly defined than in the child studies. Moreover, distributional analyses demonstrated a degree of statistical control by the interrogatives that is far beyond anything that could be extrapolated from the children's data for Stage III. As data become available for children who are at or beyond Stage IV, we will be able to further comparisons, both with Washoe and with her successors.

In his article on the development of wh-questions (1968), Brown concluded that

The derivation rules we have described for Wh questions presuppose the establishment of the major sentence constituents. The best evidence in the child's spontaneous speech that he has such constituents is his ability to make the right sort of answers to the various Wh questions addressed to him, giving noun phrases in response to *Who* and *What* questions, locatives to *Where* questions, predicates to *What-do* questions, etc.

If Washoe had been a preschool child, then by these standards her replies to the wh-questions of this sample would place her at a relatively advanced level of linguistic competence.

With the new subjects, we have traced the development of this aspect of linguistic competence. Following methods developed for testing Washoe, we began sampling replies to questions when Pili was 1½ years old and Moja was 2½. From the onset of testing, both subjects replied readily to questions as is the case with young children also (Ervin-Tripp 1970). Both were more likely to reply to wh-question ("What Pili want?" "Where my keys?") than to yes-no or intonation questions ("You sleep nice?" "You good girl?"). While each infant replied to about 80% of the wh-questions that were asked, Moja replied to 67% of yes-no questions and Pili to only 43% of yes-no questions. The difference between the subjects appeared to be a developmental one, since Moja, when retested a year later, replied to more than 90% of yes-no questions.

As with Washoe, we tested with question types that were common in everyday conversation with the subjects, and we were able to add new types of questions on successive tests. Using the same criterion for mastery that is used with children, Pili at 1½ was replying appropriately to 4 of the 6 types of questions that he was asked. This set of 4 questions included the major interrogatives *what* and *where*. Moja, in addition, was replying appropriately to questions containing the interrogative *who*. In subsequent tests for Moja we added more question types, and by 3½ Moja was working with a set of 12 question types that included all the types asked of Washoe at age 5. The younger Moja matched Washoe's performance for all except *where* questions. In addition to the question types that we had used with Washoe, Moja was replying appropriately to What Predicate questions, such as "What I do?" or "What we play?": and to the contrast between Who Subject questions and Who Object questions. For example, for "Who chase you?" Moja replied "Clayton," but for "Who Ron tickle?" she replied "Ron me." In tests at age 5 Moja added "Why?" and "How many?" to the set of question types that elicited appropriate replies. Thus, recent findings with wh-questions indicate that we can expect to reach a more advanced level of linguistic competence with the new subjects.

PRODUCTIVE VERSUS FORCED-CHOICE TESTS OF COMPETENCE

In Gardner and Gardner (1974c), we criticized Brown (1973a) for including in his evidence for the grammatical competence of human children a number of experiments based on forced-choice tests. A typical example is the study by Fraser *et al.* (1963), in which children were presented with a number of pairs of pictures. Each pair illustrated a grammatical contrast such as "The dog is biting the cat" versus "The cat is biting the dog." The pictures were untitled, of course, and the children were shown each pair in turn and asked to point to the picture that represented the sentence that the experimenter had just uttered. Fraser *et al.* added: "S sometimes pointed quickly, then reflected and corrected himself; the last definite pointing is the one we always scored (1963: 129).

The trouble with these forced-choice, apparatus-defined tests of linguistic competence is that neither the child subjects nor Fraser *et al.* nor the chimpanzee subjects of Premack and Rumbaugh *et al.* needed to have any linguistic competence whatsoever to perform the arbitrary responses that were required of them. Instead, their otherwise meaningless responses were scored by means of artificial codes and interpreted by elaborate linguistic theories. Clever Hans never spoke to his trainer, either; all he did was stamp his foot. By means of a special code, his trainer translated this binary information into an infinite system of complex answers. What all these experimenters failed to appreciate is that forced-choice apparatus introduces characteristic experimenter errors that must be controlled for by special devices (Harlow 1949). In studies of linguistic competence, the usual controls are difficult to enforce because young children and young chimpanzees seem to require living interlocutors. Gill and Rumbaugh describe in detail the deteriorating performance that followed their attempt to test Lana's ability to name M&M's and bananas when the experimenter left the test room. After 10 days of failure they concluded, "As a result of this poor performance on the test trials and the subsequent deterioration of all responding, together with the increase in emotional display, it was decided to move back into Lana's room . . ." After 3 more days of retraining with the experimenter present, the usual procedure of Rumbaugh *et al.*, "Lana achieved the established criterion in the first [sic] 100 trials, making only 7 errors" (Gill & Rumbaugh 1974:486) In the test of vocabulary reported above, we showed how easy it is to demonstrate that our subjects are intelligible, under double-blind conditions, both to their regular companions and to fluent signers who are strangers.

It is instructive to contrast the inherent weaknesses of forced-choice tests of linguistic competence with the power gained in productive tests such as the replies to wh-questions reported by Brown (1968), by Ervin-Tripp (1970), and by ourselves (1975b). In the productive tests, the verbal status of the responses is independently defined by the independent existence of a natural language. More important still, productive tests of competence leave the subject free to respond with any and all items in the verbal repertoire in any combination. This effectively eliminates the problem of Clever Hans cueing that haunts the forced-choice tests, because any hints given in a productive test must contain as much information as the correct replies. Moreover, in the productive tests any one of a large set of replies can be correct or incorrect. We should also point out that both Premack and Rumbaugh *et al.* follow a procedure in which the subjects are drilled and tested on the same day on each type of question, and sometimes the drills on one type of question go on for weeks (Gill & Rumbaugh 1974; Rumbaugh 1977). In productive tests such as those reported here, the procedure is designed to separate training from testing and to arrange for a maximum variety in the question asked within any day of testing and within any session of testing.

CONTRIBUTIONS TO DEVELOPMENTAL PSYCHOLOGY

In addition to records of sign language acquisition, this project provides records of the general intellectual development of home-reared chimpanzees, to add to the very few that are now available (Kellogg this volume). Our subjects readily adapt to the social and physical features of a human household. They use feeding utensils and toothbrushes, they play with balls and dolls and building blocks, they scribble with marking implements, and they examine themselves in mirrors. Routine household chores—washing clothes, sewing, weeding the lawn, repairs with hammers and screwdrivers—are readily imitated, perhaps because the adults who model these chores are treating them as more important matters than child care (Gardner & Gardner 1974a).

Our subjects are reared under conditions similar to those of a human child because we believe that intellectual, social, and linguistic development are intimately related. Indeed, a critical similarity to the rearing of children is the use of a system of verbal communication to foster learning about the social and physical environment that depends on language. Although these rearing conditions are required in order to compare the linguistic development of child and chimpanzee, they enable us to compare many other aspects of development, also.

This has been the topic of several studies in our laboratory, employing the chimpanzees and age-matched child controls. Among the aspects of development studied are greeting behavior and fear of strangers (Turney 1975); postural and loco-motor skills, up to the stage of unaided bipedal walking (Goyer 1976); object permanence, the first Piagetian stage in conceptual development (Wood 1976); and hand preference (Nichols 1974; Robinson 1977). For the last two studies in particular, the availability of a means of communication has been essential in order to instruct the subject comparably in the tasks they are to perform.

As the chimpanzees grow older, comparisons on more advanced aspects of conceptual development become possible. When Moja became three years old, we instituted half-hour daily school sessions for formal practice on making letters, naming and finding pictures, and motor skills (Figure 8). During school, Moja sits at a table and is given tasks involving marking implements and books, as well as sorting tasks, lacing cards, and so on. A striking development occurred some six months after school sessions began, and is illustrated in Figures 9 and 10. Figure 9 shows a typical drawing of Moja's; Figure 10 shows the sparse and different drawing that preceded it that day. Because so few lines had been made, the research assistant

Figure 8. Moja, at age 3½, using crayons during a school session. For all our subjects, playing with marking implements is a highly preferred activity.

Figure 9. A typical scribble produced by Moja, using chalk on black paper.

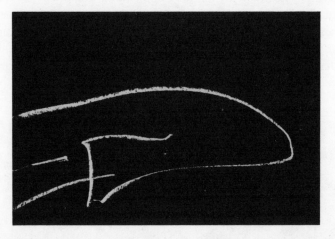

Figure 10. "Bird," the first drawing that Moja named. This drawing was made immediately before the one shown in Figure 9.

Figure 11. A drawing produced when Moja was asked to "Draw berry there." Moja chose an orange pen for this drawing. She later named the drawing "berry" when asked "What that?"

put the chalk back in Moja's hand and urged her to "Try more," but she dropped the chalk and signed "finish." The reply was unusual, and looking at the drawing the assistant noticed that it, too, was unusual in form. He then asked Moja, "What that?" and she replied, "Bird." Since that time, Moja has labeled additional drawings that she produced and she has been consistent in the form of the drawing associated with a given label, e.g., radial shapes for "flower," round forms for "berry." Moja has also drawn and then labeled figures as requested by her teacher (e.g., "Draw berry there," Figure 11), and she has replied appropriately to questions asking her to name both the artist and the subject, as in Q. "What that?" A. "Grass" and Q. "Who draw this?" A. "Moja." Here sign language allows us to explore other systems of representation, which some child psychologists have identified as the start of writing (Olivier 1974).

BUT IS IT LANGUAGE?

The results of Project Washoe presented the first serious challenge to the traditional doctrine that only human beings could have language. Before Project Washoe, the possibility of nonhuman language had always been raised in a yes-no, can-they-or-can't-they fashion, and had always been rejected categorically. A typical example is the "Straw Parrot Problem" that required countless pages of learned writing to prove and reprove that the mimicry of parrots is not language. We would call it the "straw parrot problem" because we know of no case in which the affirmative side of the question was either proposed or defended. Washoe required a radical departure from the traditional way of posing the question. She learned a natural human language and her early utterances were highly similar to, perhaps indistinguishable from, the early utterances of human children. Now, the categorical question, can a nonhuman being use a human language, must be replaced by quantitative questions; how much human language, how soon, or how far can they go? The analytical methods of structural linguistics are ill-suited to these questions. For appropriate methods of measurement we have to turn to recent advances in the study of child language.

Those who study the acquisition of language by human children have not found and do not expect to find a litmus paper test that can tell us just when this or that child has acquired his native language. What they do find is a pattern of development extending over a period of years. If the earliest utterances can be described as language, then they are best described as a primitive, childish language. Gradually and piecemeal, but in an orderly sequence, the language of the child evolves into the language spoken by the parents. This orderly sequence of development can be used as a yardstick to measure the achievements of nonhuman subjects. Because it is based on normative data, it avoids many of the speculative problems of linguistic theory. The placement of a given child at, say, Stage III, is based on comparisons between the utterances of the child and normative studies of other children. Linguistic theory enters in the explanation of why most children go through Stage III before Stage IV, or why a particular child might reach Stage III more slowly than other children, or why a particular child might exhibit a pattern that is distorted with respect to the normal pattern of Stage III children. If we can use the utterances of chimpanzee subjects to show that chimpanzees cannot progress beyond Stage III, or that the pattern of development is distorted with respect to the normal pattern of human development, then the results could be of theoretical importance. But the theoretical interpretation would depend upon the data rather than the data upon the theoretical interpretation.

The value of the comparisons that can be made between the progress of chimpanzee subjects and human children depends upon two major factors. First, the chimpanzee subjects must be taught a natural human language. Second, the chimpanzee subjects must learn this natural human language under conditions that are comparable to the conditions under which human children learn their first language. Since children require many years to master their parents' language, and since there is no reason to expect chimpanzees to acquire language faster, the conditions must be maintained and the comparisons must be made over a period of years.

The most significant results of Project Washoe were those based on comparison between Washoe and children, as in the childish or immature variants in forming signs (Gardner & Gardner 1973; Schlesinger and Meadow 1972); in the generalization of meaning of early signs (Brown this volume; Bronowski & Bellugi this volume; Gardner & Gardner 1969, 1971,1975b); in the gradual increase of length of utterances (Brown this volume; Gardner & Gardner 1974d); in the types of semantic relations expressed in the earliest combinations (Brown this volume, 1973a; Gardner & Gardner 1971); in the replies to wh-questions (Brown 1968; Ervin-Tripp 1970; Gardner & Gardner 1974b,c); and in the use of order in early sentences (Brown 1973a; Gardner & Gardner 1974b,c,d; Schlesinger 1971).

Our current project provides conditions that are closer to those of human language acquisition, hence more valid comparisons are possible. The comparisons we have made so far show that the new subjects are acquiring sign language faster than Washoe. Because the improved conditions also permit us to continue working with the subjects for a substantial number of years, we can make such comparisons for more advanced stages of language acquisition than heretofore; for example, for the further development of questioning and negation, and for sentences involving quantifiers, embedding, and the description of events in time

APES AND LANGUAGE

JANE H. HILL

INTRODUCTION

How shall anthropological linguistics assess the significance of the recent experiments with apes and language? The question is a momentous one. The answers may imply a paradigm shift, with Plato finally giving way to Darwin (Linden 1975), or perhaps "an identity crisis for *Homo sapiens*" (Gallup, Boren, Gagliaro, & Wallnan 1977: 303). At the very least, the issue poses the problem of other minds in a particularly compelling form. In linguistics and psychology, the question has stood the normal terms of the debate between empiricists and rationalists on their heads, with rationalists arguing that the evidence is inadequate, and empiricists arguing that the experiments do not teach, merely reveal (Premack 1976a). Ten years ago, the answer from most established scholars to the question, "Do other animals have language?" would have been an unequivocal "No." Chomskyan rationalism dominated American linguistics and insisted on what Lenneberg (1967) called "discontinuity theory"—the claim that "it is quite senseless to raise the problem of explaining the evolution of human languages from more primitive systems of communication that appear at lower levels of intellectual capacity" (Chomsky 1968: 59). Even continuity theorists erected formidable barriers; Hockett's lists of design features (Hockett 1960a; Hockett & Altmann 1968; Hockett & Ascher 1964) stood as a definitive statement against which a reference to the "language" of bees, birds, or dogs could be measured and found imprecise. Now all that has changed; respected students of animal communication speak of a "linguistic model" for the analysis of communication in animals as phylogenetically remote from humans as seagulls (Beer 1976, 1977), and the New York Academy of Sciences sponsors a major symposium (Harnad, Steklis, & Lancaster 1976) at which advocates of discontinuity theory are a distinct minority. The problem posed so recently by King & Wilson (1975)—the striking contrast between the similarity of humans and chimpanzees at the biochemical level and the differences at the anatomical and behavioral level—now appears to some scholars to be a false problem, at least where

JANE H. HILL • Department of Anthropology, Wayne State University, Detroit, Michigan 48202.

behavior is concerned (Gallup *et al.* 1977). Certainly the ape language experiments have been the single most important stimulus to this major shift of scientific opinion.

The experiments have been discussed from every conceivable point of view; however, most of the discussion and critique of the experiments has revolved around one question: Is what the animals do properly characterizable as "language"? Some of the experimenters have specifically rejected that issue, claiming that it is a misleading and improper one for psychology (cf. Gardner & Gardner 1971: 141). However, the question is a legitimate one for linguistics and for anthropology. It is true that there is no universally accepted definition of language against which to measure the ape experiments, and that is precisely why they are so valuable. They offer the extraordinary opportunity of a truly marginal and deviant case of what is at least very like language, and would probably be assumed to be the beginning of language should the same manifestations appear in the behavior of a human infant. Deviance has been a major tool in the analysis of language and society, because deviance "marks the limits of the rule it is a deviation from" (Pyle 1975). While we should recall that only a century or so ago even scholarly opinion held some savage speech to be less than true language, modern linguists would consider that, from the point of view of the task of definition, an investigation of the great multiplicity of human languages offers "no information except the repeated affirmation of homogeneity" (Bateson 1970: 67). Such manifestations as the usage of early childhood, pidgin languages, glossolalia, and other more or less marginal phenomena have been our only access to marginality in the past; the data from the ape experiments can be added to these materials as we triangulate in on definitions which will allow us properly to define and understand language and the role which it plays in human life, and which it has played in human evolution. This paper will attend, then, primarily to the challenges which have been made against the claim that the experimental apes manifest "language."

THE EXPERIMENTS

Experiments Using Ameslan

In the present climate of opinion, which has begun to take for granted the accomplishments of the apes, it is easy to forget the boldness of the insight which led to Allen and Beatrice Gardner's success in teaching sign language to Washoe, their first young chimpanzee subject. The Gardners were working against a background of findings which have been reviewed by Kellogg (this volume). Young, home-raised chimpanzees, while highly imitative, were never observed to reproduce human words, did not babble or "play" with vocalizations, produced only sounds which had been observed in the wild chimpanzee. Intensive attempts to teach them spoken words yielded minimal results: "the words 'mama,' 'papa,' 'cup,' and possibly 'up' represent the acme of chimpanzee achievement in the production of human speech sounds. But they were learned only with the greatest difficulty . . . [and] were sometimes confused and were used incorrectly" (this volume: 65). The Gardners reasoned that the problem lay, not in an absolute cognitive deficit in the animals, but in the unsuitability of the vocal tract configuration of the chimpanzee to the articulation of human language, a position developed particularly by Lieberman (1968). They determined that a language which could tap the formidable manual dexterity of chimpanzees might be a felicitous choice, and selected American Sign Language or Ameslan, the indigenous language of many deaf communities in the United States and Canada, as the medium for training. They began in 1966 with Washoe, a female chimpanzee about 11 months old. The Gardners wished to be able to compare Washoe's progress with that of children, so they attempted to duplicate the child's environment with toys, companions, and relative freedom. Washoe's first phase of training lasted 50 months; at the end of this time she had a vocabulary of 132 signs, which she used both singly and in combinations. She used her signs in a way very reminiscent of the usage of human children, generalizing their referential domains and making charming mistakes. Washoe's signs were tested repeatedly, both by determining that they were used aptly in communication on a regular basis, and by more demanding double-blind testing (Fouts 1972b; Gardner & Gardner 1969, 1971, 1974b, this volume).

In a second generation of experiments, the Gardners began with newborn chimpanzees, using native speakers of Ameslan as trainers. They believed that the Washoe project, a pilot experiment, had not really probed the capacities of chimpanzees deeply, so the new experiments were designed to approximate even more closely the circumstances of the human child, and to allow for longitudinal depth of many years. Early results appear to validate the revisions of the design. By 4 to 5 months of age, the infant chimpanzees showed signs comparing to the age of first appearance of signs in infants who learn Ameslan (Gardner & Gardner 1975a). As the chimpanzees mature they are exhibiting early "morphology" such as negative incorporation—the formation of negative signs by inflecting substantives and verbs with a negative component—and exhibit such new indications of true symbolic behavior as "representational" drawing (this volume).

Experiments using Ameslan with both home-raised and semicaged animals have been continued by Roger Fouts, who trained under the Gardners, and others at the Institute for Primate Studies at Norman, Oklahoma, where Washoe was taken at the end of the first stage of her training. The Oklahoma studies emphasize cross-sectional breadth rather than longitudinal depth (although they include long-term studies) and attempt to tap a broad range of different kinds of evidence, including intraspecific communication and learning (Fouts 1973, 1974, 1975a,b, 1976;

Fouts, Chown, & Goodin 1976; Fouts, Chown, Kimball, & Couch 1976; Fouts & Couch 1976; Fouts & Mellgren 1976).

Francine Patterson (Hayes 1977; Patterson 1977a, 1978) is currently replicating the Washoe project with a young female gorilla, Koko, with spectacular success. Koko has the largest vocabulary of any of the animals, at over 300 words, and displays remarkable conversational skills.

The most recent reported Ameslan experiment is by Terrace and Bever (this volume), who are working with a home-raised male chimpanzee, Nim Chimsky. At age 22 months, Nim had an active vocabulary of 30 signs and made combinations of signs up to four signs long.

Experiments Using Artificial Languages

The Ameslan experiments have special interest because they use a natural language which offers essentially infinite scope for the animals. However, in the Ameslan experiments, it is difficult and expensive to keep an exhaustive record of usage, and it is impossible to control the resources available to the animals. [For instance, Washoe learned the sign "smoke," which the experimenters did not want her to learn (Fouts & Couch 1976).] This problem has stimulated two major series of experiments using artificial languages, the first under the direction of David Premack and a more recent series directed by Duane Rumbaugh. The Premack experiments (1971a,b, 1975a, 1976a) have used four young animals, although most published reports have concerned the accomplishments of Sarah, who was 6 years old at the beginning of her training. Premack's emphasis was on determining the relationships between "the psychological capacities that underlie language" and "language in general" (1976b). The animals were kept under laboratory conditions. The language was composed of plastic tokens standing for nouns, verbs, quantifiers, connectives, and particles, which could be made into "sentences" on a magnetic board. The durable sentences avoided the problem of memory limitations, and the availability of tokens could be controlled by the experimenter. With the plastic tokens Sarah was taught analogs (or homologs, depending on one's point of view) of naming (both objects and actions), yes-no and wh-questions (answers only; Sarah was not taught to ask questions), the judgment same-different, quantifiers, dimensional classes such as color, shape, and size, metalinguistic notions such as "name of" and "color of," and logical connectives including the relation of implication.

The second major artificial language series, directed by Duane Rumbaugh at the Yerkes Regional Primate Research Center in Georgia, were conducted with Lana, who was about a year old when her training began. Lana was taught an artificial language, Yerkish (Glasersfeld 1977), which appeared as a series of "lexigrams'' inscribed on backlighted panels (the lights were turned off and on to indicate whether the lexigrams were ''in play'') on a computer terminal. The panels could be depressed to communicate with the computer; sentences thus constructed appeared on a display above the terminal. This system added to the advantages of the Premack plastic language (permanence, control over accessibility, avoiding memory problems) an additional capacity for an automatically compiled record of all usage. In addition, the system addressed a major problem of training and testing with a human trainer—inadvertent cuing and the resulting Clever Hans problem— by having training and testing done by the unfeeling computer itself (although the experiments also include interaction with human trainers; how much these interactions have polluted the system is debatable [cf. Gardner & Gardner 1975b]). From the outset, Lana had to use the terminal to obtain all food, entertainment, and companionship. She learned to construct sentences to obtain these desirables, to name things, and to ask and answer questions (Essock 1977; Gill 1977; Glasersfeld 1977; Rumbaugh 1977; Rumbaugh & Gill 1977; Rumbaugh et al. 1973).

CRITICISM OF THE EXPERIMENTS

No sooner had the earliest reports from the original Project Washoe leaked into the press than a storm of controversy began, primarily over the question of whether what Washoe did could be called ''language,'' although the Gardners in their earliest publications noted specifically that the goal of their work was to study communication, not ''language'' *per se*. The criticism revolved around a few basic problems: the problem of intention or ''pigeon ping-pong,'' the problem of symbolic capacity, the problem of syntactic productivity, and the problem of facility in acquisition. To these considerations, which primarily emphasize structure, should be added questions about communicative function, since the structure and function of language should not be separated. Criticism of the claim of language has sometimes apparently been considered by partisans of the animals to be a sort of ''speciesism,'' a denial of evolutionary continuity. Although criticism from some quarters has no doubt been contaminated with that problem, the question of the definition of language and an examination of how the experiments shed light on that definition, seems to be a process from which linguistics cannot but profit. Most linguists are charmed by the accomplishments of the experimental animals; most linguists are convinced that *Homo sapiens,* in his various biochemical, anatomical, behavioral manifestations, is firmly rooted in a mammalian matrix. However, within this context it is our scholarly responsibility to examine purported manifestations of ''language'' very closely, to see what they might tell us about language and about being human.

Pigeon Ping-Pong

The problem of intention—are the animals really using language, or does it just look that way?—is inextricably entangled with the question of the linguistic status of the systems which have been used in their training. Even the status of Ameslan itself is still somewhat controversial; however, most of the criticism has been directed at the artificial languages.

Most linguists, influenced particularly by the pioneering work of Stokoe (1960; Stokoe *et al.* 1965), now believe that Ameslan is as genuine a realization of the human *faculté de langage* as is any spoken language. Ameslan has some unusual properties, but these seem to derive primarily from the access which Ameslan speakers have to the three-dimensional spatial, visual medium as opposed to the linear medium exploited by spoken language. For instance, neither word order nor inflectional endings, major syntactic devices in spoken language, are crucial as indicators of grammatical role in Ameslan (although both occur, particularly in versions of Ameslan influenced substantially by spoken language), since the speaker can use the space around his body and the direction of his glance as syntactic devices. Ameslan can even be signed more than one sign at a time, although this may be primarily a childish device (Patterson 1978). Ameslan has many somewhat iconic elements, but the icons often appear only in hindsight or in the use of folk etymologies to facilitate teaching the signs. Certainly no one could understand Ameslan without being trained in it as one would study any foreign language, and in this respect it differs substantially from true pantomime. In any case, linguists now believe that spoken language has a good deal more iconicity than has been commonly admitted, far beyond the often-cited sound-imitation words (cf. Cooper & Ross 1975; Swadesh 1964; Wescott 1971). Syntax in spoken language may have a maplike quality which is an icon for logical paths which are derived from sensorimotor action paths (David McNeill, personal communication). Duality of patterning, often considered a crucial definitional characteristic of natural language (Hockett & Ascher 1964) can be demonstrated in Ameslan, which can be analyzed as composed of cheremes (Stokoe 1960; Stokoe *et al.* 1965) involving such basic elements as handshape, position, and direction of signing. Ameslan appears to be truly an open system; there is no limit known to what can be expressed in it. Ameslan speakers may use finger-spelling for an occasional technical word, but this is a sort of reverse of the use of mime to illustrate emotions by speakers of oral language; that is to say, each system may have areas in which its most characteristic medium is not the best suited to the message. Thus, there seems to be little doubt that Ameslan is a natural language; however, this begs the question as to whether or not we can call the very truncated version of it mastered so far by the apes a language.

With the artificial languages, however, the actual linguistic nature of the media themselves is in doubt. Roger Brown, commenting on Premack's work, notes that

it is important to remember that not English words but tokens were used. There is something mesmerizing for the native speaker about the sight of words which makes it easy for us to attribute all the linguistic knowledge we bring to such strings to the performance with tokens. (Brown 1973a: 45)

Brown considered Premack's results to be clouded by what he refers to as the "pigeon ping-pong problem" (1973a: 44)—one can teach two pigeons to bat a ball, but is it ping-pong? One can see that Sarah is manipulating tokens, but is it language? Is she perhaps "narrowly adapted," having learned "carefully-programmed language games" (1973a: 48)? Eric Lenneberg (this volume), in reference to reports of symbolic generalization by Washoe, raised this same issue when he commented that it was impossible to tell, with a chimpanzee, whether it was making a metaphor or a mistake. The Gardners (this volume) have considered that it is possible that both Sarah's and Lana's accomplishments can be interpreted most parsimoniously by imagining that they handle most of their tasks by sheer memorization, rather than by internalizing rules. They argue that since learning-set sophisticated chimpanzees are extraordinarily skillful at forced-choice problems of the type used in the Premack–Rumbaugh training paradigms, in transfer learning only first-trial performance is significant (this volume). They point out that Premack (cf. 1976a) reported transfer as a percentage of ten trials, reporting first-trial results only spottily. Harlow, in a brief review of Premack's work (1977), seems to concur that the use of the term "language" to describe the training is dubious; he describes Sarah's training as "a series of learned tasks ranging in difficulty from simple discrimination to relatively simple matching-from-sample problems" (Harlow 1977: 639), and comments that more complex learning has been shown in macaques. Harlow remarks, however, that "to describe these serialized complex learning problems as language . . . is certainly [Premack's] proper right" (1977: 639).

One of the advantages of the Ameslan experiments is that both the language system and the world in which the animals lived were relatively open compared to the limited systems and caged environments of the artificial language training. This openness, and the greater control of the animals over their own usage, limited mainly by their own aptness and not by the schemes of the trainers, allows us to infer strongly that the animals were using Ameslan in many of the ways that humans use it—to construct a world, to obtain desirables, to regulate the behavior of others. The artificial language experiments do not allow us to make the same kinds of inferences about the nature of the chimpanzee's world. They also do not allow the same kind of comparison with the maturing system in children. The languages place peculiar constraints on the animals which do not duplicate those of natural language. For instance, Yerkish (Glasersfeld 1977) does not allow any categorial sloppiness; nouns are always nouns, verbs are always verbs, and the computer is unreceptive to serendipitous innovations. Thus, the chimpanzee, even if it might have an innate grammatical ability, has the additional burden of figuring out what a psycholinguist might have thought language was like. Lana, "conversing" with her trainer to get milk in her vending machine instead of water (Gill 1977: 243), found that productions like "you put milk behind room," "you give milk in machine," and "you move milk in machine" were all rejected due to the pickiness of Yerkish grammar, which allowed only the verb "put" to take "in." Imagine the befuddlement of a child who did not get milk after three such requests! Yerkish also has a curious length constraint (the project could not afford a computer the size of a brain) which allowed only six words, plus the Yerkish initials meaning "please," "?," "yes," and "no," and the final "period." Thus, Lana could not say, after she had learned the construction "bowl of chow" (Modeled on "chow," "piece of chow"), "please you move bowl of chow behind room period," since it would have had seven words, but had to ask "? you move bowl of chow period" (Rumbaugh & Gill 1977: 182). This is precisely the type of constraint that Chomsky noted would never occur in natural language, and one must admire the chimpanzee's insight in adapting to the fact that if she wanted lots of chow at once (an end gained by constructing "bowl of"), she would have to take on faith *where* the chow would be put. That this strategy of Lana's places the semantics of the Yerkish lexigram meaning "move" under something of a cloud should be clear. Another odd element of Yerkish is the requirement that Lana use "to" before her name when she is the indirect object of a sentence (Rumbaugh & Gill 1977: 177). Since Lana did not have a directional "to," almost certainly the prototype in natural language for indirect-object "to," one cannot imagine what the particle meant to her, although she learned to use it correctly. On the other hand, Premack's artificial language did not have a "to," which mortally confused his English-speaking trainers since he insisted that the proper word order was verb/direct object/indirect object, an order requiring "to" in English. His trainers consistently erred and promoted the dative to its "toless" position after the verb; Sarah, of course, picked up this order, which eventually had to be accepted (Premack 1976a: 96). Sarah could have learned the other order had the trainers used it consistently; however, a few years of work on case and relational grammar have suggested that that "to" may not be trivial. Chimpanzees can learn to put things in any order their trainers desire, but there is very little evidence that the order requirements on the artificial languages were serving the same kinds of syntactic purposes that they serve in natural languages. Equally, we have no evidence that the set of innovations allowed in the artificial languages is a natural set, comparable to those made by the human child. In fact, the errors made by the animals in the artificial languages are not necessarily a useful source of information (unlike in the Ameslan experiments, where they have been profoundly enlightening), since they are normally made in the context of a forced-choice paradigm. In the case of Lana, many of her errors were *typing* errors due to hitting the wrong key (Rumbaugh & Gill 1977); it is extremely difficult to rigorously distinguish these errors from genuine language errors.

Fortunately we have the Ameslan experiments to complement the artificial language experiments. The Ameslan experiments show that the animals can turn systems to their own purposes and innovate in a way reminiscent of human children; the artificial languages show us that the animals can learn rigid constraints, which while not necessarily "grammatical," are certainly "rules." Although we do not yet know enough about language to construct an artificial training language that is truly "linguistic," the overall impact of the chimpanzee's performance in the artificial languages is such that most of us would observe, along with an impatient Ernst von Glasersfeld, that "if Sarah or Lana or Washoe asks for a piece of apple, she or he wants a piece of apple. And to call that a primary reinforcer, I think is nonsense" (Harnad *et al.* 1976: 606). We must recall that intentionality is a stick which can be used to beat almost all linguistic work, not only efforts with chimpanzees.

Symbolic Capacity

A cornerstone of most definitions of natural language until the early 1970s was symbolic capacity, the ability to form conceptual domains and label them with arbitrary signs. The capacity was thought to be mediated by a uniquely human brain structure, the association cortex of association cortices in the angular gyrus (Lancaster 1968), that allowed humans to integrate perceptions in different sensory modalities. However, the capacity for cross-modal association has been shown for great apes (Davenport & Rogers 1970), and the ape language experiments have stimulated major revisions in our thinking on the utility of the symbolic capacity as the benchmark for human status. The ability of apes to use artificial signs to refer has been shown consistently by the experiments. A young male chimpanzee, Ally, has been trained on spoken English words and referents, then on the association between the English word and an Ameslan sign, and was then able in testing to give the correct sign for the original referent (Fouts, Chown, & Goodin 1976). Particularly interesting are the abilities of the animals to generalize the meaning of their signs. They accept generalization readily in training. For instance, Washoe first learned the sign "more" in the context of "tickle." The sign was extended to the context of "swing," then to second helpings of food, and finally as a way to ask for repeat performances, such as somersaulting, by her companions (Gardner & Gardner 1969, 1971). Even more suggestive that the chimpanzees have the deepest sort of symbolic capacity—the capacity for "world creation"—are their spontaneous innovations. For instance, Washoe was initially taught the sign "open" for doors and containers. She herself generalized it to turning on water faucets (Gardner & Gardner 1971). Lana generalized "no" from its training usage, as the answer to Yes-No questions, to use it as a protest when people refused to give her what she wanted (Rumbaugh & Gill 1977: 170). Sarah used her name token to claim desirables, forcefully indicating something like "mine" (Premack 1976a: 104). Koko, the gorilla, generalized "straw" from drinking straws to plastic tubing, hoses, cigarettes, and radio antennae (Patterson 1977a). One of the most appealing spontaneous extensions of a sign was accomplished by Washoe, who had originally learned the sign "dirty" to refer to feces and other soil, and used it to generate insults such as "dirty monkey," of a macaque who threatened her, and "dirty Roger," to Roger Fouts when he refused her requests for desirables (this volume). Another Oklahoma chimpanzee, Lucy, has also used "dirty" in this way, suggesting that the human propensity for scatalogical insults may be susceptible to nativistic interpretation. Among the most spectacular world creations are the drawings of the 3-year-old Moja, who drew and named, spontaneously, quite a convincing "bird," and then drew a small orange "berry" on request from her trainer, who didn't quite believe "bird"! (Gardner & Gardner, this volume), thereby breaking through Premack's prediction that chimpanzees "do not, so far as is known, construct copies of existing or imaginary figures by any device" (Premack 1975b: 228). The animals are also able to use their signs productively to construct new lexical items. Washoe produced "water bird" on first seeing ducks; Lucy used "cry hurt food" for radishes, "candy fruit" for watermelon (Fouts 1975b). Lana used "coke which is orange" for Fanta orange drink (Rumbaugh & Gill 1977: 179). Koko signed "finger bracelet" for ring, "white tiger" for zebra, "eye hat" for mask, and "elephant baby" for a Pinocchio doll (Patterson 1977a). The mistakes made by the animals are particularly interesting. In the Ameslan experiments, where vocabulary is always available, errors tended to be inside semantic domains, such as "cat" for dog. Washoe would sign "baby" on tests for signs for animals, but only when small three-dimensional models, rather than photographs, were the exemplars, an unexpected extension of the original referential domain of the sign "doll" (Gardner & Gardner 1971: 161).

Premack has argued that lexical items, whether Ameslan signs or plastic symbols, are true symbols with the "power of words" for the animals (1976a,b). For Sarah, he finds that her plastic tokens were "rich representations" of their referents, as efficient for information storage and classification tasks as the actual physical referents. They could be sorted according to the same dimensional criteria as exemplars of the referent and could be used for new learning. For instance, Sarah, who knew about "chocolate," was trained on the concept "brown" with a plastic sentence "brown color-of chocolate." When asked in a test to select "brown" from a set of four colored plastic chips, she was able to do so (1976a: 202; 1976b). The animals proved to be very interested in names; Lana asks her computer to tell her the names of things (Rumbaugh & Gill 1977).

The cumulative effect of inferences from the types of performances reviewed above (there exists a much richer fund of anecdotes in the published reports than could possibly be included here) is to suggest that in the area of symbolic capacity the performances of the apes are strongly reminiscent of those of small human children. The weight of the evidence seems to belie the specter of intention; these animals are making metaphors, not mistakes. However, there remain serious questions about whether their metaphorical abilities are truly linguistic. For instance, the small size of their vocabularies is not a trivial issue. In Ameslan, 500 to 1000 words would constitute a minimal working vocabulary; none of the animals have approached that level in the modality of production, although they may be at that level in comprehension. Davidson (1976) has raised the point that, no matter how suggestive is the circumstantial evidence, the semantics of the chimpanzees' symbol systems cannot be said to be the same as the semantics of the signs into which they "translate," since the system is so truncated. The same question can be raised about the meanings of words in early child language and is a telling objection to the technique of "rich interpretation" of child language, which glosses over this problem. However, no final conclusions may be drawn in this matter, since the vocabulary is a dynamic area. None of the experiments have emphasized intensive vocabulary training, and it may be that the chimpanzees are capable of reaching levels at which the limitations on the systems will become less clear. However, symbolic capacity is one area where a tentative conclusion that the difference between the chimpanzees and human children is a quantitative, not a qualitative one seems reasonable at this time. At the very least, the symbolic capacity allows the psychologist an invaluable tool—he can probe the "folk taxonomy" of the chimpanzee, as in Fouts's work with Lucy on the classification of the world of food (1975b; Fouts & Couch 1976). He can investigate color perception, as in Essock's exemplary work (1977) with Lana. A particularly exciting possibility for animal behaviorists is the possibility of studying the indigenous chimpanzee taxonomy of vocal and gestural displays from wild animals; preliminary work here has begun in Oklahoma (Fouts & Couch 1976). We can probe the emotional life of animals, as suggested by Terrace & Bever (this volume); Koko reports poignantly on her inner states of happiness, sadness, fear, and shame (Patterson 1977a, 1978).

Productivity

The most extensive criticism of the claim of language in the experiments revolves around syntactic, rather than lexical, productivity, an issue first raised by Bronowski & Bellugi (this volume), based on early data from Project Washoe. A

major goal of the artificial language experiments was to test this capacity. But the
critics have not been satisfied. All the chimpanzees are able to use the various kinds
of signs in sequences and some of the sequences are apparently spontaneous. Linguists, though, remain unconvinced that the sequences are true "constructions,"
governed by syntactic constraints.

The Gardners have pointed out that in Washoe's case it is possible to tell
"strings" from sequences of one-word utterances following in rapid succession,
since Washoe, like many human signers, used a strategy of keeping her hands raised
in front of her chest in signing position while a sentence was being produced, and
dropping them to her sides when she had completed it (Gardner & Gardner 1971:
165). Brown, in an early review of the Washoe work, felt that the "construction-
ness" of Washoe's productions could be shown by comparing them with the constructions of children at the same age (this volume). He observed that in both cases
there was a development from shorter to longer constructions in maturation, suggesting a constraint of "complexity." In addition, Washoe (Brown 1973, this volume; Gardner & Gardner 1974b) showed the same repertoire of relations as Stage
I human children: Recurrence ("more"), action-locative, agent-action, action-
object, instrumental, agent-action-object, and agent-action-locative. The same relations characterized Koko's sequences (Patterson 1977a, 1978). Fodor *et al.* (1974:
443) noted that although these were semantic relations, they could not be shown to
have been coded as grammatical or syntactic relations. Limber (this volume) has
argued that the repertoire of semantic relations in the usage of the chimpanzees and
the Stage I child might be just those semantic relations which are inherent in any
communication system involving concatenation. Thus, the matching of the early
maturational sequence in human children has not been taken to be sufficient evidence by some psycholinguists, although it is regarded by the Gardners (this volume) as their most important result.

Critics of the claim of syntactic productivity have focused on three areas which
they require for a demonstration of syntactic coding: word order or inflection to
encode role, hierarchical constituent structure, and major strategies of recursivity
which would yield syntactic novelty, including nontrivial evidence for concatenation and nominalization, complementation, and relativization.

Many linguists now agree that the emphasis of early critics on word order, to
which Premack and Rumbaugh responded diligently in their experiments, was
rather wrong-headed. Brown (1973a) believes that word order may not be a universal syntactic strategy among children; this may be particularly true for Ameslan
learners. Washoe did exhibit some word-order preferences (Gardner & Gardner
1971: 177), and it is clear from the Rumbaugh and Premack experiments and from
some of the Oklahoma work (Fouts, Chown, Kimball, & Couch 1976) that chimpanzees can learn word order. However, as noted in the previous section, in the
artificial language context some of this word order may be of a trivial, nonsyntactic
nature.

Both the Gardners and Premack have claimed that there is evidence for a constituent hierarchy in the output of their animals. It is difficult to show constituency, particularly in short sentences. The Gardners devised an ingenious experiment in which Washoe was required to answer a great variety of wh- questions (recall that all her vocabulary was available to her at any time). She answered where, who, whose, what, etc, questions with appropriate vocabulary, suggesting a hierarchical arrangement of that vocabulary into things, directions, locations, possessors, etc. (Gardner & Gardner 1975b). These, however, are semantic, not syntactic categories, and are more like Lucy's categorizations of the world of food into "food," "fruit," "candy," "berry" (Fouts 1975b) than they are syntactic categorizations into noun phrase, verb phrase, etc, which can be shown convincingly only by one-many substitution or by movement transformations such as the passive. Premack has claimed that Sarah at least implicitly has a one-many transformation; she is able to replace the element "?" in "apple banana ? color" with "is not plural." He claims that in order to perform correctly when told "Sarah banana pail apple dish insert" (Premack 1976a: 331), she must be able to parse the immediate constituents of that sentence. He claims that she may have a node *subject* because she is able to place "plural" correctly to differentiate the answer to "apple and orange ? fruit" from "round apple ? fruit," which she could not do if she were using an order strategy (Premack 1976a: 228). Unfortunately, it may not be possible at this stage to preclude the claim that Sarah is using an order strategy, or one based in semantics, although Premack proposes arguments against such an interpretation (1976a: 330). Premack has suggested that his language is truly generative and allows novelty. Fodor *et al.* (1974) have erroneously attributed to Premack the claim that transfer learning shows "productivity," but Premack has in fact consistently distinguished "recognitional" transfer from "productivity," which he restricts to the capacity of a concept to generate new instances of itself. Examples of such a capacity are the use of tokens for "name of" and "color of" as metalinguistic elements (Premack 1976a,b). In addition, he argues that his system as used by Sarah is generative in that it displays concatenation of verbs and sentences and an implicit gapping rule, as in sentences like "Mary wash give Sarah apple." (Other evidence, though, such as Sarah's failure to delete the element "plural" in answering a question "red ? plural color" (which required placing "is," deleting "plural"), suggests that Sarah did not understand deletion as a possible process [Premack 1976a: 324].) Premack does not raise the possibility that chimpanzees will be able to learn nominalization, relativization, or complementation. The absence of these strategies (it does seem clear that concatenation, including the very interesting implicational relation, is available to the animals) may not be trivial. I have argued (1972) that the capacity for embedding sentences in other sentences through nominalization, relativization, and complementation may be a crucial defining feature of mature natural language, since it is by these techniques that we separate the world of languages from the world of context, through the "logical sloppiness" of language. For instance, in complementation, a sentence like "Sarah wants to go out" is probably derived from two propositions which directly represent concepts of context: (1.) Sarah wants X. (2.) Sarah goes out. The sentence world, created by syntactic operations, has lost the second "Sarah" and the modality of the second proposition, accomplishing a complex displacement from conceptualization. In relativization, "Sarah likes the trainer who brings chocolate" may be derived from two context-world propositions: (1.) Sarah likes the trainer. (2.) The trainer brings chocolate. If these two sentences are uttered, we must depend on context to determine that the first and second instance of "trainer" are coreferential. However, the syntax of relativization allows the creation of the sentence-world where we do not need the context to see that this is the case; coreferentiality must be satisfied (in English) because we are able to delete one instance of "trainer" and replace it with "who." Thus, a sensitive set of rules gives the logically sloppy "sentence world" from the logically particularizing world of concepts and contexts. The fact that chimpanzees do not produce this kind of syntax has been regarded as referable to a merely quantitative dimension relative to human productivity. Yet this function of syntax may be one of the most important prerequisites to the world-creating capacities of adult language, freeing the maturing child from the strict sensorimotor, action-derived semantic relations of the earliest stages of language development. It will be of great interest to see if such constructions appear in the usage of the apes as the longitudinal studies continue.

Unfortunately, the requirement that complex recursive strategies be a structural prerequisite for language eliminates young children. Other definitions, such as Katz's "effability" (1976: 37)—a natural language must have some sentence which has the same sense as the sense of every possible proposition—also exclude them. Limber (this volume) has seriously proposed that we not consider children to have language until about 3 years of age. The fault with all such criteria is that they evade the dynamic nature of language development in ontogeny, and presumably in phylogeny as well; they would require us to establish an arbitrary boundary marker— after the third candle is blown out on the third birthday cake, or at, say, 25,000 BC. Linguistics has not dealt adequately with the whole issue of potentiality, which is a problem faced not only with the early stages of child language, but also with pidgins, the "home signs" of the isolated deaf, and other marginal manifestations; we know that with maturation, change of social context, training, these can become full languages (cf. Washabaugh 1977).

Facility in Acquisition

Another criticism of the claim of language has emphasized the nature and amount of training required to elicit results from the animals. This is a crucial area because rationalist linguists claim that the human child brings to the language acquisition problem an innate device, specialized for language, which enables him to process rapidly the highly impoverished data to which he is exposed into the adult grammatical system. However, data from children, much of it collected by rationalists making that assumption, are quite incomparable with the detailed quantified materials from the apes. Rationalist linguists have simply assumed, for instance, that the purposes and concepts of the children are similar to those of adults (somehow children never remind anyone of pigeons playing ping-pong), and allow "rich interpretation" of his output. Thus, a child who says "Mommy sock" is assumed to have *intended* rich "Mommy is putting on the sock" (even when Mommy answered, "That's not Mommy's sock") (Gardner & Gardner 1974c). Such a technique would be held up to ridicule if used by workers with chimpanzees, so they are forced into a quantitative mode. When critics of Rumbaugh's work with Lana are scandalized by the fact that it took Lana 1600 trials to learn her first instance of "name of" (Rumbaugh 1977: 81), they neglect the fact that no one has ever followed a human infant around counting the number of times his interlocutors say "Mommy," "cookie," "doggy," etc. before the child first lisps these in appropriate contexts. (However it is clear, as the Gardners observe [this volume], that the fact that the 1600 trial experience took place *after* Lana had been using stock sentences containing names for many months suggests that those sentences had no semantic content for her, at least not in the usual sense.)

A number of critics (cf. Limber this volume) have commented that the Ameslan chimpanzees have to have their hands "molded" into the correct shape for the signs, rather than learning by imitation (Fouts 1972b). Some signs, though, clearly are learned by imitation, including signs like "smoke," learned by Washoe over the dead bodies of her trainers. Children, it is implied, require *no* molding. However, such a criticism neglects the characteristic babbling-imitation games which children play with parents for many months before the child utters the first word (of course I do not know that such games are played in all cultures). It is clear, in any case, that the molding technique alone is not sufficient to explain the animals' productions, since they have all been reported to use aberrant "baby" forms of the signs. For instance, Koko signs "tickle" under her arms instead of on the back of her hand; Koko is also inhibited by having a super-short brachiator's thumb and has to sign accordingly (Patterson 1978). Washoe signed "flower" with one finger rather than two (Gardner & Gardner 1971). The presence of baby words and grammar has been a major form of evidence used by rationalists to justify the assumption that the child creates grammar rather than merely learning it.

Comparison with Wild Systems: Context

An important issue raised by the experiments is why, if the experiments show that the animals have cognitive capacities adequate for simple language, we do not see anything even remotely as complex in the behavior of the wild animals. This contradiction has been used to support the interpretation of language acquisition as based on general cognitive, rather than language-specialized, capacities. However, the question of exactly what chimpanzees and gorillas do in the wild is still open. Menzel (1974; Menzel & Halperin 1975) has conducted a series of seminaturalistic experiments that have shown that chimpanzees can pass information about the quality, quantity, and direction of sources of food or unusual objects, and can even pass this information along a chain of two or three animals. (They can also lie about all these things.) Premack (1975a) has pointed out that this type of communication can probably best be explained by appealing to displays of intensity and orientation, rather than by symbols, as long as all the animals share the same affective stance toward their world (which he assumes is the case). Premack believes that this is crucially different from the situation in the language experiments, where it was possible for two parties in communication to disagree, but still exchange information. Observations of wild chimpanzees (Goodall 1968; Kortlandt 1967) have shown that they have a fair-sized repertoire of gestures, but this repertoire has not been shown to be anywhere near the size used in the Ameslan experiments. (Some signs observed in the experiments may be like those seen in the wild, and may even be "natural" signs.) Marler (1969) has shown that the chimpanzee vocal repertoire is complexly graded, with similar systems found only in other great apes and in terrestrial monkeys such as macaques. Green (1975) has suggested that most of the immense variety and complexity of the vocal communication system in macaques can be accounted for on a scale which incorporates only intensity of arousal and affect antithesis. If so, then the gradation may constitute a trivial form of productivity which does not violate the claim of an upper limit on the number of possible messages in vertebrate communication proposed by Smith (1969) and Wilson (1972). True compounding, the most primitive productive technique, has, as far as I know, been recorded only from wolves and coyotes (Fox 1971: 34), where combinations of play, aggression, and submission have been recorded when animals of different dominance feed together in close proximity, a situation which is particularly interesting because it is, of course, a characteristic context of social hunter societies. In fact, the most suggestive examples of signlike gesturing in chimpanzees are seen in the context of food sharing and in the reassurance by dominant animals of inferiors (Goodall 1968). The interpretation of wild communication systems is an area where emphases are rapidly shifting; Beer (1976, 1977) has suggested that the traditional ethological paradigm must be replaced by a *linguistic* paradigm which explains variation in displays as governed by context, as phonemes mean something different depending on what word they are in. Increasingly, students of animal behavior are attending more to questions which have been previously disallowed as too susceptible to anthropomorphization; perhaps the most notable recent instance of this trend is the work of Griffin (1976). Thus, scholars who have argued that there is nothing we can call "language" in the wild should probably hedge their bets. My own point of view (cf. 1972) is that much elaboration in the gestural modality is unlikely; over the long run there would be selection against it in favor of elaboration in the vocal-auditory modality, since this is less involved with other behaviors. Chevalier-Skolnikoff (1976) has suggested, however, that the developmental sequence in the vocal-auditory modality is arrested at an early stage in great apes; they lack a fully developed sensorimotor or imitation series in this modality. Hewes (1973b) has suggested that the gestural system did elaborate, and that it is in this elaboration that we must seek the origins of language. However, Marler (1975) and Green (1975) have suggested that there are areas of vocal-auditory behavior which are relatively affect-free and might be susceptible to evolutionary pressure toward elaboration. A major lesson of the ape experiments is that they show that it takes very little change in context to shift behavior from something rather less like language—the wild system, elaborate as it is—to something rather more like language—the behavior observed in the experimental animals. This result seems to belie the claim that language is an all-or-nothing, emergent *faculty* of humans. Hewes (1973a) has suggested that my arguments that we attend critically to the homelike contexts of the Ameslan experiments (1972) somehow imply that these contexts are inappropriate. On the contrary, they are entirely appropriate if one wishes to push the behavior of the animal over a boundary into the something which is "more like language." They show us in a dramatic way the crucial importance of context, and by extension the role of environment in evolution.

Communicative Function

Perhaps the most important problem with the reports of the ape language experiments for the anthropological linguist lies in their failure to attend explicitly to the question of communicative function. This was a particularly dramatic flaw in Premack's experiments, which he himself has discussed: "a failure to interdigitate the motivational pressures that make language valuable with the cognitive program teaching it" (1976a: 153) is considered by him to be an important problem. Only a few communicative possibilities (primarily answering questions) were made available to Sarah. However, when the problem of communicative function was raised by Mounin (this volume), Premack replied that he was not interested in communicative function, but in intelligence only (1976b). This is an excellent illustration of the capacity of the "comparative intelligence" model (cf. Hill 1974) of the comparison of human language with communication systems in other animals to mislead us. Communication systems do not function simply to pass along ever more richly analyzed information; they form societies as well. The point made by Mounin, and missed by Premack, is that it is quite impossible, from a structuralist point of view, to separate the cognitive side of language from its instrumental, pragmatic side; to do so leaves one with a system which lacks structural significance, and renders most of what is done in the laboratories trivial, a set of "games," to use Brown's (1973a) term. Even where communication was an important purpose of the experiments, as in the Gardners' work, the reports have de-emphasized it in favor of intensive attention to structure. The only exception in the published reports so far is a brief survey by Miles (1976) on the repertoire of speech acts used by the Oklahoma chimpanzee, Ally.

Halliday, in his brilliant analysis of the development of language functions in children (1975), has shown that from a very early stage the use of language to learn about the world—its cognitive aspect, Halliday's "mathetic" function—cannot be separated from the interactional or pragmatic function. Knowledge and society are inseparable, as the child cannot use language to learn without using it for interaction. The earliest stages of functional development may be almost entirely pragmatic, as the child uses his protolanguage to regulate those around him, to acquire desirables, and to invite interaction. Those protofunctions in which the child uses objects as foci for interaction develop into the more mature mathetic function, inextricably linked to the interactional and pragmatic aspects of language. Thus, when chimpanzees are not encouraged to ask questions, but only to answer, the pragmatic function is sufficiently truncated that a full development of the mathetic function is highly unlikely. The function of giving information—asked of chimpanzees in test after test—appears relatively late in human ontogeny as the maturing child learns the sophisticated strategies by which discourses are regulated. Thus, in Halliday's terms the "conversations" which Lana held with her trainers, where she was required to innovate in order to acquire desirables, were probably manifestations of the archaic instrumental and regulatory capacities. The only discourse constraint upon the conversations was the primitive one of sticking to the subject (Gill 1977; Rumbaugh & Gill 1977: 174)—the desire for food and companionship—and this can be accounted for parsimoniously through an appeal to affect alone.

Reports of the experiments have not attended formally to the possibility that the chimpanzees have input into the communication dynamic, and hence into the society being constructed. The chimpanzees have often been seen as exclusively receptive; the interest in their output has been in its structural complexity or lexical variety. However, input from the chimpanzees is plain from many anecdotes. Premack was forced to restructure sessions with Sarah to allow her a period for "creative writing," or she would steal the plastic tokens and play with them (1976a: 16; 1976b). Lana managed to create a type of expression—the polite imperative—which was not anticipated in Yerkish grammar; this creation is passed over very briefly by Rumbaugh, who sees it as an accident of the system:

So far as the computer program was concerned, ? was the functional equivalent of PLEASE, and hence the two lexigrams coupled together were redundant. Eventually when making requests of people, Lana came to use it by itself (e.g. ? TIM MOVE INTO ROOM PERIOD) (Rumbaugh & Gill 1977: 171).

A review of the texts shows that Lana used the ? imperative, a well-known "politeness" form, almost all the time with trainers, and tended to use "please," the original Yerkish imperative, only with the machine. When the machine was recalcitrant, though, she would shift to the ? imperative (Rumbaugh & Gill 1977: 171). Handelman (1976) noted that status differentiations communicated by the experimenters might severely constrain the chimpanzees' communicative behavior. This is clear in Premack's and Rumbaugh's studies, where the trainers controlled access to the lexicon, and probably lies behind the fact that Lana's trainers blithely accepted the lexigram ? as a marker of the imperative as well as in its originally intended domain as the mark of yes-no questions, while rejecting "? Tim move milk in machine period" because "move" did not take "in"! Thus, we may see that communication is not a side issue, even in experiments designed to study intelligence. Premack is willing enough to appeal to communication dynamics when it is convenient, to explain the decrement in Sarah's performance on double-blind tests with a strange tester (1976a: 35; 1976b), but he is not willing to consider seriously that it may be a structurally significant component of all his other experiments as well.

A number of issues need to be raised in assessing the meaning of the experiments in functionalist terms. We should understand more precisely the purposes of the humans involved in the process. This will involve us in the ethnography of modern science and in Western attitudes toward language to determine why, within the context of the elaborate status maneuverings of modern psychology, someone might wish to be seen talking to a chimpanzee. We will have to ask why the chimpanzees answered, apparently with some eagerness. We should investigate the "pidginness" of the communication systems used. For instance, in the early Ameslan experiments the very poor control over the language by the experimenters has been considered something of an accident. However, the communicative context, of extreme status differentiation and narrow purpose, is particularly appropriate for a pidgin language. It may be unlikely that either the experimenters or the animals will develop the trappings of "full" language, even if cognitive capacities are adequate, unless all parties in the system consider each other equal. Not only is the situation reminiscent of language contact with a pidgin, it is highly ritualized, due to the constraints of psychological custom. Bloch (1975) has suggested that highly ritualized messages, as in religious liturgy and political oratory, constrain their respondents to highly ritualized replies, which can be altered only by revolutionary changes in the system; thus, we cannot dismiss ritualization as a possible serious constraint on chimpanzee production. The allocation of the various codes in the experiments is equally interesting. We know that the chimpanzees are multilingual (Fouts, Chown, & Goodin 1976), as are their trainers, but no formal attention has been paid to how the chimpanzees and trainers use the several available codes.

Even Premack's highly controlled studies were polluted by vocal English, since his trainers often said the English "translation" of the tokens aloud as they placed them on the board (1976a: 39; 1976b: 559). Premack's message board is, as he points out (1976a), a highly interesting element of the system, since to use it the two parties to a communication do not face each other as they put their sentences on the board; they use the board instead, rather like floating messages in bottles.

Unfortunately, we do not have enough material to speculate usefully on the purposes of the humans or the chimpanzees, on the nature and allocation of the codes, or on the organization of the speech communities which are being constructed through the experiments. It is a fundamental claim of modern sociolinguistics (cf. Hymes 1975) that the structures of codes and their functions are interconnected and that we will never understand the one without attending to the other. If we are to take what the chimpanzees are doing seriously enough to call it language, we must attend to both kinds of considerations and do the ethnographies of communication of the experimental communities with as much care as we would give to those of the Navajo or the French.

CONCLUSIONS

The range of issues which should be raised in an examination of the ape language experiments is without limit; I have reviewed only a fragmentary few of the possible problems. The whole question of what the experiments suggest about the origins of language has been neglected here, although a major effort of the experiments has been to stimulate new interest in this question. I have not been able to discuss the problem of the physiological basis of the apes' behavior, particularly in neurology, although this is an area which could be explored by techniques similar to those used with humans, such as tachistoscopy; Malmi (this volume) has reviewed this question. The experiments suggest many possibilities for the construction of an experimental, as opposed to an observational, study of animal behavior, as in the color experiments of Essock (1977) or the studies of the distinctive features of wild chimpanzee vocalization (Fouts & Couch 1976). Anthropologists should watch the impact of the experiments on the development of ethics in animal experimentation. Currently, two equally high-toned, but radically opposed, positions have crystallized; on the one hand, a position which holds that great apes may no longer be used as experimental animals or in zoo displays but must be left in the wild or raised like children (cf. Hayes 1977); on the other hand, a position which sees the experiments as revealing whole new orders of utility for the great apes as experimental subjects, particularly as human surrogates in studies of aphasia, retardation, and other language problems (cf. Rumbaugh 1977). Already the analogy between the language competencies of the apes and of the human right hemisphere has led to exciting work with aphasics and retarded children using the artificial languages for training (Davis & Gardner 1976; Gazzaniga 1975; Premack 1976b; Premack & Premack 1974). However, it seems clear that the experiments, while holding out the possibility of apes as particularly refined human surrogates, at the same time cast serious doubts on the ethics of using them in that role.

Linguists, the present author included, have not always been warm in their welcome of the great apes into the community of language-learning beings, but there is no question of the brilliance of the experiments. At once they remind us of our fundamental continuity with other animals, and of the beauty and richness of human science. It is unlikely that any of us will in our lifetimes see again a scientific breakthrough as profound in its implications as the moment when Washoe, the baby chimpanzee, raised her hand and signed "come-gimme" to a comprehending human.

ACKNOWLEDGMENTS

I would like to thank Robert Most and Deborah Keller-Cohen for bibliographical assistance, and Allen and Beatrice Gardner, Roger Fouts, and Francine Patterson for graciously furnishing unpublished material.

LINGUISTICALLY MEDIATED TOOL USE AND EXCHANGE BY CHIMPANZEES (PAN TROGLODYTES)*

E. SUE SAVAGE-RUMBAUGH, DUANE M. RUMBAUGH,
and SALLY BOYSEN

*The "fundamental linguistics situation" is that of cooperation
between two individuals, A and B, in which A stimulates B to do something
impossible for A to accomplish, but which is adaptive with respect to A.*

—Paraphrase of Bloomfield by Crawford (1941: 259)

It is surprising that, given the enduring interest in the ability of apes to communicate symbolically with one another, all language work with apes has hitherto concentrated upon their ability to demonstrate language functions and syntax in interaction with human beings (Fouts 1974a; Gardner & Gardner 1969, 1971; Patterson 1977; Premack 1976a; Rumbaugh 1977; Terrace & Bever this volume). As Steklis and Harnad (1976) point out, the true adaptive function of language lies in the ability it confers upon man to transmit *specific* information in an abstract, context-free form. Previous work with apes has, for the most part, placed a highly sophisticated and competent, language-using human being in the role of either the receiver or the transmitter in every linguistic interchange. This essentially allows the chimpanzee to "fill in the blanks" while the nature of the interchange is basically structured by the human being. Anecdotal reports on chimpanzees learning American Sign Language (ASL) have suggested that the animals are communicating with one another through signs (Fouts 1973, 1974a; Gardner & Gardner 1978); however, what is not clear in any of these cases is (1) whether the chimpanzees were indeed gesturing to one another and not to the humans who were nearby, (2) whether the ASL gestures

*Supported by grants from the National Institute of Child Health and Human Development (HD–06016) and Animal Resources Branch (RR–00165), of the National Institutes of Health. The authors gratefully acknowledge Janet Lawson's work with the chimpanzees, which facilitated the research. They also thank Harold Warner, Victor Speck, Tom Smith, Sarah Ennis, Judy Sizemore, Susan Wilson, Cheryl Meuleners, and Jenny Campbell for important assistance and colleagueship.

E. SUE SAVAGE-RUMBAUGH AND SALLY BOYSEN • Yerkes Regional Primate Research Center, Emory University, Atlanta, Georgia 30322. DUANE M. RUMBAUGH • Department of Psychology, Georgia State University, Atlanta, Georgia 30303.

353

used were essentially different from the nonverbal gestures reported for wild chimpanzees, (3) whether the recipient of the gesture altered his behavior in specific response to the gesture, or (4) whether the animals were capable of reversing transmitter's and receiver's roles.

The nonverbal communication of nonhuman primates has traditionally been characterized as species-specific, emotionally based, and nonintentional (Lancaster 1968; Myers 1976). However, the complex use of gesture reported by van Lawick-Goodall (1968b, 1973) in common chimpanzees, and by Savage-Rumbaugh, Wilkerson, and Bakeman (1977) in pygmy chimpanzees suggests that these animals might be capable of expressing actions that they wish others to perform.

Menzel & Halperin (1975) have demonstrated that captive chimpanzees in an open-field enclosure can provide one another with information concerning the nature and location of hidden objects. This information is communicated by the rate of locomotion and other accompanying indices of excitement exhibited by the "leader," the animal previously shown the location of the hidden item. Although Menzel suggests that the leader is aware of the communicative value of his signals, there is no indication that any leader intentionally varied the types of signals emitted as a function of whether the hidden object was a food or a toy. Rather, it appears that when a chimp observes the hiding of a preferred food, he is simply more anxious to retrieve it than in the case of a toy. Overt gestures were not observed.

Crawford (1937, 1941) also found it difficult to elicit overt gestures between chimpanzees required to work cooperatively to obtain a food reward. When both animals achieved an understanding of the task at hand (mutually pulling to draw in heavy objects, sequential lever operation, etc.), they simultaneously set about working on the problem, often without even glancing at one another. While their behavior was coordinated and cooperative, the author in some cases questioned whether or not the animals were in fact aware of this. He was able to elicit gestures from one animal by distracting the other with a novel object, but when gestures were used to implore aid, they did not vary with the type of aid required. This indicates that the animals were not conveying specific information about what they wanted one another to do. Since gestures did differ widely among individuals, this would suggest that there was no common gestural code. Rather, when gestures were used, the initiator of the gesture expected the recipient to understand why he was gesturing, based on the context of the gesture.

These studies demonstrate that as long as the animals share similar expectancies, complex acquired information can be exchanged with minimal awareness and intentionality. Similar conclusions can be drawn about other experiments of this sort (e.g., Mason & Hollis 1962). Information transmitted in this fashion lacks the specificity of symbolically coded information. Whenever there is a simple situation (in which the animals either do or do not work together on the one task available), or when there is a simple linear relationship between degree of arousal and intensity of a set of nonverbal signals, the specificity of symbolic communication is not needed. Had Crawford's experiments been modified so that several cooperative tasks were simultaneously available to the animals, but with only one resulting in a reward and only one animal *knowing* which the rewarded task was, gestures of greater specificity would have been called for. If, in addition, pointing had somehow been precluded, symbolic coding of some sort would have been required. Menzel (1972) has attempted such a manipulation in his work. The leader was shown the food but was not permitted to lead the others to it, only to convey to them its location (if possible). The chimpanzees did not succeed in this task.

The present authors have recently reported a study in which two chimpanzees, Sherman and Austin, employed a learned symbol system to indicate to one another the contents of a container (Savage-Rumbaugh, Rumbaugh & Boysen 1978). In the study, only one animal was allowed to see the container, whose contents consisted of one of eleven possible foods and drinks. In order to obtain this item, both animals had to correctly identify it in their request. It was arranged so that this could be accomplished only if the animal who saw the container being baited transmitted this information to the second animal. The transmission of this information occurred with 95% accuracy unless the animals were prevented from using their learned symbol system, in which case their accuracy dropped to 15%.

The present study extends this work in a variety of directions. In an attempt to move away from a vocabulary linked directly to food, tools that could be used to obtain food were introduced to the animals. Prior studies of tool use in nonhuman primates have tended to focus upon the acquisition and transmission of these skills. In the present study, although the animals did have to learn to use the tools, and mutual observation did play an important role in this process, the main interest was in establishing certain symbolic behaviors associated with tool use.

Tool use has repeatedly been linked to the emergence of language by individuals interested in the origins of language and culture in the human species (Beck 1974; Holloway 1969; see Steklis & Harnad 1976 for a review of these ideas). Some support for such a position comes from the fact that because apes have considerable difficulty, compared to human children, in learning to use one object to manipulate another (Köhler 1927; McGrew 1977; Menzel 1972; Redshaw 1978), a cognitive skill that has been implicated as an important precursor of language (Bates 1976). Object/object manipulation also tends to be accompanied by another similar skill, often late in appearance or absent in the ape, namely, using another individual to manipulate an object (Bates 1976; Redshaw 1978). These observations would suggest that both tool use and language may share a similar cognitive base (but cf. Steklis & Harnad 1976).

Although most higher primates readily manipulate a variety of objects (Parker 1974), reported instances of (a) object use to obtain a specific goal are relatively infrequent (Beck 1972, 1973b; Cooper & Harlow 1961; van Lawick-Goodall 1968;

McGrew 1974, 1977; McGrew, Tutin, & Midgett 1975; Menzel 1972, 1973b, c; Rumbaugh 1970). Still rarer are (b) cooperative object use, (c) the use of one moveable object on another movable object, and (d) the use of another individual as an intermediary to obtain an object (Bierens de Haan 1931; Beck 1973a; Crawford 1937, 1941; Köhler 1927; Parker & Gibson 1977; Yerkes & Yerkes 1929). In the study reported below, (a) and (c) are clearly and reliably exhibited by our chimpanzee subjects, together with a symbolically mediated manifestation of (d), all in a context of a high degree of cooperativity and interdependence (b). These performances constitute the first documented instance of a symbolically mediated exchange of goods and information in a nonhuman species.

PHASE I: REQUESTING TOOLS

The subjects of this study were two chimpanzees *(Pan troglodytes)*, Austin and Sherman, 3.5 and 4.5 years of age, respectively. They were laboratory-born animals; at the initiation of the study reported below, both had a working vocabulary of 22 symbols. By the end of this study their vocabulary had increased to 56 symbols. [These vocabulary counts are approximate, due to problems inherent in the determination of levels of semanticity and word acquisition (Rumbaugh & Savage-Rumbaugh 1978). For an account of the training these animals had received prior to the present study, see Savage-Rumbaugh & Rumbaugh 1978, 1979, in press, and Savage-Rumbaugh *et al.* 1978].

The communication system used in the training of these animals is that devised by Rumbaugh and associates (Rumbaugh 1977). It basically consists of a large board covered with keys, each distinctively marked with a geometric symbol or lexigram together with a colored background that serves to code a semantic category. The geometric patterns themselves are produced from arbitrary combinations of nine basic elements. Each lexigram is the functional equivalent of an English word, and the elements of the lexigrams are analogous to the phonemes of spoken words. Every depression of the lexigram on the keyboard is monitored by a DEC PDP–8 computer. It is important to note that although the numerous computer-linked keyboards in the laboratory constitute the principal symbolic medium, symbol use is not restricted to such keyboards. Portable units with lexigram plates are used by the animals during their domestic activities, such as working around the sink, preparing food, and even going outdoors. Throughout the duration of the present study, all word-keys in the animals' vocabulary were continuously lighted and available.

The first phase of this study consisted of teaching the chimpanzee (C) to request one of six simple tools: *key, sponge, wrench, stick, money,* or *straw*. Tool names were taught by making the acquisition of various foods dependent upon tool use, and in order to use a tool, C had first to obtain it. This could be accomplished by requesting the tool with the keyboard symbols. The tools and their functions are listed below:

Tool:	Function:
Key	To unlock a variety of padlocks placed on boxes, doors, etc.
Money	To operate a vending device for food.
Straw	To obtain liquids by threading through small holes in the tops of containers and the wall.
Stick	To dip pudding, yogurt, etc., from containers out of reach, and to push food out of a long hollow tube mounted horizontally on the wall
Sponge	To dip into a vertically-mounted, narrow hollow tube; to soak up liquid from flat surfaces.
Wrench	To unscrew bolts from the keyboard and from small bolted doors mounted in various locations in the rooms.

Tools were introduced by permitting C to watch as food was placed at a particular tool site.[1] Initially, C was allowed to observe while a human experimenter (E) used a tool to extract food from the site. When C noted the rebaiting of the site, he typically took the tool from E and attempted to use it himself, gesturally and vocally soliciting aid if he could not orient the tool properly. When C was thoroughly acquainted with the use of the tool, the site was again baited, and as C reached for the tool, E stated at the keyboard "This stick (sponge, straw, key, etc.)." The animals were already accustomed to requesting many things through the keyboard, and they quickly responded "Give stick." E replied, "Yes, give Sherman (Austin) stick," and handed him the tool.

Following this initial tool naming, E attempted to avoid providing further help and instead encouraged C to deduce which tool was needed at any given time as well as the correct symbol to be used in requesting that tool. C could ask for any tool at any time. He could also ask for the food located at the tool site, to go out of the room, to play, to tickle, etc., since all words in the animal's vocabulary were always available, as were symbols for words not yet learned. Whenever C requested a tool, E searched through a kit containing tools and various other objects until the requested tool was located; it was then given to C. If the request was inappropriate to the problem, C was allowed to attempt to use the inappropriate tool or to request again, as he wished. At first, frequent attempts were made to use the inappropriate tool: keys were inserted into long tubes, locks were poked with

_____

[1]Consequently, with each new tool there was associated one or more tool sites.

sticks, sponges were twisted on bolts, etc. Later, the inappropriate tool was immediately refused by C, and another tool was requested. Thus, tool labels were not
initially learned as "names" *per se,* but as complex functional activities

For purposes of comparison, we also attempted to teach the animals to name
or label a variety of objects. These were objects that the animals encountered every
day in a number of contexts. No particular problem or use for the object was provided during training; C was simply asked its name and rewarded with food or social
praise if correct. C was allowed to play with the object as much as he wished and
to use it in any way. Similar name-training procedures have been employed with
chimpanzees by Gardner & Gardner (1971), Fouts (1973), Premack (1972, 1976a),
and Gill & Rumbaugh (1974). Although these procedures varied in specific details,
they all involved the common feature of requiring C to label an item presented or
indicated by E. Items used in the present naming task were: *box, bowl, key, blanket,* and *lock. Key* was used in this task prior to its introduction into the tool-use
task.

Items were introduced one at a time in both tasks. When C was working well
with the first item, a second one was introduced, and so on, with additional items
(at their respective sites in the case of the tool task). Tool problems and specific
items to be named were always presented in a random order, with the exception
that when confusion persisted on a given item, work would concentrate on that one
for three or four consecutive trials. In the naming task, due to difficulties experienced by the animals, no more than three randomly selected items were involved in
any session. In the tool task, the baiting of the various sites was randomized as of
the initial use of each tool's respective site. Once a tool was introduced, it was
always available in the animal's tool kit and was always provided when requested,
regardless of the appropriatenesssof the request. Concurrently with this training,
the chimpanzees were using the keyboards throughout the day in a variety of communicative contexts not specifically related to the present study. It was neither
feasible nor critical to administer a fixed number of trials per day on any task.
Rather, each task was worked on as long as the animals remained interested and
participated cooperatively. If the animals tired of the task, or experienced frequent
consecutive failures (as often occurred in the naming task), they simply refused to
continue to participate. Animals always received full food rations, irrespective of
whether they had worked on any given training task.

Object naming would appear *prima facie* to be the simpler of the two tasks,
since it only required that C attend to some perceptual characteristics of an object
and reliably associate a label with those characteristics. In comparison, the tool
task required that the animal (1) note that food had been placed at a particular site,
(2) shift attention from the food itself to its physical surroundings, (3) determine
what sort of tool would be needed to extract the food, and (4) recall the tool's name
and use it in a communicative request.

Contrary to expectation, the object-naming paradigm proved the more difficult task. Table I displays the number of trials administered on various items in both paradigms. *No* item was learned to a criterion of nine consecutive correct responses (with all exemplars randomized) in the naming paradigm. No item even approached this criterion. By contrast, all tool symbols were learned to this criterion, some quite readily.

Initially, the animals appeared to understand the principle of object naming. As long as only one or two objects were used simultaneously, performance was quite good. However, as additional objects were added and as the exemplars changed from trial to trial, performance dropped noticeably on both new and practiced names. In addition, when the animals began to do well within a session, considerable intersession forgetting persisted. The motivation and attention of the animals were poor, and it was frequently difficult to sustain more than cursory visual orientation toward the object to be named. The acquisition of object names in this manner appeared to be of an associative nature, readily subject to interference, confusion of alternatives, and intersession forgetting.

It is noteworthy that had a weaker criterion of initial acquisition been employed, such as *one* correct spontaneous usage per day for 15 consecutive days (Gardner & Gardner 1971), all object names, regardless of training paradigm, would have easily met it. The production of single names was typically accurate. However, when alternatives were repeatedly presented randomly, simple associations

Table I. Naming and Tool Use

	Naming paradigm				
	Number of trials[a]				
	Box	Bowl	Key	Blanket	Lock
Sherman	215	215	966	704	77
Austin	not attempted	not attempted	220	390	50

	Tool-use paradigm					
	Number of trials[b]					
	Wrench	Straw	Stick	Sponge	Money	Key
Sherman	59	9	88	21	70	49
Austin	54	55	41	34	21	20

[a] Until name-training on item was discontinued, due either to C's lack of progress or his lack of willingness to cooperate. Neither animal reached criterion with any object.

[b] From introduction of tool name until the tool was correctly requested on nine consecutive correct presentations with all tools randomly sequenced.

broke down rapidly and an item spontaneously named correctly on the first presentation would frequently be named incorrectly on the fourth or fifth presentation.

These problems were essentially nonexistent in the tool-use situation; learning was rapid and intersession forgetting was minimal. Randomly changing the problem while increasing the number of tools did not result in confusion with respect to previously learned tool names. Introduction of new tools did require closer attention to the specific physical properties of each of the preceding tools, however, and some confusion regarding the function of each new tool did invariably occur, with the animals attempting to use it inappropriately. Persistent errors could be traced to a misperception of, or an actual overlap among, the functions of various tools. For example, wrench-stick errors became common when Sherman discovered that a wrench vigorously shoved into a tube would occasionally succeed in extracting food that E placed therein. Thus, the wrench could, by virtue of a novel behavior, function as the stick was intended to function. Wrench-stick labeling errors persisted until the tube was modified to prevent this solution. As Table I shows, the total number of trials, or instances of use, required to learn and appropriately employ the tool symbols was quite small.

Although the names of the tools were well known when the animals reached the criterion of Table I, it is important to note that even at this point some mistakes continued to occur, particularly as other new words were introduced, or if several days lapsed without use of the tool name. Opportunities to employ these words were accordingly provided daily for two to three months beyond the time of reaching the initial acquisition criterion (of Table I) up to a point when performance remained almost error-free across days, and daily practice could be omitted without affecting accuracy as long as the staff working with the animals remained the same. (Changes in personnel inevitably resulted in performance decrement at all stages of training.)

PHASE II: SEPARATION OF THE TOOL FROM ITS FUNCTION: THE DEVELOPMENT OF NONFUNCTIONAL NAMING CAPACITY

This phase of training concentrated exclusively upon the testing and refinement of the relatively more abstract skills of naming and receptive comprehension and compliance, as divorced from the context of function. The correct response now became either simply naming the tool displayed by E, or selecting the tool requested by E. E's requests that C name a tool or that C hand E a tool specified by name were from the outset randomized across all trials. The tools were not used to obtain food. They were instead returned to the tool kit, and the animal received social praise or a token with which food could be obtained from a vending machine.

The purpose of this assessment and training was to ensure that the animals were able to respond to the physical-perceptual properties of the tool and to link these with the corresponding symbol, apart from the context of use. Without this ability, a word such as "stick" might function as a communicator, but only in a specific need-related context. In other words, the use of "stick" (or any other tool name) would not arbitrarily represent the properties of "stickness." It would instead be linked with the act of putting a long, narrow object into a hole, to dip or extract. In such a case, "stickness" would be a combination of something like a noun and something like a verb, and would be linked to the actual physical activity of stick use. If our tool names had been functioning at this level, the animals could not be expected to employ them to communicate with one another, because the lexigram words would only be functioning as performatives (Greenfield & Smith 1976), with their use limited to occasions when food was seen at a given tool site.

The receptive capacity to decode the statements of other individuals, to comprehend and to comply with those statements, is also a necessary skill for interanimal communication. Being able to use a word does not automatically imply being able to understand it when it is used by others. Even if the recipient does comprehend the referent of the word, he may not understand that other individuals can also use it, and that he must attend, comprehend, and alter his behavior in a cooperative fashion if such communication is to be productive. This part of the language acquisition process has been overlooked or taken for granted in much of the chimpanzee language work, because interest has centered upon what the animal could say, not what it could do in response to statements of others. However, receptive capacity is critical to any complete form of language use, and it is obviously an essential prerequisite to interanimal communication.

The testing of simple naming capabilities was conducted by holding up a tool and asking through the keyboard, "What this?" E pointed toward the tool and then gestured toward C's keyboard to encourage him to reply. The animals had previously been asked to name foods and photographs of food, and they responded to the question as though they understood that a naming or designating response was required. When confronted with this task, Sherman's initial replies were tool names, although the names he used did not coincide with the tool being displayed. He tended to pick a tool name randomly or to cycle through all tool names, as though attempting to guess the correct alternative. He did not attend closely to the displayed tool, but instead seemed to interpret the act of holding up a tool as a signal to guess again. Austin, however, did attend to the tools closely from the beginning and seemed to comprehend that a correspondence was to be drawn between the tool displayed and the tool to be indicated at the keyboard. It is unclear why Austin initially generalized from food-naming skills more rapidly than Sherman. However, as indicated in Table II, Sherman spontaneously began assigning appropriate tool names after the first 96 trials. There was, at this time, a distinct change in his performance, reflected in his careful observation of the tools prior to

Table II. Naming and Receptive Skills with
Tool Symbols after Functional Training

	Proportion and percent correct			
	Sherman		Austin	
Naming				
First 96 trials[a]	67/96	70%	85/96	89%
next 69 trials	67/69	97%	64/69	93%
Receptive				
First 73 trials	24/73	33%	59/73	81%
next 57 trials	45/57	79%	48/57	84%

[a]The 165 naming trials and 130 receptive trials have been
partitioned (96:69 and 73:57, respectively) in terms of the
points at which Sherman's performance radically
improved. In contrast to Sherman's performance, Austin's performance was at a high level from the beginning
of both tests.

responding. During this naming task, all tools were always present in front of C,
who was to name the tool that E indicated by pointing or holding. The test for
receptive skills was similar to that for naming skills, in that the tools were not
employed for any specific purpose. By means of C's keyboard, E would state (for
example) "Give sponge," and then direct C's attention toward the tool kit, encouraging him with pointing gestures to hand over the named tool. The animals had been
asked to proffer food items and photographs of food items in the past, and they
readily comprehended that they were being requested to hand an object to E. As
had occurred with the tool naming, Austin attended closely to E's statements and
did well from the beginning of this task. Sherman again experienced initial difficulty,
tending to cycle through various tools regardless of the one indicated at the keyboard. His behavior again altered quite suddenly, after 73 trials on this task, and the
altered response performance was again accompanied by a noticeable orientation
of the attentional response and the emergence of a careful search pattern. The type
of difficulty and recovery pattern exhibited by Sherman has tended to occur with
all the chimpanzees on the project whenever they are confronted with a task that
they do not yet completely understand, but already possess the requisite skills to
solve, once they do achieve understanding.

Daily practice on the naming and receptive tasks continued, as did practice on
the functional task, until the animals used these names in an almost errorless fashion across a number of days. Practice on all tasks ran jointly for approximately one
month and was accompanied by a variety of other activities (tickling, naming of
slides, going to various locations, etc.) also discussed at the keyboard, but not relevant to this study.

"Blind" tests were then administered to ensure that C's ability to request, name, and choose tools requested by E was not in any way cued by E. If it was, interanimal communication regarding tools would surely fail. The blind tests were conducted in a slightly different manner for each task. In the functional task, E stood outside C's room so that he was not visible from the keyboard; he observed C's request on projectors outside the room and then handed C any tool requested. In the naming task, E again stood outside the room and held up a tool so that it was visible to C through a lexan wall, although neither C nor E could see one another. C's responses were again monitored on projectors located outside the room.

During the test of receptive skills, E_1 sat in the room with his back toward the keyboard and projectors, while E_2 stood outside the room out of C's view and used the keyboard to indicate, through the projectors, which tool was to be chosen. C then noted this information on his projectors, chose a tool from the tool kit, and carried it to E_1. This choice was verbally reported by E_1 to E_2, who recorded the score. Both animals did very well on these tasks (Table III), thus demonstrating that their abilities were not dependent upon cues from E. It should be noted that the scores reported are not based on multiple administration of these tests; each test was administered only once.

Although the blind tests in Table III demonstrated that the animals were extremely competent with the use of these words in these contexts, such tests do not reveal the conceptual stages through which the animals progressed during their acquisition of these words. A brief description of this acquisition process is appropriate at this point, so that the reader will not be left with the impression that the acquisition of these words was merely a matter of repetitive practice.

When training described above began, C's vocabulary consisted of food and drink names and a few simple verbs ("give," "tickle," etc.). Once the first tool word, "key," was learned, it was used whenever C wanted to obtain food that was not directly accessible. In other words, the food would be named or requested if it was in the direct possession of E, whereas the word "key" would be used when

Table III. "Blind" Tests

	Proportion and percent correct	
	Sherman	Austin
Functional task[a]	29/30 = 97%+	29/30 = 97%+
Naming task[b]	30/30 = 100%+	30/30 = 100%+
Receptive task[c]	27/30 = 90%+	27/30 = 90%+

[a]C requests tool with E out of the room. E reads request on projectors outside the room.
[b]Question posed through projectors. Object to be named held by E, who is outside the room and cannot see the C.
[c]E_1 informs C of requested tool through symbols on projectors. Tool is selected by C from tool kit and is given to E_2 who does not know which tool was requested.

food had been placed in an adjacent locked room. With the introduction of stick, the second tool word, "key–stick" errors appeared often, although "food–tool" errors almost never occurred. The animals now had two words to use when food was not directly accessible, and at first they tended to confuse them. Subsequently, however, they distinguished these words, as well as additional tool words, by using "key" whenever food was in the adjacent room and "stick" whenever food was in a container just beyond a hole in their own room. "Key" and "stick" became the food from a particular location. Consequently, even when C could correctly request all his tools, he could not name them. In fact, he gave no indication of understanding what the request to name his tools meant.

When a second tool site and a second tool function were introduced for each tool, C could select the correct tool from his tool kit, but could not request it at the keyboard. For example, when the key was first used to open a box instead of a door, both Sherman and Austin chose the key from their tool boxes and tried to fit it into the lock on the box. However, when they were required to request the key from E, they could not do so, because "key," at that point, represented what was said when food was in the adjacent room, not in the box. In other words, the semantic properties of the word "key" were initially linked to the location of food in space, not to the functional properties of *keyness*. Once each tool had been used in a second site, however, the meaning of "key" began to focus on the object itself as the single common element, because the location was no longer redundant with the chosen tool. At this point, more tool sites and tool functions were introduced, and the proper tool was requested immediately via the keyboard. Another chimpanzee, Lana, did not experience such a difficulty. She began to learn tool names after she had learned many other language tasks (Rumbaugh 1977); this permitted her to approach the acquisition of tool names with more sophistication. She immediately transferred tool requests to other sites, even sites which had to be recalled from memory while the request was made. She was also able to name all her tools correctly when first asked to do so.

Thus, as tools came into use for diverse functions in different locations, with all settings linked by a common symbol, the properties covarying with these symbols were sorted out from the rest of the environmental events associated with the tools. At this point it became possible to attempt to require C to name the tools apart from their use, since the symbol was now more closely associated with the item itself. Whenever C experienced naming difficulties, as he often did initially, E could help him recall the symbol by demonstrating *in vacuo* the use of the tool. For example, if C had trouble recalling "wrench," E could demonstrate that it unscrewed bolts, although there was no food to be obtained by this demonstration. While this type of prompting did not need to be continued long, it helped to bridge the gap between these two very different ways of using these words. Because different ways of using the same words were often perceived as totally different tasks by the animals, this type of link was important. Once this level of comprehension was reached, the animals readily used these words for a variety of keys with a variety of uses (Boysen & Savage-Rumbaugh 1978).

PHASE III: INTERANIMAL COMMUNICATION

When the animals had acquired these individual skills, we wondered if they would be able to combine them in order to communicate to one another the need for a particular tool. They were now capable of asking us for the proper tools, and if we requested one, they could search among their own and find the correct one. The question now became: if only one chimp had access to tools, but could not obtain food with them, and if the other one could obtain food but had no tools, could they perceive the necessity of requesting tools from one another? Would the animal who had access to the tools watch the actions of the other at the keyboard? Would they willingly hand objects back and forth? Would they understand that they could make symbolic requests to one another, as well as to their human experimenters? These were important language adjuncts and communicative coordinators, which they had not had to learn as long as they communicated with human beings who always attended closely to all that was said and who consistently responded and provided food and tools rather than absconding with such things for themselves. If these animals could attend to one another, coordinate their communication, and exchange roles of tool-requester and tool-provider—if they could comprehend the function and *intentionality* of their communications and, through joint symbolic communication, *share* their access to tools and the food obtained through tool use—then, by all definitions of human culture, they would surely have taken a large step.

If only one animal were given access to the tools and the other animal were given access to a food source requiring a tool, then it would be necessary for the animals to communicate, provided the animal who needed the tool could not simply walk over and take it. The latter solution was eliminated by placing the animals in separate rooms with a large window between them. Both rooms were equipped with keyboards and projectors that were clearly visible from the window of the adjacent room. A cover could be lowered in front of the window to cut off vision while food was placed at various tool sites in one room without the knowledge of the animal in the adjacent room. The window could then be uncovered to permit communication between the chimpanzees. All tool sites were painted with opaque paint so that the chimpanzee in the other room could not simply look through the window, see where the food had been placed, and infer that a particular tool was needed. This also meant that the tool requester had to remember where he had seen food placed, because once the food was deposited at a given site (for example, in the bottom of a long hollow tube), it was no longer visible.

One animal was chosen as the requester (C_r) of the tool, and food was placed in his room with the window between the rooms covered. The window was then uncovered and the animals were permitted to use their keyboards in any fashion they chose. Initially C_r, who had seen the food placed in his room, readily requested a tool. However, the first few requests were directed, by gaze pattern, toward E and not toward the tool provider chimpanzee (C_p). E responded by demonstrating to C_r, with an open-hand gesture, that he had no tools; then, by pointing, he drew C_r's attention to C_p in the adjacent room, who did have the tools. Often, C_p would initially observe the request and even pick up the correct tool, but would drop it and begin to play once it became evident that C_r was directing the communication toward E. However, after five or six trials the animals seemed to begin to comprehend that they could request tools from one another and that C_p would comply with requests. Initially E encouraged the cooperative behavior by gesturing in order to direct one C's attention to the other. By the second day the animals quickly anticipated the nature of the task and their respective roles in it. (Roles had been systematically alternated from the outset.) C_p began to rush hurriedly to the window to observe the request of C_r, who would look at him to see that his request had been noted. Often, if C_p appeared inattentive, C_r would draw his attention to the request by repeating it or by pointing to the projectors where the requested tool was displayed. Initially, E would give some of the food obtained with the tool to C_p; however, C_r eventually came to do this on his own.

The performance of the animals during the exchange of various tools is shown in Figures 1 through 6. The results, presented in terms of groups of trials per day, are shown in Table IV. *All* trials during all instances of interanimal communication concerning tools are shown in this table. The sequence for the baiting of all tool sites was randomly generated daily by the computer. Even on the first day, joint communication was attempted, and the performance of the animals was far above chance. Chance performance with six tools, any of which could be either erroneously requested (1/6) or provided (1/6) would be (1/6 × 1/6) or 2.8%. As Table IV indicates, the accuracy of the animals increased steadily across days. The rapidity with which their accuracy rose, coupled with their previous scores on blind tests of naming, functional, and receptive skills would suggest that the improvement in performance was not due to increasing symbol facility, but rather to a refinement of their ability to (1) reverse communicative and cooperative roles, (2) attend to the communications of another chimpanzee, and (3) direct toward one another communications coordinated in time and space with the other's readiness to receive those messages. Prior to this point, they had never observed one another use tools or employ tool symbols to request tools, and thus had no reason to presume that the other animal knew and used such symbols or would cooperate with requests for tools.

As Table IV indicates, without E present during the request and tool-transfer portions of the task, performance remained unchanged. Earlier, E's presence had

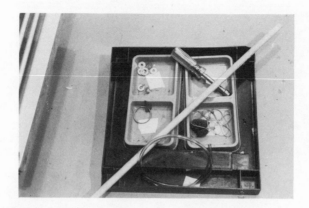

Figure 1. Tool kit: money (washers), socket wrench, key, stick, sponge, straw (plastic tubing).

been necessary to help both Cs to coordinate attention and action and to remind them of the task at hand. Without this help and structure they tended to play and to become easily distracted, like preschool children.

The increase in accuracy also suggested that perhaps the animals had learned nonverbal signals during this period that enabled them to transmit information through a channel other than the symbols on the keyboard. Perhaps their use of the keyboard merely reflected the continuance of behaviors that they had been conditioned to emit by E, although these behaviors were no longer functional in the communication. Observation of trials on which Cs were in error suggested that this was not the case, however. For example, on one trial Sherman requested "key" erroneously when he needed a wrench. He then watched carefully as Austin searched the tool kit. When Austin started to pick up the key, Sherman looked over his shoulder toward his keyboard, and when he noticed the word "key," which he had left displayed on the projectors, he rushed back to the keyboard, depressed "wrench," and tapped the projectors to draw Austin's attention to the new symbol he had just transmitted. Austin looked up, dropped the key, picked up the wrench, and handed it to Sherman.

These kinds of observations were verified by posing the same problem for the animals with the keyboards turned off, so that any information transfer regarding the needed tool had to occur through some other medium. On the first trial of this control test, Sherman, who needed a sponge, rushed to the keyboard and tried once to press sponge. When it did not light, he then went to the window and looked at Austin. Austin did nothing. E pointed toward Austin's tool kit and encouraged him to hand Sherman a tool. Austin handed him a straw. Sherman grabbed the straw, stared at E, held the straw toward her, shook it vigorously, and threw it at her feet. The keyboard was then turned back on and the correct tool was immediately requested and given. This procedure continued for the remainder of the control

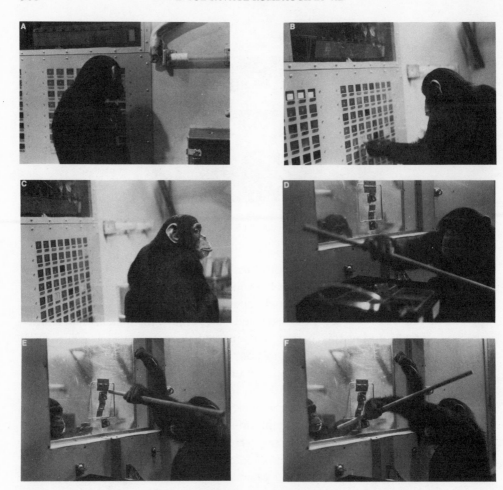

Figure 2. A: Austin looks on as food is placed inside a long, narrow tube. He will need a stick to push it out, as it is inserted beyond the reach of his finger. (Sherman's view to this point is blocked by a cover over the window.) *B:* Austin goes to the keyboard and states "stick." (The implied verb, "give," is often dropped when Sherman and Austin communicate with one another.) *C:* Austin looks to see if Sherman is attending. *D:* Sherman (right) is selecting the stick from the tool kit, so Austin (left) approaches. *E,F:* Sherman passes the stick through the open window to Austin.

test. The animals' performance on trials with the keyboard turned off was only 10% correct, while on trials with the keyboard turned on it was at a high level: 97% correct. Clearly, the keyboard symbols served a critical role in the transfer of information regarding the necessary tool.

On trials without the keyboard, the animals adopted a strategy of either cycling through all the tools or repeatedly offering the same tool. It did not appear that they were attempting to innovate gestures or other means of communication, although

Figure 2 (cont.). G: Austin orients the stick toward the end of the tube. *H:* Austin inserts the stick in order to remove the food. *I:* Sherman (right) watches Austin closely through the window. *J:* Austin gives Sherman a portion of the food obtained with the stick (see text for description of controls).

they would perhaps do so if this type of situation continued for a longer period of time. Iconic gestures have been devised repeatedly by the animals, and by the experimenters, as an adjunct to the abstract symbols available on the keyboard. The use of such gestures often serves as an intermediate link between symbol and event.

INTERANIMAL COMMUNICATION AND ITS SIGNIFICANCE IN THE PERSPECTIVE OF EARLIER APE–LANGUAGE STUDIES

It is not possible to place the present study in proper perspective relative to prior research in this field without first reviewing briefly the quality and content of the earlier work. This is because, in the past, tests showing that chimpanzees can do certain things suggestive of linguistic capacity have tended to be controlled for cueing only in the simpler cases, while the more complex skills have typically been reported without any such controls at all. We view this as a major deficit in work with apes.

Figure 3. A: Sherman watches as food is dropped into a box. The box is then padlocked, and Sherman will need a key to open it. (Austin's view to this point is blocked by a cover over the window.) *B:* Sherman states "Give key." The symbols for this statement are visible on the projectors above the keyboard. *C:* Austin watches ánd reaches into the tool kit for the key. *D:* Sherman (left) approaches and Austin hands him the key. *E:* Sherman goes back to the box and inserts the key into the padlock. *F:* Sherman then twists the key to open the lock.

Man became fascinated with the communicative potential of apes soon after these animals first began to appear in Europe (La Mettrie 1748—published in 1912) and America (Yerkes 1943). The important studies by Furness (1916), the Kelloggs (1933—see Kellogg & Kellogg 1967), and the Hayeses (1954a) were all motivated by such an interest and helped to clarify a number of problems associated with teaching language to apes. These studies demonstrated that vocal modality was extremely difficult for apes to use because of insufficient voluntary control over

Figure 3 (cont.). *G:* Austin watches carefully as Sherman opens the lock. *H:* Sherman (left) takes half of the food back to Austin (right). He tastes Austin's portion on the way. *I:* Sherman (left) passes part of food to Austin (see text for description of controls).

vocal output, and because their vocal repertoire was not well suited to the production of many human speech sounds. These initial conclusions regarding the inappropriateness of the vocal modality were later confirmed by du Brul (1958) and by Lieberman, Crelin, and Klatt (1972).

The studies by Crawford (1937, 1941) and Hayes (1954a) revealed something even more significant: the existence of a fairly elaborate gestural communicative capacity. The gestures they reported were complex and apparently intentional, and it can be surmised that had these researchers been innovative enough to translate these gestures into ASL signs and to attribute intentionality to the ape, the sign-language capacity of the chimpanzee would have been discovered many years before the report by Gardner and Gardner (1969). For example, the Hayeses (1954a) reported that Viki

places our hands on the objects she wants us to manipulate and often moves our hands in a manner suggesting the action to be performed. For instance, if she wants to go outside, she leads us to the drawer where the door key is kept and places our hand on the drawer pull. If we don't open it promptly, she gives our wrist a tug. When the drawer is open, she puts our hand to the key, and when we grasp it, she moves our hand to the key hole. If we continue to lag, she moves our hand until the key enters the key hole and finally twists our wrist to

Figure 4. A: Austin observes banana slices being placed behind a small, bolted door. (Sherman's view to this point is blocked by a cover over the window.) *B:* At the keyboard, Austin asks, "Give wrench." *C, D:* Austin approaches the window as he sees Sherman pick up the tool. *E:* Sherman (right) hands the wrench through the window to Austin. *F:* Austin carefully places the wrench on the bolt. *G:* Austin then turns the wrench to unscrew the bolt.

indicate the unlocking movement. . . . This kind of communication may be said to involve "iconic signs" whose meanings are related in an obvious and logical way to their physical character. . . . We have observed some behavior in Viki which suggests that chimpanzees may readily convert such signs into "symbols" whose meanings have an essentially arbitrary relationship to their physical character. (Hayes & Hayes 1954a: 229)

Many of the signs used by ASL-trained chimpanzees are also iconic in nature (Fouts 1974; Gardner & Gardner 1971; Patterson 1978; Terrace & Bever this vol-

Figure 4 (cont.). *H:* Sherman watches closely as Austin works to open the door. *I:* After obtaining some food for himself, Austin goes to the window. *J:* Austin hands food through to Sherman (left) (see text for description of controls).

ume), and it is often difficult to ascertain from reports alone whether the behavior of these chimpanzees is in fact significantly different from Viki's, or is merely the writer's interpretative ASL-to-English gloss, which creates a difference in the mind of the reader. The repertoire of hand motions is definitely larger in ASL-tutored apes, but virtually all reports of informative messages are anecdotal in nature (Mounin this volume) and are rendered contextually appropriate by rich interpretations made after the fact.

Project Washoe

Is there sufficient evidence to conclude that chimpanzees tutored in ASL comprehend the nature, function, and symbolic power of the ASL signs that they use? This question has been raised repeatedly in numerous forms by scholars from a variety of disciplines (Bronowski & Bellugi this volume; Mounin this volume; Limber this volume) since the first reports of Washoe's progress were released (Gardner & Gardner 1969). The issue is complicated by the fact that a grasp of the basic nature of the signing chimpanzee's communicative ability tends to elude all who

Figure 5. A: Sherman watches as juice is poured into a long tube. He needs a sponge tied to a string to obtain the juice. (Austin's view to this point is blocked by a cover over the window.) *B, C:* Sherman goes to the keyboard and asks, "Give sponge." *D:* Sherman looks toward Austin while finishing his request.

have not worked and interacted extensively with such animals. This is because the reports by the Gardners and others have repeatedly remained vague on many important points. We strongly agree with Mounin (this volume) when he states:

> in order to have communication (in the linguistic sense of the term specifying human communication), (a) there must be someone transmitting and someone receiving, (b) the transmitter must be aware that his target is the receiver, (c) the receiver must be aware of being the transmitter's target, and (d) the receiver must be capable of becoming the transmitter by using the channel (as a rule, using the same code or one equivalent to it). . . . (Mounin this volume: 166)

The writings of Gardner & Gardner (1969, 1971) suggest that the criteria listed by Mounin were indeed met by the relationship between Washoe and her trainers; however, they present no evidence, other than anecdotal, in support of this. We have argued elsewhere (Savage-Rumbaugh & Rumbaugh, in 1979, in press) that the

Figure 5 (cont.). E: Austin (right) has picked up the sponge, so Sherman approaches and Austin gives it to him. *F:* Sherman inserts the sponge into the tube. *G:* After squeezing out some juice, Sherman dips the sponge into the tube again and takes it, filled with juice, to Austin (see text for description of controls).

ability to name pictures and the ability to produce an iconic gesture repeatedly, under the same or similar circumstances, are not evidence of symbolic communicative capacity. The issue of iconicity is an important one, not because an icon cannot function as an arbitrary symbol, but because, given the weak acquisition criteria employed by the Gardners (1969) and others who work with signing apes (Fouts 1974a; Patterson 1977), it is impossible to tell whether the chimpanzee is simply imitating or echoing, in a performative sense, the action or object, or whether the animal is indeed attempting to relay a symbolic message. The high degree of iconicity in ASL (Bellugi & Klima 1976) could permit the animal to recall the proper hand motion simply by observing either the object or the form of the action. If the animal has previously been encouraged to emit such iconic hand movements in the presence of a wide variety of objects, then "spontaneous" signs can be interpreted more parsimoniously as a preparatory "set" to gesture when attention is drawn to those objects or actions. The object/action itself would then

Figure 6. A: Austin observes juice being poured into a container outside the room. (Sherman's view to this point is blocked by a cover over the window.) *B:* Austin states, "Give straw" at the keyboard. *C:* Austin (left) approaches Sherman, who is sitting by the tool kit in the adjacent room. *D:* Sherman picks up the straw and hands it through the opening to Austin.

prompt the particular sign because of the iconic relationship between the sign and referent. As Steklis and Harnad (1976) so aptly point out:

It is not the act itself, the sign or manual gesture, which is the proposition. Initially, a gestural act may just be a short-circuited iconic sequence, which it makes no sense to deny [or disbelieve] as long as it continues to retain some isomorphism with its referent and the referent is present for the verification . . . As soon. . . as the gestures are divorced from their immediate referents and are intended and relied upon to transmit information about those referents, *propositionality* is involved. (p. 451)

We believe that there is no evidence, other than richly-interpreted anecdote, to suggest that Washoe and other signing apes are producing anything more than short-circuited iconic sequences.

Further interpretative difficulties arise because the correct message is never specified in advance, and it is unclear whether or not the trainers have recently used a similar sign or series of signs in the same context. Also, the animals are rarely required to use signs of related semantic content (such as a variety of specific food names) across a series of randomly-sequenced trials. Can Washoe, for example, repeatedly name an apple, a drink, an orange, and a banana, while eating ice cream?

Figure 6 (cont.). E: Because the liquid in the straw cannot easily be passed to Sherman, he is allowed to enter Austin's room and use the straw with Austin. *F:* Austin (left) taps Sherman's hand to signal that it is his turn. *G:* Sherman passes the straw to Austin and watches closely as Austin finishes the juice (see text for description of controls).

Table IV. Interanimal Communication re Tool Transfer

Day	1[a]	2[a,d]	3[a,d]	4[a]	5[b]	6[c]
		Proportion and percentage correct				
Total correct	32/47 68%	38/50 76%	28/40 70%	27/30 90%	55/60 92%	3/30 10%
Sherman	36/47 77%	45/50 90%	32/40 80%	28/30 90%	57/60 95%	2/30 7%
Austin	43/47 91%	41/50 82%	34/40 85%	29/30 97%	58/60 97%	1/30 3%
Requesting errors[e]	12	9	8	1	3	
Providing errors[e]	3	5	6	2	2	27

[a]Experimenter not blind.
[b]Experimenter blind or absent.
[c]Deactivated keyboard.
[d]Due to two instances of errors in both requesting and providing on the same trial, error scores are correspondingly higher than those reflected in total scores on these days.
[e]Errors represent totals for both chimps, since roles were systematically alternated.

Can she do this when there are additional controls for trainer cueing, and when the actions of peeling the orange, peeling the banana, and drinking are not demonstrated, and hence cannot remind her of the iconic hand motions involved in the sign (Savage-Rumbaugh & Rumbaugh 1979)? These kinds of competence issues have not been considered by the Gardners. While competence tests of this sort would not assure that Washoe was extending the meaning of a food sign and applying it to a novel context, such tests would at least reveal that Washoe could competently use the sign apart from the context of the original acquisition. By contrast, the Gardners simply state that signs occur in a new context. There is, however, no evidence that such occurrences cannot be accounted for by deferred imitation, or error interpreted in a novel manner by the experimenters.

Thus, the question of whether signing chimpanzees (or any chimpanzees who have learned an arbitrary symbol system) comprehend the nature, function, and symbolic power of the symbols they use becomes a question of *awareness* and *intentionality*. These are issues which psychologists have long sought to avoid, and it is perhaps concentration upon performance, as opposed to competence, which led the Gardners and others to conclude that sign use could be equated with semantic comprehension.

The questions of intentionality and awareness are now being reopened, as more individuals in a variety of fields attempt to deal with nonverbal purposive behavior, behavioral linguistic precursors displayed by human children, and pragmatic aspects of language itself (Bates 1976; Bruner 1975; Glasersfeld 1976; Greenfield & Smith 1976; Griffin 1976, 1978; Premack & Woodruff 1978; Grice 1969; Lyons 1972; Mounin this volume).

Project Sarah

Criticisms of the Gardners' work have, in the past, often been refuted by citing Premack's (1976a) findings (with the chimp Sarah) and vice versa (Hewes 1971; Sarles 1976). Thus, where there is an absence of sound experimental evidence for one ape, the training of another is cited to strengthen the argument that apes as a species are competent symbol-users. Setting aside the logical weaknesses of such arguments, one might ask whether Sarah has given better evidence than Washoe of comprehending the communicative nature and symbolic function of her signs, which consist of plastic chips. Unlike the Gardners, Premack (1972, 1976a) has not ignored this question. He has first of all taken the approach that it is as impossible to determine what the plastic chips really mean to Sarah as it is to determine what words really mean to a human being. This position, then, enables him to attempt to demonstrate that Sarah does comprehend the symbolic function of at least two words, and then to imply, by analogy, that she probably uses the rest of her vocabulary symbolically.

The main evidence for Sarah's symbolic representational capacity comes from a single match-to-sample paradigm with the word "apple," followed by one replication with the word "caramel." The replication differs from the original test in terms of a counterbalanced presentation order for word and object. Sarah's first task was to choose (with an apple present) from two items, the one that most closely matched the real apple on the basis of color, shape, presence or absence of a stem, and shape versus presence of a stem. After Sarah had made these four choices, the apple was removed and a blue triangle (Premack's symbol for "apple") was placed in front of Sarah. She was again given the same four choices, and she chose as she had when the apple was present.

Premack states that it is unlikely that Sarah's choices with the blue triangle present were controlled by her ability to recall the choices she had made when the apple had been present. However, this type of recall is well within the capacity of the chimpanzee. In a test of this specific possibility, Lana (one of our chimps) has demonstrated an ability to recall her choices on ten consecutive one-trial discrimination problems with 100% accuracy. Thus, Sarah should have been able to recall four such choices. It is important to note that physical matching of the sample (the blue triangle) with one of the alternatives was precluded because no blue objects or triangles were provided. Thus, it seems reasonable to conclude that if Sarah could no longer solve the problems by matching, she would return to her original choices, particularly since she had been rewarded for all of them. Had Premack included the control of a blue or triangular object for use in matching, and had Sarah still chosen the alternative most like a real apple, then there would be reason to conclude that Sarah was responding in a representational manner.

Premack repeated this test with the word "caramel," presenting the word first (instead of the real candy) in the initial matching problem "to rule out the unlikely possibility that on the former test she had simply memorized the alternatives for apple and then applied them to the (word) 'apple' (1976a: 171)." However, Premack does note that iconicity existed in some symbols, so it would be important to determine whether the shape and color of the symbol "caramel" (which were not reported), might have allowed Sarah to match on a dimension of physical similarity. In addition, the data for Sarah's performance on the most critical trials, the first presentation of each of the four alternatives with the word "caramel" as the samples, are not analyzed separately. It is reported that, with the word as the sample, she made five errors, but not whether any of these errors occurred on the first four critical trials. After these first four trials, Sarah could presumably have solved the problem by matching the real food to one of the alternatives and then successfully dealt with all further presentations by recalling her earlier choices.

These two tests have been interpreted repeatedly as evidence for symbolic-representational capacity in the chimpanzee (Cohen 1977; Mistler-Lachman & Lachman 1974; Mounin this volume); however, the above-described methodological shortcomings and lack of controls (e.g., for cueing), as well as the fact that both studies together involved only eight pairs of items, suggest that the decisive tests for symbolic capacity have yet to be performed. Questions as to whether or not such a capacity is necessary to solve the various problems could and should be raised for the entire gamut of tests presented to Sarah. The suspicion lingers that perhaps

the illusion of language [could] come from the fact that the coded messages are always "paraphrased" in English. Could the meanings of all Sarah's messages not be reduced to two: (a) if I put two, three, . . . eight well-defined plates in such an order, I will receive a fruit, a chocolate, or a candy, etc.; (b) if I put two, three, . . . eight well-defined plates in any order other than the first, I will not receive a fruit, a chocolate, or a candy, etc.? (Mounin this volume: 174).

However, since present space does not permit a thorough consideration of all the work done to date, a brief discussion of the nature of Sarah's communicative behavior must suffice. Does Sarah give evidence of comprehending that she is communicating with her teachers or that they are communicating with her, as opposed to simply solving a set of problems? Communication is a universal feature of language, but it was not initially the direct object of interest in the studies on Sarah. Her ability to communicate was clearly limited by the plastic chips available to her at any one time. Since she only worked on one problem at any given point, she did not have the option of communicating anything other than a correct or incorrect answer to that problem. This makes it difficult to understand how Sarah could come to realize that the plastic chips could be used to communicate desires and to control or orient the behavior of others. It is also not clear that she viewed her trainers' use of these chips as a means of transmitting information or controlling her behavior.

   Premack (1976a) attempts to demonstrate that Sarah can transfer readily from the productive to the receptive or comprehensional mode, thereby implying that she understands the communicative value of the symbols and the communicative intent of the trainer sufficiently to permit her to reverse the roles of sender and receiver. In one task, Sarah was asked to "take" one of five foods for which she was reported to have names. She had previously been required to use those food names in a problem task in which she chose from among two or more food symbols to complete the sentence "Mary give X." She was never required to switch back and forth between these tasks. It is important to note that in both types of sentences (*give* versus *take*) Sarah's main objective was to obtain food. In the first case she had to present a symbol that matched the food in the trainer's hand; in the second, she was to eat the food indicated by the symbol already presented. Since her goal (to eat the food) was similar in both cases, there is no reason to believe that she comprehended that the roles of receiver and transmitter were being reversed. It is more reasonable to conclude that she was simply aware that different responses were required of her in order to obtain food. This fact should have been apparent to her, for in one task she was presented with plastic tokens and no food, and in the other, she was presented with no tokens and a choice of foods. In addition, the trainer indicated by gesture that Sarah was not to take just any food; this was accomplished by passively guiding Sarah's hand through the proper motions. There is no communicative role-reversal required in this task. Sarah only needed to recall an association between the plastic chip and the food to solve the problem. We have found that such simple associative connections are formed at a fairly primitive level of "wordness" and do not require comprehension of either the symbolic or communicative value of a word (Savage-Rumbaugh & Rumbaugh 1979). (It should once again be noted that no controls for cueing were employed. In the receptive task, cueing would be very likely to occur, since the chimpanzee would tend to move her hand toward a food and then look to the trainer for permission to take it, if not for cues as to correctness.)

Thus, while there are abundant descriptions of very interesting and suggestive phenomena, there are few firm data, collected in controlled, blind-test situations, to support the contention that either Washoe or Sarah are employing symbolically-mediated, abstract communication similar to that involved in human langauge.

CONCLUSION

In contrast to the single-animal approaches described above, which were basically designed to demonstrate that chimpanzees *could,* through use of a nonverbal medium, acquire something like a human language, the goal of the present project was quite different. Our main emphasis was on (1) the value of the chimpanzee as an animal model for the investigation of basic language acquisition processes, and (2) the application of these findings to a companion project concerned with mentally-retarded children (Parkel, White, & Warner 1977; Parkel & Smith 1978; Parkel 1977). It was clear, therefore, from the outset that effort was to be primarily directed toward the development of functional, symbolic communication between animals. Not only did such skills have to be demonstrable to a degree sufficient to conclude that they were not based on chance, but they had to be reliably deployable between animals. This called for levels of comprehension, receptive sophistication, role-reversal, and cooperativity not previously required in ape language research. While earlier criteria were perhaps adequate for demonstrating that the animal was not responding by chance or solving the task solely through simple conditioning, they were not sufficient for the purpose of elucidating the processes that *were* employed or for determining the level of mastery.

Techniques for ascertaining the existence of reliable comprehension and for assessing the receptive and productive use of individual symbols have proven essential to this work. These techniques have also shown that earlier criteria for single-word acquisition are far from sufficient, and they raise significant questions concerning prior claims of novel sentence constructions by chimpanzees. Chimpanzees have certainly combined symbols in ways that seem interpretable to human beings. But since we do not understand exactly how children come to combine words, it has been assumed too often that we must likewise remain agnostic as to how chimpanzees come to do this and thus, that all we can do is record and present observed word combinations. The fact that the chimpanzee's comprehension of single words may be quite different from that of the child, due in part to weak experimental criteria for determining wordness, suggests that chimpanzees may come to string signs together for reasons quite different from those of the child. Thus, without a better understanding of the phenomenon of single-word acquisition, the interpretations of "novel combinations" remain intrinsically speculative.

A number of significant factors in initial language acquisition and use by chimpanzees have been demonstrated in the present study. First, as found in earlier work with the food-oriented vocabulary (Savage-Rumbaugh & Rumbaugh 1978), word acquisition was not found to be an all-or-none phenomenon, in the sense that a word is either known or not. Rather, there appear to exist various levels of "wordness," or competence with particular symbols; these reflect the chimpanzee's comprehension of a given symbol and of communicative symbol-use in general. Simple usage criteria prove to be inadequate to distinguish these various levels of comprehension (Rumbaugh & Savage-Rumbaugh 1978). Such criteria are also inadequate for determining receptive capacity. One must look at the ways in which the animal *cannot* use symbols, as well as the ways in which he can: the initial limitations of the chimpanzee's concept and use of tool names in the present study have provided some insight into the factors that determine and promote evolving, complex forms of word usage. Similarly, since productive ability alone was not found to promote interanimal exchange, the importance of the dynamics of communicative role-reversal and receptive capacity was emphasized and clarified.

The behavior of the animals at every point in training has made apparent the links between the prelinguistic, sensorimotor understanding of various forms of action and communication, and the symbolic encoding of those actions, objects, and desires. It was necessary that the chimpanzees understand how a sponge worked, and that they use it motorially, before they could reliably attach a symbol to it. It was also necessary that the functions of a sponge be differentiated from those of other tools which could also be used to obtain liquids, and that the sponge be used in a variety of locations, so that the invariant properties of "spongeness" could become motorially and perceptually distinguished from its surroundings. It was also important to observe other individuals using a sponge, and to share in the liquid thereby obtained, in order to appreciate the need to cooperate with the symbolically-encoded requests of others. From this type of interanimal communication, there have emerged spontaneous sharing of resources between animals, spontaneous requests for food (Savage-Rumbaugh *et al.* 1978) and objects of interest, and spontaneous cooperation with such requests. Prior to the training that promoted learning of tool function and cooperative sharing of the obtained foods, mutual requests for aid were not observed between the animals. They did not use gestures to request that another animal open a lock, insert a stick into a hole, go into another room and retrieve a discarded tool, etc. Now, they regularly employ prelinguistic gestures of this sort quite spontaneously in their interactions with one another.

This observation suggests that "language instruction" is too simple a term to account for the learning, for it is not merely proficient and accurate symbol-use that the animals seem to have acquired. Rather, a concept of role-reversal and cooperative, communicative behavior directed toward a common goal seems to be emerging. The relationships among internal representation (symbolization of objects), tool-use, and interanimal communication observed in this study suggest that the coordinated emergence of these skills in human evolution may not have been accidental. Instead, as Holloway (1969) has suggested, they may have been a set of interlinked abilities functioning in concert to promote a greater and greater symbolizing capacity.

IS PROBLEM-SOLVING LANGUAGE?*

H. S. TERRACE

Opinion has shifted during the last decade about the ability of chimpanzees to learn language. Recent projects have reversed earlier failures to establish communication between man and chimpanzee by bypassing the troublesome (for chimpanzees) vocal medium of language. Nevertheless, just what a chimpanzee can learn about language remains controversial. Through the media of American Sign Language (e.g., Gardner & Gardner 1969) and the artificial visual languages invented by Premack (1970a) and by Rumbaugh and Glasersfeld (1973), chimpanzees have been taught to produce and comprehend far greater vocabularies than were thought possible following the unsuccessful attempts of the Hayeses (1951), the Kelloggs [1933 (1967); this volume], and others to communicate with chimpanzees via vocal languages. What is at issue is whether a chimpanzee can master relationships between words—in particular, relationships as expressed in sentences.

Premack's ambitious book, *Intelligence in Apes and Man* (1976a) adds an important new dimension to the controversy about a chimpanzee's linguistic ability. On the one hand, Premack continues to ask the question he has posed in numerous research papers: what kinds of evidence show that a chimpanzee can learn certain rudiments of human language? But as the title of his book reveals, Premack is not content to limit himself to that rather complex question. Having accepted his own data as evidence that a chimpanzee can master many aspects of language, Premack speculates about the intellectual skills implied by the linguistic performance of the chimpanzees he has studied over a 10 year period. Why, Premack asks, is a chimpanzee more successful at learning language skills than a rat or a pigeon?

Premack addresses this question with both empirical and conceptual arguments. First, he cites evidence that chimpanzees have learned something about language. From his interpretation of those data, he derives a conceptual framework for relating language to more general intellectual functions. *Intelligence in Apes and*

*Preparation of this review was supported in part by grants from NIMH (5R01 MH29293) and the Harry Frank Guggenheim Foundation.

H. S. TERRACE ● Department of Psychology, Columbia University, New York, New York 10027.

Man provides the most comprehensive presentation to date of training techniques and data on language learning by chimpanzees. It also provides the most comprehensive set of hypotheses about what a chimpanzee might or might not be doing when it uses language.

In many ways, the most interesting and readable sections of *Intelligence in Apes and Man* are Premack's conclusions about the intellectual bases of language. These conclusions, however, are only as valid as the data on which they are based. For that reason alone, it is appropriate to begin our evaluation of Premack's hypotheses about intelligence and language by examining their empirical foundations.

TEACHING THE LANGUAGE

Premack's basic training procedures have been described extensively in earlier papers, as well as in *Intelligence in Apes and Man*. In this review, it should suffice to summarize a few representative examples of some of the "atomic constituents" of language that Premack has selected for study. The "words" of Premack's artificial language were plastic chips of different colors and shapes. Only those few chips that signified individual chimpanzees or trainers resembled their referents. The other chips provided no information about what they signified. An obvious advantage of a language consisting of tangible words is that performance is not limited by the chimpanzee's ability to remember words that are not physically present. Instead, performance would appear to be limited only by the subject's ability to grasp the semantic and syntactic complexities of a particular problem.

Premack not only invented a vocabulary of plastic words, he also provided an exact set of rules for using these words. In devising these rules, Premack

made no attempt to simulate the "natural approach," mainly because the natural approach is far from well-defined and it is difficult to simulate an ill-defined condition. Instead, . . . [he] . . . attempted to devise the most efficient training procedure possible, without regard to whether it did or did not simulate the human one. (p. 20)

Some of these training procedures are illustrated in the following examples:

1. *Production of four-word sentences.* All communication took place on a "language board" placed between the chimpanzee and the trainer. Words were placed on the board either by the trainer or by the chimpanzee. The trainer could also place various incentives on the board. In teaching a sequence such as "Mary give apple Sarah," training began with an exchange of the word "apple" for a real apple. A plastic blue triangle (the word "apple") was placed within easy reach of Sarah, Premack's star student. The apple itself was out of reach. When Sarah picked up

the blue triangle and handed that word to the trainer, she was given the apple. Later, Sarah had to discriminate between two pieces of plastic, one the word "apple," the other a word for a less desired incentive. Variations of this procedure were used to train a vocabulary describing different incentives.

The next step was to train Sarah to "write" the words "give apple," in that order. That did not prove very difficult. However, attempts to teach Sarah to differentiate between the consequences of writing "give apple" versus "insert apple" proved unsuccessful. During the next stage of training, Sarah had to write "give apple Sarah" in order to obtain an apple. As an aid to learning her name, Sarah was given a necklace from which was suspended the plastic word signifying "Sarah." Trainers and other chimpanzees were likewise identified. In an earlier paper, but not in *Intelligence in Apes and Man,* Premack (1971a) described a procedure in which Sarah had to choose between the names of two recipients, when another chimpanzee (Gussie) was present. If Sarah wrote "give apple Gussie," the apple was given to Gussie (and not to Sarah). Through contrasts of this type, Sarah presumably learned the difference between "Sarah" and "Gussie."

The addition of the donor's name completed the sequence, "Mary give apple Sarah." It is difficult to determine from the information provided in *Intelligence in Apes and Man* just how well Sarah performed at this stage of training. Premack writes that the "use of the donor name was optional at this early stage. . . . Later, when several agents were present in the same session, use of the donor's name was made obligatory" (p. 102). However, no data were presented concerning Sarah's performance when the donor's name was obligatory. Premack does report that two other juvenile female subjects, Peony and Elizabeth, were required to produce sequences of the variety *Donor give X recipient.* Neither subject performed well when offered choices of different words for donors, objects, and recipients. They also balked at performing four-word sequences. For that reason, they were returned to two-word sequences of the type *give X.*

2. *Comprehension of the hierarchical structure of a sentence.* Hierarchical structure is one of the hallmarks of human sentences. The clear advantage of arranging words according to a hierarchical, as opposed to a linear rule, is that certain words can dominate other words. For example, in the instruction "Sarah apple pail banana dish insert," "insert" dominates two phrases, "apple pail" and "banana dish." Premack argued that Sarah demonstrated comprehension of the hierarchical structure of a sentence by carrying out instructions of this type.

Compliance with these and similar instructions was trained as follows. First, Sarah was required to perform the actions called for by sentences of the following variety: "Sarah banana pail insert." "Sarah apple pail insert." "Sarah banana dish insert." "Sarah apple dish insert." In each case, Sarah was given two empty receptacles (a pail and a dish) and two fruits (an apple and a banana). From these basic sequences, the instructions were changed progressively until the target sentence was reached. First, two sentences were simply linked together, for example: "Sarah banana pail insert Sarah cracker dish insert." Then, the agent's name was omitted: "Sarah banana pail insert cracker dish insert." Finally, the target instruction was achieved by deleting the first "insert": "Sarah banana pail cracker dish insert."

After learning to carry out the instruction "insert," Sarah was given sequences requesting her to "take" (remove) the contents of the receptacles in front of her. Having learned to follow the instruction "take," Sarah was given a series of instructions in the presence of two pairs of receptacles. One pair was empty. Each member of the other pair contained a banana and an apple. An apple and a banana were also placed between the two pairs of receptacles. On some trials, Sarah was instructed to "insert"; on others, she was asked to "take." Sarah made only one error on the 10 trials in which the action required of her changed from trial to trial.

3. *Prepositions.* Sarah, Elizabeth, and Peony were trained on comprehension and production problems that used the word "on." In a typical comprehension problem, Sarah was given red and green cards. The trainer wrote instructions such as "red on green" or "green on red." Reward was provided when Sarah arranged the cards as described in the instructions. At a later stage of training, Sarah was given production problems in which she was required to write "red on green," "yellow on red," and so on, in accordance with the configuration of colored cards arranged by the trainer. On these trials, Sarah was provided with the words "on," "red," "yellow," and "green." Peony and Elizabeth were given similar problems in which object names ("keys," "clay," "shoes," and so on) were used instead of colors names. Peony and Elizabeth were subsequently required to choose among three alternatives (a sponge, clay, and monkey chow) in executing an instruction such as "sponge on clay." This type of problem forced the subject to attend both to the object names and to the order of the object names in the trainer's instructions.

If only two choices were provided in a comprehension problem, the subject need pay attention only to the first word. That word identifies which object is to be placed in the top position. With three choices, the topmost object can be identified in the same way. However, identification of the bottom object requires the subject to match the name of the bottom object with one of the two remaining choices.

4. *Mapping of predicates onto various arguments.* Premack considered problems dealing with same-different judgments, properties of objects, and causal inference as problems in which the subject was asked to apply a particular predicate to different arguments. For example, the predicates "same," "name of," "color of," "if then," derive their meanings only to the extent that they can be applied to various arguments and to the extent that these predicates are perceived as descriptive relational terms:

To acquire these predicates in a generalized sense that is indispensable for language, the subject must in each case respond to a relation between relations. For example, to acquire "same-different" the subject must first recognize that the relation between, say, apple and apple is same$_1$; likewise that the relationship between say, banana and banana is same$_2$; and finally that same$_1$ is the same as same$_2$. It must make a comparable judgment in the case of other predicates:

1. The relation between, say, "apple" and apple is the same as that between say, "banana" and banana, i.e., name of$_1$ is the same as name of$_2$.

2. The relation between say, red and apple is the same as that between, say, yellow and banana, i.e., color of$_1$ is the same as color of$_2$.

3. The relation between, say, dropping a glass and the glass breaking is the same as that between, say, tipping the glass and the water spilling, i.e., $\supset_1$ is the same as $\supset_2$. (p. 133)

(a) *Same-different*. Before teaching the words "same" and "different," Premack verified that his subjects could sort familiar objects in simple matching and oddity tasks involving cups, spoons, keys, paperclips, and so on. Same-different training was introduced once a subject could match novel items (or select the odd item from a set of two similar items and one dissimilar item). When like items were presented, the response of putting the word "same" near, or between, them was reinforced. When dissimilar objects were presented, the word "different" was reinforced.

At a later stage of training, a symbol meaning "?" was placed between two objects. If the objects were the same, the required response was to replace "?" with "same." If the objects were dissimilar, "different" was the correct answer. This paradigm was used in a number of ways. In some instances, the problem was posed as a question in the following formats: "object, same, ?," or "object, different, ?" The correct solution of the problem was to replace "?" with the appropriate object. The choices were an object identical to the object presented in the question or a dissimilar object. In another variation, problems of the following variety were presented:

Object A *same* Object A?
Object A *different* Object A ?
Object A *same* Object B?
Object A *different* Object B?

In each case, the subject's choices were the words, "yes" and "no."

(b) *Property names*. After Sarah learned the words for "red," "yellow," "round," "square," "large" and "small," along with the names of various objects, she was taught to use plastic chips meaning "color of," "shape of," "size of," and "name of." This was accomplished through training on frames of the following variety:

? color of apple
?color of banana   where the alternatives are
? not color of apple   *"red"* and *"yellow"*
? not color of banana, etc.

Red color of ?   where the alternatives are
Yellow color of ?, etc.   *"apple"* and *"banana"*

? red color of apple   where the alternatives are
? orange color of apple   *"yes"* and *"no"*

round ? ball   where the alternatives are
round ? square, etc.   *"shape of"* and *"not shape of"*

After "color of," "shape of," "size of," and "name of" were taught separately, a series of problems was presented in which questions concerning the name of, shape of, and color of different objects were given in the same session. The range of Sarah's performance was 83 to 100% correct.

(c) Causal inference. Premack offered two types of evidence that a chimpanzee can perceive causal relationships between various events. In one type of problem, the trainer presented the subject with a pair of objects in two different states, for example, a whole apple and a piece of an apple. The task was to place between the two objects the instrument that was responsible for causing it to change from one state to the other. As was the case with other types of problems, the first stage of training consisted of errorless trials. A knife was placed on the training table within easy reach. All Sarah had to do was to move the knife between the whole and the cut apple. When the array consisted of a blank piece of paper and a paper with scribbles on it, Sarah was required to place a crayon (the only other object on the training table) between the marked and unmarked pieces of paper. During the next phase of training, two instruments were present on each trial. For example, when the chimpanzee was shown a dry sponge and a wet sponge, it had to select between a crayon and a container of water and place one of those objects between the two test objects. On transfer tests, novel pairs of objects were presented, for example, a sponge marked with a crayon and an unmarked sponge. The choices on this trial might consist of a container of water and a crayon. Sarah, Peony, and Elizabeth performed at typical levels of accuracy on these tests (75 to 95% correct).

Causal inference was also studied in problem sets made up of conditional sentences. In one type of problem, Sarah was shown a pair of sentences providing instructions as to how she might obtain a piece of chocolate. For example: "Sarah take apple if then Mary give Sarah chocolate." "Sarah take banana if then Mary no give Sarah chocolate." Sarah preferred chocolate to both apples and bananas. In order to obtain a piece of chocolate, Sarah had to choose the alternative stipulated by the instructions. An English paraphrase of these instructions would read: when presented with a choice between an apple and a banana, Sarah will be given a piece of chocolate if she takes an apple. If she chooses the banana, she is given nothing.

In a more complicated version of this type of problem, Sarah was confronted with sentences such as: "Mary give cracker Debby if then Sarah eat yellow fruit." "Debby give cracker Mary if then Sarah eat red fruit." On each trial, Sarah had to choose between an unnamed yellow (cantaloupe) and a red (strawberry) fruit. Though not stated explicitly in the text, it seems reasonable to assume that Sarah was allowed to eat the fruit only if she followed the instructions.

Premack recognized that Sarah did not have to attend to the verbal argument in this last type of problem (e.g., "Mary give cracker Debby"). In the first place, the antecedent was always carried out. Thus, the verbal specification was redundant with the action. In actual fact, Sarah could solve the problem without paying any attention to either the antecedent or to the action. All she had to do was to discriminate the piece of plastic describing the color of the fruit she was allowed to eat.

To determine whether Sarah could relate the action of one of her trainers, as specified by a particular verbal argument, with the action specified in the consequence, Sarah was shown simultaneously a set of six conditional sentences. These sentences referred to possible exchanges between two of the three trainers present. The exchange could go either way, for example, "Mary give X Debby"; "Debby give X Mary"; "John give X Debby"; "Debby give X John," and so on. Following each of these types of argument was a conditional instruction, such as "if then Sarah take red" or "if then Sarah take green." The sight of an array of six long conditional sentences caused Sarah to leap away from the language board to the other side of her cage, where she remained until the board was erased.

A simpler version of the same problem was tried with only two trainers. Depending on whether Mary gave green or red to Debby, or vice versa, Sarah was supposed to choose a candy or a cracker. On each trial, she was shown only two sentences. One stated when she was supposed to choose the candy, the other when she should choose the cracker. Sarah made two errors on eight trails of this type of conditional problem.

LANGUAGE OR DISCRIMINATION LEARNING?

The validity of these and related examples of problem solving as exemplars of language invites careful scrutiny. One issue is the general procedure used to train and test each exemplar. With but few exceptions, all the problems presented during each training session were of the same nature. Also, within a session, the number of answers was quite limited. In most cases, only two alternatives were available, and these alternatives were restricted to a small subset of contrasts, for example same versus different, yes versus no, red versus blue (or yellow or green), and so on.

The homogeneous nature of the questions posed during any one session, along with the restricted range of possible answers, increases the likelihood that nonlinguistic contextual cues contributed to the performance of Premack's subjects. In at least one instance, it was possible to solve a problem without paying any attention to the critical word. Consider, for example, training designed to convey the meaning of the preposition *on*. In these problems, *on* was never contrasted with any other preposition. The format of each trial was the same: either the trainer placed one object on top of another or Sarah was required to do so. The only contribution of language to the solution of this problem was a lexical one: to identify which object was to be placed on top of the other object (red on green, clay on key, and so on). In fairness to Premack, I should mention an earlier article in which he did refer to Sarah's use of the prepositional term *in front of* (Premack 1971a). But even this fact was presented in a footnote that provided no information about whether *in front of* was contrasted with other prepositions or about how well Sarah performed.

Another factor to keep in mind is the chimpanzee's motivation for solving the problems posed by Premack and his trainers. In each case, some tidbit of food is provided following each correct response or at the end of a run of correct responses. Premack does not seem terribly concerned about this state of affairs. He argues that reinforcement is a performance variable and not necessary for learning *per se*. Premack also asserts that he could have obtained the same results using social reinforcement. As far as I could tell, however, no evidence was presented to support this conjecture.

In view of the many training procedures Premack describes, it is often frustrating not to be able to learn from the text just what took place and when. No mention is made of the actual sequence of problems given to each chimpanzee. Only in a few cases is it possible to infer which problems were presented earlier and which problems were presented later in the course of each subject's training. This problem could have been obviated by presenting a table describing the sequence and the number of each type of session, as experienced by each subject.

Also lacking are lists of the actual words mastered by Sarah, Elizabeth, and Peony. In the legend of Figure 3.3, Premack states that what is shown is "a major portion of a lexicon of 130 words" (p. 75). However, a count of the words shown in Figure 3.3 reveals only 48 words, at least two of which are repeated. Nor does it help to read on page 156 that Figure 3.3 shows how "the negative particle was appended to the head of one of the instruction". None of the symbols shown in Figure 3.3 was identified.

What is at issue, though, is the contents of Sarah's, Peony's, and Elizabeth's vocabularies, and not their actual size. It would be of interest to compare the nature of their vocabularies with that of chimpanzees from other contemporary studies. It would also be of interest to know whether each word was mastered in both the production and the comprehension modes. It does not suffice to assert parenthetically at the end of the book that "after sufficient training, every word taught the chimpanzees in either production or comprehension transferred to the other mode" (p. 354). For example, after the unsuccessful attempt to teach *insert* productively, it is stated that this verb was taught in the comprehension mode. However, no information is provided as to when and how *insert* was taught productively. Here again it would have been helpful to look at a simple table, showing when each word was learned and in what mode.

The ambiguous nature of some of Premack's data gives rise to yet another type of frustration. For example, in Table 4.1, performance is characterized by reference to the number of incorrect responses that occurred in a set of problems. In Tables 6.1 and 12.1, performance is characterized by the number of correct responses. However, neither the neighboring text nor the title of Table 15.1 specifies what Table 15.1 (a three-page table!) shows: correct or incorrect responses? Given the marginal nature of the performance shown by many of the entries of Table 15.1, it is especially unsettling not to be sure what kind of data it presents.

SENTENCES OR ROTELY LEARNED SEQUENCES?

The most serious drawback of *Intelligence in Apes and Man* is to be found in Premack's interpretations of his data, rather than in their fragmentary nature. The basic problem is whether Premack has correctly interpreted how his subjects perceived the symbols and sequences of symbols they were required to use.

1. *Is "Mary give apple Sarah" a sentence?* As far as one can tell from Premack's description of the procedures used to train four-symbol sequences, the trainer's name was never contrasted with the names of other trainers during the same session. While "Mary," "Amy," "Debbie," were alternated as choices between sessions, they were never contrasted with one another within the same session. Furthermore, since each trainer wore his or her name symbol on a necklace, all the subject had to do in order to solve this type of problem was to start the sequence by matching the symbol on the language board with that worn by the trainer. Even though, as Premack observed, the subject could have learned to associate the symbols worn by the trainers with the trainers themselves, there is no evidence that they did so. From the chimpanzee's point of view, the symbols referring to each teacher's name could well have been nonsense words.

As mentioned earlier, the meaning of "give" was not distinguished from that of other verbs. As such, it could have amounted to another nonsense word that had to be placed after the first nonsense word. On some trials, two recipient names were provided. Throughout each of these problem sets, however, only one recipient was present. From the subject's point of view then, one alternative was consistently correct. The subject need not have used the fourth symbol of the sequence to refer to itself. Of the symbols used in training four-symbol sequences, Sarah, Elizabeth, and Peony appear to have learned the meanings of only those symbols that referred to the objects they requested. But even in the case of objects, only two choices of object names were provided on each trial.

These considerations suggest that the sequences glossed by Premack, in general form, as "trainer give X recipient" (p. 81), could just as well be described as the rotely-learned sequence ABXD in which substitutions of the meaning of X varied with the object on hand. Premack is clearly aware that only

the object class was explicitly mapped. . . . Action, recipient, and donor classes were not mapped in the same explicit fashion, although the training session did provide information that an appropriately inferential subject might have used to arrive at the referents of the other particles. (p. 82)

Premack recognizes that only the meaning of the symbols for objects was learned. Yet, throughout *Intelligence in Apes and Man,* he refers to the symbols of the three remaining elements of the sequence *trainer give X recipient* as if Sarah, Elizabeth, and Peony had in fact learned their meanings.

Premack's poetic licence in attributing meanings to each of the symbols glossed as *trainer give X donor* is clearly revealed by considering the performance of another organism on a similar problem. If a pigeon performed a sequence ABXC, where X referred to different incentives, it would seem far-fetched to refer to that sequence as "trainer give grain R–42." That type of performance is easy to obtain. Pigeons were trained to peck the sequence $A \rightarrow B \rightarrow C \rightarrow D$, where A,B,C, and D were different colors, at levels of accuracy comparable to that reported by Premack in the case of "four-word sentences" (Straub, Seidenberg, Bever, & Terrace 1979). On each trial, A,B,C, and D were presented simultaneously in different physical arrays. We have yet to try to extend this performance to ABXC problems (where X_1 could refer to one type of grain, X_2 to a different type of grain, X_3 to water, X_4 to the opportunity to attack another pigeon, and so on). If a pigeon could learn such a sequence (a not unlikely outcome), one wonders what is to be gained by assigning names to each member of that sequence.

Similar problems of interpretation arise in connection with Sarah's performance on comprehension problems in which a single verb refers to actions that are to be repeated (e.g., "Sarah apple pail banana dish insert"). Premack asks, "Did syntax play any role or were the compound sentences processed on an exclusively semantic basis?" (p. 329).

After rejecting a semantic interpretation of his instructions, Premack concluded that Sarah did use a syntactic rule:

In processing such sentences . . . the functional effect of the rules, although not necessarily the rules themselves, can be described by grouping the words in a sentence by parentheses and brackets. . . . The bracketing emphasizes the following features of her performance.

She recognized the word "Sarah" applied to the whole sentence and not just to the first clause; that is, she did not confine her behavior to the first clause but carried out the whole instruction. By the same token, she must have recognized that the word "insert" applied across the sentence to all appropriate items mentioned in the sentence, and not just to the second clause. Additionally she divided the food and container words appropriately, using a container word as a break and grouping together all food words that occur above any container word (p. 330)

In these tests, Sarah was the only chimpanzee present. That context, along with a long history of working with problems that required the execution of all aspects of an instruction, is sufficient to ensure the result that "she . . . recognized [that] the word 'Sarah' applied to the whole sentence and not just to the first clause." To conclude otherwise would imply that Sarah would not have performed in the same manner if "Sarah" was omitted. No evidence of that unlikely outcome was presented.

Even the most difficult instruction put to Sarah required only one kind of action (insert or take). In view of the progression from compound to condensed instructions, it would be surprising if she had *not* recognized that the verb "applied across the sentence, to all appropriate items mentioned in the sentence, and not just to the second clause."

What Sarah appears to have learned in this type of problem is what Premack rejected as a "semantic" rule: operate on all of the objects listed before the name of a container in the manner specified by the verb at the end of the sequence. Unless Sarah could decode an instruction in which she had to perform two actions, where at least one of the actions had to be repeated, it seems gratuitous to conclude that Sarah had learned a hierarchical rule of syntax.

PREPOSITIONS

The major difficulty in concluding that Premack's chimpanzees learned the meaning of *on,* is that *on* does not appear to have been contrasted with other prepositions. Since there is no evidence that a chimpanzee can learn the meaning of a preposition from its context, there is no basis for concluding that Sarah, Elizabeth, and Peony interpreted *on* as a preposition. These considerations apply to both comprehension and production tasks. Accordingly, Premack's conclusion that Sarah, Elizabeth, and Peony showed "successful transfer of 'on' from comprehension to production" (p. 128) seems unwarranted.

PREDICATES

The evidence that chimpanzees master predicates is most convincing in the case of *same* and *different* and least convincing in the case of causal relationships. More so than any other set of words of the vocabularies of Sarah, Elizabeth, and Peony, *same* and *different* appear to have been used in a wide variety of situations. But even in this instance, evaluation is difficult. Consider, for example, Peony's performance with *same* and *different* when she worked with them in a new format. Instead of having Peony put *same* or *different* between a pair of similar or dissimilar objects, she was invited to use these symbols without configurational constraints. She was simply presented with three objects (two similar and one different) and two words (*same* and *different*). Premack noted that Peony's performance on the first such test was "profoundly reassuring as to the nontrivial character of pongid intelligence":

1. She puts spoons together (did nothing with the word "same," as though it was redundant) and put the word "different" on top of the piece of clay.
2. She put clay next to the word "different," the two spoons together and the word "same" on top of them.
3. She put two pieces of clay together and the word "same" on top of them (and did nothing with "different" and the spoon).
4. She wrote out in a linear fashion "clay same clay different spoon," that is, A same A different B.

On subsequent lessons, she was given a large variety of objects in the same configurationless way. Most of the forms that appeared in the first lesson disappeared, including, regretably, her linear format: A same A different B. By and large, she settled on only one of the several forms that had appeared on her first lesson; she superimposed the two like objects insofar as possible and placed the word "same" on top of them; then she placed the word "different" either on or alongside the odd object. One form she did not use was to bring together the two unlike objects and place the word "different" on them, perhaps because to have done so would have left no way to deal with the remaining object and the word "same". Peony's behavior suggested that to her "same" meant an object with a twin, whereas "different" meant the condition in which an object did not have a twin; this was a construction altogether compatible with the terms of her training. (p. 145)

Without question, Peony, Sarah, and Elizabeth mastered the use of *same* and *different* in the standard configuration in a variety of problem sets. What is not clear is the reliability of Peony's use of *same* and *different* in novel configurations.

Performance on problems with the more difficult *if-then* predicate provided little evidence to justify Premack's conclusion that his subjects understood the conditional relationship expressed in the instructions. With but one exception, the subject could discover what to do by attending only to the consequence of the conditional sentence (the second clause). It was unnecessary to understand the contents of the argument (the first clause). In the one test in which it was necessary to attend to the argument, Sarah made two mistakes in a set of eight problems. With such meager information, it is not possible to decide if that was a reliable result.

More reliable data were obtained with problem sets on causal inference. On these problems, the subject had to place the appropriate instrument between an object in two different states and thereby indicate what caused the object to change from one state to another. Performance was reliably accurate when novel objects and instruments were presented. Accordingly, Premack concluded that his subjects expressed an understanding of how the instrument caused a particular consequence (for example, how a crayon, as opposed to a glass of water, caused an unmarked sponge to become a marked sponge).

About this performance, Premack remarked:

Simple as this outcome is, it can be given a stronger interpretation than may first meet the eye. The visual sequences are infinitely ambiguous: each can be coded in indefinitely many ways, such as red-blank-red, one-blank-two, round-blank-flat, large-blank-small. Not only the test items but also the three alternatives are subject to indeterminately many codings. Knife, for instance, need not be read as knife (instrument that cuts) but can be coded as sharp, metal, long, shiny, etc., and the same holds for the other alternatives. The subjects evidently did not code the sequences or alternatives in this way, for they consistently chose alternatives compatible with only one coding, viz.: how do you change the object from the intact to the terminal stage? With what instrument do you produce the change? Because the subjects read the sequences in a specific and consistent way—finding the same question in each of the sequences—I infer that they have a schema, a structure that assigns an interpretation to an otherwise infinitely ambiguous sequence. (p. 337)

Adult educated human observers will probably agree with Premack that a "knife . . . need not be read as a knife (instrument that cuts) but can be coded as sharp, metal, long, shiny, etc." However, the critical question is, can a *chimpanzee* code the knife and other so-called instruments in the manner described by Premack? Without any training to code objects according to sharpness, shininess, and so on, there is no reason to expect the chimpanzee to do so. In suggesting that his subjects refrained from coding instruments in these irrelevant manners, Premack appears to have taken for granted a competence on the part of chimpanzees for which there is no evidence.

How Premack's subjects solved these problems is suggested by the following thought experiment. Suppose that the set of causal inference problems was given without each object in its original state. That is, the subject would see only identical pairs of objects such as two damp sponges, two apples, each showing a crayon mark, and so on. It would not be at all surprising if the chimpanzee performed as well under this condition as it did when the object was presented in its original state. Without such tests, it is not clear whether Premack's data support his conclusion that the chimpanzee inferred a *causal* relationship between an instrument and a particular state. A more parsimonious interpretation is that as a result of extensive drill, the subjects learned to associate certain so-called instruments with various states, e.g., crayons with crayon marks, wet objects with water, pieces of objects with knives, and so on.

SYNTAX

One of the most perplexing chapters of *Intelligence in Apes and Man* is the chapter entitled "Syntax", a chapter in which Premack concluded that a chimpanzee can learn some rudimentary rules of grammar. Premack recognized that his method did not encourage his subjects to generate new sentences by combining separately learned phrases. However, in evaluating Sarah's performance, Premack observed that "there are at least five different cases sifted through her record in which Sarah comprehended (and in a few cases produced) sentences formed by a process more demanding than that of combining phrases" (p. 319). We have already considered one of these five cases: "hierarchical organization". Let us consider briefly a few of the other examples of Sarah's purported syntactical competence.

ATTRIBUTION

The instructions in this instance were in the form *Sarah take X*. Initially, *X* was a property, for example, *red*. Thus, when Sarah was instructed *Sarah take red*, where the choices were a red and a green dish, she was rewarded for taking the red dish. In the same series, it appears as if Sarah was also given instructions of the type: *Sarah take dish* where the choices were a dish and a pail. For Premack, the critical test was Sarah's performance following the instructions: *Sarah take red dish*, where the alternatives were red and green dishes and red and green pails. No details of this test are provided nor are any data presented. Nor is there any entry in the index under "attribution" that would help the reader to dig out the relevant information from other sections of the book. Let us assume that Sarah did perform reliably on this test and ask, would it be valid to conclude that

in taking the red dish, instead of either the green dish or the red/green pail, [Sarah] demonstrated comprehension of the attribute form. Her accomplishment went beyond that of a child, for although she had been taught "Sarah take dish" and "Sarah take red," she had never been taught "red dish." Unlike the sentence a child produces at stage II, Sarah comprehended a sentence involving a unit that had no history of independent occurrence. (p. 320)

The question at issue is whether Sarah regarded *red* as an attribute of *dish* or whether she solved the problem by matching *red* and *dish* to the objects on hand in two separate operations. Her correct performance when instructed *Sarah take red/ green,* and confronted with red and green dishes or pails, clearly suggests that she could match the symbols *red* and *green* to the objects on hand. In following the instructions *Sarah take red dish,* one must assume that Sarah was encouraged to choose only one object from the set of four objects with which she was confronted (red and green pails, and red and green dishes).

Sarah could have followed one of two strategies that would have enabled her to solve this problem without any understanding of the attributional relation between *red* and *pail.* All she had to do was first to attend to the dishes and then to the red object, or first to the red object and then the dish. So long as this type of problem can be solved by relating symbols to the objects on hand, one at a time, the interpretation that Sarah spontaneously exhibited attribution seems gratuitous.

The further interpretation that this type of performance shows evidence of understanding an "actor-action-attribute-object form" (p. 320) seems even more far-fetched. The evidence that Sarah encoded *Sarah* as an exemplar of a class of symbols meaning "actor" is virtually nil. Since *take* appears to be the only verb in this series of problems, it seems foolhardy to interpret *take* as an action. And, as mentioned earlier, it seems doubtful that Sarah truly distinguished between an object and its attribute in this set of problems.

FROM DEMONSTRATIVE PRONOUN TO DEMONSTRATIVE ADJECTIVE

Sarah was taught the so-called demonstrative pronouns *this* and *that* in situations in which she was instructed to take a near *(this)* or a far *(that)* object. She was also taught to produce sequences of the form *Give Sarah this* in order to request a near object and *Give Sarah that* in order to request a far object. Premack observed:

When required to produce "Give Sarah this cookie" vs. " . . . that cookie," she made only three errors in fifteen trials, with none on the first five trials. She wrote the incorrect "give Sarah cookie this" almost as often as " . . . this cookie" but there was no reason why her word order should have been correct. Sentences of that kind have never been modeled for her. Her own production of the demonstrative adjective form on that occasion was her first experience with the form. (pp. 320–321)

Here again, the details of training are scanty. In this instance, however, one can refer back to a slightly more elaborate description of the procedure (p. 282):

Two cookies were placed on the table, one notably larger than the other, with the larger one closer to Sarah on some trials and closer to the trainer on other trials. She was given the words "give," "Sarah," "cookie," "this," and "that." Without further training she was required to write either "give Sarah this cookie" or "give Sarah that cookie" depending on the location of the desired cookie.

Having just been trained on a series of problems in which she was required to produce sequences of the form *Give Sarah this* and *Give Sarah that,* it is not surprising that Sarah added the additional symbol for the incentive when that symbol was provided. The interpretation of this performance pivots on just what Premack meant by saying that Sarah, "without further training was required to write either *give Sarah this cookie* or *give Sarah that cookie. . . ."* Just how did Sarah know what was required? Though Premack successfully rebutted Clever Hans' interpretations of performance on other types of problems, in this instance the trainer could have cued Sarah's performance. Did the trainer end a trial and code Sarah's performance as an error if she wrote "give Sarah cookie?" If he did not, and instead encouraged Sarah to choose another symbol, the sequences "give Sarah cookie this" or "give Sarah cookie that" are not surprising, especially in view of her training to write "give Sarah this" or "give Sarah that" in the preceding problem set. Since "this" and "that" had just been used to refer to objects, it seems premature to refer to them as demonstrative adjectives.

CONJUNCTION

In arguing that Sarah learned the "major recursive form" of conjunction, Premack states:

In the beginning she requested separate fruits with separate sentences, say, "Mary give Sarah apple" and "Mary give Sarah banana," but subsequently requested separate fruits with a single sentence "Mary give Sarah apple banana." . . . [The] use of conjunction was impressive . . . because it was invented. . . . no aspect of conjunction was taught Sarah or the other subjects. (p. 321)

The reader is invited to judge the validity of Premack's claim that Sarah (and Elizabeth and Peony as well) "invented" the conjunctive form by reading Sarah's training on "conjunction reduction":

Before we taught Sarah an explicit marker for "and," we invited her to engage in an implicit form of conjunction reduction on her own. In previous drills on simple sentence reduction, a piece of food had been placed in front of her along with a small set of words, her task being simply to request the food—for instance, to write "Mary give apple Sarah" when the food was apple and "Mary give banana Sarah" when it was banana, etc. After many such drills, we invited Sarah to behave conjunctively, by placing before her pieces of two different fruits and giving her the usual set of words, including names for both fruits. On the first eight trials, she responded in keeping with her previous training, writing "Give apple Sarah," and "Give banana Sarah," her usual individual sentences. On the ninth trial, however, she changed her approach and wrote, "Mary give Sarah apple orange," for the first time requesting both items in one sentence. On a subsequent lesson when given three items per trial, she requested all three of them, writing for example, "Give banana apple orange Sarah." (p. 243)

The initial conjunction of *apple* and *orange* could well have occurred by chance, then been strengthened by virtue of a dual reward. It is also the case that putting together a sequence in which it is possible to use all of the object names is an easier task than one that requires the subject to choose only one object name.

INTELLECTUAL FOUNDATIONS OF LANGUAGE OR PROBLEM SOLVING?

Particularly in the last chapter of *Intelligence in Apes and Man* (entitled Mechanisms of Intelligence: Preconditions for language), Premack's mixture of creative hypotheses, scanty data, and overly rich interpretations of that data detract from the soundness of his conclusions about the intellectual basis of language. The longest section of this chapter considers the ability of an organism to engage in casual inference. For the sake of brevity, I will focus on this section (though similar problems arise in Premack's discussion of intentionality, representational capacity, map reading, multiple internal representation, mnemonic capacity, and second-order relations).

The evidence of causal inference to which Premack seems to assign the most weight is the data on object$_{state\ 1}$-object$_{state\ 2}$-instrument matching that was described earlier. Premack remarks that "Because the subjects read the sequences in a specific and consistent way—finding the same question in each of the sequences—I infer that they have a schema, a structure that assigns an interpretation to an otherwise infinitely ambiguous sequence" (p. 337). Recall that the problems posed by such sequences could have been solved by symbolic matching. Knives could be matched with multiple portions of an object, crayons with crayon marks, and so on. Given the existence of convincing data on symbolic matching in pigeons, it seems doubtful that this kind of problem solving is unique to chimpanzees and man.

Premack acknowledges that another line of evidence supporting the chimpanzee's ability to communicate about causal inference is indirect. The data come from problem sets in which Sarah was required to write sequences of the type *trainer give X Sarah*:

At an early stage of training, when first being taught word order, Sarah wrote 409 three- or four-word sentences requesting that one of several fruits be given her. Although she made 76 errors of word order (and many more errors of word choice) in doing so, only three times did she begin a sentence with the name of an object, mistakenly putting the object in the agent's role, as in for example, "orange give Sarah apple." The infrequency of this kind of error is compatible with the view that she divides the word as we do, assigning different functions to objects and agents. (p. 338)

The problem with Premack's interpretation is not that the data regarding Sarah's perception of objects and agents was indirect but that a simpler explanation is readily available. The sheer number of trials in simpler forms of this problem, in which the object name was always required to appear last, would seem to account for Premack's observation equally well. Before three- and four-word sequences were trained, two sequences of the type *give X* were trained. Unfortunately, no data are provided in *Intelligence in Apes and Man* regarding the number of *give X* trials to which Sarah was subjected, nor are any data provided on the number of *give Sarah X* trials that Sarah experienced before being trained on four-symbol sequences (where *give Sarah X* sequences were also acceptable). It is at this stage of training that Premack reports only three instances in 409 sequences in which the object was placed first.

Having exhausted his own data on causal inference, Premack turns to data from a natural situation in which chimpanzees were observed to hunt termites by first selecting straws from nearby plants, then inserting them in the termite mound and finally "fishing out" the termites:

One finds in the chimpanzee's technology evidence of at least three aspects of its intelligence: planning, memory, and inference.

1. In selecting the straw at a distance from the mound, the chimpanzee appears to be able to plan. If planning is correctly said to depend on the ability to hold in mind a representation of one's objective, then there is no question but that the chimpanzee should be able to plan. Its ability to generate and use—in highly determinate ways—internal representations has been amply demonstrated in the present research.

2. The individual chimpanzee is said to collect from over 25 different mounds. Its ability to remember the locations of the mounds, and the orifices concealed in each of them, testifies to the ape's long-term memory. Our own evidence confirms this capacity most dramatically in the animal's ability to identify the "anatomy" of various fruits.

3. I have not been in a position to interrogate the field chimpanzee, but if I were, I would ask it whether or not it knows what goes on in a termite mound. Can it infer what happens between the time when it inserts an empty straw into the mound and withdraws it full of termites? Fishing is after all a cognitively special activity, a kind of black-box technology in which input is related to output by a hidden middle. We could use the visual causality tests described earlier to interrogate the chimpanzee, and find out whether or not it knows the content of the hidden middle. The three pictures would consist of a chimp fishing, a blank frame, and a chimp holding a laden straw, about to eat the termites. The chimp's task would be the usual one of selecting the missing picture. The alternatives could include: (1) termites with their feet or antenna caught in cracks in the straw; (2) termites caught while using the straw essentially as a bridge to cross little streams inside the mound; (3) termites either attacking the straw or trapped by their own curiosity, and carried out even while exploring the straw, etc. Let us assume that the third alternative is the correct one. Whether it is or not, if the chimpanzee chose it consistently, this would demonstrate that it could identify the fishing situation as one that induced curiosity, and could picture another species responding to the situation in the same way it would. But perhaps this exceeds pongid intelligence. (pp. 340–341)

This and other passages of *Intelligence in Apes and Man* illustrate how Premack's creative imagination can isolate instances in the chimpanzee's environment that *may* reveal important aspects of intelligence. Just what does this example say about the uniqueness of this performance in chimpanzees and how does it contribute to language? There is sufficient evidence available from lower forms that could just as readily serve as a basis for arguing that those forms have representations of their worlds. Hummingbirds can remember where they last gathered food from a comparable number of alternative food sites. The hypothetical experiment Premack poses is an interesting one, but, as we have seen earlier, it is not clear what it would demonstrate about causal inference as opposed to symbolic matching.

In opting for an artificial, as opposed to a natural language, the reader will recall that Premack "attempted to devise the most efficient training procedure possible, without regard to whether it did or did not simulate the human one." By claiming indifference to the degree to which his language simulated human language, Premack may have hoped to discourage comparison between the linguistic achievements of his subjects and those of a human child. Yet the very title of Premack's book, and the numerous comparisons drawn between chimpanzee and human language, invite the reader to ask just what did Sarah and her companions learn about language?

Despite Premack's frequent claims to the contrary, a careful reading of *Intelligence in Apes and Man* reveals that Premack himself is aware that many exemplars of language in the chimpanzee fall considerably short of simulating human language. These contradictions weaken the impact of this provocative book. We have already seen that Premack's own analysis of the semantic limitations of the four-symbol sequences he tried to teach Sarah, Elizabeth, and Peony did not stand in the way of referring to such sequences as sentences and to referring to the elements of these sequences by their English glosses.

Premack also seems to be aware of the limitations of his teaching procedures. In discussing Sarah's ability to ignore irrelevant words during one phase of her training, he writes:

Lessons were typically devoted to a small well-defined topic. A search of extensive data reveals perhaps a dozen lessons that pick up several topics and shifted freely from one to the other, but most lessons dealt with only one topic. . . . Always the lesson concerned a well-defined set of alternatives. The words and even the sentence to each lesson also made up a well-defined set. The boundedness of the lesson, in both its verbal and non-verbal alternatives, could not but have helped Sarah discover the topic of the lesson. (p. 127)

Unfortunately, Premack did not pursue the implications of training his subjects on homogeneous sets of problems whose solutions were selected from a minimal set of alternatives. Were he to do so, he would have to recognize fundamental differences between Sarah's use of language and that of a child's. In learning a natural language, a child experiences a large variety of utterances, as expressed by its parents and siblings. That variety of linguistic input is matched by the variety of words the child has available in responding to another person's utterance or in generating spontaneous utterances. When a child says, "ball red" upon picking up a ball, its use of language differs fundamentally from that of Sarah who may have produced the corresponding sequence in response to a question such as "ball?" The issue here is not simply a question of the spontaneity of the child's utterance. Of equal importance are three other factors: the range of alternative words available to the child and the chimpanzee, the kind of training needed to produce the utterance in question, and the motivation for making that utterance.

One can anticipate Premack's reaction to the concern that Sarah's use of language was rather limited from his reaction to the objection that sequences containing the symbol ? were not really questions. One answer to this objection is especially telling:

She answered but did not ask questions. This would be a serious objection if she failed to ask questions when given an opportunity to do so. The omission was in the training program, however, not in the subject. In the beginning we could not find a simple condition in which to make the test; then we were diverted from the matter by other issues and ended up by (conveniently) forgetting it. Moreover, on more than one occasion, apparently bored by excessive drill, she stole all the words before her—retired to the floor where her position was less vulnerable—and wrote out and then answered all the questions taught her. Hunching over the plastic pieces, passionately arraying them in sequences taught her, she offered one of the few displays not only of her ability to learn but of her ardent desire to use what she had learned. The display also bespeaks one of our main failures, a failure to interdigitate the motivational pressures that make language valuable with the cognitive program teaching it. In fact, we did not locate Sarah or the other objects in a world, such as the child's, where increasing command of language provided increased command of the world. (p. 153)

Here again, Premack recognized the limitations of his program. At the same time, he nonchalantly ducks the issue by noting that his subjects were not provided with the opportunity to ask questions, or the motivation to use language to increase their command of the world. Premack seems to assume that by simply arranging the appropriate training sequences and by somehow changing the motivation, Sarah would ask a question and would advance beyond the rote learning that gave rise to the dubious behavior of stealing the words from the language board and then writing out and answering all of the questions put to her by her trainer.

MORGAN'S CANON REVISITED

The attitude that anything is possible, if one uses the appropriate training procedure, is expressed all too frequently in *Intelligence in Apes and Man*. Regrettably, it begs the questions that Premack set out to answer in the first place. If Premack wants to study language in the chimpanzee, its performance on his exemplars must be compared with some sort of reference performance. In order to go beyond the interpretations given to the solutions of animal problem tasks, the burden of proof is on Premack to show that those interpretations are inadequate. A chimpanzee's closeness to man does not exempt it from C. Lloyd Morgan's observation that an animal's behavior should be interpreted at the simplest level, unless there is compelling evidence that an explanation involving higher processes is needed. Such evidence cannot be obtained gratuitously by assigning to Sarah's symbols whatever meanings might be appropriate in English or by arguing that Sarah would exhibit different features of language if she were exposed to certain untried protocols.

An optimistic assessment of Premack's approach to language is that it is incomplete but viable, so long as the problems he posed are presented in a more elaborate form. Specifically, one would want to see how performance fares when (1) the heterogeneity of the questions posed to a chimpanzee is increased so that, within a single problem set, a subject might encounter same-different questions, name-of questions, causal inference questions, instructions thorough enough to test for an understanding of syntax, etc.; (2) the available answers span many categories of words; (3) each word used by the trainer and by Sarah is contrasted with at least one other word from that word category; and, (4) the referents of each of those words are not present.

In short, one would want to see even a modest attempt at synthesizing, within a single session, some of the so-called "atomic constituents" that Premack had identified in his analysis of language.

Whether problem solving, no matter how elaborate, can ever simulate language is, of course, an empirical question. *Intelligence in Apes and Man* provides a stimulating sketch as to how one might undertake such an approach. In view of the creative effort that went into this research and its interpretation, it is unfortunate that it falls short of providing clear answers to the many stimulating questions it poses, both about language and intelligence.

LOOKING IN THE DESTINATION FOR WHAT SHOULD HAVE BEEN SOUGHT IN THE SOURCE*

THOMAS A. SEBEOK

Gibt es nicht gelehrte Hunde?
Und auch Pferde, welche rechnen
Wie Commerzienräthe? Trommeln
Nicht die Hasen ganz vorzüglich?

—Heinrich Heine, *Atta Troll* (Cap. V, Quatr. 15)

The notorious but unimpeachably corroborated case of Pavlov's mice raises, in capsule form, a variety of fascinating issues with far-reaching ramifications in several directions, but with particularly serious implications, several of which are well worth restating and pondering further (cf. Sebeok 1977b), both for the foundations and research methodology of contemporary semiotics.

The facts, as reconstructed by Gruenberg (1929:326–327), Zirkle (1958), and Razran (1959) are straightforward enough. Pavlov, convinced that acquired characters could be inherited, thought at one time that this process might be demonstrated by inducing conditioned reflexes in mice and then counting the conditioning trials required through successive generations. His expectation, in conformity with the Lamarckian model of information transmission then, as later, favored in the

---

*The substance of this paper was delivered during a conference on Language and Psychotherapy, organized by the Institute for Philosophy of Science, Psychotherapy, and Ethics, at Wagner College, on April 17, 1977. Several leading themes developed there were later touched upon in different lectures and seminars given, during the fall of 1977, at the University of Kansas (week of October 10), Texas Tech University (October 17), and the University of Texas at Dallas (October 18). Some were also presented, in synoptic form, under the title "Natural Semiotics," at the 76th Annual Meeting of the American Anthropological Association, suited to the context of an all-day symposium on the "Semiotics of Culture: Towards a New Synthesis in World Anthropology" (co-organized by Drs. Jean Umíker-Sebeok and Irene Portis Winner, and held in Houston, December 1). The illustrations appearing here were first used in the book from which this article is reprinted, *The Sign and Its Masters* (Sebeok 1979b). Different versions have been published, in English and in French, in *Dîogenes/Diogène:* cf. Sebeok 1978.

THOMAS A. SEBEOK ● Research Center for Language and Semiotic Studies, Indiana University, Bloomington, Indiana 47405.

USSR (Razran 1958), was that the numbers would significantly decrease. Accordingly, he caused an assistant of his, one Studentsov (who appears in the history of science solely as an obscure although, for present purposes, emblematic figure confined to this single episode), to conduct a series of experiments over five generations of mice, the astounding results of which the collaborator then reported to the 1923 Soviet Physiological Conference, as expressed by the following dramatically cascading figures (rounded out later by Pavlov himself): 300, 100, 30, 10, and 5.

The intellectual milieu in which Pavlov worked, and, of course, the very assumptions he brought to the investigation of the problem, accounts for his remissness in not instantly questioning the results, let alone repudiating the conclusions, obtained and announced by his "over-zealous assistant" (Razran 1959:916). "It seems reasonable to assume," Razran continues, "that Pavlov would not have been so gullible if he had not shared the Lamarckian predisposition, common to Russian bioscientists—and to the intelligentsia in general—even before the Revolution, and if he had reviewed critically the general evidence on the topic." Only in 1929 did this uncompromisingly empirical scientist, whose honesty was never in doubt, indeed who, in a famous lecture, as far back as April 23, 1921, on the basic qualities of mind deemed indispensable to a scientist, put in a leading place exceptional facility in constructing scientific hypotheses—the capacity, that is, "to get behind the facts," as he used to say (Frolov 1938: 256)—set forth publicly an alternative

Figure 1. Ivan Pavlov in his study. From the author's files.

hypothesis to explain the astonishing data emanating from his laboratory. As related in Gruenberg's *The Story of Evolution* (1929: 327, n.1), "in an informal statement made at the time of the Thirteenth International Physiological Congress, Boston, August 1929, Pavlov explained that in checking up these experiments it was found that the apparent improvement in the ability to learn, on the part of successive generations of mice, was really due to an improvement in the ability to teach, on the part of the experimenter! And so this 'proof' of the transmission of modifications drops out of the picture, at least for the present."

This little tale of self-deception—a variant of what Merton (1948) has dubbed the self-fulfilling prophecy, a phenomenon which was later most creatively and ingeniously explored by Rosenthal (e.g., 1976: 136–137) and Rosenthal and Jacobson (e.g., 1968: 36), but which is perhaps best known by the tag Clever Hans Fallacy—evokes certain urgent lines of inquiry which continue to be neglected by semioticians, as well as most other students of human and animal behavior, at their peril. The issue is such an important one because the Clever Hans effect informs, in fact insidiously infects, all dyadic interactions whatsoever, whether interpersonal, or between man and animal,[1] and by no means excepting the interactions of any living organism with a computer.[2]

In what follows, I will confine my observations to the three salient features suggested by the Pavlov episode which seem to me to be especially instructive for a general theory of signs. The first of these has to do with the notion of deception, especially within or at the perimeter of the academy, and the importance of being able to recognize different kinds and degrees thereof, ranging from out-and-out fraud for financial gain (say, royalties) and preferment (in the form, for example, of a doctorate), as in a scintillating instance of fictional ethnography, cleverly unwrapped by de Mille (1976; cf. Truzzi 1977). Imposture is sometimes alleged

[1]In view of the now hardly controvertible fact, underlined once again by Hediger (1974: 27–28), that the Clever Hans effect in "animals is only explainable by the continually repressed fact that the animal—be it horse, monkey or planarian—is generally more capable of interpreting the signals emanating from humans than is conversely the case," it is irksome to repeatedly come across reports fatuously stating that "in order to avoid the results of suggestion [certain] investigators decided to use animals rather than humans as their experimental organisms" (this in reference to mice, in a test of "laying-on-of-hands" healing, as reported by Rhine [1970: 316–317]).

[2]Cf. Weizenbaum's (1977) telling remark about the "power of . . . [his] computer program [being] no more and no less than the power to deceive," and the constant, inevitable, yet apparently discounted intrusion of Clever Hans cues into the Lana experiment intended to be conducted by means of an "impersonal" computerized system—see, e.g., Rumbaugh (1977: 159, 161), and the acerb comment on this project by Gardner and Gardner (this volume: 317–318) alleging that Rumbaugh's "results . . . presented thus far are more parsimoniously interpreted in terms of such classic factors as Clever Hans cues. . ." The Gardners claim that, to the contrary, testing procedures they themselves developed rule out this and kindred alternative interpretations. The procedures they refer to presumably involve the "double-blind" design, adapted from psychopharmacological researches. The objectivity of this method, however, though comforting, is altogether illusory; see, e.g., Tuteur 1957–1958. So what we have here is a blatant case of, paraphrasing Cervantes, the pot calling the kettle black.

where facts remain forever bafflingly insubstantial while nonetheless mortally dam-
aging, as in the melodramatic Paul Kammerer scandal made famous anew by Koestler (1971): was the principal in the case deliberately trying to perpetrate a swindle,
or was he an ingenuous yet suicidal victim of his own Lamarckian tendencies, or
did he have a Studentsov in his lab, and, if so, was this putative staff member
doctoring critical specimens of *Alytes obstetricans* either to please or to discredit
(124) his master? No possibility can be entirely excluded, just as we shall never
know whether Claudius Ptolemy is the most successful fraud in the history of science, as Robert R. Newton recently argued, or the greatest astronomer of antiquity,
as Owen Gingerich reaffirms. The question hinges on whether Ptolemy systematically invented or doctored earlier astronomers' data in order to support his own
theories, whether he was unknowingly deceived by a dishonest assistant, or
selected, for pedagogical purposes, just the data which happened to agree best with
his theory (Wade 1977).

From a semiotic point of view, the deliberate exercise of fraud and deceit—the
traditional confidence game or, as this is known to its practitioners, the con—is less
interesting than self-deception and its farflung consequences. For centuries, of
course, one very special and continuing form of the con has been perpetrated upon
marks by an operator using a tame, trained, domesticated animal, such as a horse,
as his or her pivotal prop. A celebrated equine in point, popularized in a ballad
published on November 14, 1595, was Morocco (see Fig. 2), "Maroccus Extaticus,
or Bankes [John Banks', the operator's] Bay Horse in Trance," whose astonishing
feats, suspected of verging on magic, were graphically portrayed, in 1602, by Jean
de Montlyard, Sieur de Melleray, in a long note (transcribed in Halliwell-Phillipps
1879: 31–36) to a French translation of the *Golden Ass of Apuleius*. If contemporary
accounts are to be believed, both Banks and Morocco were burned upon orders of
the Pope, as alluded to by Ben Jonson in one of his *Epigrams:* "Old Bankes the
juggler, our Pythagoras,/ Grave tutor to the learned horse, Both which,/ Being,
beyond sea, burned for one witch . . ." (1616). Pepys witnessed just such a horse,
operated for profit nearly a century later, as noted in his *Diary* for September 1,
1668: "So to the Fair, and there saw several sights; among others, the mare that
tells money, and many things to admiration; and, among others, come to me, when
she was bid to go to him of the company that most loved a pretty wench in a corner.
And this did cost me 12 *d.* to the horse, which I had flung him before, and did give
me the occasion to baiser a might belle fille that was in the house that was exceedingly plain, but forte belle." And Christopher (1970, Ch. 3) entertainingly relates
the adventures of "the most discussed animal marvel of recent times," a mare
named Lady, and her operator, Mrs. Claudia Fonda.[3] Although Dr. Joseph Banks

---

[3]As recently as 1975, one still finds books on communication between man and horse imbued with the
  Clever Hans Fallacy. Thus, Blake (1975, Ch. 10) devotes an entire chaotic chapter to "telepathy in horse
  language." He describes, no doubt accurately, his experiences with a horse, Weeping Roger (Ch. 7), but

Figure 2. John Banks with Morocco, the wonderful horse of 1595. In this woodcut, the steed is stomping out the numbers on a pair of rolled dice. From Robert Chambers, *Book of Days: A Miscellany of Popular Antiquities in Connection with the Calendar including Anecdote, Biography, and History Curiosities of Literature and Oddities of Human Life and Character* (London and Edinburgh: W. & R. Chambers, 1869).

Rhine declared Lady "the greatest thing since radio,"[4] and claimed that she possessed ESP, the skillful conjuror and historian of magical entertainment exposed the technique used by Mrs. Fonda, an obvious trick—obvious, that is, to mentalists—sometimes employed by mediums and known as pencil reading.

Christopher's key sentence reads (45): "If Dr. Rhine was interested in testing for ESP, he should have ignored the horse and studied Mrs. Fonda." He is restating here a basic principle, explicitly recognized already in 1612 by a certain Samuel Rid, the author of a wondrously sophisticated instructional manual, or how-to-do-it

---

goes on to imply an absurd explanation: "I discovered that I could direct [this stallion] where I wanted to go just by thinking it. I would steer him to the left or right or straight ahead simply by visualizing the road. This was the first time I had consciously experienced telepathy with a horse" (p. 126). Elsewhere (p. 94), he remarks, "I was always at one with him." Plainly, all the constituents for a Clever Hans setup are present, but Blake still finds it necessary to resort to ESP instead of the correct semiotic explanation, which he apparently knows nothing of.
[4]Perhaps echoing Upton Sinclair's (1930: 4) technologically puerile yet by virtue of that very fact endearing simile, comparing ESP to "some kind of vibration, going out from the brain, like radio broadcasting." This imagery has its ultimate source in Democritus.

Figure 3. A: Presenting Clever Hans. *B:* Testing Clever Hans. *C:* Hans blindfolded. The place of attachment of the blindfold was tightly closed by an impenetrable flap which reached under Hans's neck. From Karl Kroll, *Denkende Tiere* (Leipzig, 1912).

book, of whom, alas, nothing further is known. This book, *The Art of Juggling*, ought to be made required reading for all would-be semioticians; here I will reproduce only a brief passage of commentary on the exploits of a performing horse, presumably Morocco:

As, for ensample, His master will ask him how many people there are in the room? The horse will paw with his foot so many times as there are people. And mark the eye of the horse is always upon his master, and as his master moves, so goes he or stands still, as he is brought to it at the first. As, for ensample, his master will throw you three dice, and will bid his horse tell you how many you or he have thrown. Then the horse paws with his foot whiles the master stands stone still. Then when his master sees he hath pawed so many as the first dice shews itself, then he lifts up his shoulders and stirs a little. Then he bids him tell what is on the second die, and then of the third die, which the horse will do accordingly, still pawing with his foot until his master sees he hath pawed enough, and then stirs. Which, the horse marking, will stay and leave pawing. And note, that the horse will paw an hundred times together, until he sees his master stir. And note also that nothing can be done but his master must first know, and then his master knowing, the horse is ruled by him by signs. This if you mark at any time you shall plainly perceive.

Let me underscore Rid's last sentence: "This if you mark at any time you shall plainly perceive." The point is that, until the advent of Oskar Pfungst in 1907 (1965 [1911]), no scientist that we know of had the insight to ask an animal—in this instance, Clever Hans, the horse of Herr von Osten—a question to which the inquirer himself did not know the answer. It turned out that, no matter how severely skeptical the audience, whether unschooled or expert, was, it was the observer who had involuntarily and unknowingly signed to the observed to stop tapping at the precise instant where the message destination—alive to the correct answer—expected the message source to cease emitting. "This," Polanyi (1958: 169–170) says, "is how they made the answers invariably come out right" (continuing: "this is exactly also how philosophers make their descriptions of science, or their formalized procedures of scientific inference, come out right").

Actually, Lord Avebury, in the 1880s, came very close to rediscovering the correct solution in his experiments with Van, his black poodle, supplemented by his casual inspection of other dogs, some score of years before Pfungst, who himself regarded Van "as a predecessor of our Hans" (1965: 178). Avebury had the right attitude to begin with, "that hitherto we have tried to teach animals, rather than to learn from them: to convey our ideas to them, rather than to devise any language or code or signals by means of which they might communicate theirs to us" (Lubbock 1886: 1089). He sensitively discerned that when a dog—or a chimpanzee (see Thomson 1924: 132) for that matter—is taught how to "count," the operator need not, in fact, ordinarily does not, "*consciously* give the dog any sign, yet so quick [is] the dog in seizing the slightest indication that he [is] able to give the correct answer. . . . Evidently, the dog seize[s] upon the slight indications unintentionally given" (Lubbock 1886: 1091).

Figure 4. Lord Avebury. Courtesy of N. H. Robinson, Librarian, The Royal Society in Britain.

Avebury, furthermore, shrewdly connected these observations "with the so-called 'thought-reading,'" one variant of which, commonly known as "muscle reading," came eventually to be investigated in painstaking detail by three prominent Berkeley psychologists, Edward C. Tolman among them. "Muscle reading" was shown to crucially hinge on the performer's perception of motor signs of an exceedingly delicate character, signs, moreover, "unintentionally" communicated to him "by each of the persons who acted as his guide" (Stratton 1921; discussed further in Sebeok 1977b).[5] It is established by now beyond serious doubt that the

---

[5]On "muscle reading" as explanation for other pseudo-occult phenomena, such as the movement of a Ouija board, table tipping, and automatic writing, see Gardner (1957: 109), who speaks of the "unwitting translation of thoughts into muscular action . . ." See also Vogt and Hyman (1959, Ch. 5). Regarding the most flashy of contemporary "psychics," the Israeli stage-performer Uri Geller, see Marks and Kammann (1977: 17), who similarly conclude that "parsimony dictates the choice of normal explanations for the phenomena described. . . . Geller's procedures allow him to use ordinary sensory channels and ordinary motor functions." Incidentally, James ("The Amazing") Randi, a top-flight Canadian conjurer,

working ingredient of many other mind-reading acts—much in the manner of the children's game of Hot and Cold—consists of unwitting and inadvertent nonverbal signs transmitted from audience to "psychic"; nor is this surprising, "since people constantly pass nonverbal signals to each other through such things as changes in their tones of voice and body movements. In fact, this nonverbal communication forms the basis of a well-known magic act. One performer, for example, asks to have his check in payment for a show hidden in the auditorium in full view of his audience. He then comes on stage and finds the check by reading the nonverbal cues of the audience as he wanders closer to or farther from the check" (Kolata 1977: 283, interviewing Persi Diaconis, who is both a prominent mathematician and magician; the identical illusion is discussed, in his somewhat hokey style, by Kreskin [George Kresge, Jr.] [1973: 80–84], describing how "I concentrate on reading every direction, every clue, and sensitize myself to hear or see any supportive factors beyond the perceived thought. . . . It can be likened to a highly stimulating game of charades . . ."). This example is far from insignificant, since, as Diaconis emphasizes, it suggests an enormous problem area of "how much usable information is being transmitted in this way and what the best guessing strategy is," which arises in many contexts other than parapsychology—in fact, whenever and wherever organisms interact.

As to the mental operation of guessing, it was none other than Peirce (1929: 269–270) who had emphasized that "its full powers are only brought out under critical circumstances," a claim he went on to substantiate in a colorful extended narrative of a true personal incident in which the great philosopher was metamorphosed into a master sleuth (for the full story, see: 267–282). As one of his editors summarizes the anecdote, it concerned "the theft of [Peirce's] coat and a valuable watch from his stateroom on a Boston to New York boat. He says that he made all the waiters stand in a row and after briefly talking with each, but without consciously getting any clue, he made a guess as to which one was guilty. The upshot of the story is that after many difficulties, and by making more successful guesses, he proved that his original guess had been correct" (Peirce 1935–1966: 40 n. 15; cf. 7.45). What Peirce attempted to do by talking briefly with each man in turn was, as he put it, "to detect in my consciousness some symptoms of the thief" (Peirce 1929: 281). His expectation was that the crook would emit some unwitting index, but Peirce also stressed that his own perception of telltale signs, while he held himself "in as passive and receptive a state" as he could, had to be unconscious, or, to use a preferred term he suggested, unself-conscious, "a discrimination below the surface of consciousness, and not recognized as a real judgment, yet in very

---

has publicly duplicated all of Geller's feats. Concerning Peirce's disapprobation of telepathy, "with its infrequency and usual deceptiveness" (Peirce 1935–1966: 7.686), and of kindred psychic doctrines and claims, see his extended if apparently incomplete essay on "Telepathy and perception" (597–686).

Figure 5. C. S. Peirce: an official Coast and Geodetic Survey photograph; he would not have been quite forty at the time. Photo obtained courtesy of Max H. Fisch. Peirce Edition Project, Indiana University, Purdue University, Indianapolis.

truth a genuine discrimination . . ." (ibid.: 280).[6] He mentions two conjectural principles that may furnish at least a partial explanation for his successful application of "this singular guessing instinct. I infer in the first place," he concluded, "that man divines something of the secret principles of the universe because his mind has developed as a part of the universe and under the influence of these same secret

[6]The incidents happened on Saturday, June 2, 1879. Peirce recovered his watch on the following Tuesday. The two culprits were committed for trial on Wednesday, June 25. Peirce's detective procedure is compared in detail, in Sebeok and Umiker-Sebeok (1979), with the famous "method" of Sherlock Holmes, wherein the similarity is accounted for by virtue of their common roots in Natural Semiotics (including medical). Kreskin (1975: 27–28) incidentally sketches a stage illusion, *Guilty,* which unfolds precisely according to the strategy devised by Peirce, in applying which, Kreskin claims, it is "impossible for the 'guilty' person not to give himself away . . ." For a flagrant case of real life abuse of "telepathy" in law enforcement, seemingly motivated by social prejudice, see Posinsky (1961).

principles; and secondly, that we often derive from observation strong intimations of truth, without being able to specify what were the circumstances we had observed which conveyed those intimations" (1929: 281–282). In Peirce's incomparably insightful fashion, the first principle adduced provides the ultimate evolutionary rationale for the workings of the Clever Hans effect, while the other addresses its specifically semiotic roots.

The work on deception by illicit communication in the laboratory recently adumbrated by Pilisuk and his collaborators (1976) surely is on the right track, but merely scratches the surface of deception as a pervasive fact of life characteristic of experimental studies of human and animal behavior; Rosenthal (1976: 156), for instance, admits that "deception is a necessary commonplace in psychological research," although I believe that he tends to substantially underestimate (1976: 388) potentially harmful consequences, particularly in the context of placebos, which may have decided toxic effects and even the power to produce gross physical change (see, e.g., Beecher 1955: 1606), as well as of the dubious role of double-blind "controls."[7]

The first general lesson of the Pavlov episode thus boils down to this: be ever on the lookout against deception, but beware, above all, of self-deception. The second moral is expressly methodological, and may be best understood in a semiotic frame. It has been formulated, as we saw, in more or less the same way by Rid, Avebury, Pfungst, Christopher, and stated perhaps most comprehensively in the title of this article. Pfungst (1965: XXX) and his chief, the eminent psychologist Carl Stumpf, distilled the essence of their investigation by recognizing and admitting that the Hans Commission made the initial mistake of "looking for, in the horse, what should have been sought in the man." In physics, one speaks of couplings between the observer and the observed, and keeps asking how the former affects the latter. In psychological jargon, the experimenter becomes a proxy for "man," while "horse" can stand for any subject, whether human or animal (Rosenthal 1976). In anthropological, folkloristic (Fine and Crane 1977), and even linguistic (Sebeok 1977b), field work, we are concerned with the distorting influence of elicitor upon native informant. In a clinical setting, we are interested in what the agentive physician's (or quack's or shaman's) personality and paraphernalia contribute toward the healing of the patient/client (Sebeok 1977b, 1979, chap. 10). In the argot of the con, the police want to know how does the operator "take" the mark? All these dynamic/dyadic relationships between living systems have specific commonalities ultimately modeled on or, more exactly, programmed after the one universal dependence relationship which must be both basic and paradigmatic: the cybernetic cycle that prevails between mother (or other caretaker) and child. The nature of this

[7]See note 2; I intend to return in much more depth elsewhere to these complex semiotic topics, which I had occasion to discuss but briefly before (Sebeok 1977b and 1978).

system, "in which one partner assumes the functions of sensor and regulator for the other one and vice versa," was first outlined by Thure von Uexküll (1978), in an attempt to account for the efficaciousness of placebos. All of us are assumed to be reliving and reiterating the early months of our extrauterine existence, when gesturing, posturing, vocalizing, and eventually articulating wicca words like "mama" produced something from nothing—milk and toys, for instance—"out of the blurry, remote world of the adult gods," in Wagoner's (1976: 1598) apt and evocative conceit.

Although von Uexküll states his hypothesis in exceedingly fruitful semiotic terms, it is likewise in obvious conformity with psychoanalytic theory, which suggests consideration of this primal program as a reactivation of a pivotal early experience and one which "may be a permanent available pattern of social interchange in human life, which is not confined only to child–mother or patient–doctor relationship" (von Uexküll 1978). Plausible as this formula appears, it nevertheless leaves open the question most often asked about the seemingly miraculous placebo effect and comparable forms of therapy—say, the "laying on of hands" (currently taught at the graduate level at the New York University School of Education, Health, Nursing and Arts Professions) or "mother's kiss or voodoo drums, leeches, purgatives, poultices, or snake oil" (Moertel et al. 1976: 96)—or, indeed, the workings of one's belief in Christian Science (Sebeok 1979): namely, how are the semiotic agencies and habiliments transmuted into physiologically operational mechanisms? The answer was foreshadowed in Janet's (1925: I:43–53) discussion of the value of miraculous methods of treatment, from the shrine of Aesculapius to Lourdes. Cannon's classic article (1942) on the cause of "voodoo" death notwithstanding,[8] much fascinating research remains to be done at the borders of the sign science with the life science before this problem can be wholly resolved.

Another area of role-demand where the von Uexküll paradigm is palpably manifested is in hypnotic and posthypnotic responsiveness. As in the placebo effect— for, as Paul Sacerdote emphasizes, "hypnosis may be in many ways the most powerful of placebos" (Holden 1977: 808)—the audience, or, using a semiotic term with a broader charge, context (cf. Fisher 1965: 85), serves at least four functions that combine to reinforce the realization and maintenance of so-called hypnotic behavior; these were conveniently summarized by Sarbin and Coe (1972: 96–97), but may be assigned to a wider category of effects sometimes called *artifacts*, such as

---

[8] Francis Huxley, who professes to believe in the existence of ESP (1967: 282), and appears perversely unaware of Cannon's highly significant study of a quarter of a century earlier than his, nevertheless gropes toward an analysis of voodoo in semiotic terms: "Is it . . . possible that symbols," he asks, meaning icons, "by containing the field of relationships and providing the ground of consciousness, are responsible for what we call ESP?" (302). Discussion of the etiology of voodoo death continues in anthropological and other circles; for a summary of the recent literature and latest interpretations, see Lex 1974.

Figure 6. Bernadette of Lourdes. From the author's files.

increased motivation and role-playing, in contrast to *essence,* which, if it really exists, refers to what is more or less vaguely known as "an altered state of consciousness," or sometimes "cortical inhibition" or "dissociation" (Orne 1959). Artifacts are systematic errors stemming from specifiable uncontrolled conditions—a bouquet of subtle cues emanating from both the experimental procedure and the experimenter. Thus investigation has revealed that the paraphonetic features selected by the source—viz., forceful or lethargic tone of voice—constitute a ruling variable which must, if feasible, be carefully controlled (Barber and Calverley 1964). The hypnotized subject exhibits the behavior which he thinks the hypnotist expects of him, or, more accurately, what he thinks hypnosis is. The phrase "demand characteristics" is applied to this invigorating idea in the history of hypnosis research (Sheehan and Perry 1976), to which Jaynes's (1977: 385) notion of the "collective cognitive imperative" corresponds exactly.

The intimate mutual gaze of lovers furnishes one example among many of how this fundamental paradigm is played out in young adulthood: the reason why both a boy and a girl spend so much time peering closely into each other's eyes is that "they are unconsciously checking each other's pupil dilations. The more her pupils expand with emotional exictement, the more it makes his expand, and vice versa"

(Morris 1977: 172). The pupil response is, as a rule, unknowingly emitted as well as, even more often, unknowingly perceived (Janisse 1977; Sebeok 1977c). "Hip dudes," metaphoric "cats," wear dark glasses, or "shades," like the Chinese jade dealers of yore, to conceal their excited pupil dilation and thus to project a cool look—one that demands heightened participation of their "transparent" interlocutors (Gump 1962: 229).[9]

Semiotics, which is commonly defined as the study of any messages whatsoever, whether verbal or not, must be equally concerned with the processes of generation and encoding on the part of the most various sources, whether human or not; with the transmission of any string of signs through all possible channels; and with the successive processes of decoding and interpretation on the part of the most various destinations, whether human or not. What the Pavlov tale reminds us of is the peculiar force of the linkage joining any source with any destination. In marveling at the accomplishments of animals—especially hand-reared dolphins in the 1960s and the great apes in the 1970s—trained to engage in two-way communication with man, attention to the behavior of the human has all too often been either shunted aside by deliberate misdirection (imposture) or ignored in innocence (self-deception). Thus, many people cherish the belief that police dogs are infallible as trackers, enabling them to recognize the trail of a stranger after getting the scent. However, in one historic experiment (Katz 1937: 8–10), it turned out that the man in charge of the police dogs had provided unwitting cues: in other words, "it was not the dog guiding the man, but the man guiding the dog owing to his preconceived opinion about the result to be expected."[10]

Those who stage-manage the circus antics of apes have known for centuries what scientists who aim to instill manually encoded and visually decoded verbal communicative skills in such animals have still scarcely grasped. It is widely imagined, for example, that imitation of the human model in learning situations of this sort is critical. On this issue, Hachet-Souplet (1897: 83–84, 91), author of the standard textbook on dressage, quotes Buffon: "'Le singe, ayant des bras et des mains, s'en sert comme nous, mais sans songer à nous; la similitude des membres et des

---

[9]See Umiker-Sebeok (1979) for a detailed treatment of the elaborate semiotic code for partial or total eye concealment by means of eyeglasses and other devices in American culture.

[10]One side effect of this 1913 experiment was a decisive improvement in the training of police dogs and in their consequent accuracy in tracking. Katz's conclusion is, of course, equally applicable to any "muscle reading" act. The performer may have a spectator take hold of his hand believing that "he is being led by the magician, but actually the performer permits the *spectator to lead him* by unconscious muscular tensions" (Gardner 1957: 109). The best muscle readers, like the famous Eugen de Rubini (whose case I discussed in 1977b), may dispense with physical contact altogether, relying on far more elusive guiding cues, such as tremors of the floor, faint sounds of feet, movements of arms and clothing, and/or those made by changes in breathing (Rinn 1950: 531). The workings of several variants of the Clever Hans theme were known to scientists of the stature of Michael Faraday (table turning) and Michel Eugene Chevreul (the magic pendulum) by at least the early 1850s (Hansel 1966: 33–34).

organes produit nécessairement des mouvements qui ressemblent aux nôtres; étant conformé comme l'homme, le singe ne peut se mouvoir que comme lui; mais se mouvoir de même n'est pas agir pour imiter'. . . . Du reste,'' the canny author concludes, ''le public se laissait prendre à cette ruse innocente.'' When the subject patently fails to imitate the trainer, this imperfection, too, is reinterpreted to fit with the anticipated design. Patterson (1977a), for instance, instructed Koko to smile for a photograph. Her gorilla signed ''frown'' or ''sad.'' The psychologist's explanation of this contrary behavior was not at all that Koko responded erroneously; Patterson's preconception of her design constrained her to assert that negative occurrences of this sort ''demonstrate [the ape's] grasp of opposites.'' With dialectic unfolded in this vein how can you lose?[11]

What actually happens, as Hediger (1974: 40) keeps patiently repeating, is that whoever poses the question about the linguistic accomplishment of apes

> often already has preconceived ideas about the outcome of the experiments, indeed, he must frequently have possessed such ideas before being able to set up the experiment in the first place. Another factor is the choice of suitable experimental animals. It is up to him to choose a suitable species and individuals, treat and prepare them in a definite way. In this, the 'context' . . . , are already included many possibilities of influence by channels still largely unknown to us.''

The modish mirage of the Pathetic Fallacy, or the attribution of human characteristics to objects in the natural world, especially to the speechless creatures populating it, reinforced in ways that are more or less well understood (cf. Sebeok 1977b), is so powerful that observers are not uncommonly prone to report a Barmecidal feast of signs where the more candid among them admit to having perceived none. Thus Stokoe (1977: 1)—a leading expert on Ameslan—remarks about some infant chimpanzees: ''These baby chimps sign as they move—very rapidly; and we often found that we had seen a sign or a sequence of two or three signs without consciously realizing that we had in fact seen it.'' Stokoe's encounter with baby Dar and bantling Tatu is disturbingly reminiscent of my own experience in the early 1960s with dolphins in Miami's long defunct Communication Research Institute. In that laboratory, *Tursiops* was being trained to mimic the speech of a human investigator by standard operant conditioning technique. Numerous rumors and some reports were put in circulation to the effect that the animals, especially Elvar and Chee-Chee, were indeed capable of reproducing words ''appropriately.'' Of Elvar,

---

[11] Parents who act on the assumption that their child is bright appear to proceed in just this way. I recently observed an infant of seventeen months being fed beef. Her mother interrogated her, ''What's this?'' The daughter replied, ''Chicken.'' The mother observed, ''She loves to tease me!'' She then followed this remark up with a further unsubstantiated general comment: ''She enjoys making a game out of oppositions.'' Bingham (1971) has shown that preverbal children are addressed in a carefully accommodated register by mothers who judge that *their* infants have the capacity to understand quite a bit, but not by mothers who set a lower estimate on *their* infants' capacity.

it was avouched, for instance (Lilly 1963: 114): "He does not reproduce a word in 'tape-recorder' fashion or in the fashion of a talking bird. In one's presence he literally [*sic*] analyzed acoustic components of our words and reproduced various aspects in sequence and separately." Perhaps mistaken for a mark, I was permitted to observe one training session, and later to listen to recordings of several previous sessions. I heard only random dolphin noises, no dialogue. My puzzled demurral was countered by the assertion that these coastal porpoises articulated much too fast for their emissions to be interpreted by the human ear unaided: understanding presupposed analysis by means of the sound spectrograph and oscillographic methods. It was shown a decade or so afterward that the papers published in scientific journals by this Florida research group "provided no solid evidence in support of [such] speculations" (Wood 1973: 91). The project was, accordingly, scratched, in 1968, altogether. The long shadow of Clever Hans darkened that undertaking from the start, as is perfectly patent from a sentence the principal had printed in italic type: "*And he* [i.e., Elvar] *first did it* [i.e., spoke] *when and only when we believed he could do it and somehow demonstrated our belief to him*" (Lilly 1963: 114).

Stokoe (1977: 1) drew another methodological conclusion from his observation in Nevada, or, more accurately, the lack of it: videotape or film "can never be an adequate substitute for trained live observation . . ." This, however, holds only if the inevitability of voluntary and involuntary influence upon the animals being experimented upon is objectively and critically recognized and assessed at every turn, and if all conceivable media of communication between men and animals are kept in constant view. Concerning the work Stokoe describes, and the like, it is not enough to exclaim in awe on what the chimps do; the real challenge is to uncover—the relationship being reciprocal—what the University of Nevada team, for one, is up to.

The use of recording devices is no panacea, of course. As F. J. J. Buytendijk's scrutiny of a film of a fight between a mongoose and a cobra established, the reaction time of their coordinate exchange of some messages is so short that it can neither be viewed by human observers nor re-viewed even in slow motion. This is explicable in terms of the concept of zero signifier (Sebeok 1976: 118). "These dissimilar combatants behaved part of the time like a pair of dancers, in which each anticipated the other's next movement" (Hediger 1974: 38), that is, their reaction time was reduced to naught. Hediger believes that something similar takes place in the circus, for example, between a skilled trainer causing a panther sitting on a pedestal to strike out with a forepaw and withdrawing in exquisite accord with that movement, or a springboard acrobat adjusting his leap to the blow of the elephant's foot at the other end of the plank. In his keen observations on movement coordination, or microsynchrony, in human social interaction, Kendon (1977: 75) has noted the same kind of foreknowledge: "The precision with which the listener's movements are synchronized with the speaker's speech means that the listener is in some way able to anticipate what the speaker is going to say . . ."

Hediger's mention of channels focuses attention on yet a third dimension of the Pavlov yarn. It is insufficient to shift one's attention back from the destination to the source. It is essential to consider, as well, the means whereby the two are conjoined. Although the visual, auditory, tactile, and chemical mechanisms of rodent communication, for instance, are understood to a degree (Eisenberg and Kleiman 1977: 637–649), no one, least of all the principals, had the slightest idea how, precisely, Studentsov unwittingly disciplined Pavlov's mice; neither was Rosenthal (1976: 178) able to determine to his satisfaction how his "bright" and "dull" rat subjects were differentially educated by his naïve students: "we cannot be certain of the role of handling patterns as the mediators of the experimenters' expectancies, nor of whether such other channels as the visual, olfactory, and auditory were involved." As to this, we can but reiterate Hediger's query and observation (1974: 39): "How many channels exist between man and animal? We know little more today than we did half a century ago, i.e., that many other channels exist besides those of optic and acoustic question and answer. On account of the inadequacy of our sense organs and the apparatus at our disposal, such channels remain for the moment unknown. It is known, however, that many apparently quite objective laboratory experiments have given, and continue to give, false results for the very reason that many experimenters believe themselves aware of and able to control all the channels of communication existing between those conducting the experiment and the animal involved."

The following principles deserve, in consequence, attentive consideration:

1. Any form of physical energy propagation can be exploited for communication purposes (Sebeok 1972: 40, 67, 124).

2. Channel selection is governed and constrained by the source encoder's sensorium. The source decoder will generate an acceptable reproduction of the source output if endowed (at least in part) with a correspondingly functioning sensorium.

3. It is reasonable to assume that messages are routinely transmitted between organisms through hitherto undiscovered or as yet scarcely discerned channels. One arresting case in point is the electrical channel, "a new modality" (Hopkins 1977: 286), the multifaceted communicative functions of which are in the process of being actively disclosed.

4. The range of each of man's sense organs is significantly exceeded by those of a host of other animals. Hediger (1974: 32) cites Pfungst as having demonstrated that the horse is capable of perceiving movements in the human face of "less than one fifth of a millimetre." Pierce and David (1958: 102–103) relate amusingly how a trio of electronics experts learned about the ultrasonic stridulation of crickets, drawing from this story the moral "that we hear only what we can hear, and that

Figure 7.A: Georges Roux's conception of Dr. Johausen luring his "informants" to him by making music.

there may be a great many obvious differences among sounds which must forever escape our ears," wisely adding: "to some degree we hear what we expect to hear." Parallel comparisons can be adduced, *mutatis mutandis,* about the human eye, to say nothing of the olfactory field.

5. Man has invented a variety of technical aids to enhance the ineffectualness of his channel capacity. However, such intensifying equipment "has frequently been shown to have been a [further] source of error . . ." (Hediger 1974: 30).

6. Before resorting to cheap ad hoc paranormal rationalizations, a sophisticated, if time-consuming, research program must be conducted to pin down the mechanism actually at work in each instance. Elegant and exhaustive investigations of this character are illuminatingly inventoried in Vogt's and Hyman's (1959, Ch. 6) psychophysical exegesis of the movement of the dowsing rod in water witching (cf. Gardner 1957: 101–113). The contrary is illustrated by the widely publicized case of Rosa Kuleshova (Pratt 1973: 63), who was reputed to be capable of "seeing," particularly reading, through her fingertips. Astute press-agentry led to a global rash of other reported "dermooptical" manifestations (Sebeok 1977b), in the early 1960s, all of which turned out to be phony (Gardner 1966). "X-Ray Eye Act" is the profes-

Figure 7(cont.). B: Georges Roux's portrayal of Sa Majesté Msélo-Tala-Tala after the scientist lost his language capacity. From Jules Verne, *Le Village aérien* (Paris: Collection Hetzel, 1901).

sional designation of hoaxes of this nature,[12] where the performer can easily open his or her eyes and is able to look down both sides of the nose; blindfold magic can be achieved with seemingly impenetrable coverings like bread dough, silver dollars, wads of cotton, powder puffs, folded paper, sheets of metal, adhesive tape, and, of course, a variety of cloth shields.

The small but influential segment of mankind that can afford leisure for the contemplation of such matters longs to establish communication links in two opposite directions: with the rest of animate existence (plant forms, involving phytosemiosis, as well as animal forms, involving zoosemiosis), in the matrix of which our lives lie inalienably embedded; and with supposititious extraterrestrial civilizations. Leaving unearthly aspirations and efforts aside (see, e.g., Ponnamperuma and Cameron 1974: 213–215 for selected references to "interstellar communication lan-

---

[12] In part no doubt inspired by Jules Romains, the French writer, who was obsessed with "paroptic" vision, or "eyeless sight." His book on this subject (Romains 1920; American version, 1924) was widely read in the postwar years here and throughout Europe.

guages''), one can confidently assert that the fundamentals of code-switching between our species and not a few others are adequately understood, not just intuitively—that kind of comprehension was the imperative semiotic prerequisite for domestication—but also scientifically, thanks, in the main, to Hediger's brilliantly creative lifelong spadework (cf., *inter alia,* Hediger 1974, and the references given in Sebeok 1976: 219–220). Two-way zoosemiotic communication is thus not at issue, but such communication between man and animalkind by *verbal* means is quite another matter. The fascinating paradox of language-endowed speechless creatures has been iteratively resolved in myth and fiction, but not in reality. That search, for a resolution of the authentic kind, has lately taken a disturbingly pseudoscientific turn. An account of the socioeconomic reasons for this craze, interesting though it may be, of ''humanizing'' pets, quasi-feral terrestrial and marine mammals, and an occasional tame bird,[13] falls outside of the scope of this article.

Leo Szilard's satirical story ''The Voice of the Dolphins'' (1961) and Robert Merle's thriller *The Day of the Dolphin* (1967) are chimerical treatments of the same theme in what may well be called the Decade of the Dolphin. In the 1970s, writers have, fittingly, emerged from the brine. Peter Dickinson's ''chimpocentric'' tale of detection, *The Poison Oracle* (1974), where the action hinges on the linguistic capacity of an ape, and John Goulet's affecting book *Oh's Profit* (1975), the hero of which, a gifted young signing gorilla, is pitted against the merciless forces of a singularly sinister coalition of linguists, are modern transfigurations of Jules Verne's diverting (if today seldom read) parodic science fiction pastiche, *The Great Forest* (originally published with his *Le Village aérien,* in 1901). This work was inspired by the genuine, if eccentric, exploits of Richard L. Garner, who, in 1892, left America on a field trip for Gabon, where he lived in Libreville for two years. He then proceeded upcountry, where he was sheltered at a mission of the Fathers of the Holy Ghost, located on the banks of the Ogowe. In due course, he published (Garner 1892) a book on the ''speech'' of monkeys. His studies were themselves an odd mishmash of valuable observations, pure inventions, and colorful humbug: ''Peut-être a-t-on souvenir de l'expérience à laquelle voulut se livrer l'Américain Garner dans le but d'étudier le langage des singes et de donner à ses théories une démonstration expérimentale,'' Verne questions tongue-in-cheek, and then goes on to invent a lunatic proto-ethologist, one Dr. Johausen (obviously Garner, but in Teutonic guise), who journeys to Central Africa to seek out ''le prétendu langage

---

[13]Chauvin-Muckensturm (1974: 207) explicitly compares the drumming code she imparted to her Greater Spotted Woodpecker to the man–monkey performances variously shaped by the Gardners and Premack, stressing that *''le bec est au moins l'égal de la main du chimpanzé.''* This woodpecker is French. It will not have escaped notice, however, that the happily defunct myth of dolphin discourse, as well as the currently continuing promotion of primates to the status of a putatively (Limber this volume) productive *animal loquens,* have been confined, so far without a single exception, to the United States.

des singes." Predictably, he finds just what he was expecting to find—speaking monkeys—but with a difference: "Ce qui les distingue essentiellement des hommes [est qu'ils] ne parlaient jamais sans nécessité." In passing, Verne makes some exceedingly prescient observations about language and cognition, intelligence and verbal propensity, and animal communication in general. The story ends with an ironic twist: Johausen's expectations are indeed fulfilled, and he even rises to become the ruler of the beasts, Sa Majesté Msélo-Tala-Tala, but the cost he has to pay for his achievement is enormous: that price is the loss of his most precious possession, his own language, which is to say, his humanity: "Il est devenu singe . . .'' Thus, in an unending cycle, does Pop Art burlesque scientific lore while Big Science apes *(le mot juste)* the presentiments of Pop Culture—no less in today's ecologically remorseful USA than in yesteryear's Lysenko-ridden USSR.

The road from Russian rodents to American apes is paved with good intentions, but for an innocent onlooker, trained in the sign science, at least three signposts pointing to a need for ventilation loom behind and ahead, each beckoning to as yet insufficiently explored byways at the dangerous intersection of two synergetic causes of error: the Clever Hans Fallacy and the Pathetic Fallacy. The trio of problems that seem, from a semiotic point of view, to cry out for immediate, impartial, intensive investigation are: the destructive pitfall of self-deception; the predominance, in dyadic encounters, of the source over the destination; and the paucity of accurate knowledge about the multiplicity and range of natural channels connecting both extremities of the communication chain.

HUMAN LANGUAGE AND OTHER SEMIOTIC SYSTEMS*

I would like to begin with a familiar and elementary conceptual distinction between two quite different questions: (1) what is a *human language?* (2) what is a *language?*

The first question, what is a human language, belongs in principle to the natural sciences. It is analogous to such questions as: what is the nature of the human visual system, or the human system of locomotion? There are some quite reasonable research strategies aimed at answering this question. One approach, which happens to interest me particularly, begins with the observation that knowledge of language—what is sometimes called *competence*—develops through a series of stages reaching a mature *steady state,* some time before puberty, after which changes are quite marginal to the system. We may therefore assume that the organism passes from an *initial state* to a *final state* through some interaction with the environment. The final state attained distinguishes a speaker of Japanese from a speaker of English. The initial state distinguishes a human from a rock or a bird or a chimpanzee, which do not attain the same steady state given comparable experience. We may assume that the initial state is a uniform species property (to first approximation), genetically determined as a part of the human biological endowment. We may think of this initial state as a function that maps evidence available into state attained, just as we can think of a genotype as a function that maps a course of experience into a phenotype. We can then proceed to characterize the mature steady state attained and the experience required to attain it, and by so doing, can propose ideas concerning the initial state. The range of this function is the class of human languages—or, more precisely, the class of grammars of human languages, since what is actually acquired in the steady state is a finite grammar, neurally represented in some manner, that specifies phonetic, semantic and structural prop-

*Paper delivered on February 16, 1978, in a symposium on the Emergence of Language: Continuities and Discontinuities, held at the annual meeting of the American Association for the Advancement of Science.

NOAM CHOMSKY • Department of Linguistics and Philosophy, Massachusetts Institute of Technology, Cambridge, Massachusetts 02173.

erties of an infinite class of linguistic expressions. While problems arise in pursuing this research program and refining the concepts that enter into it, the general line of inquiry seems clear and has been profitably pursued. In a comparable way, we may study some physical organ of the body to determine its intrinsic nature, say, the visual system of a cat.

The second question, what is a language, is not, as it stands, a question of science at all—just as the question, what is a visual system, or a system of locomotion, is not as it stands a question of science. Rather, these questions pose problems of conceptual analysis. To determine whether music, or mathematics, or the communication system of bees, or the system of ape calls, is a *language,* we must first be told what is to count as a *language.* If by *language* is meant *human language,* the answer will be trivially negative in all these cases. If by *language* we mean *symbolic system,* or *system of communication,* then all these examples will be languages, as will numerous other systems—e.g., style of walking, which is in some respects a conventional culturally-determined system used to communicate attitude, etc. If something else is intended, it must be clarified before inquiry can begin.

There is still another, perhaps related, question that might in principle be profitably investigated. Suppose that two biological systems have been characterized to some level of success, with a partial account of the initial and final state—for example, the visual systems of a cat and an insect. We may then ask whether anything can be learned about one of these systems from the results attained in the study of the other. Note that this is not a question of science in the sense in which the question, what is the cat's visual system, is a question of science. But it may be a question relevant to science. One can raise questions about homology or evolutionary origin, though considerable caution is naturally in order. Gross functional or phenomenal similarities can mean very little, as is well known. Suppose, for example, that I meet a biologist who is studying the flight of birds, and I suggest to him the following line of inquiry: "What is 'flying'? It is an act in which some creature rises into the air and lands some distance away, with the goal of reaching some remote point. Humans can 'fly' about 30 feet, chickens about 300 feet, Canada geese far more. Humans and chickens 'cluster' (only an order of magnitude difference), as compared with geese, eagles, and so on. So if you are interested in the mechanisms of bird flight, why not pay attention to the simpler case of human 'flight'." The biologist is not likely to be too impressed.

I don't mean to suggest that all variants of this third question are necessarily that silly. They are not. For example, Richard Gregory (1970) has suggested that the human language capacity "has its roots in the brain's rules for ordering retinal patterns in terms of objects," that is, "in a take-over operation, in which man cashed in on" the development of the visual system in higher animals. While the suggestion seems to me dubious, it is by no means outlandish. Speculation about homology and evolutionary origins is legitimate, observing proper caution.

Let us turn briefly to the first question: what is a human language? We may study a cognitive structure such as human language (analogously, a bodily organ) along several dimensions: (a) structural principles; (b) physical mechanisms; (c) manner of use; (d) ontogenetic development; (e) phylogenetic development; (f) integration into a system of cognitive structures. Currently, results of any depth or complexity fall primarily along dimensions (a) and (d), to my knowledge, and in particular relate to a particular aspect of (d), namely, the character of the initial state.

In standard terminology, a characterization of the mature state attained in the case of a particular language is a *grammar* of that language: a system of rules and principles that specifies the properties of its expressions. The term *universal grammar* is used to refer to a system of principles to which any grammar must conform as a matter of biological necessity, thus, a characterization of a crucial aspect of the initial state. One theory of language development, which I believe is very plausible, postulates that the actual grammar representing knowledge attained results by fixing certain parameters in the genetically-determined universal grammar and adding further specifications and articulation within a narrowly circumscribed range. Universal grammar, then, determines the essential nature of human language; each grammar specifies a particular instance.

Consider the six dimensions of inquiry just mentioned. Along dimension (a), the most elementary property of human language is that it involves a denumerable infinity of functionally distinct expressions, in contrast to systems that involve continuity (like the dance of bees) or strict finiteness (calls of apes). Evidently, such a system must be based on a finite system of rules that specify the properties of the denumerable infinity of expressions—i.e., a grammar. In the case of human language, these principles crucially involve a hierarchy of phrases, abstractly represented, and structure-dependent rules operating on these phrases. Recursive embedding of several types is the basic device for constructing new phrases. For example, a simple clause such as "he read the book" can be embedded in a verb phrase such as "to buy the book that he read", which can appear within another clause, as in "I wanted to buy the book that he read", which can be embedded in an adjective phrase within a further clause as in "they were so surprised that I wanted to buy the book that he read that they gasped", etc. The structures formed by various processes of recursive embedding are assigned phonetic and semantic representations by further rules. These devices provide for the range of expression characteristic of all human languages, allowing us to denote previously unexamined or newly-imagined objects, actions, properties, events, etc., and to form propositions of various sorts. These are the most basic and elementary properties of human language; they account for the traditional observation that human language is a system for the infinite use of finite means.

Moving to less elementary properties, we enter the study of universal grammar, with its various hypotheses as to how grammars are organized and constructed and the principles that govern these highly special systems. It is here that the most significant results have been obtained, in my opinion. As I mentioned before, these results bear directly on the essential nature of human language and the basis for its growth in the individual, what is called, with misleading connotations, *langauge learning*.

I cannot try to outline such results here, but perhaps a very simple example will illustrate what has proven to be a very productive direction of inquiry. Consider the rule forming reciprocal constructions such as "the men saw each other". A child learning English, or someone learning English as a second language, must learn that *each other* is a reciprocal expression; that is an idiosyncratic fact about the grammar of English. Given that it is a reciprocal expression, it must have an antecedent, for example: *the men* in "the men saw each other", which has the meaning, roughly: "each of the men saw the other(s)". But it is not always so easy to find the antecedent. For example, sometimes it can be in another clause, as in "the candidates wanted each other to win", where *each other* appears in a subordinate clause as the subject of *win* and its antecedent, *the candidates,* appears in the main clause. Sometimes, however, the reciprocal cannot find its antecedent outside of its clause, as in "the candidates wanted me to vote for each other", which does not have the perfectly sensible meaning "each of the candidates wanted me to vote for the other". We might assume that the antecedent must be the 'nearest noun phrase', but this is false as we can see from such sentences as "the candidates hurled insults at each other".

Now it is difficult to imagine that children learning English receive specific instruction about these matters, or even that they are provided with relevant experience. In fact, we find that while children make many errors in language learning, they never make such mistakes as these: they never assume, until corrected, that "the candidates wanted me to vote for each other" means that each candidate wanted me to vote for the other. In fact, relevant experience is never presented for most speakers of English, just as no pedagogic grammar would ever point out these facts. Somehow, this is information that the child himself brings to the process of language learning. Or, in other words, some general principle of universal grammar is applied to permit the proper choice of antecedent, not an entirely trivial matter, as these examples suggest. This principle is one of several providing a basic framework within which knowledge of language develops as the child progresses to the mature state of knowledge; it is on a par with the factors that determine that he will have binocular vision. As we turn to the investigation of such principles and their interaction, we begin to approach the richness of structure of the language faculty, one element of our biological endowment.

Along dimension (b) (physical mechanisms in which the structural principles are realized) little is known. It seems clear that lateralization plays a crucial role and that there are special language centers, perhaps linked to the auditory and vocal systems. It is a particularly striking fact that very severe peripheral defects can be overcome in acquisition and use of language, a biological function of humans that matures even under serious conditions of deprivation.

With regard to the functions of language (dimension [c]), there are only general observations and elaborate taxonomy, but little in the way of explanatory theory. Human language is characteristically used for free expression of thought, for establishing social relations, for communication of information, for clarifying one's ideas, and in numerous other ways. While some describe its essential purpose as "communication," there is, to my knowledge, no substantive formulation of this proposal with empirical content; it can be sustained only if the term "communication" is used in so loose a sense as to deprive the proposal of any interest. Crucially, there is no basis for the belief that human language is used "essentially" for "instrumental ends"— to obtain some benefit.

The study of ontogenetic development (dimension [d]) involves in the first place universal grammar, which determines the basic framework within which growth of language in the individual proceeds; and, beyond that, whatever can be learned about onset, character, relative order, etc., of language growth. Universal grammar has been studied along the lines suggested earlier, and there are, I believe, a number of promising theories involving principles of some explanatory power. Actual language development has been studied primarily at the very early stages, and this study has quite often been concerned with properties of language about which very little of substance is known even at the mature stage (e.g., naming, social conditions of language use). Given these limitations, results of any generality or explanatory force are meager. Although this work is of considerable potential importance, it is necessary to regard it with some caution. Crawling precedes walking, and there are incipient motions prior to bird flight, but one must be cautious in drawing conclusions about the character of some biological system from observation of its early manifestations.

With regard to evolutionary development of language (dimension [e]), evidence is slight. It seems reasonable to assume that evolution of the language faculty was a development specific to the human species long after it separated from other primates. It also seems reasonable to suppose that possession of the language faculty conferred extraordinary selectional advantages, and must be a primary factor in the remarkable biological success of the human species, that is, its proliferation. It would be something of a biological miracle if we were to discover that some other species had a similar capacity but had never thought to put it to use, despite the remarkable advantages it would confer, until instructed by humans to do so—rather as if we were to discover in some remote area a species of bird that had the capacity of flight but had never thought to fly. There seems little indication that this biological miracle has occurred. Upright posture has also no doubt been a factor in human biological success, and, as has often been remarked, though dogs and horses can be taught to walk, in a sense, we do not therefore conclude that this capacity is part of their biological endowment, as bipedal locomotion is part of the endowment of the human species.

As for the interaction of the language faculty and other cognitive systems (dimension [f]), investigation is premature, pending deeper analysis of other related systems. The question arises at central points in the study of language, particularly with regard to studies of word meaning, which appears to involve other systems of knowledge and belief in an intimate and perhaps inseparable way, a matter of much recent discussion. It has even been proposed—quite plausibly, in my personal view—that the study of word meaning is not, properly speaking, part of the study of language at all, but rather concerns other cognitive systems that are connected in part to language through some sort of "labeling." To the extent that this assumption is valid, the study of semantics of human language will be concerned with compositional properties, that is, the ways in which the meaning of a phrase relates to the meaning of its parts.

These remarks relate to the first question, namely: what is a *human language?* Let us now turn to the second: what is a *language?*

When we ask whether some other system, say music, or the various systems taught to apes, are *languages,* we are not raising a question of science, as already noted; at least, not until the concept *language* is specified. Although *human language* is a biological property, which can be investigated as we would study *human vision,* the same is not true of *language* (or *vision,* or *locomotion,* etc.). The question whether other systems are "like" human languages is a question about the usefulness of a certain metaphor—although, as I mentioned before, it is possible to imagine that some day useful questions about homology or evolutionary history might be raised. It is important to stress this point of logic, since only confusion and empty debate can result if it is ignored. For example, when one reads in the popular press that "there is no longer a very serious argument to be made as to whether the great apes have access to language" (H. T. P. Hayes 1977), one wonders at once what is the question about which there is no longer a serious argument. It is true that there is no serious argument as to whether the great apes have access to human language, just as there is no argument as to whether humans have access to bird flight, despite superficial similarities. With regard to *language* in some other, as yet unspecified sense, there is and cannot be any serious argument, because there is no clear question to argue about.

How useful are the metaphors and analogies? In the nature of the case, this is a difficult question to answer. For example, Straub *et al.* (1978) have recently reported success in training pigeons to tap four colored buttons in sequence to obtain food. Suppose that we label these buttons, successively, "please," "give," "me," "food." Do we now want to say that pigeons have been shown to have the capacity for language, in a rudimentary way? This is much like the question whether humans can fly, almost as well as chickens though not as well as Canada geese. The question is not clear or interesting enough to deserve an answer. What we might try to do is to investigate capacities for serially-ordered behavior, symbolization, and so on, in pigeons, dogs, chimpanzees, and other organisms, and to ask how or in what respects these systems compare with the biologically-given human language faculty. Investigating the six dimensions discussed briefly a moment ago in connection with the language faculty, it seems to me that there is very little similarity between any of these systems and human language, at least in the areas where anything significant is known about human language—and it is only here, naturally, that the question is of any real interest.

With regard to structural principles (dimension [a]) and the related matter of universal grammar (which relates crucially to dimension [d]), the systems taught to apes and other species differ from human language at the most primitive and elementary level. As far as has been reported, they are strictly finite systems in principle (apart, perhaps, from trivial devices such as conjoining), with no significant notion of phrase and no recursive rules of embedding or structure-dependent operations. Thus, they belong to a category of systems entirely unlike human language from the point of view of their essential formal structure. Similarly, elements basic to the semantics of human language, such as modality and propositional attitude, description and presupposition, aspect and anaphora and quantification, and so on, seem to be entirely lacking. As we turn to less elementary properties of human language, there seem to be no comparable features in the systems taught to other species. Similarities have been noted at a very general level—e.g., "use of symbols" in reference or to evoke action, serial order, perhaps some kind of limited substitution in frames. Human language also manifests these properties; however, there is little reason to suppose that they are specific to human language, to humans, or even to primates.

One might suspect that humans suffering severe language deficit and perhaps even lacking the neural capacity for language might be comparable to higher apes in ability to learn certain conventional symbol systems. Results of this nature, concerning the training of severely aphasic patients to communicate using a symbol system other than natural language, have been reported (Velletri-Glass *et al.* 1973; H. Gardner *et al.* 1976; Davis & Gardner 1976). This work indicates that "severely aphasic patients can use a system of visual symbols to act upon command, answer questions, and describe actions in their immediate situational environment" and perhaps "to express their own wants and feelings", possibly with some limited productivity (Gardner *et al.* 1976). However, "the ability to handle the more purely computational aspects of a symbolic system . . . is particularly devastated . . ." and spontaneity is at a minimum (Davis & Gardner), and analogies to performances of apes have been noted (Velletri-Glass *et al.*). Perhaps one might interpret this work as suggesting that apes are in this regard similar to humans lacking the "organ of language," much as birds with their wings clipped can "fly" in the sense in which humans can 'fly', though again, it is doubtful what significance any of these vague analogies may have.

With regard to dimensions (b), (e), and (f) (physical mechanisms, phylogenetic development, and interaction), so little is known that there seems little purpose in pursuing analogies. It appears that human language involves neural structures that may not exist in the same form in other primates, and the evolutionary history is surely quite separate. As for language use (dimension [c]), here, too, such elementary and primitive uses of language as telling a story, requesting information merely to enhance understanding, expressing an opinion or a wish (as distinct from an instrumental request), monologue, casual conversation, and so on, all typical of very young children, seem utterly unrelated to the functions of the ape systems, which appear to be strictly instrumental and thus quite unlike human language, as has been reported by Rumbaugh and Gill (1976b) and others. Perhaps it will turn out that their functional properties are similar to those of the alternative systems taught to global aphasics. Here, too, since current understanding does not extend much beyond extensive taxonomy, it is not clear that there is much point in pursuing analogies.

With regard to the one remaining dimension, namely, ontogenetic development (d), quite apart from the apparently vast and qualitative differences at the level of principles of universal grammar, human language is acquired effortlessly and without training, in sharp contrast to the systems taught to apes. But since the results attained are qualitatively so different, there seems little point in pursuing the differences in mode of acquisition further. Perhaps I should stress that the question of the precise role of experience in triggering and shaping language growth is, naturally, an important one. Obviously, experience has a shaping effect—English is not Japanese—and presumably some social interaction is necessary to trigger the process, as has been found in the case of other biological systems. Since crucial experiments are ruled out on ethical grounds, we cannot proceed to answer these questions directly through experiment. Mention should be made of some recent work indicating that deaf children with zero linguistic input have nevertheless spontaneously developed a sign system remarkably like human language—one 3-year-old in this group is reported to have produced "sentences" up to 13 signs in length (Feldman et al. 1977; cf. also Goldin-Meadow & Feldman 1977), and it would be quite interesting to know how such systems might develop past these early stages.

As already noted, the usefulness of some analogy is difficult to assess, and it is doubtful that the question merits much attention. To me it seems that the analogies observed between human language and the systems taught to apes are not particularly suggestive, at least in the domains where something nontrivial is known about human language. Some researchers on ape symbol systems have criticized the tendency to draw such analogies much more harshly. For examples, the Gardners, in an article reviewing such work (this volume), argue that virtually all of it apart from their own is undermined by a false analogy, in that researchers have labeled the symbols taught to apes with values derived from human languages, as in the hypothetical pigeon example mentioned earlier, then mistakenly concluding that the symbol correlated by the human researcher with a term of human language is being used by the ape with the properties of its human language correlate. (They also make other criticisms that do not concern me here.) The Gardners argue that when such illegitimate assumptions are dropped, no "linguistic" abilities are demonstrated in the work they criticize. They believe that this difficulty is overcome in their own work, since they are teaching their chimpanzees an actual human language, namely, Ameslan (American sign language). But their claim obviously begs the question. True, they are teaching Washoe (*et al.*) signs correlated with symbols of Ameslan, just as the researchers they criticize are teaching Lana, Sarah, etc., signs correlated with symbols of natural language. In each case, the correlation is imposed by the researcher and the question is whether it is a valid one; that is, are the apes using symbols with properties of their natural language correlates? The question arises in exactly the same form under the Gardners's approach (in which the correlation is based on visual similarity or even identity) as in the work of Premack, Rumbaugh, etc., which they criticize.

The Gardners argue further that their chimpanzees use symbols in a manner comparable to that of very young children, from which they conclude that the chimpanzees are exhibiting the first stages of "language development" exactly as the children are. Again, the argument is fallacious. As has often been remarked, we know that the children are exhibiting "incipient human language behavior" only because of the later stages achieved. A child may "flutter his arms" more or less in the way a fledgling flutters its wings, but we cannot conclude from this fact that the child is exhibiting "incipient flight motions."

In short, insofar as the Gardners's criticism of other work in this respect is valid, it applies in exactly the same form to their own work. To use symbols correlated by the investigator with symbols of Ameslan (even if the correlation is one of visual matching) is not necessarily to use Ameslan, as they mistakenly conclude; to act in a manner resembling the early stages of the manifestation of some system of another species is not necessarily to be producing incipient manifestations of this system, obviously. Once again, we are back to the question whether certain analogies are useful and suggestive.

The same article contains other curious examples of reasoning. Thus, the Gardners cite my·remark that "as far as we know, possession of human language is associated with a specific type of mental organization, not simply a higher degree of intelligence," and conclude erroneously that "this view argues from surface dissimilarities to the conclusion that there are fundamentally different sets of laws for different types of behavior." That is incorrect. Nothing is implied in the remark they cite concerning "different sets of laws." Consistently with this remark, there might be no "general laws of behavior" at all, or there might be the same laws for all behavior, applying to different systems of mental organization. The question is left entirely open, and of course has nothing to do with the commitment to approaching language as a natural biological phenomenon. Perhaps their mistaken conclusion derives from assumptions they seem to be making about natural biological systems that are quite odd, to say the least. Thus, they argue that "if a form of behavior such as human language appears to be different in character from other forms of human and animal behavior, then we do not abandon the search for general laws [i.e., 'fundamental laws' of behavior], instead we question the adequacy of existing observations." Apparently, they believe that the discovery that two systems are organized and function on the basis of quite different principles implies abandonment of the search for "general laws" (it may lead to the denial that there are "general laws of behavior" of any significance, but that is quite another matter). It is difficult to understand what point they think they are making. Surely it is obvious that human language not only appears to be but is different in character from certain other forms of human and animal behavior, which are in turn different in character from one another: say, spinning a web, building a nest, fishing, tying one's shoes, etc. To accept these truisms is not to abandon the search for "general laws," though, contrary to the Gardners's assumption, there is little reason to suppose that there are fundamental laws of *behavior* that "combine and recombine" to yield "the rich diversity of behavior that we observe" in these and other cases. In fact, it is quite difficult to make any sense of the proposals that they take as axiomatic for "a science of psychology".

Conceivably, there are fundamental, general, and nontrivial laws of behavior that have some significant empirical coverage and explanatory force, though the existing literature proposes none, to my knowledge. But this assumption is by no means a necessary consequence of a commitment to study behavior as a natural, biological phenomenon. *A fortiori,* such a commitment does not lead us to insist that all behavior of all organisms falls under general "laws of behavior," in any significant sense of this notion (as distinct from biological or physical laws). Other fallacies abound in the discussion of these issues, for example, the belief that the principles of evolutionary biology commit us to some doctrine of "continuity" of a sort that has never been made at all clear but is commonly invoked. The initial commitment is an entirely reasonable one, but some care must be taken in drawing specific consequences from it.

Suppose that future studies reveal that apes or some other species can, contrary to natural expectations, acquire a language sharing significant properties with human language in some domain where something is known about the latter so that analogies might become suggestive. What would this teach us about the nature or evolutionary origins of human language? The answer is: very little. Similarly, if some enterprising athletic coach were to discover a previously untried motion of the arms that enabled humans to "fly" as well as chickens, that fact in itself would not be very helpful to the biologist studying the mechanisms of bird flight and their genetic basis or evolutionary origins. True, if apes were discovered to have a latent "human language capacity," or something sufficiently like it so that analogies become significant, then experimental programs would no doubt be devised of a sort that are excluded on ethical grounds in the case of humans—I do not mean to suggest that ethical questions would not arise or do not already arise, in the case of intrusive experimentation with other species. Thus, scientists might feel that it is permissible to proceed with design of contrived environments, ablation studies, fetal neurosurgery, and other investigations that might be expected to yield understanding of the nature of this latent capacity and its physical basis. But apart from these possibilities (putting to the side the ethical issues), the discovery that some other species had the capacity to attain knowledge of a rudimentary human language and the ability to use it in a quasi-human way would, in itself, leave the scientific problems concerning human language exactly where they now stand. We would now face a dual problem, replacing the single problem to which research is currently addressed: what is the nature of human language in its mature and initial stage—ultimately, a question of human biology. Under these purely hypothetical circumstances, the question of the nature of human language would remain as it is, but we would now face the additional problem of explaining why this capacity remains latent in this other species, despite the selectional advantages that would no doubt be conferred by its exercise. These speculations are, needless to say, entirely academic as far as we know.

Summarizing, recent work seems to confirm, quite generally, the not very surprising traditional assumption that human language, which develops even at very low levels of human intelligence and despite severe physical and social handicaps, is outside of the capacities of other species, in its most rudimentary properties, a point that has been emphasized in recent years by Eric Lenneberg, John Limber, and others. The differences appear to be qualitative; not a matter of "more or less," but of a different type of intellectual organization, so it appears. As for the value of analogies between human language and other systems (e.g., alternative systems taught to aphasics, or ape systems), one must reserve judgment, just as in other cases of biological analogy, say jumping and flying. As I have already mentioned, it is not very surprising that careful study should confirm the traditional belief that there are striking qualitative differences between humans and other species in "capacity for language," given the enormous selectional advantages conferred by language—for humans, in a period of human evolution that is recent in evolutionary terms.

For reasons that are unclear to me, this topic has aroused considerable emotion, at least in popular discussion. For this reason, perhaps I should emphasize a point that should be obvious without specific mention. The study of symbolic systems taught to apes will no doubt prove rewarding in mapping their intellectual capacities, and perhaps it will ultimately help to locate more exactly the cognitive systems that are specific to humans, a contribution of some potential interest. But it is not obvious to me that one should expect this work to shed any significant light on what appear to be specific, biologically fixed human faculties such as the language capacity, or, conversely, that the study of human language will prove enlightening for the study of ape communication or intelligence—just as the study of broad jumping probably has little to offer concerning the flight of birds, or conversely.

DO YOU SPEAK YERKISH? THE NEWEST COLLOQUIAL LANGUAGE WITH CHIMPANZEES

H. HEDIGER

The desire to enter into direct linguistic communication with animals is very ancient. It has lived in mankind for thousands of years. Whereas until recently this dream has been realized only in fairy tales, today man is trying to achieve it with the most modern and scientific means. It is true that ethology has reached astonishing results, whereby it has become possible to comprehend and communicate to a certain degree with many species of animals. To be mentioned in this connection are the great experiments of the Nobel Prize winner Karl von Frisch, who has discovered surprising facets of the language of bees, details of which today are known to every schoolchild. Also to be remembered are the discoveries of Karl von Frisch's former student, Martin Lindauer, who succeeded in understanding the "negotiations" of swarming bees concerning their next residence so well that he was able—on his bicycle—to arrive at the new location before the bees did. Another ethologist, Peter Marler, understands so thoroughly the language of the chaffinch, which consists of about twenty signals, that he can attract the bird to a simulated female, or make him flee from a cat or hide from a bird of prey—all this through the replay of the appropriate calls with sensitive instruments. An Italian ethologist, Floriano Papi (1969), has deceived glowworms by using an electric instrument to emit the light signals of female glowworms, thus attracting the males.

Such deceptive signals, for example to attract game, have been used by hunters for hundreds if not thousands of years, in order to get close enough to a fox or a good roebuck—signals like the imitation of the suffering cry of a wounded hare to attract a fox, or the rutting call of a female roebuck to attract the male deer. Some scholars even maintain that human language itself was invented mainly for reasons of deception, that is for lying. Until recently, it was believed that lying represented a doubtful peculiarity of humans. However, many cases of the use of deceptive signals have been discovered in the animal world, especially among certain predatory female glowworms, which send out the normally harmless female signals of another species and devour the misguided males of that species that respond by

H. HEDIGER ● Emeritus, University of Zurich, Switzerland.

approaching them (Lloyd 1965). It has been discovered among many birds, and, among mammals, as the polar fox, that, in view of an especially attractive morsel, will emit alarm sounds in order to cause all nearby fellows of the same species to flee so that he can have the food to himself.

The above-mentioned dream of mankind with regard to man–animal communication is, however, far more grand. Man not only wishes to record animal signals and understand them at the animal level and make use of this through playback and control of the reaction; he also wants to enter into a more intricate dialogue, into an actual conversation with the animal in the hope of receiving from the animal detailed information about its world. In hoping to achieve this, man entertains completely illusory conceptions just as he did thousands of years ago. He overlooks the fact that animals not only have communication systems completely different from man's—although these may be studied to a large extent—but also that animals live in completely different surroundings, in totally different spheres of interest. To put it simply, the animal is not interested in what we read in the newspaper, in what we hear on the radio or see on television. Our libraries, our sciences, our mathematics, technology, philosophy, or theology do not interest the animal at all. This fact should have been clear, at the latest, half a century ago, when the worldwide rumors about hoof-speaking horses and thinking dogs were uncovered and clarified in at least one sense—neither horses nor dogs can philosophize or calculate square roots.

However, the intense desire to enter into direct dialogue with animals proves to be so strong that new paths toward its realization are being sought and in a naive way are worked at for short periods of time. And this always at a considerable cost and ever increasing effort! With all this, these pioneers are not people who have had from an early age intimate and practical contact with animals of all levels of the zoological system. They are people who either know only a few laboratory animals or who only at a later age discovered animals at all. This fact is practically universal among this group of people. The fanatical promoter of the hoof-speaking horse, Wilhelm v.Osten, was originally a schoolteacher who discovered the fantastic capability of his experimental animals only after his retirement. John C. Lilly, who believed he could enter into direct conversation with dolphins, was a normal neurologist who at best had to do with white rats or with brain-deficient monkeys before he became interested in interspecific communication. His experiments with costly instruments at expensive institutes have led to nothing. The man–dolphin dialogue did not materialize. The whole thing proved to be a great bluff.

Today, it is mainly psychologists, linguists, electronic engineers, computer specialists, and the like, who occupy themselves with linguistic communication with animals, above all with chimpanzees. But even they often lack the overall knowledge of the zoological system in that they have little personal experience with living representatives of all species. Many of the scientists working today on interspecific communication know their animal "guineapigs" only from the scientific literature, as figures out of a zoology book or laboratory report. This same fact appears very clearly in the Lana Project, which I will discuss at greater length below. This project is based on the optimistic idea that man will be able to talk with chimpanzees in Yerkish.

In order to understand this highly modern method, we should become acquainted briefly with a few of the preceding experiments. First of all, the one made by the psychologist couple Hayes, who in a heroic way accepted the literally impossible task of adopting a newborn chimpanzee in the real sense of the word, i.e., to consider it as their child—Viki—and to treat it subsequently as such. Viki Hayes did not tolerate baby sitters and, for years, demanded of its stepparents an almost inhuman readiness to make sacrifices. In 1951, I made a pilgrimage to this unique family in Orange Park, Florida, in order to convince myself on the spot of the actual facts. A book by Viki's stepmother, Cathy Hayes, *An Ape in Our House* (1951)—which did not get the attention it really deserves—contains the details of Viki's development, especially her linguistic development. This classic document is discouraging in that—as in all other such experiments by others—Viki did not, despite all the love, care, and attention given her, learn to speak more than three words. She learned, like all of her predecessors, only simple words like "Mama," "Papa," and "cup" (drinking), and, even with this very modest vocabulary, there were sometimes confusions.

What makes this strenuous Viki experiment scientifically valuable and so sympathetic in a human way is, among other things, the fact that it was always readily admitted that during long private sessions the Hayes had tried very hard to teach Viki more words, by patiently pronouncing and actively shaping the chimp's mouth and lips. Without success. Other authors with other apes never succeeded either. This finally led to the decisive realization that apes do not lack the physical means to talk (larynx, tongue, lips, etc.), but they lack the necessary specialized organs of the brain, the speech center, i.e., the Broca's Area and Wernicke Area.

This fact gave another couple, R. Allen and Beatrice T. Gardner, the idea of trying out a new method of teaching language to a chimp by by-passing the insufficient speech center. They used a sign language, the ASL or AMESLAN (American Sign Language), the official gesture language of the American deaf. Their first conversational partner was the eleven-month-old, by now famous chimpanzee Washoe, whom I also visited after it moved to its new home with Professor Lemmon and Dr. Fouts, in Norman, near Oklahoma City. Washoe was, at the time of my visit, six years of age. She was a friendly, reliable animal that could be taken along in a car. When, during a drive in the country, we stopped along the roadside, near a grazing cow, and asked her what that was, she willingly made the sign for cow: she stretched thumb and small finger from the closed left hand and held it sideways to the forehead (where the cow has its horns). Or, when a soft drink machine appeared during a stop at a gas station, she spontaneously made the sign for drink, which is also familiar to persons who do not practice ASL: she held the stretched out thumb of her fisted hand against her mouth and indicated with this that she wished something to drink, preferably Coca-Cola. Washoe has learned well over 100 signs, but this method was not without problems: the signs were not always clearly definable. Therefore, as long as they were not documented on film, they left quite a margin of interpretation to the experimenter. Besides that, this method had the prerequisite of a direct contact between man and animal. This means an additional source of mistakes. One then never knows for sure if the animal reacts to the official signs or if it realizes what it is supposed to do by watching the face of the experimenter.

In order to eliminate such difficulties, David and Ann Premack tried to enter into a more controlled type of contact with another, by now well-known chimpanzee named Sarah, by teaching her the meaning of word-like plastic symbols. For example, there was a symbol for "Give me," for "bananas," "drink," etc. These symbols could be fixed magnetically in a planned sequence onto a vertical board, e.g., "Mary give apple Sarah." But even with this method, the risk of giving involuntary signs is present, even though the experimenters really tried not to influence Sarah at work in any way, as should be done in such experiments.

Finally, the Lana Project claims to have reached the highest degree of objectivity and of noninfluence over the animal (Rumbaugh 1977). Lana is, so to speak, the youngest colleague of Viki, Washoe, and Sarah. Her name is composed of the first letters of LANguage ANAlogue Project. This demanding research project offers a special, previously unattained level of objectivity inasmuch as in this man–animal dialogue experiment, it is not man and chimpanzee that face one another, but chimpanzee and computer, i.e., an absolutely objective authority, a machine. More rigorous experimental conditions cannot, apparently, be demanded.

Nevertheless, it is virtually impossible to push an animal into a sequence of experiments with a printed set of directions for use. The machine needs to be explained to the animal somehow or other, i.e., the animal has to get to know it before it can use it correctly; it needs to gather experience with it. In other words, this type of experiment is always necessarily preceded by a breaking-in period, by instructions from the leader or leaders of the experiment. By doing this, an inevitable, and often decisive, contact is established between man and animal. This preparatory phase is nonetheless generally very little appreciated by the experimentalists. These preparations are not important to them. They are more interested in the subsequent experimental results which are statistically comprehensive and, for them, indispensable.

From the standpoint of animal psychology, this preparatory phase *per se* deserves special attention. It determines correct interpretation of the final phase of training, i.e., of the experiments to be evaluated. During this phase of initial training, of becoming familiar with the experimental apparatus, the animal also has the extensive possibility of getting acquainted with the leader of the experiment, the interpretation of his movements, his expressions, appearance, moods, etc. One should remember, moreover, that the animals used for experiments are often extremely sophisticated, cunning observers in front of whom we practically "nakedly sit in a showcase" (before whom we cannot hide anything). This is especially so in the case of highly organized animals like chimpanzees. Up to now, no one has been able to determine exactly what such an animal can read from our face, smell, gestures, and sounds, or what they know about us at all. Human ethology and semiotics are only just beginning to understand the amazing abundance of such unwitting signals.

It is assumed nowadays that in experiments concerning dialogues with animals, nonverbal communication needs to be carefully taken into account. For example, this is the case when attempts are being made to teach chimpanzees language with the aid of plastic symbols or—and this is the latest experiment—to get them to learn Yerkish. Yerkish is thought of as a dialogue with a computer. Thus, at least theoretically, with this robot of a partner, all involuntary human signs like facial expression, etc., should be eliminated. Yerkish is a completely new language created at the desk of linguists and other specialists. It consists of symbolic figurative signs which may be typed or played back on a kind of keyboard with picture slides. As in the experiments of the Gardners, Fouts, Premack, and others, one has to do with communication through signs.

In the Lana Project, one deals with signs that are fed into a computer and are registered by it. By doing this, one not only expects an unsurpassable objectivity but also a considerable saving of effort. The attempts at communication with the language of the deaf or with plastic tokens were very time-consuming. They worked only as long as the leader of the experiment occupied himself with the animals. The computer, on the other hand, works day and night, as Rumbaugh emphasizes with a certain pride. The whole apparatus, Lana's living room with the keyboards, measures $2.1 \times 2.1 \times 2.1$ meters. But one has to realize that in these narrow quarters Lana was not active 24 hours a day, even though all her desires like food and drink, blankets, opening of window, music, and film performances, were fulfilled only via computer through activation of the respective key. Even a chimpanzee needs to sleep once in a while, and this generally longer than man: 12 hours compared with eight hours.

Everything that Lana asserted through her own natural language and all that she did otherwise for herself was neither observed nor recorded. This did not interest the computer. Through the partially transparent walls of her cage, however, Lana had the possibility of observing what went on outside her pen, as well as the (unsuspecting) experimentalists and keepers. She undoubtedly could draw certain conclusions and behave accordingly. The room had to be cleaned periodically. During that time, contacts with humans were not to be excluded. During her training, man had to teach Lana how to activate the keys, to push Lana's finger down onto the keys, etc. The ideal condition for all such experiments would be complete isolation of the experimental animal from the leader of the experiment. This is completely illusory, since each higher living being becomes neurotic under complete isolation and therefore would no longer react naturally or normally. One has only to imagine how big, how tremendous the difference is between the natural habitat and natural family unit of the chimpanzee and his isolated existence in a desolate computer experimental cage. Despite the unavoidable contact with man which Lana so intensely desired, she relatively soon reached a condition in which she became fed up with all the experimenting, so that she received one month's holiday. Lana's need for social contact became so acute that the leader of the experiment decided to give her the company of a young orangutan. But the disappointment was great. Orangutans react, as is well known, considerably more slowly than chimpanzees. The orangutan proved to be an undesirable hindrance which soon caused Lana to use the computer only to request the most necessary food and drink. This meant removing the orangutan. Only outside of the 24 hours that were originally planned for the experiment was Lana allowed to have contact with other primates in order to play.

In other ways as well, one could notice that the Lana experimental team did not include enough people (if any) who had the above-mentioned basic, practical experience in dealing with animals. For example, the whole apparatus was suited not to a chimpanzee but rather to the gentle touch of a secretary or a laboratory student. Chimpanzees do not remain babies forever. Keeping in mind how demanding and costly such projects are, one should always remember that a half-grown chimpanzee excels man physically several times. Thus, it was unavoidable that the valuable keyboard was maltreated with terrible blows of hands and feet, and became dirtied with food and feces. It was also exposed to Lana's chisellike fingernails and teeth. For this reason, the whole installation had later on to be rebuilt in a much more solid way. Anyone with just a little bit of practical experience with apes could have foreseen this and could have asked from the beginning for a sturdier construction.

Any experienced person would have predicted a pessimistic outcome with certainty, by analogy to the unsuccessful dolphin project. He would have foreseen that the greatly conceived Lana Project would come to a disappointing end. This is not only due to the naivité of the goal as seen from an animal expert's point of view, but simply because practically every chimpanzee ceases, at the latest by the age of ten, to submit to man's orders. For every practician, this is an old truth which has been confirmed a hundred times in zoos and circuses. The same thing manifested itself with the famous artist chimpanzee Kongo of the London Zoo. Some of his paintings obtained fantastic prices, but he finally gave up his career as an artist upon reaching the age of five. Washoe too, at the age of six, developed a temper and obstinancy which at times turned into aggression. After a few bites, her guardians had to equip themselves with electric sticks whereby they could dish out strong shocks as penalty stimuli. Today, of course, all kinds of drugs are readily available in order to subdue the temper and aggression, which are really only a normal biological development toward individual independence. The question arises, however, whether one can expect significant reactions from an experimental animal which is under the influence of strong drugs.

If things stay as they are, it looks as though the dream of a dialogue with chimpanzees based on the computerized Yerkish[1] will soon come to an end. The Lana Project is very ambitious, like every project which aims at direct linguistic communication between man and animal. One has imagined until now unheard of possibilities to help numberless brain-deficient children by the Yerkish method. One hopes to enter into a dialogue, into a conversation, with them also. Apparently, with Lana, this has already proved to be a success. And what does the perceptive reader believe to be the first piece of this new decisive conversation between man and animals? Of course, Coca-Cola, the lovable popular abbreviation, "Coke"!

However, there is still something basic to think about: in the Lana Project also, we are dealing with experiments of animal psychology. The experimental animal is supposed to learn signs (in this case computer signals) in order to use them in a way that makes sense, that is, the sense the leader of the experiment desires. Experience has shown it always to be extremely difficult to decide if the animal actually understands and uses the signs that are presented to it, or if it reacts to involuntary signs (like expressions or the movements of the leader of the experiment). It is for this reason that the long-time editor of the *Journal of Animal Psychology*, Otto Koehler, has for many decades warned of this dangerous source of mistakes: "Leaders of experiments should not believe in advance what the results of the experiment are to teach them." From the beginning, the members of the team involved with the Lana Million Dollar Project have succumbed to this strong temptation. They either never heard of Koehler's justified warning or they did not wish to know about it. From the very beginning, they did not expect an objective result out of their quest but emphatically hoped for the success of their bold, if not fantastic, project. This in itself is basically a dangerous starting point. We have to ask ourselves very seriously if the Lana Project will reach the same fate as the alleged dialogue with dolphins?

---

[1]The designation Yerkish for the newly invented man–chimpanzee language has, by the way, been derived from Robert M. Yerkes (1876–1956), who was one of the most well-known researchers on apes. Nobody knows if he would have agreed to the daring Lana Project.

REFERENCES[1]

ALTMANN, STUART A. "The Structure of Primate Social Communication." In *Social Communication Among Primates,* edited by Stuart A. Altmann. Chicago: University of Chicago Press, 1967, pp. 325–362.

AMMAN, J. C. *Surdus Loquens.* London: Sampson Low, Marsten Low & Seale, 1692.

———. *Dissertatio de Loquelo.* Amsterdam: North-Holland. (Originally published 1700.)

APFELBACH, R. "Electrically Elicited Vocalizations in the Gibbon *Hylobates lar* (Hylobatidae) and Their Behavioral Significance." *Zeitschrift für Tierpsychologie* 30 (1972): 420–430.

BARBER, THEODORE XENOPHON, and DAVID SMITH CALVERLEY. "Effect of *E*'s Tone of Voice on 'Hypnotic-Like' Suggestibility." *Psychological Reports* 15 (1964): 139–144.

BASTIAN, JARVIS R. "Primate Signalling Systems and Human Languages." In *Primate Behavior: Field Studies of Monkeys and Apes,* edited by Iroen DeVore. New York: Holt, Rinehart & Winston, 1965, pp. 585–606.

BATES, ELIZABETH. *Language and Context, The Acquisition of Pragmatics.* New York: Academic, 1976.

BATESON, GREGORY. "The Message of Reinforcement." In *Language Behavior,* edited by J. Akin, A. Goldberg, G. Meyers and J. Stewart. The Hague: Mouton, 1970, pp. 62–72.

BATTISON, ROBIN. "Phonological Deletion in American Sign Language." *Sign Language Studies* 5 (1974): 1–19.

BATTISON, ROBIN, and I. K. JORDON. "Cross-Cultural Communication with Foreign Signers: Fact and Fancy." *Sign Language Studies* 10 (1976): 53–68.

BECK, B. "Tool Use in Captive Hamadryas Baboons." *Primates* 13 (1972): 277–295.

———. "Cooperative Tool Use by Captive Hamadryas Baboons." *Science* 182 (1973): 594–597. (a)

———. "Observation Learning of Tool Use by Captive Guinea Baboons *(Papio papio)*." *American Journal of Physical Anthropology* 38 (1973): 579–582. (b)

———. "Baboons, Chimpanzees, and Tools." *Journal of Human Evolution* 6 (1974): 509–516.

BEECHER, HENRY K. "The Powerful Placebo." *Journal of the American Medical Association* 159 (1955): 1602–1606.

BEER, C. G. "Some Complexities in the Communication Behavior of Gulls." In *Origins and Evolution of Language and Speech,* edited by S. R. Harnad, H. D. Steklis, and J. Lancaster. New York: Annals of the New York Academy of Sciences, 280 (1976): 413–432.

———. "What Is a Display?" *American Zoologist* 17 (1977): 155–165.

BELLUGI, URSULA. "The Acquisition of Negation." Doctoral thesis, Harvard University, 1967.

BELLUGI, URSULA, and ROGER BROWN, eds. *The Acquisition of Language. Monographs of the Society for Research in Child Development,* 29 (1964): 5–191.

BELLUGI, URSULA, and SUSAN FISHER. "A Comparison of Sign Language and Spoken Language." *Cognition* 1 (1972): 173–200.

BELLUGI, URSULA, and EDWARD S. KLIMA. "Aspects of Sign Language and Its Structure." In *The Role of Speech in Language,* edited by James F. Kavanagh and James E. Cutting. Cambridge, Mass.: M.I.T. Press, 1975, pp. 171–203. (a)

---

[1]References for the editors' introduction are on pp. 56–59.

————. "Perception and Production in a Visually Based Language." *Annals of the New York Academy of Sciences* 263 (1975): 225–235. (b)

————. "Two Faces of Sign: Iconic and Abstract." In *Origins and Evolution of Language and Speech,* edited by S. R. Harnad, H. D. Steklis, and J. Lancaster. New York: Annals of the New York Academy of Sciences, 280 (1976): 514–538.

BELLUGI, URSULA, EDWARD S. KLIMA, and P. A. SIPLE. "Remembering in Signs." *Cognition 3* (1975): 93–125.

BEM, D. J. "Self-Perception Theory." In *Advances in Experimental Social Psychology,* vol. 6, edited by L. Berkowitz. New York: Academic, 1972, 1–62.

BENACERRAF, PAUL, and HILARY PUTNAM, eds. *Philosophy of Mathematics: Selected Readings.* Englewood Cliffs, N.J.: Prentice-Hall, 1964.

BENVENISTE, ÉMILE. *Problèmes de linguistique générale.* Paris: Gallimard, 1966.

BEVER, T. G. "The Cognitive Basis for Linguistic Structures." In *Cognition and the Development of Language,* edited by John R. Hayes. New York: Wiley, 1970, pp. 279–362.

BEVER, T. G., J. R. MEHLER, and V. V. VALIAN. "Linguistic Capacity of Very Young Children." Unpublished manuscript, n.d.

BIERENS DE HAAN, J. "Werkzeuggebrauch und Werkzeugerstellung bei einem niederen Affen *(Gebu hypoleucus Humb.)."* *Zeitschrift für Vergleichende Physiologie* 13 (1931): 639–695.

BINGHAM, N. E. "Maternal Speech to Pre-Linguistic Infants: Differences Related to Materna ments of Infant Language Competence." Unpublished paper, Cornell University, 1971.

BLAKE, HENRY N. *Talking with Horses: A Study of Communication between Man and Horse.* London: Souvenir, 1975.

BLAKEMORE, C. B., and G. ETTLINGER. "Cross-Modal Transfer of Conditional Discrimination Learning in the Monkey." *Nature* 210 (1966): 117–118.

BLOCH, MAURICE. Introduction to *Political Language and Oratory in Traditional Society,* edited by M. Bloch. London: Academic, 1975, pp. 1–28.

BLOOM, LOIS. *Language Development: Form and Function in Emerging Grammars.* Cambridge, Mass.: M.I.T. Press, 1970.

————. *One Word at a Time: The Use of Single Word Utterances before Syntax.* The Hague: Mouton, 1974.

BOLLES, ROBERT C. "Learning, Motivation, and Cognition." In *Handbook of Learning and Cognitive Processes.* Vol. 1. Edited by William K. Estes. New York: Wiley, 1975, pp. 249–280.

BONVILLIAN, J. D., K. E. NELSON, and V. D. CHARROW. "Language and Language-Related Skills in Deaf and Hearing Children." *Sign Language Studies* 12 (1976): 211–250.

BORNSTEIN, M. "Color Vision and Color Naming." *Psychological Bulletin* 80 (1973): 257–285.

BORNSTEIN, M., W. KESSEN, and S. WEISKOPF. "Categories of Hue in Infancy." *Science* 191 (1976): 201–202.

BOWERMAN, MELISSA. "Brief Comparison of Finnish I and English I." Unpublished paper, Harvard University, 1969. (a)

————. "The Pivot-Open Class Distinction." Unpublished paper, Harvard University 1969. (b)

————. *Early Syntactic Development: A Cross Linguistic Study with Special Reference to Finnish.* London: Cambridge University Press, 1973.

BOYSEN, S., and E. SUE SAVAGE-RUMBAUGH. "Form vs. Function in Chimpanzee Communication." Paper presented at the Midwest Animal Behavior Society, Purdue, 1978.

BRAINE, MARTIN D. S. "The Ontogeny of English Phrase Structure: The First Phase." *Language* 39 (1963): 1–14.

BRONOWSKI, JACOB. "Human and Animal Languages." In *To Honor Roman Jakobson,* Vol. 1. The Hague: Mouton, 1967, pp. 374–394.

————. *The Ascent of Man.* Boston: Little, Brown, 1973.

BROWN, ROGER. "The Development of Wh Questions in Child Speech." *Journal of Verbal Learning and Verbal Behavior* 7 (1968): 277–290.

————. *Psycholinguistics. Selected Papers by Roger Brown.* New York: Free Press, 1970.

————. *A First Language: The Early Stages.* Cambridge, Mass.: Harvard University Press, 1973. (a)

————. "Development of the First Language in the Human Species." *American Psychologist* 28 (1973): 97–106. (b)

BROWN, ROGER, and URSULA BELLUGI. "Three Processes in the Acquisition of Syntax." *Harvard Educational Review* 34 (1964): 133–151. (a)

——. "Three Processes in the Child's Acquisition of Syntax." In *New Directions in the Study of Language,* edited by Eric H. Lenneberg. Cambridge, Mass.: M.I.T. Press, 1964, pp. 131–161. (b)

BROWN, ROGER, and COLIN FRASER. "The Acquisition of Syntax." In *Verbal Behavior and Learning: Problems and Processes,* edited by Charles N. Cofer and Barbara S. Musgrave. New York: McGraw-Hill, 1963, pp. 158–197.

BROWN, ROGER, and ROBERT J. HERRNSTEIN. *Psychology.* Boston: Little, Brown, 1975.

BROWN, ROGER, and ERIC H. LENNEBERG. "A Study in Language and Cognition." *Journal of Abnormal and Social Psychology* 49 (1954): 454–462.

BROWN, ROGER, COURTNEY CAZDEN, and URSULA BELLUGI. "The Child's Grammar from I to III." In *Minnesota Symposia on Child Psychology,* vol. 2, edited by Jane P. Hill. Minneapolis: University of Minnesota Press, 1969, pp. 28–73.

BRUNER, JEROME S. "Nature and Uses of Immaturity." In *The Growth of Competence,* edited by Kevin J. Connolly and Jerome S. Bruner. New York: Academic, 1974, pp. 11–48.

——. "The Ontogenesis of Speech Acts." *Journal of Child Language* 2 (1975): 1–19.

CAMPBELL, C. B. G., and W. HODOS. "The Concept of Homology and the Evolution of the Nervous System." *Brain, Behavior and Evolution* 3 (1970): 353–367.

CANNON, WALTER B. "'Voodoo' Death." *American Anthropologist* 44 (1942): 169–181.

CARROLL, JOHN B. "Review of Stokoe 1960." *Exceptional Children* 28 (1961): 113–116.

CASSIRER, ERNST. *The Philosophy of Symbolic Forms.* vol. 1, *Language.* New Haven: Yale University Press, 1953.

CHAO, YUEN REN. "The Cantian Idiolect: An Analysis of the Chinese Spoken by a Twenty-Eight-Month Old Child." In *Semitic and Oriental Studies.* University of California Publications in Semitic Philology XI. Berkeley: University of California Press, 1951, pp. 27–44.

CHAUVIN-MUCKENSTURM, BERNADETTE. "Y-a-t-il Utilisation des Signaux Appris Comme Moyen de Communication Chez le Pic Epeiche?" *Revue du Comportement Animal* 9 (1974): 185–207.

CHEVALIER-SKOLNIKOFF, SUZANNE. "The Ontogeny of Primate Intelligence and Its Implications for Communicative Potential: A Preliminary Report." In *Origins and Evolution of Language and Speech,* edited by S. R. Harnad, H. D. Steklis, and J. Lancaster. New York: Annals of the New York Academy of Sciences, 280 (1976): 173–211.

CHOMSKY, NOAM. *Syntactic Structures.* The Hague: Mouton, 1957.

——. *Aspects of the Theory of Syntax.* Cambridge, Mass.: M.I.T. Press, 1965.

——. *Language and Mind.* New York: Harcourt, Brace & World, 1968.

——. *Language and Mind.* Enlarged edition. New York: Harcourt Brace Jovanovich, 1972.

CHOWN, W. B., ROGER S. FOUTS, and L. T. GOODIN. "Productive Competence in a Chimpanzee's Comprehension of Commands." Master's thesis, University of Oklahoma, 1974.

CHRISTOPHER, MILBOURNE. *ESP, Seers & Psychics.* New York: Thomas Y. Crowell, 1970.

CHURCH, JOSEPH. "The Ontogeny of Language." In *The Ontogeny of Vertebrate Behavior,* edited by Howard Moltz. New York: Academic, 1971, pp. 451–479.

CLARK, EVE V. "On the Acquisition of the Meaning of 'Before' and 'After'." *Journal of Verbal Learning and Verbal Behavior* 10 (1971): 266–275.

COHEN, G. *The Psychology of Cognition.* New York: Academic, 1977.

COOPER, L., and Harry F. HARLOW. "Note on a Cebus Monkey's Use of a Stick as a Weapon." *Psychological Reports* 8 (1961): 418.

COOPER, W. E., and J. R. ROSS. "World Order." In *Papers from the Parasession on Functionalism,* edited by R. E. Grossman, L. J. San, and T. J. Vance. Chicago: Chicago Linguistic Society, 1975, pp. 63–111.

COVINGTON, V. C. "Juncture in American Sign Language." *Sign Language Studies* 2 (1973): 29–38.

COWEY, A., and L. WEISKRANTZ. "Demonstration of Cross-Modal Matching in Rhesus Monkeys, *Macaca mulatta.*" *Neuropsychologia* 13 (1975): 117–120.

CRAWFORD, M. P. "The Cooperative Solving of Problems by Young Chimpanzees." *Comparative Psychological Monographs* 14 (1937): 1–88.

——. "The Cooperative Solving by Chimpanzees of Problems Requiring Serial Responses to Color Cues." *Journal of Social Psychology* 13 (1941): 259–280.

CROMER, R. F. "Development of Temporal Reference during the Acquisition of Language." Doctoral dissertation, Harvard University, 1968.

DAVENPORT, RICHARD K. "Cross-Modal Perception in Apes." In *Origins and Evolution of Language and Speech*, edited by S. R. Harnad, H. D. Steklis, and J. Lancaster. New York: Annals of the New York Academy of Sciences, 280 (1976): 143–149.

DAVENPORT, RICHARD K., and C. M. ROGERS. "Intermodal Equivalence in Stimuli in Apes." *Science* 168 (1970): 279–280.

DAVENPORT, RICHARD K., C. M. ROGERS, and I. S. RUSSELL. "Cross-Modal Perception in Apes." *Neuropsychologia* 11 (1973): 21–28.

———. "Cross-Modal Perception in Apes: Altered Visual Cues and Delay." *Neuropsychologia* 13 (1975): 229–235.

DAVIDSON, D. Introduction to *Origins and Evolution of Language and Speech*, edited by S. R. Harnad, H. D. Steklis, and J. Lancaster. New York: Annals of the New York Academy of Sciences, 280 (1976): 18–19.

DAVIS, L., and H. GARDNER. "Strategies of Mastering a Visual Communication System in Aphasia." In *Origins and Evolution of Language and Speech*, edited by S. R. Harnad, H. D. Steklis, and J. Lancaster. New York: Annals of the New York Academy of Sciences, 280 (1976): 885–897.

DE MILLE, RICHARD. *Castaneda's Journey: The Power and the Allegory*. Santa Barbara: Capra, 1976.

DENNIS, W. *Children of the Creche*. New York: Appleton-Century-Crofts, 1973.

DESCARTES, RENÉ. *Discourse on Method and Meditations*, translated by L. Lafleur. Indianapolis: Bobbs-Merrill, 1960. (Originally published 1637, 1640.)

DEWSON, J. H., and A. C. BURLINGAME. "Auditory Discrimination and Recall in Monkeys." *Science* 187 (1975): 267–268.

DINGWALL, WILLIAM ORR. "The Species-Specificity of Speech." In *Developmental Psycholinguistics: Theory and Applications*, edited by Daniel P. Data. Washington, D.C.: Georgetown University Round Table on Languages and Linguistics 1975, 1975, pp. 17–62.

DOBRZECKA, C., G. SZWEJKOWSKA, and J. KONORSKI. "Qualitative versus Directional Cues in Two Forms of Differentiation." *Science* 153 (1966): 87–89.

DRACHMAN, GABRIEL. "Adaptation in the Speech Tract." In *Papers from the Fifth Regional Meeting of the Chicago Linguistic Society*, edited by R. J. Binnick, A. Davison, G. Green, and J. L. Morgan. Chicago: Department of Linguistics, University of Chicago, 1969.

DU BRUL, E. L. *Evolution of the Speech Apparatus*. Springfield, Ill: Charles C Thomas, 1958.

EIBL-EIBESFELDT, IRENÄUS. "The Expressive Behaviour of the Deaf-and-Blind Born." In *Social Communicaton and Movement: Studies of Interaction and Expression in Man and Chimpanzee*, edited by Mario Von Cranach and Ian Vine. New York: Academic, 1973, pp. 163–194.

EISENBERG, JOHN F. "Mammalian Social Systems: Are Primate Social Systems Unique?" *Symposia IVth International Congress of Primatology* 1 (1973): 232–249.

EISENBERG, JOHN F., and DEVRA G. KLEIMAN. "Communication in Lagomorphs and Rodents." In *How Animals Communicate*, edited by Thomas A. Sebeok. Bloomington: Indiana University Press, 1977, pp. 634–654.

ELDER, J. H. "Auditory Acuity of the Chimpanzee." *Journal of Comparative Psychology* 17 (1934): 157–183.

———. "The Upper Limit of Hearing in Chimpanzee." *American Journal of Physiology* 112 (1935): 109–123.

ERVIN-TRIPP, SUSAN. "Discourse Agreement: How Children Answer Questions." In *Cognition and the Development of Language*, edited by John R. Hayes. New York: Wiley, 1970, pp. 79–107.

ESSOCK, SUSAN M. "Color Perception and Color Classification." In *Language Learning by a Chimpanzee*, edited by D. M. Rumbaugh. New York: Academic, 1977, pp. 207–224.

ESTES, WILLIAM K. "The State of the Field: General Problems and Issues of Theory and Metatheory." In *Handbook of Learning and Cognitive Processes*, vol. 1, edited by William K. Estes. New York: Wiley, 1975, pp. 1–24.

———. (ed.). *Handbook of Learning and Cognitive Processes*, vol. 2, New York: Wiley, 1976.

ETTLINGER, GEORGE. "Analysis of Cross-Modal Effects and Their Relationship to Language." In *Brain*

Mechanisms Underlying Speech and Language, edited by Frederic L. Darley and Clark H. Millikan. New York: Grune & Stratton, 1967.

——. "Interactions between Sensory Modalities in Nonhuman Primates." In *Behavioral Primatology. Advances in Research and Theory,* vol. 1, edited by Allan M. Schrier. Hillsdale, N.J.: Lawrence Erlbaum, 1977, pp. 71–104.

FALK, D. "Comparative Anatomy of the Larynx in Man and the Chimpanzee: Implications for Language in Neanderthal." *American Journal of Physical Anthropology* 43 (1975): 123–132.

FANT, L. J. *Ameslan: An Introduction to American Sign Language.* Silver Spring, Md.: National Association of the Deaf, 1972.

FARRER, D. N. "Picture Memory in the Chimpanzee." *Perceptual and Motor Skills* 25 (1967): 305–315.

FELDMAN, H., S. GOLDIN-MEADOW, and L. GLEITMAN. "Beyond Herodotus: The Creation of Language by Linguistically Deprived Deaf Children." In *Action, Gesture and Symbol: The Emergence of Language,* edited by A. Lock. New York: Academic, 1977.

FINE, GARY ALAN, and BEVERLY J. CRANE. "The Expectancy Effect in Anthropological Research: An Experimental Study of Riddle Collection." *American Ethnologist* 4 (1977): 517–524.

FISHER, SEYMOUR. "The Role of Expectancy in the Performance of Posthypnotic Behavior." In *The Nature of Hypnosis,* edited by Ronald E. Shor and Martin T. Orne. New York: Holt, Rinehart & Winston, 1965, pp. 80–88.

FODOR, JERRY A. *Psychological Explanation.* New York: Random House, 1968.

FODOR, JERRY A., T. G. BEVER, and MERRILL F. GARRETT. *The Psychology of Language. An Introduction to Psycholinguistics and Generative Grammar.* New York: McGraw-Hill, 1974.

FOSSEY, DIANE. "Vocalizations of the Mountain Gorilla *(Gorilla gorilla beringei)."* *Animal Behaviour* 20 (1972): 36–53.

——. "Observations on the Home Range of One Group of Mountain Gorillas *(Gorilla gorilla beringei)."* *Animal Behaviour* 22 (1974): 568–581.

FOUTS, ROGER S. "The Acquisition and Testing of Gestural Signs in Four Young Chimpanzees *(Pan troglodytes)."* Paper presented at the annual meeting of the Animal Behavior Society, Reno, 1972. (a) (See Fouts 1973 for published version.)

——. "Use of Guidance in Teaching Sign Language to a Chimpanzee *(Pan troglodytes)."* *Journal of Comparative and Physiological Psychology* 80 (1972): 515–522. (b)

——. "Acquisition and Testing of Gestural Signs in Four Young Chimpanzees." *Science* 180 (1973): 978–980.

——. "Language: Origins, Definitions, and Chimpanzees." *Journal of Human Evolution* 3 (1974): 475–482.

——. "Capacities for Language in Great Apes." In *Socioecology and Psychology of the Primates,* edited by Russell H. Tuttle. The Hague: Mouton, 1975, pp. 371–390. (a)

——. "Communication with Chimpanzees." In *Hominisation und Verhalten,* edited by G. Kurth and I. Eibl-Eibesfeldt. Stuttgart: Fischer, 1975, pp. 137–158. (b)

——. "Field Report: The State of Apes." *Psychology Today* 7(8) (1975): 31–54. (c)

——. "Comparison of Sign Language Projects and Implications for Language Origins." In *Origins and Evolution of Language and Speech,* edited by S. R. Harnad, H. D. Steklis, and J. Lancaster. New York: Annals of the New York Academy of Sciences, 280 (1976): 589–591.

FOUTS, ROGER S. and J. B. COUCH. "Cultural Evolution of Learned Language in Chimpanzees." In *Communication Behavior and Evolution,* edited by M. E. Hahn and E. C. Simmel. New York: Academic, 1976, pp. 141–161.

FOUTS, ROGER S., and R. L. MELLGREN. "Language, Signs, and Cognition in the Chimpanzee." *Sign Language Studies* 13 (1976): 319–346.

FOUTS, ROGER S., W. CHOWN, and L. T. GOODIN. "The Use of Vocal English to Teach American Sign Language (ASL) to a Chimpanzee: Translation from English to ASL." Paper presented at the Southwestern Psychological Association Meeting, Dallas, 1973.

——. "Transfer of Signed Responses in American Sign Language from Vocal English Stimuli to Physical Object Stimuli by a Chimpanzee *(Pan)."* *Learning and Motivation* 7 (1976): 458–475.

FOUTS, ROGER S., W. B. CHOWN, G. KIMBALL, and J. B. COUCH. "Comprehension and Production of

American Sign Language by a Chimpanzee *(Pan)*." Paper presented at the 21st International Congress of Psychology, Paris, 1976.

FOUTS, ROGER S., R. L. MELLGREN, and W. B. LEMMON. "American Sign Language in the Chimpanzee: Chimpanzee-to-Chimpanzee Communication." Paper presented at the Midwestern Psychological Association meeting, Chicago, 1973.

FOX, M. J., and B. P. SKOLNICK. *Language in Education: Problems and Prospects in Research and Training.* New York: Ford Foundation, 1975.

FOX, MICHAEL W. *Behavior of Wolves, Dogs, and Related Canids.* New York: Harper & Row, 1971.

FRANÇOIS, FRÉDÉRIC. "La Description Linguistique." In *Le Langage,* edited by André Martinet. Paris: Gallimard, 1968, pp. 171–282.

FRASER, COLIN, URSULA BELLUGI, and ROGER BROWN. "Control of Grammar in Imitation, Comprehension and Production." *Journal of Verbal Learning and Verbal Behavior* 2 (1963): 121–135.

FREEDMAN, D. *Human Infancy: An Evolutionary Perspective.* Hillsdale, N.J.: Lawrence Erlbaum, 1975.

FRIEDMAN, LYNN A. "Space, Time and Person Reference in American Sign Language." *Language* 51 (1975): 940–961.

FRISHBERG, N. "Arbitrariness and Iconicity: Historical Change in American Sign Language." *Language* 51 (1975): 696–719.

FROLOV, IURII P. *Pavlov and His School.* London: Paul, Trench, Trubner, 1938.

FURNESS, W. H. "Observations on the Mentality of Chimpanzees and Orangutans." *Proceedings of the American Philosophical Society* 65 (1916): 281–290.

GALLUP, G. G., JR. "Chimpanzees: Self-Recognition." *Science* 167 (1970): 86–87.

GALLUP, G. G., JR., J. L. BOREN, G. J. GAGLIARO, and L. B. WALLNAN. "A Mirror for the Mind of Man, or Will the Chimpanzee Create an Identity Crisis for *Homo Sapiens?"* *Journal of Human Evolution* 6 (1977): 303–314.

GARCIA, J., B. K. McGOWAN, F. R. ERVIN, and R. A. KOELLING. "Cues: Their Relative Effectiveness as a Function of the Reinforcer." *Science* 160 (1968): 794–795.

GARDNER, BEATRICE T., and R. ALLEN GARDNER. "Teaching Sign Language to a Chimpanzee. Part I: Methodology and Preliminary Results. Part II: Demonstrations." *Psychonomic Bulletin* 1(2) (1967): 36.

———. "Teaching Sign Language to a Chimpanzee." *Science* 165 (1969):664–672.

———. "Development of Behavior in a Young Chimpanzee." In *Eighth Summary of Washoe's Diary.* Reno: University of Nevada, Department of Psychology, 1970.

———. "Two-Way Communication with an Infant Chimpanzee." In *Behavior of Nonhuman Primates,* vol. 4, edited by Allan M. Schrier and Fred Stollnitz. New York: Academic, 1971, pp. 117–183.

———. "Communication with a Young Chimpanzee: Washoe's Vocabulary." In *Modeles du Comportement Humain,* edited by R. Chauvin. Paris: Centre National de la Recherche Scientifique, 1972, pp. 241–264.

———. "Teaching Sign Language to the Chimpanzee Washoe." (16-mm sound film, transcript of the soundtrack available from the authors on request) University Park: Pennsylvania State University Psychological Cinema Register, 1973.

———. "Behavioral Development of the Chimpanzee, Washoe." Film presented at the American Psychological Association meeting, New Orleans, 1974. (a)

———. "Comparing the Early Utterances of Child and Chimpanzee." In *Minnesota Symposium in Child Psychology,* vol. 8, edited by A. Pick. Minneapolis: University of Minnesota Press, 1974, pp. 3–23. (b)

———. Review of *A First Language: The Early Stages* by Roger Brown, 1973. *American Journal of Psychology* 87 (1974): 729–736. (c)

———. "Teaching Sign Language to a Chimpanzee. VII: Use of Order in Sign Combinations." *Bulletin of the Psychonomic Society* 4 (1974): 264. (d)

———. "Early Signs of Language in Child and Chimpanzee." *Science* 187 (1975): 752–753. (a)

———. "Evidence for Sentence Constituents in the Early Utterances of Child and Chimpanzee." *Journal of Experimental Psychology: General* 104(3) (1975): 244–267. (b)

———. "Emergence of Language." Paper presented at the meeting of the American Association for the Advancement of Science, Washington, D.C., 1978.

GARDNER, HOWARD, EDGAR ZURIF, T. BERRY, and E. BAKER. "Visual Communication in Aphasia." *Neuropsychologia* 14 (1976): 275–292.

GARDNER, MARTIN. *Fads and Fallacies in the Name of Science*. New York: Dover, 1957.

———. "Dermo-Optical Perception: A Peek Down the Nose." *Science* 151 (1966): 654–657.

GARNER, RICHARD L. *The Speech of Monkeys*. New York: Charles L. Webster, 1892.

GAZZANIGA, MICHAEL S. "Brain Mechanisms and Behavior." In *Handbook of Psychobiology*, edited by Michael S. Gazzaniga and C. B. Blakemore. New York: Academic, 1975, pp. 591–605.

GESCHWIND, NORMAN. "Disconexion Syndromes in Animals and Man." *Brain* 88 (1965): 237–294; 585–644.

———. "Anatomy and the Higher Functions of the Brain." *Boston Studies in the Philosophy of Science* 4 (1969): 98–136.

GILL, TIMOTHY V. "Conversations with Lana." In *Language Learning by a Chimpanzee*, edited by D. M. Rumbaugh. New York: Academic, 1977, pp. 225–246.

GILL, TIMOTHY V., and DUANE M. RUMBAUGH. "Mastery of Naming Skills by a Chimpanzee." *Journal of Human Evolution* 3 (1974): 483–492.

GLASERSFELD, ERNST VON. "The Development of Language as Purposive Behavior." In *Origins and Evolution of Language and Speech*, edited S. R. Harnad, H. D. Steklis, and J. Lancaster. New York: Annals of the New York Academy of Sciences, 280 (1976): 212–226.

———. "The Yerkish Language and Its Automatic Parser." In *Language Learning by a Chimpanzee*, edited by D. M. Rumbaugh. New York: Academic, 1977, pp. 91–130.

GOLDIN-MEADOW, SUSAN, and H. FELDMAN. "The Development of Language-Like Communication without a Language Model." *Science* 197 (1977): 401–403.

GOODALL, JANE. "Chimpanzees of the Gombe Stream Reserve." In *Primate Behavior: Field Studies of Monkeys and Apes*, edited by Irven DeVore. New York: Holt, Rinehart & Winston, 1965, pp. 425–473.

———. "Behavior of Male and Female Chimpanzees." Paper presented at the L. S. B. Leakey Memorial Lectures, Philadelphia, 1974.

GOODMAN, M. "Toward a Geneological Description of the Primates." In *Molecular Anthropology*, edited by M. Goodman and R. E. Tashian. New York: Plenum, 1976, pp. 321–353.

GOYER, M. "Comparison of the Postural Development of Home-Reared Chimpanzees, Lab-Reared Chimpanzees and Human Infants." Paper presented at the meeting of the Western Psychological Association, Los Angeles, 1976.

———. "The First Year: A Comparison of the Development of Three Home-Reared Chimpanzees." Master's thesis, University of Nevada, Reno 1977.

GREEN, S. "Variation of Vocal Pattern with Social Situation in the Japanese Monkey *(Macaca fuscata)*: A Field Study." In *Primate Behavior*, vol. 4, edited by Leonard A. Rosenblum. New York: Academic, 1975, pp. 1–102.

GREENFIELD, PATRICIA M., and JOSHUA H. SMITH. *The Structure of Communication in Early Language Development*. New York: Academic, 1976.

GRÉGOIRE, ANTOINE. *L'Apprentissage du Langage: Les Deux Premières Années*. Paris: Librairie E. Droz, 1937.

GREGORY, RICHARD. "The Grammar of Vision." *The Listener*, Feb. 19, 1970, 242–244.

GRETHER, W. "Chimpanzee Color Vision." *Journal of Comparative and Physiological Psychology*, 29 (1940): 167–192.

GRICE, H. P. "Utterer's Meanings and Intentions." *Philosophical Review* 78 (1969): 147–177.

GRIFFIN, DONALD R. *The Question of Animal Awareness. Evolutionary Continuity of Mental Experiences*. New York: The Rockefeller Press, 1976.

———. "Prospects for a Cognitive Ethology." *The Behavioral and Brain Sciences* 4 (1978): 527–538.

GRUENBERG, BENJAMIN C. *The Story of Evolution: Facts and Theories on the Development of Life*. Garden City, N.Y.: Garden City Publishing, 1929.

GUMP, RICHARD. *Jade: Stone of Heaven*. Garden City, N.Y.: Doubleday, 1962.

GUNDERSON, K. *Mentality and Machines*. Garden City, N.Y.: Anchor, 1971.

GUTTMAN, N. Review of *Biological Foundations of Language* by Eric H. Lenneberg, 1967. *Journal of the Acoustical Society of America* 43 (1968): 178–179.

HACHET-SOUPLET, PIERRE. *Le Dressage des Animaux et les Combats des Bêtes, Révélation des Procédés Employés par les Professionels pour Dresser le Chien, le Singe, le Cheval, l'Éléphant, les Bêtes Féroces, etc.* Paris: Firmin Didot, 1897.

HALLIDAY, MICHAEL A. K. *Learning How to Mean.* New York: Elsevier, 1975.

HALLIWELL-PHILLIPPS, JAMES O. *Memoranda on Love's Labour's Lost, King John, Othello, and on Romeo and Juliet.* London: James Evan Adlard, 1879.

HAMILTON, J. "Hominid Divergence and Speech Evolution." *Journal of Human Evolution,* 3 (1974): 417–424.

HANDELMAN, D. Comment. *Current Anthropology* 17 (1976): 9–10.

HANSEL, CHARLES E. M. *ESP: A Scientific Evaluation.* New York: Scribner's, 1966.

HARLOW, Harry F. "The Formation of Learning Sets." *Psychological Review* 56 (1949): 51–65.

———. Review of *Intelligence in Ape and Man* by David Premack, 1976. *American Scientist* 65 (1977): 639–640.

HARNAD, STEVAN R., HORST D. STEKLIS, and JANE LANCASTER, eds. *Origins and Evolution of Language and Speech.* New York: Annals of the New York Academy of Sciences 1976, 280.

HARRISSON, B. *Orang-utan.* London: Collins, 1962.

HAYES, CATHERINE H. *The Ape in Our House.* New York: Harper & Row, 1951.

HAYES, HAROLD T. P. "The Pursuit of Reason." *New York Times Magazine* June 12, 1977, pp. 21–23, 73, 75–79.

HAYES, KEITH J., and CATHERINE H. HAYES. "Vocalization and Speech in Chimpanzees" (16-mm sound film). University Park: Pennsylvania State University Psychological Cinema Register, 1950.

———. "The Intellectual Development of a Home-Raised Chimpanzee." *Proceedings of the American Philosophical Society* 95 (1951): 105–109.

———. "Imitation in a Home-Raised Chimpanzee." *Journal of Comparative and Physiological Psychology* 45 (1952): 450–459.

———. "Picture Perception in a Home-Raised Chimpanzee." *Journal of Comparative and Physiological Psychology* 46 (1953): 470–474.

———. "The Cultural Capacity of Chimpanzees." *Human Biology* 26 (1954): 288–303. (a)

———. "The Mechanical Interest and Ability of a Home-Raised Chimpanzee" (16-mm silent film). University Park: Pennsylvania State University Psychological Cinema Register, 1954. (b)

HAYES, KEITH J., and CATHERINE H. NISSEN. "Higher Mental Functions of a Home-Raised Chimpanzee." In *Behavior of Nonhuman Primates,* vol. 4, edited by Allan M. Schrier and Fred Stollnitz. New York: Academic, 1971, pp. 59–115.

HEDIGER, H. "Communication between Man and Animal." *Image Roche* 62 (1974): 27–40.

HEIDER, F. *The Psychology of Interpersonal Relations.* New York: Wiley, 1958.

HESS, L. *Christine the Baby Chimp.* London: Bell, 1954.

HEWES, GORDON W. "Conversations with Chimpanzees: Recent Studies of the Capacity of Apes to Acquire and Use Human Language." *Sociolinguistics Newsletter* 2(4) (1971): 3–5.

———. "Pongid Capacity for Language Acquisition." *Symposia of the Fourth International Congress of Primatology,* vol. 1. Basel: Karger, 1973, pp. 124–143. (a)

———. "Primate Communication and the Gestural Origin of Language." *Current Anthropology* 14 (1973): 5–24. (b)

———. "The Current Status of Gestural Origin Theory." In *Origins and Evolution of Language and Speech,* edited by S. R. Harnad, H. D. Steklis, and J. Lancaster. New York: Annals of the New York Academy of Sciences, 280 (1976): 482–504.

———. "Language Origin Theories." In *Language Learning by a Chimpanzee,* edited by D. M. Rumbaugh. New York: Academic, 1977, pp. 3–53.

HILL, JANE H. "On the Evolutionary Foundations of Language." *American Anthropologist* 74 (1972): 308–317.

———. "Possible Continuity Theories of Language." *Language* 50 (1974): 134–150.

HILL, O. *Evolutionary Biology of Primates.* London: Academic, 1972.

HINDE, ROBERT A., ed. *Non-Verbal Communication.* London: Cambridge University Press, 1972.

HOCKETT, CHARLES F. *A Course in Modern Linguistics.* New York: Macmillan, 1958.

———. "Logical Considerations in the Study of Animal Communication." In *Animal Communication,* edited by W. E. Lanyon and W. N. Tavolga. Washington, D.C.: American Institute of Biological Sciences, 1960, pp. 392–430. (a)

———. "The Origin of Speech." *Scientific American* 203 (1960) 89–96. (b)

———. "The Problem of Universals in Language." In *Universals of Language,* edited by Joseph H. Greenberg. Cambridge, Mass.: M.I.T. Press, 1963, pp. 1–22.

HOCKETT, CHARLES F., and STUART A. ALTMANN. "A Note on Design Features." In *Animal Communication,* edited by Thomas A. Sebeok. Bloomington: Indiana University Press, 1968, pp. 61–72.

HOCKETT, CHARLES F., and R. ASCHER. "The Human Revolution." *Current Anthropology* 5 (1964): 135–168.

HOEMANN, H. W. "The Transparency of Meaning of Sign Language Gestures." *Sign Language Studies* 7 (1975): 151–161.

HOFFMEISTER, R. J., D. F. MOORES, and R. L. ELLENBERGER. "Some Procedural Guidelines for the Study of the Acquisition of Sign Languages." *Sign Language Studies* 7 (1975): 121–137.

HOLDEN, CONSTANCE. "Pain Control with Hypnosis." *Science* 198 (1977): 808.

HOLLOWAY, R. "Culture: A Human Domain." *Current Anthropology* 10 (1969): 395–412.

HOOFF, J. A. R. A. M. VAN. *Aspects of the Social Behavior and Communication in Human and Higher Nonhuman Primates.* Rotterdam: Bronder-Offset, 1971.

HOPKINS, CARL D. "Electric Communication." In *How Animals Communicate,* edited by Thomas A. Sebeok. Bloomington: Indiana University Press, 1977, pp. 263–289.

HOYT, A. M. *Toto and I: A Gorilla in the Family.* New York: Lippincott, 1941.

HUMBOLDT, WILHELM VON "Über das vergleichende Sprachstudium in Beziehung auf die verschiedenen Epochen der Sprachentwicklung." *Gesammelte Schriften,* vol. 4. Berlin: Königlich-Preussische Akademie der Wissenschaften, 1905.

HURVICH, L. and D. JAMESON. "Opponent Processes as a Model of Neural Organization." *American Psychologist* 29 (1974): 88–102.

HUXLEY, FRANCIS. "Anthropology and ESP." In *Science and ESP,* edited by John R. Smythies. New York: Humanities, 1967, chap. 13.

HYMES, DELL H. *Foundations in Sociolinguistics.* Philadelphia: University of Pennsylvania Press, 1975.

JACOBSEN, CARLYLE, MARION M. JACOBSEN, and JOSEPH G. YOSHIOKA. *Development of an Infant Chimpanzee during Her First Year (Comparative Psychological Monographs 9, no. 41), 1932, pp. 1–94.

JAKOBSON, ROMAN, C. G. M. FANT, and MORRIS HALLE. *Preliminaries to Speech Analysis.* Cambridge, Mass.: M.I.T. Press, 1969.

JAMES, WILLIAM. *Principles of Psychology.* New York: Dover, 1890.

JANET, PIERRE. *Psychological Healing: A Histological and Clinical Study.* New York: Macmillan, 1925. (Originally published 1919.)

JANISSE, MICHEL PIERRE. *Pupillometry: The Psychology of the Pupillary Response.* New York: Halsted, 1977.

JAYNES, JULIAN. "The Origin of Consciousness." In *A Symposium in Honor of Robert MacLeod,* edited by David Krech and H. Levin. Ithaca, N.Y.: Privately printed, 1973.

———. *The Origin of Consciousness in the Breakdown of the Bicameral Mind.* Boston: Houghton Mifflin, 1977.

JERISON, HARRY J. *Evolution of the Brain and Intelligence.* New York: Academic, 1973.

JORDON, I.K. "A Referential Communication Study of Signers and Speakers Using Realistic Referents." *Sign Language Studies* 6 (1975): 65–103.

JORDON, I. K., and R. BATTISON. "A Referential Communication Experiment with Foreign Sign Languages." *Sign Language Studies* 10 (1976): 69–80.

KATZ, DAVID. *Animals and Men: Studies in Comparative Psychology.* London: Longmans, Green, 1937.

KATZ, JERROLD J. "A Hypothesis about the Uniqueness of Natural Language." In *Origins and Evolution of Language and Speech,* edited by S. R. Harnad, H. D. Steklis, and J. Lancaster. New York: Annals of the New York Academy of Sciences, 280 (1976): 33–41.

KEARTON, C. *My Friend Toto: The Adventures of a Chimpanzee and the Story of His Journey from the Congo to London.* London: Arrowsmith, 1925.

KELEMAN, GEORGE. "The Anatomical Basis of Phonation in the Chimpanzee." *Journal of Morphology* 82 (1948): 229–256.

———. "Structure and Performance in Animal Language." *Archives of Otolaryngology* 50 (1949): 740–744.

KELLEY, HAROLD H. "Attribution Theory in Social Psychology." In *Nebraska Symposium on Motivation,* vol. 15, edited by D. Levine. Lincoln: University of Nebraska Press, 1967, pp. 192–238.

KELLOGG, WINTHROP N., and LOUISE A. KELLOGG. "Comparative Tests on a Human and a Chimpanzee Infant of Approximately the Same Age." (16-mm silent film) University Park: Pennsylvania State University Psychological Cinema Register, 1932. (a)

———. "Experiments upon a Human and a Chimpanzee Infant after Six Months in the Same Environment." (16-mm silent film) University Park: Pennsylvania State University Psychological Cinema Register, 1932. (b)

———. "Some Behavior Characteristics of a Human and a Chimpanzee Infant in the Same Environment." (16-mm silent film) University Park: Pennsylvania State University Psychological Cinema Register, 1933. (a)

———. "Some General Reactions of a Human and a Chimpanzee Infant after Six Months in the Same Environment." (16-mm silent film) University Park: Pennsylvania State University Psychological Cinema Register, 1933. (b)

———. "Facial Expressions of a Human and a Chimpanzee Infant Following Taste Stimuli." (16-mm silent film) University Park: Pennsylvania State University Psychological Cinema Register, 1945.

———. *The Ape and the Child: A Study of Environmental Influence Upon Early Behavior.* New York: Hafner, 1967. (Originally published in 1933 by Whittlesey House.)

KENDLER, HOWARD H., and TRACY S. KENDLER. "From Discrimination Learning to Cognitive Development: A Neobehavioristic Odyssey." In *Handbook of Learning and Cognitive Processes,* vol. 1, edited by William K. Estes. New York: Lawrence Erlbaum, 1975, pp. 191–247.

KENDON, ADAM. *Studies in the Behavior of Social Interaction.* Lisse: Peter de Ridder, 1977.

KING, J. E., and J. L. FOBES. "Evolutionary Change in Primate Sensory Capacities." *Journal of Human Evolution* 3 (1974): 435–443.

KING, M. C., and A. C. WILSON. "Evolution at Two Levels in Humans and Chimpanzee." *Science* 186 (1975): 107–116.

KINNEY, ARTHUR E., ed. *Rogues, Vagabonds, and Sturdy Beggars.* Barre, Mass: Imprint Society, 1973.

KLATT, D. H. and R. STEFANSKI. "How Does a Myna Bird Imitate Human Speech?" *Journal of the Acoustical Society of America* 55 (1974): 822.

KLIMA, EDWARD S., and URSULA BELLUGI. "The Signs of Language in Child and Chimpanzee." In *Communication and Affect,* edited by T. Alloway, L. Krames, and P. Pliner. New York: Academic, 1972, pp. 67–96.

KOESTLER, ARTHUR. *The Case of the Midwife Toad.* London: Hutchinson, 1971.

KÖHLER, WOLFGANG. *The Mentality of Apes.* London: Routledge & Kegan Paul, 1927.

———. "Methods of Psychological Research with Apes." In *The Selected Papers of Wolfgang Köhler,* edited by Mary Henle. New York: Liveright, 1971, pp. 197–223. (Originally published 1921.)

KOHTS, N. *Infant Ape and Human Child (Instincts, Emotions, Play and Habits)* Scientific Memoirs of the Museum Darwinian, Moscow, 1935.

KOLATA, GINA BARI. "Mathematics and Magic: Illumination and Illusion." *Science* 198 (1977): 282–283.

KORTLANDT, A. "How Do Chimpanzees Use Weapons when Fighting Leopards?" *Yearbook of the American Philosophical Society.* Philadelphia: The American Philosophical Society, 1965.

———. "Experimentation with Chimpanzees in the Wild." In *Progress in Primatology,* edited by D. Starck, R. Schneider, and H. J. Kuhn. Stuttgart: Fischer, 1967, pp. 208–224.

———. Comment. *Current Anthropology* 14 (1973): 13–14.

KRESKIN [GEORGE KRESGE, JR.]. *The Amazing World of Kreskin.* New York: Random House, 1973.

KUMMER, HANS. *Social Organization of Hamadrayas Baboons: A Field Study.* Chicago: University of Chicago Press, 1968.

LAMETTRIE, JULIEN OFFRAY DE. *Man a Machine.* Chicago: Opencourt, 1912. (Revised edition, 1948.)

LANCASTER, JANE B. "Primate Communication Systems and the Emergence of Human Language." In *Primates. Studies in Adaptation and Variability,* edited by Phyllis C. Jay. New York: Holt, Rinehart & Winston, 1968, pp. 439–457.

LAWICK-GOODALL, JANE VAN. "A Preliminary Report on Expressive Movements and Communication in Gombe Stream Chimpanzees." In *Primates. Studies in Adaptation and Variability,* edited by Phyllis C. Jay. New York: Holt, Rinehart & Winston, 1968, pp. 313–374. (a)

———. "The Behaviour of Free-Living Chimpanzees in the Gombe Stream Reserve." *Animal Behaviour Monograph* 1 (1968): 161–311. (b)

———. "Tool-Using in Primates and Other Vertebrates." *Advances in the Study of Behavior* 3 (1970): 195–249.

———. *In the Shadow of Man.* Boston: Houghton Mifflin, 1971.

———. "Cultural Elements in a Chimpanzee Community." In *Precultural Primate Behavior,* edited by Emil W. Menzel. Basel: Karger, 1973, pp. 144–184.

LEES, ROBERT B. *The Grammar of English Nominalizations.* Bloomington: Indiana University Press, 1960.

LENNEBERG, ERIC H. "Understanding Language without Ability to Speak: A Case Report." *Journal of Abnormal Social Psychology* 65 (1962): 419–425.

———. *Biological Foundations of Language.* New York: Wiley, 1967.

———. "The Natural History of Language." In *The Genesis of Language, a Psycholinguistic Approach,* edited by Frank Smith and George A. Miller. Cambridge, Mass.: M.I.T. Press, 1968, pp. 215–252.

———. "On Explaining Language." *Science* 164 (1969): 635–643.

———. "Brain Correlates of Language." In *The Neurosciences: Second Study Program,* edited by Francis O. Schmitt. New York: Rockefeller University Press, 1970.

LEOPOLD, WERNER F. *Speech Development of a Bilingual Child: A Linguist's Record,* vol. 3, *Grammar and General Problems in the First Two Years.* Evanston, Ill.: Northwestern University Press, 1949. (a)

———. *Speech Development of a Bilingual Child: A Linguist's Record,* vol. 4, *Diary from Age Two.* Evanston, Ill.: Northwestern University Press, 1949. (b)

LEWIS, MICHAEL, and LOUISE CHERRY. "Social Behavior and Language Acquisition." In *Interaction, Conversation, and the Development of Language,* edited by Michael Lewis and Leonard A. Rosenblum. New York: Wiley, 1977, pp. 227–245.

LEX, BARBARA W. "Voodoo Death: New Thoughts on an Old Explanation." *American Anthropologist* 76 (1974): 818–823.

LIBERMAN, A. M., F. S. COOPER, D. P. SHANKWEILER, and M. STUDDERT-KENNEDY. "Perception of the Speech Code." *Psychological Review* 74 (1967): 431–461.

LIEBERMAN, PHILIP. *Intonation, Perception and Language.* Cambridge, Mass.: M.I.T. Press, 1967.

———. "Primate Vocalizations and Human Linguistic Ability." *Journal of the Acoustical Society of America* 44 (1968): 1574–1584.

———. *On the Origins of Language.* New York: Macmillan, 1975. (a)

———. "The Evolution of Speech and Language." In *The Role of Speech in Language,* edited by James F. Kavanagh and James E. Cutting. Cambridge, Mass.: M.I.T. Press, 1975, pp. 83–106. (b)

LIEBERMAN, PHILIP, E. S. CRELIN, and D. H. KLATT. "Phonetic Ability and Related Anatomy of the Newborn and Adult Human, Neanderthal Man, and the Chimpanzee." *American Anthropologist* 74 (1972): 287–307.

LIEBERMAN, PHILIP, D. H. KLATT, and W. H. WILSON. "Vocal Tract Limitations on the Vowel Repertoires of Rhesus Monkey and Other Non-Human Primates." *Science* 164 (1969): 1185–1187.

LILLY, JOHN C. "Productive and Creative Research with Man and Dolphin." *Archives of General Psychiatry* 8 (1963): 111–116.

LIMBER, JOHN. "The Genesis of Complex Sentences." In *Cognitive Development and the Acquisition of Language,* edited by Timothy E. Moore. New York: Academic, 1973, pp. 169–185.

———. Essay Review of *Language by Eye and Ear,* by James Kavanagh, ed. *Harvard Educational Review* 44 (1974): 336–343.

———. "Unravelling Competence, Performance, and Pragmatics in the Speech of Young Children." *Journal of Child Language* 3 (1976): 309–318.

LINDBLOM, B. E., and J. SUNDBERG. "Neurophysiological Representation of Speech Sounds." Paper presented at the 15th World Congress of Logopedics and Phoniatrics, Buenos Aires, August 1971.

LINDEN, EUGENE. *Apes, Men and Language*. New York: Dutton, 1975.

LINTZ, G. D. *Animals Are My Hobby*. New York: McBride, 1942.

LLOYD, J. E. "Aggressive Mimicry in *Photurus:* Firefly Femmes Fatales." *Science* 149 (1965): 653–654.

LOVELL, K., and E. M. DIXON. "The Growth of the Control of Grammar in Imitation, Comprehension and Production." *Journal of Child Psychology and Psychiatry* 5 (1965): 1–9.

LUBBOCK, JOHN [LORD AVEBURY]. "Note on the Intelligence of the Dog." *Report of the Fifty-fifth Meeting of the British Association for the Advancement of Science*. London: Murray, 1886, pp. 1089–1091.

LUCHSINGER, R., and G. E. ARNOLD. *Voice–Speech–Language. Clinical Communicology: Its Physiology and Pathology*. Belmont, Cal.: Wadsworth, 1965.

LYONS, JOHN. "Human Language." In *Non-Verbal Communication,* edited by R. A. Hinde. London: Cambridge University Press, 1972, pp. 49–85.

MANNING, AUBREY. *An Introduction to Animal Behavior*. Reading, Mass.: Addison-Wesley, 1972.

MARKS, DAVID, and RICHARD KAMMANN. "The Nonpsychic Powers of Uri Geller." *The Zetetic* 1(2) (1977): 9–17.

MARLER, PETER. "Communication in Monkeys and Apes." In *Primate Behavior: Field Studies of Monkeys and Apes,* edited by Irven DeVore. New York: Holt, Rinehart & Winston, 1965, pp. 544–584.

———. "Vocalizations of Wild Chimpanzees: An Introduction." *Proceedings of the Second International Congress of Primatology,* vol. 2, 1969, pp. 94–100.

———. "On the Origin of Speech from Animal Sounds." In *The Role of Speech in Language,* edited by James F. Kavanagh and James E. Cutting. Cambridge, Mass.: M.I.T. Press, 1975, pp. 11–37.

———. "An Ethological Theory of the Origin of Vocal Learning." In *Origins and Evolution of Language and Speech,* edited by S. R. Harnad, H. D. Steklis, and J. Lancaster. New York: Annals of the New York Academy of Sciences, 280 (1976): 386–395. (a)

———. "Social Organization, Communication and Graded Signals: The Chimpanzee and the Gorilla." In *Growing Points in Ethology,* edited by P. P. G. Bateson and R. A. Hinde. London: Cambridge University Press, 1976, pp. 239–280. (b)

MARLER, PETER, and JANE VAN LAWICK-GOODALL. "Vocalizations in Wild Chimpanzees" (sound film). New York: Rockefeller University Film Service, 1971.

MARLER, PETER, and RICHARD TENAZA. "Signalling Behavior of Wild Apes with Special Reference to Vocalization." In *How Animals Communicate,* edited by Thomas A. Sebeok. Bloomington: Indiana University Press, 1978, pp. 965–1033.

MARSHACK, ALEXANDER. "Some Implications of the Paleolithic Symbolic Evidence for the Origin of Language." In *Origins and Evolution of Language and Speech,* edited by S. R. Harnad, H. D. Steklis, and J. Lancaster. New York: Annals of the New York Academy of Sciences, 280 (1976): 289–311.

MASON, W. A. "Information Processing and Experimental Deprivation: A Biologic Perspective." In *Early Experience and Visual Information Processing in Perceptual and Reading Disorders,* edited by Francis A. Young and Donald B. Lindsley. Washington, D.C.: National Academy of Sciences, 1970.

———. "Environmental Models and Mental Modes: Representational Processes in the Great Apes and Man." *American Psychologist* 31 (1976): 284–294.

MASON, W. A., and JOHN H. HOLLIS. "Communication between Young Rhesus Monkeys." *Animal Behaviour* 103 (1962): 211–221.

MAYR, ERNST. *Principles of Systematic Zoology*. New York: McGraw-Hill, 1969.

———. "Behavior Programs and Evolutionary Strategies." *American Scientist* 62 (1974): 650–659.

MCADAM, D. W., and H. A. WHITAKER. "Language Production: Electroencephalographic Localization in the Normal Human Brain." *Science* 172 (1971): 499–502.

MCCALL, ELIZABETH A. "A Generative Grammar of Sign." Unpublished master's thesis, University of Iowa, 1965.

McCarthy, Dorothy. "Language Development in Children." In *Manual of Child Psychology*, edited by Leonard Carmichael. New York: Wiley, 1954, pp. 492–630.

McGrew, William C. "Tool Use by Wild Chimpanzees in Feeding upon Driver Ants." *Journal of Human Evolution* 3 (1974): 501–508.

———. "Socialization and Object Manipulation of Wild Chimpanzees." In *Primate Bio-Social Development: Biological, Social, and Ecological Determinants*, edited by S. Chevalier-Skolnikoff and F. E. Poirier. New York: Garland, 1977, pp. 261–287.

McGrew, William C., C. E. G. Tutin, and P. S. Midgett. "Tool Use in a Group of Captive Chimpanzees. I. Escape." *Zeitschrift für Tierpsychologie* 37 (1975): 145–162.

McIntire, M. L. "The Acquisition of American Sign Language Hand Configuration." *Sign Language Studies* 16 (1977): 247–266.

McNeill, David. "Explaining Linguistic Universals." Paper presented at the 19th International Congress of Psychology, London, 1969.

———. *The Acquisition of Language. A Study of Developmental Psycholinguistics*. New York: Harper & Row, 1970.

Mellgren, R., Roger S. Fouts, and W. B. Lemmon. "American Sign Language in the Chimpanzee: Semantic and Conceptual Functions of Signs." Paper presented at the Midwestern Psychological Association meeting, Chicago, 1973.

Menzel, Emil W. "Spontaneous Invention of Ladders in a Group of Young Chimpanzees." *Folia Primatologica* 17 (1972): 87–106.

———. "Chimpanzee Spatial Memory Organization." *Science* 182 (1973): 943–946. (a)

———. "Further Observations on the Use of Ladders in a Group of Young Chimpanzees." *Folia Primatologica* 19 (1973): 450–457. (b)

———. "Leadership and Communication in Young Chimpanzees." In *Proceedings of the Fourth International Congress of Primatology*, vol. 1, Basel: S. Karger, 1973, pp. 192–225. (c)

———. "Leadership and Communication in Young Chimpanzees." In *Precultural Primate Behavior*, edited by Emil W. Menzel. Basel: S. Karger, 1973, pp. 172–204. (d)

———. "A Group of Young Chimpanzees in a One-Acre Field." In *Behavior of Nonhuman Primates*, vol. 5, edited by Allan M. Schrier and Fred Stollnitz. New York: Academic, 1974, pp. 83–153.

Menzel, Emil W., and S. Halperin. "Purposive Behavior as a Basis for Objective Communication between Chimpanzees." *Science* 189 (1975): 652–654.

Menzel, Emil W., and Marcia K. Johnson. "Communication and Cognitive Organization in Humans and Other Animals." In *Origins and Evolution of Language and Speech*, edited by S. R. Harnad, H. D. Steklis, and J. Lancaster. New York: Annals of the New York Academy of Sciences, 280 (1976): 131–142.

Merton, Robert K. "The Self-Fulfilling Prophecy." *Antioch Review* 8 (1948): 193–210.

Miles, L. W. "The Communicative Competence of Child and Chimpanzee." In *Origins and Evolution of Language and Speech*, edited by S. R. Harnad, H. D. Steklis, and J. Lancaster. New York: Annals of the New York Academy of Sciences, 280 (1976): 592–597.

Miller, N. E. "Liberalization of Basic S-R Concepts: Extensions to Conflict Behavior, Motivation and Social Learning." In *Psychology: A Study of a Science*, vol. 2, edited by Sigmund Koch. New York: McGraw-Hill, 1959, pp. 196–292.

Miller, R. E., J. H. Banks, Jr., and N. Ogawa. "Communication of Affect in 'Cooperative Conditioning' of Rhesus Monkeys." *Journal of Abnormal and Social Psychology* 64(5) (1962): 343–348.

Miller, Wick, and Susan Ervin. "The Development of Grammar in Child Language." In *The Acquisition of Language (Monographs of the Society for Research in Child Development)*, edited by Ursula Bellugi and Roger Brown, 1964, pp. 9–34.

Mistler-Lachman, J. L., and R. Lachman. "Language in Man, Monkeys, and Machines." *Science* 186 (1974): 871–872.

Moertel, Charles G., William F. Taylor, Arthur Roth, and Francis A. J. Tyce. "Who Responds to Sugar Pills?" *Mayo Clinic Proceedings* 51 (1976): 96–100.

Moore, Timothy E. Introduction to *Cognitive Development and the Acquisition of Language*, edited by Timothy E. Moore. New York: Academic, 1973, pp. 1–8.

MORRIS, DESMOND. *Manwatching: A Field Guide to Human Behavior.* New York: Abrams, 1977.

MORRIS, RAMONA, and DESMOND MORRIS. *Men and Apes.* New York: McGraw-Hill, 1966.

MOUNIN, GEORGES. Review of *Non-Verbal Communication,* by Robert A. Hinde, 1972. *Journal of Linguistics* 1 (1974): 197–206.

MOWRER, O. H. "The Psychologist Looks at Language." *American Psychologist* 9 (1954): 660–694.

MOYNIHAN, M. "Control, Suppression, Decay, Disappearance and Replacement of Displays." *Journal of Theoretical Biology* 29 (1970): 85–112.

MUCHMORE, E., and W. W. SOCHA. "Blood Transfusion Therapy for Leukemic Chimpanzee." *Laboratory Primate Newsletter* 15(3) (1976): 13–15.

MYERS, R. "Comparative Neurology of Vocalization and Speech: Proof of a Dichotomy." In *Origins and Evolution of Language and Speech,* edited by S. R. Harnad, H. D. Steklis, and J. Lancaster. New York: Annals of the New York Academy of Sciences, 280 (1976): 745–757.

NEISSER, U. *Cognitive Psychology.* New York: Appleton-Century-Crofts, 1967.

NELSON, KATHERINE. *Structure and Strategy in Learning to Talk (Monographs of the Society for Research in Child Development* 38, serial no. 149), 1973, pp. 1–137.

NICHOLS, S. "Methodology for Testing Left- and Right-Handedness in Human and Chimpanzee Infants." Paper presented at the Western Psychological Association, San Francisco, 1974.

NOTTEBOHM, FERNANDO. "Neural Lateralization of Vocal Control in a Passerine Bird." *Journal of Experimental Zoology* 179 (1972): 35–50. (a)

———. "The Origins of Vocal Learning." *The American Naturalist* 106 (1972): 116–140. (b)

OLIVIER, F. "Le Dessin Enfantin Est-Il une Ecriture?" *Enfance* 22 (1974): 183–216.

OMARK, D. "Peer Group Formation in Children." Unpublished doctoral dissertation, Committee on Human Development, University of Chicago, 1972.

ORNE, MARTIN T. "The Nature of Hypnosis: Artifact and Essence." *Journal of Abnormal and Social Psychology* 58, (1959): 277–299.

OSGOOD, CHARLES E. *Method and Theory in Experimental Psychology.* New York: Oxford University Press, 1953.

———. "A Behavioristic Analysis of Perception and Language as Cognitive Phenomena." In *Contemporary Approaches to Cognition: A Symposium Held at the University of Colorado.* Cambridge, Mass.: Harvard University Press, 1957, pp. 75–118.

———. "Toward a Wedding of Insufficiencies." In *Verbal Behavior and General Behavior Theory,* edited by Theodore R. Dixon and David L. Horton. Englewood Cliffs, N.J.: Prentice-Hall, 1968, pp. 495–519.

PAPI, FLORIANO. "Light Emission, Sex Attraction and Male Flash Dialogues in a Firefly, *Luciola lusitanica (Charp.)." Monitore Zoologico Italiano (N.S.)* 3 (1969): 135–184.

PARKEL, DOROTHY A. "The Use of Visual Symbols in an Experimental Language Training Program for the Severely and Profoundly Retarded." Paper presented at the National Invitation Conference on Communication Research in Mental Retardation and Learning Disabilities, Columbia, Maryland, 1977.

PARKEL, DOROTHY A., and S. T. SMITH. "Application of Computer Assisted Language Designs." In *Nonspeech Communication,* edited by Richard Schiefelbusch. University Park, Md.: University Park Press, 1978.

PARKEL, DOROTHY A., ROYCE A. WHITE, and HAROLD WARNER. "Implications of the Yerkes Technology for Mentally-Retarded Human Subjects." In *Language Learning by a Chimpanzee,* edited by D. M. Rumbaugh. New York: Academic, 1977, pp. 273–283.

PARKER, C. E. "The Antecedents of Man the Manipulator." *Journal of Human Evolution* 3 (1974): 493–500.

PARKER, S. T. "Piaget's Sensorimotor Period Series in an Infant Macaque: A Model for Comparing Unstereotyped Behavior and Intelligence in Human and Nonhuman Primates." In *Primate Bio-Social Development: Biological, Social, and Ecological Determinants,* edited by S. Chevalier-Skolnikoff and F. E. Poirier. New York: Garland, 1977, pp. 43–112.

PARKER, S. T., and K. R. GIBSON. "Object Manipulation, Tool Use and Sensorimotor Intelligence as Feeding Adaptations in Cebus Monkeys and Great Apes." *Journal of Human Evolution* 6 (1977): 623–641.

PATTERSON, FRANCINE G. "Linguistic Capabilities of a Young Lowland Gorilla." Paper presented at a symposium of the American Association for the Advancement of Science, "An Account of the Visual Mode: Man versus Ape," Denver, 1977. (a)

———. "The Gestures of a Gorilla: Language Acquisition in Another Pongid Species." In *Perspectives on Human Evolution,* edited by David Hamburg, Jane Goodall, and R. E. McCown. Menlo Park, Calif.: Benjamin, 1977. (b)

———. "The Gestures of a Gorilla: Sign Language Acquisition in Another Pongid Species." *Brain and Language* 5 (1978): 72–97.

PEIRCE, CHARLES S. "Guessing." *The Hound and Horn* 2 (1929): 267–282.

———. "Order of Nature." In *Meaning and Knowledge,* edited by Ernest Nagel and Richard B. Brandt. New York: Harcourt, Brace & World, 1965. (Originally published 1878.)

———. *Collected Papers of Charles Sanders Peirce,* edited by Charles Hartshorne, Paul Weiss, and Arthur W. Burks. Cambridge, Mass.: Harvard University Press, 1935–1966. (References are to volumes and paragraphs, not pages.)

PEPYS, SAMUEL. *The Diary of Samuel Pepys,* edited by R. Latham and M. Williams. Berkeley: University of California Press, 1970. (Originally published 1912.)

PFUNGST, OSKAR. *Clever Hans, The Horse of Mr. von Osten.* New York: Holt, 1911.

———. *Clever Hans (The Horse of Mr. von Osten),* edited by Robert Rosenthal. New York: Holt, Rinehart & Winston, 1965.

PIAGET, JEAN. *The Construction of Reality in the Child.* New York: Basic, 1954. (Originally published 1937.)

———. *Six Psychological Studies.* New York: Random, 1967.

PIERCE, JOHN R., and EDWARD E. DAVID, JR. *Man's World of Sound.* Garden City, N.J.: Doubleday, 1958.

PILISUK, MARK, BARBARA BRANDES, and DIDIER VAN DER HOVE. "Deceptive Sounds: Illicit Communication in the Laboratory." *Behavioral Science* 21 (1976): 515–523.

PLOOG, DETLEV W. "The Behavior of Squirrel Monkeys *(Saimiri sciureus)* as Revealed by Sociometry, Bioacoustics, and Brain Stimulation." *Social Communication among Primates,* edited by Stuart A. Altmann. Chicago: University of Chicago Press, 1967, pp. 149–184.

PLOOG, DETLEV W., and M. MAURUS. "Social Communication among Squirrel Monkeys: Analysis by Sociometry, Bioacoustics and Cerebral Radio-Stimulation." In *Comparative Ecology and Behavior of Primates,* edited by R. Michael and J. Crook. London: Academic, 1973, pp. 211–233.

PLOOG, DETLEV W., and T. MELNECHUK. "Primate Communication." *Neuroscience Research Program Bulletin* 7(5) (1969): 419–510.

———. "Are Apes Capable of Language?" *Neuroscience Research Program Bulletin* 9 (1971): 600–700.

PLUTCHIK, ROBERT. "Emotions, Evolution, and Adaptive Processes." In *Feelings and Emotions,* edited by Magda B. Arnold. New York: Academic, 1970, pp. 3–24.

POIRIER, F. E. *Fossil Man: An Evolutionary Journey.* St. Louis: Mosby, 1973.

POLANYI, MICHAEL. *Personal Knowledge: Towards a Post-Critical Philosophy.* Chicago: University of Chicago Press, 1958.

PONNAMPERUMA, CYRIL, and A. G. W. CAMERON, eds. *Interstellar Communication: Scientific Perspectives.* Boston: Houghton Mifflin, 1974.

POSINSKY, S. H. "The Case of John Tarmon: Telepathy and the Law." *Psychiatric Quarterly* 35 (1961): 165–166.

PRATT, J. GAITHER. *ESP Research Today: A Study of Developments in Parapsychology since 1960.* Metuchen, N.J.: Scarecrow, 1973.

PREMACK, ANN J., and DAVID PREMACK. "Teaching Language to an Ape." *Scientific American* 227(4) (1972): 92–99.

PREMACK, DAVID. "A Functional Analysis of Language." Invited address, Div. I, American Psychological Association, Washington, D.C., 1969. (Mimeographed prepublication copy.)

———. "A Functional Analysis of Language." *Journal of the Experimental Analysis of Behavior* 14 (1970): 107–125. (a)

———. "The Education of Sarah, a Chimp." *Psychology Today* 4 (1970): 55–58. (b)

———. "Language in Chimpanzee?" *Science* 172 (1971): 808–822. (a)

————. "On the Assessment of Language Competence in the Chimpanzee." In *Behavior of Nonhuman Primates,* vol. 4, edited by Allan M. Schrier and Fred Stollnitz. New York: Academic, 1971, pp. 186–228. (b)

————. "Some General Characteristics of a Method for Teaching Languages to Organisms That Do Not Ordinarily Acquire It." In *Cognitive Processes of Nonhuman Primates,* edited by L. Jarrad. New York: Academic, 1971, pp. 47–82. (c)

————. "Teaching Language to an Ape." *Scientific American* 227 (1972): 92–99.

————. "On the Origins of Language." In *Handbook of Psychobiology,* edited by Michael S. Gazzaniga and Colin Blakemore. New York: Academic, 1975, pp. 591–605. (a)

————. "Putting a Face Together." *Science* 188 (1975): 228–236. (b)

————. *Intelligence in Ape and Man.* Hillsdale, N.J.: Lawrence Erlbaum, 1976. (a)

————. "Mechanisms of Intelligence: Preconditions for Language." In *Origins and Evolution of Language and Speech,* edited by S. R. Harnad, H. D. Steklis, and J. Lancaster. New York: Annals of the New York Academy of Sciences, 280 (1976): 544–561. (b)

————. "On the Study of Intelligence in Chimpanzees." *Current Anthropology* 17 (1976): 516–521. (c)

PREMACK, DAVID, and ANN PREMACK. "Teaching Visual Language to Apes and Language-Deficient Persons." In *Language Perspectives—Acquisition, Retardation, and Intervention,* edited by Richard L. Schiefelbusch and L. L. Loyd. Baltimore: University Park Press, 1974, pp. 347–376.

PREMACK, DAVID, and A. SCHWARTZ. "Preparations for Discussing Behaviorism with Chimpanzee." In *Genesis of Language,* edited by Frank Smith and George A. Miller. Cambridge, Mass.: M.I.T. Press, 1966, pp. 295–335.

PREMACK, DAVID, and GUY WOODRUFF. "Does the Chimpanzee Have a Theory of Mind?" *The Behavioral and Brain Sciences,* 4 (1978): 515–526.

PRESTRUDE, A. M. "Sensory Capacities of the Chimpanzee: A Review." *Psychological Bulletin* 74 (1970): 47–67.

PRIBRAM, KARL H. *Languages of the Brain.* Englewood Cliffs, N.J.: Prentice-Hall, 1971.

————. "Language in a Sociobiological Frame." In *Origins and Evolution of Language and Speech,* edited by S. R. Harnad, H. D. Steklis, and J. Lancaster. New York: Annals of the New York Academy of Sciences, 280 (1976): 798–809.

PYLE, C. "Pragmatics." Paper presented at the 50th annual meeting of the Linguistic Society of America, San Francisco, 1975.

RADINSKY, L. "Primate Brain Evolution." *American Scientist* 63 (1975): 656–663.

RALEIGH, MICHAEL, J., and FRANK R. ERVIN. "Human Language and Primate Communication." In *Origins and Evolution of Language and Speech,* edited by S. R. Harnad, H. D. Steklis, and J. Lancaster. New York: Annals of the New York Academy of Sciences, 280 (1976): 539–541.

RAZRAN, GREGORY. "Pavlov and Lamarck." *Science* 128 (1958): 758–760.

————. "Pavlov the Empiricist." *Science* 130 (1959): 916–917.

REDSHAW, M. "Cognitive Development in Humans and Infant Gorillas." *Journal of Human Evolution* 7 (1978): 133–141.

RESTLE, FRANK. "Cognitive Structures." In *Cognitive Theory,* vol. 1, edited by Frank Restle, R. M. Shiffrin, N.J. Castellan, H. R. Lindman, and D. B. Pisoni. New York: Wiley, 1975. (a)

————. *Learning: Animal Behavior and Human Cognition.* New York: McGraw-Hill, 1975. (b)

REYNOLDS, P. C. "Language and Skilled Activity." In *Origins and Evolution of Language and Speech,* edited by S. R. Harnad, H. D. Steklis, and J. Lancaster. New York: Annals of the New York Academy of Sciences, 280 (1976): 150–166.

REYNOLDS, VERNON and F. REYNOLDS. "Chimpanzees of the Budongo Forest." In *Primate Behavior: Field Studies of Monkeys and Apes,* edited by Irven DeVore. New York: Holt, Rinehart & Winston, 1965, pp. 368–424.

RHINE, LOUISA E. *Mind Over Matter: Psychokinesis.* New York: Macmillan, 1970.

RIESEN, A. H. "Comparative Perspectives in Behavior Study." *Journal of Human Evolution* 3 (1974): 433–434.

RIESEN, A. H., and E. F. KINDER. *Postural Development in Infant Chimpanzees.* New Haven: Yale University Press, 1952.

RINN, JOSEPH F. *Sixty Years of Psychical Research.* New York: Truth Seeker, 1950.

ROBINSON, BRYAN W. "Vocalization Evoked from Forebrain in *Macaca Mulatta*." *Physiology and Behavior* 2 (1967): 345–354.

———. "Anatomical and Physiological Contrasts between Human and Other Primate Vocalizations." In *Perspectives on Human Evolution,* vol. 2, edited by Sherwood L. Washburn and Phyllis C. Dolhinow. New York: Holt, Rinehart & Winston, 1972, pp. 438–443.

———. "Limbic Influences on Human Speech." In *Origins and Evolution of Language and Speech,* edited by S. R. Harnad, H. D. Steklis, and J. Lancaster. New York: Annals of the New York Academy of Sciences, 280 (1976): 761–771.

ROBINSON, C. E. "The Development of Hand Preference in Children and Young Chimpanzees." Doctoral dissertation, Psychology Department, University of Nevada, Reno 1977.

ROMAINS, JULES. *La Vision: Extrarétinienne et le Sens Paroptique; Recherches de Psychophysiologie Expérimentale et de Physiologie Histologique.* Paris: Nouvelle Revue Française, 1920.

ROSCH, ELEANOR H. "On the Internal Structure of Perceptual and Semantic Categories." In *Cognitive Development and the Acquisition of Language,* edited by Timothy E. Moore. New York: Academic, 1973, pp. 111–144.

ROSENTHAL, ROBERT. *Experimenter Effects in Behavioral Research.* 2nd ed. New York: Appleton-Century-Crofts, 1976.

ROSENTHAL, ROBERT, and LENORE JACOBSON. *Pygmalion in the Classroom: Teacher Expectation and Pupils' Intellectual Development.* New York: Holt, Rinehart & Winston, 1968.

ROZIN, P., S. PORITSKY, and R. SOTSKY. "American Children with Reading Problems Can Easily Learn to Read English Represented by Chinese Characters." *Science* 171 (1971): 1264–1267.

RUMBAUGH, DUANE M. "Learning Skills of Anthropoids." In *Primate Behavior: Developments in Field and Laboratory Research,* vol. 1, edited by Leonard A. Rosenblum. New York: Academic, 1970, pp. 1–70.

———. "Evidence of Qualitative Differences in Learning Processes among Primates." *Journal of Comparative and Physiological Psychology* 76 (1971): 250–255.

———. ed. *Language Learning by a Chimpanzee. The LANA Project.* New York: Academic, 1977.

RUMBAUGH, DUANE M., and TIMOTHY V. GILL. "Language, Apes, and the Apple Which-Is Orange, Please." In *Proceedings from the Symposia of the Fifth Congress of the International Primatological Society,* edited by S. Kondo, M. Kawai, A. Ehara, and S. Kawamura. Tokyo: Japan Science, 1975, pp. 247–257.

———. "Language and the Acquisition of Language-Type Skills by a Chimpanzee *(Pan)*." *Psychology in Progress: An Interim Report* 270 (1976): 90–135. (a)

———. "The Mastery of Language-Type Skills by the Chimpanzee *(Pan)*." In *Origins and Evolution of Language and Speech*, edited by S. R. Harnad, H. D. Steklis, and J. Lancaster. New York: Annals of the New York Academy of Sciences, 280 (1976): 562–578. (b)

———. "Lana's Mastery of Language Skills." In *Language Learning by a Chimpanzee,* edited by D. M. Rumbaugh. New York: Academic, 1977, pp. 165–192.

RUMBAUGH, DUANE M., and ERNST VON GLASERSFELD. "Reading and Sentence Completion by a Chimpanzee." *Science* 182 (1973): 731–733.

RUMBAUGH, DUANE M., TIMOTHY V. GILL, ERNST VON GLASERSFELD, HAROLD WARNER, and PIER PISANI. "Conversations with a Chimpanzee in a Computer-Controlled Environment." *Biological Psychiatry* 10 (1975): 627–641.

RUMBAUGH, DUANE M., ERNST VON GLASERSFELD, HAROLD WARNER, PIER PISANI, TIMOTHY V. GILL, JOSEPHINE V. BROWN, and CHARLES L. BELL. "A Computer-Controlled Language Training System for Investigating the Language Skills of Young Apes." *Behavior Research Methods and Instrumentation* 5 (1973): 385–392.

RUMBAUGH, DUANE M., and E. SUE SAVAGE-RUMBAUGH. "Chimpanzee Language Research: Status and Potential." *Behavior Research Methods and Instrumentation* 10 (1978): 119–131.

SACKETT, G. P. "Monkeys Reared in Isolation with Pictures as Visual Input Evidence for an Innate Releasing Mechanism." *Science* 154 (1966): 1468–1472.

SARBIN, THEODORE R., and WILLIAM C. COE. *Hypnosis: A Social Psychological Analysis of Influence Communication.* New York: Holt, Rinehart & Winston, 1972.

SARICH, V. M. "The Origin of the Hominids: An Immunological Approach." In *Perspectives on Human*

Evolution, vol. 1, edited by Sherwood L. Washburn and Phyllis C. Jay. New York: Holt, Rinehart & Winston, 1968, 94–121.

SARICH, V. M., and J. E. CRONIN. "Molecular Systematics of the Primates." In *Molecular Anthropology,* edited by M. Goodman and R. E. Tashian. New York: Plenum, 1976, pp. 141–170.

SARLES, HARVEY B. "The Study of Language and Communication across Species." *Current Anthropology* 10 (1969): 211–215.

———. "Could a Non-h?" In *Language Origins,* edited by Roger W. Wescott. Silver Spring, MD.: Linstock, 1974, pp. 219–238.

———. Comment. *Current Anthropology* 17 (1976): 334–335.

SAVAGE, E. SUE, and DUANE M. RUMBAUGH. "Communication, Language, and Lana: A Perspective." In *Language Learning by a Chimpanzee,* edited by D. M. Rumbaugh. New York: Academic, 1977, pp. 287–309.

———. "Symbolization, Language and Chimpanzees: A Theoretical Reevaluation Based on Initial Language Acquisition Processes in Four Young *Pan Troglodytes.*" *Brain and Language* 6 (1978): 265–300.

———. "Initial Acquisition of Symbolic Skills via the Yerkes Computerized Language Analog System." In *Language Intervention from Ape to Child,* edited by Richard L. Schiefelbusch and John H. Hollis. Baltimore, Md.: University Park Press, 1979, pp. 277–294.

———. "Language Analog Project, Phase II: Theory and Tactics." In *Children's Language,* vol. 2, edited by Keith Nelson. New York: Gardner, in press.

SAVAGE-RUMBAUGH, E. SUE, DUANE M. RUMBAUGH, and SALLY BOYSEN. "Symbolic Communication between Two Chimpanzees *(Pan Troglodytes).*" *Science* 201 (1978): 641–644.

SAVAGE-RUMBAUGH, E. SUE, B. J. WILKERSON, and R. BAKEMAN. "Spontaneous Gestural Communication among Conspecifics in the Pygmy Chimpanzee *(Pan paniscus).*" In *Progress in Ape Research,* edited by Geoffrey Bourne. New York: Academic, 1977, pp. 97–116.

SAVIN, H. B. "What the Child Knows about Speech When He Starts to Learn to Read." Paper presented at the Conference on the Relationships between Speech and Learning to Read, Elkridge, Md., 1971.

SAVIN, H. B., and T. G. BEVER. "The Nonperceptual Reality of the Phoneme." *Journal of Verbal Learning and Verbal Behavior* 9 (1970): 295–302.

SCHACHTER, S., and J. E. SINGER. "Cognitive, Social and Physiological Determinants of Emotional State." *Psychological Review* 69 (1962): 379–399.

SCHALLER, GEORGE B. *The Mountain Gorilla.* Chicago: University of Chicago Press, 1963.

SCHLESINGER, HILDE S., and KATHERINE P. MEADOW. *Deafness and Mental Health: A Developmental Approach.* Berkeley: University of California Press, 1972.

SCHLESINGER, I. M. "The Grammar of Sign Language and the Problems of Language Universals." In *Biological and Social Factors in Psycholinguistics,* edited by John Morton. Urbana: University of Illinois Press, 1970, pp. 98–121.

———. "Production of Utterances and Language Acquisition." In *The Ontogenesis of Grammar,* edited by Dan I. Slobin. New York: Academic, 1971, pp. 63–101.

SCHLESINGER, I. M., B. PRESSER, E. COHEN, and T. PELED. "Transfer of Meaning in Sign Language." *Working paper No. 12,* Department of Psychology, Hebrew University, Jerusalem, 1970.

SCHULTZ, A. H. *The Life of Primates.* New York: Universe, 1969.

SEBEOK, THOMAS A. *Perspectives in Zoosemiotics.* The Hague: Mouton, 1972.

———. *Contributions to the Doctrine of Signs* Lisse: Peter de Ridder Press/Bloomington: Indiana University Research Center for Language and Semiotic Studies, 1976.

———. "Displaying the Symptoms." *Times Literary Supplement* (no. 3, 939) September 9, 1977, p. 1075. (a)

———. "Ecumenicalism in Semiotics." In *A Perfusion of Signs,* edited by Thomas A. Sebeok. Bloomington: Indiana University Press, 1977, pp. 180–206. (b)

———. "Zoosemiotic Components of Human Communication." In *How Animals Communicate,* edited by Thomas A. Sebeok. Bloomington: Indiana University Press, 1977, pp. 1055–1077. (c)

———. "Mieux vaut cherchez à la source qu'à l'embouchure." *Diogène* 104 (1978): 116–142.

———. *The Sign and Its Masters.* Austin: University of Texas Press, 1979.

SEBEOK, THOMAS A., and JEAN UMIKER-SEBEOK. " 'You Know My Method': A Juxtaposition of C. S. Peirce and Sherlock Holmes." *Semiotica* 26(nos. 3/4) (1979) 203–250.

SEGAL, E. "Toward a Coherent Psychology of Language." In *Handbook of Operant Behavior,* edited by Werner K. Honig and J. E. R. Staddon. New York: Prentice-Hall, 1977.

SELIGMAN, M. E. P. "On the Generality of the Laws of Learning." *Psychological Review* 77 (1970): 406–418.

SHANTHA, TOTADA R., and SOHAN L. MANOCHA. "The Brain of Chimpanzee *(Pan troglodytes).*" In *The Chimpanzee,* vol. 1, edited by Geoffrey Bourne. Basel: S. Karger, 1969, pp. 187–368.

SHEEHAN, PETER W., and CAMPBELL W. PERRY. *Methodologies of Hypnosis: A Critical Appraisal of Contemporary Paradigms of Hypnosis.* Hillsdale, N.J.: Lawrence Erlbaum, 1976.

SIMPSON, G. G. *Principles of Animal Taxonomy.* New York: Columbia University Press, 1961.

SINCLAIR, UPTON. *Mental Radio.* Pasadena, Cal.: Upton Sinclair, 1930.

SINCLAIR-DE-ZWART, HERMIONE. "Language Acquisition and Cognitive Development." In *Cognitive Development and the Acquisition of Language,* edited by Timothy E. Moore. New York: Academic, 1973, pp. 9–25.

SKINNER, B. F. *Verbal Behavior.* New York: Appleton-Century-Crofts, 1957.

SLOBIN, Dan I. "The Acquisition of Russian as a Native Language." In *The Genesis of Language,* edited by Frank Smith and George A. Miller. Cambridge, Mass.: M.I.T. Press, 1966, pp. 129–148.

———. *Psycholinguistics.* Glenview, Ill.: Scott, Foresman, 1971.

SMITH, FRANK, and GEORGE A. MILLER, eds. *The Genesis of Language.* Cambridge, Mass.: M.I.T. Press, 1966.

SMITH, W. JOHN. "Messages of Vertebrate Communication." *Science* 165 (1969): 145–158.

SNOW, CATHERINE E. "Mothers' Speech to Children Learning Language." *Child Development* 43 (1972): 549–565.

SNOW, CATHERINE E., and CHARLES FERGUSON, eds. *Talking to Children.* Cambridge: Cambridge University Press, 1977.

STEBBINS, WILLIAM C. "Hearing." In *Behavior of Nonhuman Primates,* vol. 3, edited by Allan M. Schrier and Fred Stollnitz. New York: Academic, 1971, pp. 159–192.

STEKLIS, HORST D., and STEVAN R. HARNAD. "From Hand to Mouth: Some Critical Stages in the Evolution of Language." In *Origins and Evolution of Language and Speech,* edited by S. R. Harnad, H. D. Steklis, and J. Lancaster. New York: Annals of the New York Academy of Sciences, 280 (1976): 445–455.

STOKOE, WILLIAM C., JR. *Sign Language Structure: An Outline of the Visual Communication Systems of the American Deaf* (Studies in Linguistics, Occasional Papers 8). Buffalo, N.Y.: University of Buffalo Press, 1960.

———. "Linguistic Description of Sign Language." *Georgetown University Monograph Series on Languages and Linguistics* 19 (1961): 243–250.

———. *Semiotics and Human Sign Languages.* The Hague: Mouton, 1972.

———. "The Shape of Soundless Language." In *The Role of Speech in Language,* edited by James F. Kavanagh and James E. Cutting. Cambridge, Mass.: M.I.T. Press, 1975, pp. 207–228.

———. "First Hand Reporting from the Field . . ." *Signs for Our Times* 46, (1977): 1.

STOKOE, WILLIAM C., JR, DOROTHY C. CASTERLINE, and CHARLES G. CRONEBERG. *A Dictionary of American Sign Language on Linguistic Principles.* Washington, D.C.: Gallaudet College Press, 1965.

STRATTON, GEORGE M. "The Control of Another Person by Obscure Signs." *Psychological Review* 28 (1921): 301–314.

STRAUB, R. O., M. S. SEIDENBERG, T. G. BEVER, and H. S. TERRACE. "Representation of a Sequence by Pigeons." Unpublished manuscript, Department of Psychology, Columbia University, 1978.

———. "Serial Learning in the Pigeon." *Journal of the Experimental Analysis of Behavior,* 32 (1979): 137–148.

STRUHSAKER, THOMAS T. "Auditory Communication among Vervet Monkeys *(Cercopithecus aethiops).*" In *Social Communication among Primates,* edited by Stuart A. Altmann. Chicago: University of Chicago Press, 1967, pp. 281–324.

SWADESH, MORRIS. "Linguistic Overview." In *Prehistoric Man in the New World,* edited by J. D. Jennings and E. Norbeck. Chicago: University of Chicago Press, 1964, pp. 527–558.

TERRACE, H. S., L. PETITTO, and T. G. BEVER. "Project NIM Progress Report II." Unpublished manuscript, Columbia University, 1976.

THOMAS, R. K., and S. R. KERR. "Conceptual Conditional Discrimination in *Saimiri sciureus.*" *Animal Learning Behavior* 4 (1976): 333–336.

THOMSON, J. ARTHUR. "Zoology (Animal Behaviour)." In *The Life-Work of Lord Avebury (Sir John Lubbock).* London: Watt, 1924, pp. 115–156.

TOULMIN, STEVEN. "Brain and Language: A Commentary." *Synthèse* 22 (1971): 369–395.

TRUZZI, MARCELLO. Review of *Castaņeda's Journey: The Power and the Allegory* by Richard de Mille, 1976. *The Zetetic* 1(2) (1977); 86–87.

TURNEY, T. H. "Greeting Behavior in Infant Home-Reared Chimpanzees." Doctoral dissertation, Psychology Department, University of Nevada, Reno, 1975.

TUTEUR, WERNER. "The 'Double-Blind' Method: Its Pitfalls and Fallacies." *American Journal of Psychiatry* 114 (1957–1958) 921–922.

TWEENEY, R. D., G. W. HEIMAN, and H. W. HOEMANN. "Psychological Processing of Sign Language: The Effects of Visual Disruption on Sign Intelligibility." *Journal of Experimental Psychology and Genetics* 106 (1977): 255–268.

UEXKÜLL, THURE VON. "Terminological Problems of Medical Semiotics." Unpublished manuscript, 1978.

UMIKER-SEBEOK, JEAN. "Face Work in American Magazine Advertisements." Unpublished manuscript, Indiana University, 1979.

VARTANIAN, A. *La Mettrie's l'Homme Machine: A Study in the Origins of an Idea.* Princeton, N.J.: Princeton University Press, 1960.

VELLETRI-GLASS, A., M. S. GAZZANIGA, and D. PREMACK. "Artificial Language Training in Global Aphasics." *Neuropsychologia* 11 (1973): 95–104.

VENDLER, Z. *Res Cogitans.* Ithaca, N.Y.: Cornell University Press, 1972.

VOGT, EVON A., and RAY HYMAN. *Water Witching U.S.A.* Chicago: University of Chicago Press, 1959.

VYGOTSKY, L. S. *Thought and Language.* Cambridge, Mass.: M.I.T. Press, 1962.

WADE, NICHOLAS. "Scandal in the Heavens: Renowned Astronomer Accused of Fraud." *Science* 198 (1977): 707–709.

WAGONER, DAVID. "The Literature of Legerdemain." *Times Literary Supplement* No. 3, 902, December 24, 1976, pp. 1598–1599.

WALL, C. *Predication: A Study of its Development.* The Hague: Mouton, 1974.

WARDEN, C. J., and L. H. WARNER. "The Sensory Capacities and Intelligence of Dogs with a Report on the Ability of the Noted Dog, 'Fellow' to Respond to Verbal Stimuli." *Quarterly Review of Biology* 3 (1928): 1–28.

WARNER, HAROLD, CHARLES L. BELL, DUANE M. RUMBAUGH, and TIMOTHY V. GILL. "Computer-Controlled Teaching Instrumentation for Linguistic Studies with Great Apes." *IEEE Transactions on Computers* C–25 (1976): 38–43.

WARREN, J. M. "Learning in Vertebrates." In *Comparative Psychology: A Modern Survey,* edited by Donald A. Dewsbury and Dorothy A. Rethlingshafer. New York: McGraw-Hill, 1973.

——. "Possibly Unique Characteristics of Learning by Primates." *Journal of Human Evolution* 3 (1974): 445–454.

WARREN, J. M. and K. AKERT, eds. *The Frontal Granular Cortex and Behavior.* New York: McGraw-Hill, 1964.

WASHABAUGH, W. "The Iconic and Analog in Sign." In *Aspects of Nonverbal Communication* (Publications in Southwestern English 6), edited by Walburga von Raffler-Engel and B. Hoffer. San Antonio, Tex.: Trinity University, 1977, pp. 75–120.

WASHBURN, SHERWOOD L. *The Study of Human Evolution.* Eugene: University of Oregon Press, 1968.

WASHBURN, SHERWOOD L., and SHIRLEY C. STRUM. "Concluding Comments." In *Perspectives on Human Evolution,* vol. 2, edited by Sherwood L. Washburn and Phyllis C. Dolhinow. New York: Holt, Rinehart & Winston, 1972, pp. 469–471.

WECHSLER, D. "Intelligence Defined and Undefined: A Relativistic Appraisal." *American Psychologist* 30 (1975): 135–139.

WEIR, RUTH H. *Language in the Crib.* The Hague: Mouton, 1962.

WEIZENBAUM, JOSEPH. "Computers as 'Therapists'." *Science* 198 (1977): 354.

WESCOTT, ROGER W. "Linguistic Iconism." *Language* 47 (1971): 416–428.

WHITAKER, HARRY A. "On the Representation of Language in the Brain." *UCLA Working Papers in Phonetics 12,* Los Angeles: University of California, 1969.

WHORF, BENJAMIN L. *Language, Thought, and Reality,* edited by John B. Carroll. New York: Wiley, 1956.

WICKELGREN, W. "Memory." In *Psychology and the Handicapped Child,* edited by J. A. Swets and L. L. Elliot. Washington, D.C.: U.S. Government Printing Office, 1974.

WIENER, NORBERT. *Cybernetics.* New York: Wiley, 1948.

WILSON, EDWARD O. "Animal Communication." *Scientific American* 227 (1972): 52–60.

WINOGRAD, T. "Understanding Natural Language." *Cognitive Psychology* 3 (1972): 1–191.

WITMER, LIGHTNER. "A Monkey with Mind." *The Psychological Clinic* (Philadelphia): 3 (1909): 179–205.

WOLFE, JOHN B. "Effectiveness of Token-Rewards for Chimpanzees." *Comparative Psychological Monographs* 12 (5) (1936): 1–72.

WOOD, FORREST G. *Marine Mammals and Man: The Navy's Porpoises and Sea Lions.* Washington, D.C.: Robert B. Luce, 1973.

WOOD, S. "Early Cognitive Development in Child and Chimpanzees: Object Permanence." Paper presented at the XXIst International Congress of Psychology, Paris, 1976.

WOODWARD, JAMES C. "A Transformational Approach to the Syntax of American Sign Language." In *Semiotics and Human Sign Language,* edited by W. C. Stokoe, Jr. The Hague: Mouton, 1972, pp. 131–153.

———. "Inter-Rule Implication in American Sign Language." *Sign Language Studies* 3 (1973): 47–56.

———. "Implicational Variation in American Sign Language." *Sign Language Studies* 5 (1974): 20–30.

———. "Signs of Change: Historical Variation in American Sign Language." *Sign Language Studies* 10 (1976): 81–94.

WRANGHAM, R. W. "Behavioural Ecology of Chimpanzees in Gombe National Park, Tanzania." Doctoral thesis, Cambridge University, 1975.

WRIGHT, C. *Philosophical Discussions.* New York: Franklin, 1971. (Originally published 1877)

YERKES, ROBERT M. *The Mental Life of Monkeys and Apes: A Study of Ideational Behavior (Behaviour Monographs* 3), 1916, pp. 1–145.

———. *Almost Human.* New York: Century, 1927.

———. *Chimpanzees. A Laboratory Colony.* New Haven: Yale University Press, 1943.

YERKES, ROBERT M., and B. W. LEARNED. *Chimpanzee Intelligence and Its Vocal Expression.* Baltimore, Md.: Williams and Wilkins, 1925.

YERKES, ROBERT M., and A. W. YERKES. *The Great Apes.* New Haven: Yale University Press, 1929.

YOUNG, FRANCIS A., and DONALD N. FARRER. "Visual Similarities of Non-Human and Human Primates." In *Medical Primatology,* edited by Edward I. Goldsmith and J. Moor-Jankowski. Basel: S. Karger, 1971, pp. 316–328.

ZEEMAN, E. C., and O. P. BUNEMAN. "Tolerance Spaces and the Brain." in *Towards a Theoretical Biology: An I.U.B.S. Symposium,* vol. 1, edited by C. H. Waddington. Chicago: Aldine 1968, pp. 140–151.

ZHINKIN, N. I. "An Application of the Theory of Algorithms to the Study of Animal Speech. Methods of Vocal Intercommunication between Monkeys." In *Acoustic Behaviour of Animals,* edited by René-Guy Busnel. Amsterdam: Elsevier, 1963, pp. 132–180.

ZIRKLE, CONWAY. "Pavlov's Beliefs." *Science* 128 (1958), 1476.

INDEX OF NAMES